THE CORPSE ON BOOMERANG ROAD

Telluride's War on Labor 1899-1908

BY MARYJOY MARTIN

WESTERN REFLECTIONS PUBLISHING COMPANY®

Montrose, CO

ISBN 1-932738-02-9

Library of Congress Control Number: 2004105377

Cover and text design: Laurie Goralka Design

First Edition
Printed in the United States of America

Western Reflections Publishing Company®
219 Main Street
Montrose, CO 81401
www.westernreflectionspub.com

To Wayne Kirchgasser

who holds the breaker boys in his heart,

To Ralph C. Barney

who always wondered where Will went,

And to Nakoa Martinez

who is our hope for the future:

This book is yours,

with love.

(S.D.G.)

TELLURIDE REGIONAL MAP

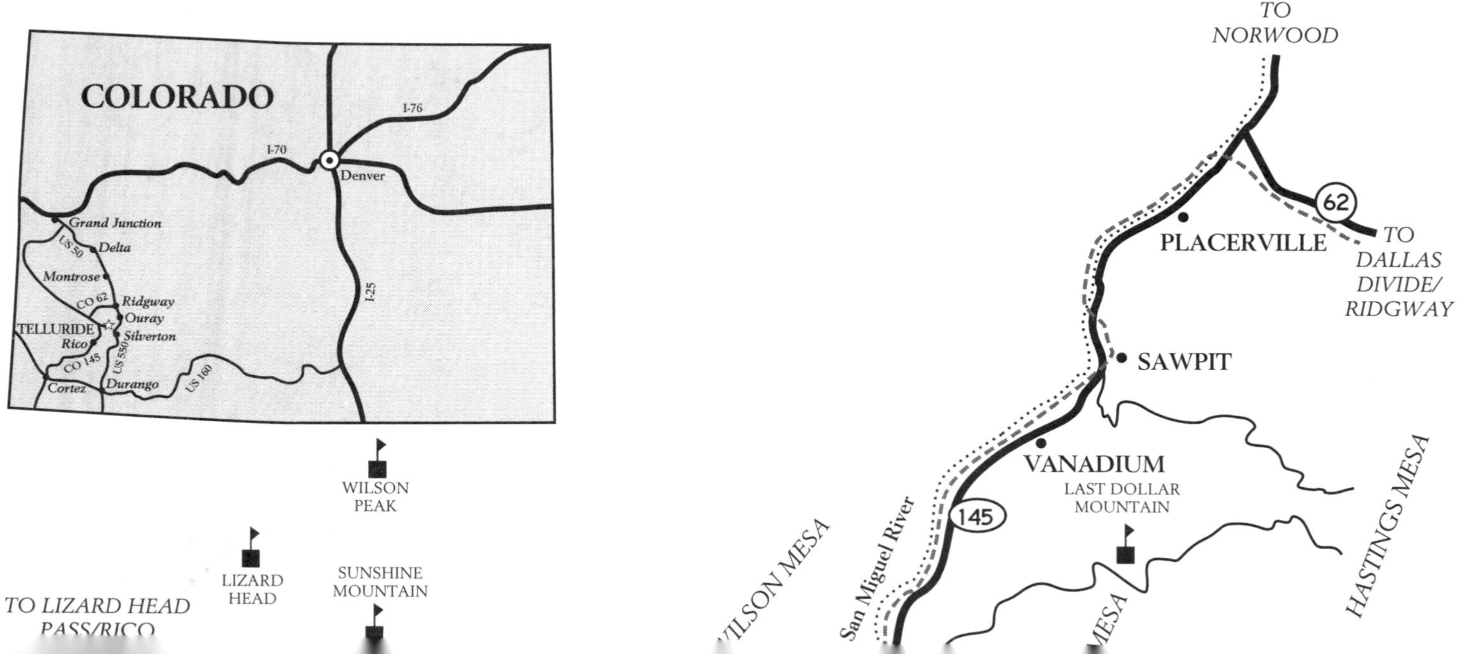

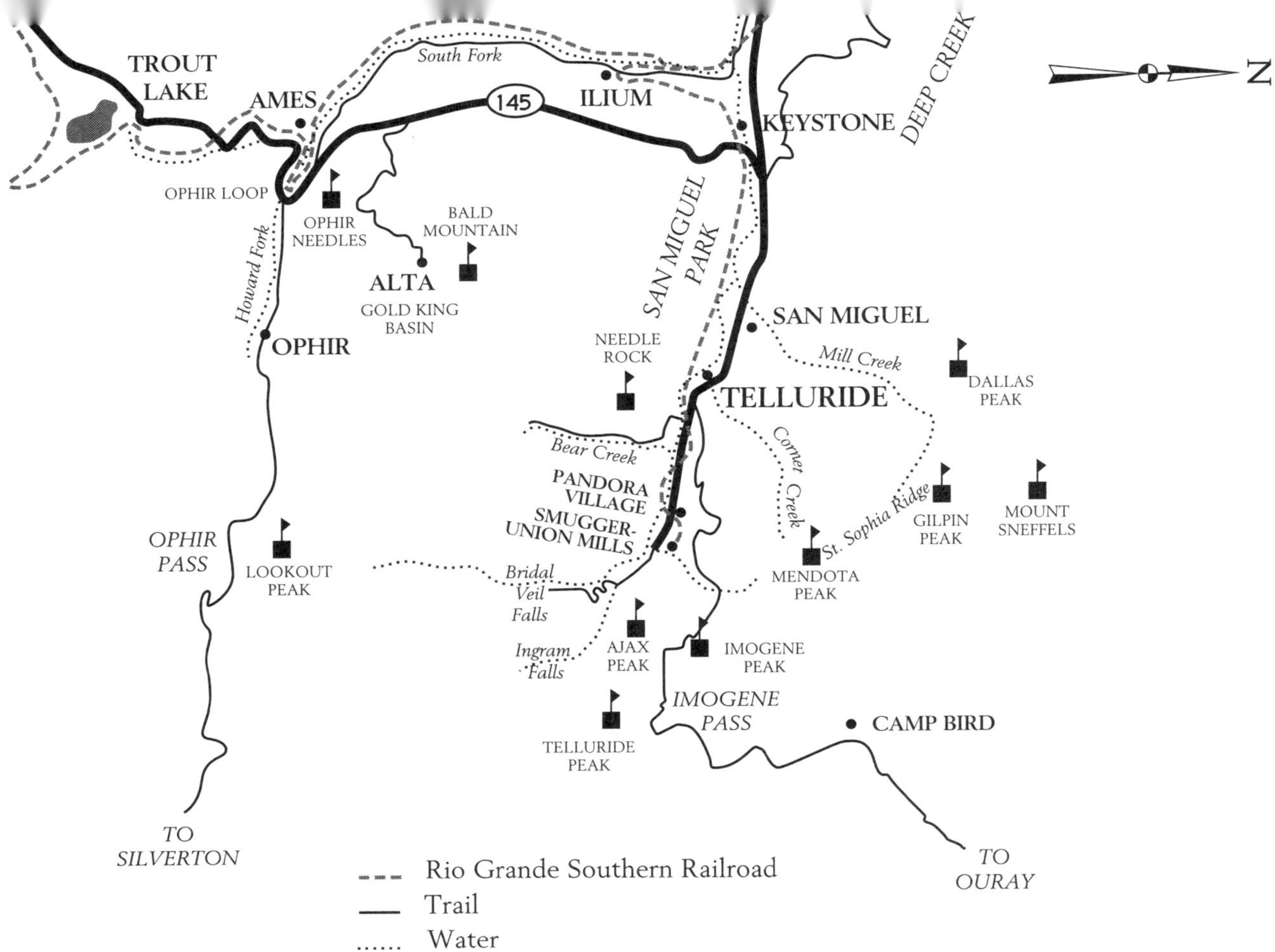

N
TROUT LAKE
AMES
South Fork
145
ILIUM
KEYSTONE
DEEP CREEK
OPHIR LOOP
OPHIR NEEDLES
BALD MOUNTAIN
ALTA
GOLD KING BASIN
Howard Fork
OPHIR
SAN MIGUEL PARK
SAN MIGUEL
NEEDLE ROCK
Mill Creek
DALLAS PEAK
TELLURIDE
Bear Creek
PANDORA VILLAGE
SMUGGER-UNION MILLS
Cornet Creek
St. Sophia Ridge
GILPIN PEAK
MOUNT SNEFFELS
OPHIR PASS
LOOKOUT PEAK
Bridal Veil Falls
MENDOTA PEAK
Ingram Falls
AJAX PEAK
IMOGENE PEAK
IMOGENE PASS
CAMP BIRD
TELLURIDE PEAK
TO SILVERTON
TO OURAY
Rio Grande Southern Railroad
Trail
Water

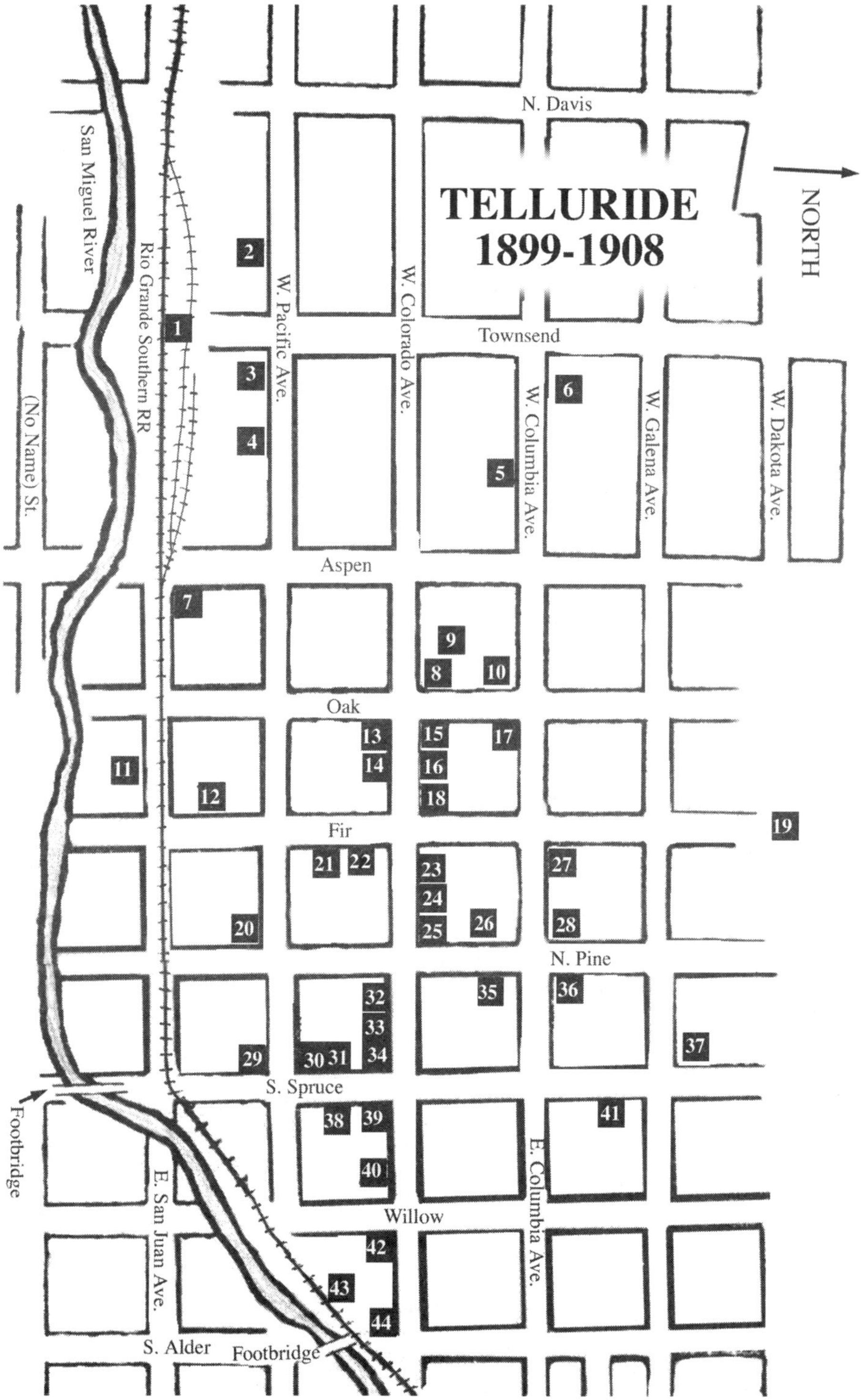
TELLURIDE
1899-1908
NORTH
N. Davis
Townsend
Aspen
Oak
Fir
N. Pine
S. Spruce
Willow
S. Alder
San Miguel River
Rio Grande Southern RR
(No Name) St.
W. Pacific Ave.
W. Colorado Ave.
W. Columbia Ave.
W. Galena Ave.
W. Dakota Ave.
E. San Juan Ave.
E. Columbia Ave.
Footbridge
Footbridge
1
2
3
4
5
6
7
8
9
10
11
12
13
14
15
16
17
18
19
20
21
22
23
24
25
26
27
28
29
30
31
32
33
34
35
36
37
38
39
40
41
42
43
44

(Division between north and south streets is Colorado Ave.; between east and west avenues is Pine Street.)

1. Rio Grande Southern Depot
2. Telluride Iron Works
3. Swedish-Finn Temperance Hall
4. Finn Hall
5. First Congregational Church
6. Telluride Public School
7. Tomboy Gold Mines shipping station
8. San Miguel County Courthouse
9. County Jail
10. Oscar Downtain residence
11. Oscar Carpenter residence
12. Red Men's Hall & Opera House
13. *Telluride Daily Journal*
14. *San Miguel Examiner*
15. Sheridan Building
16. New Sheridan Hotel
17. First M. E. Church
18. First National Bank
19. Hall's/Allen's Hospital
20. The Big Swede (dance hall)
21. Rogers' Livery
22. Telluride Electric Light & Power
23. Cosmopolitan Restaurant
24. Bank of Telluride
25. Van Atta's store
26. Victoria House Hotel
27. City Hall & Fire Department
28. Miners' Union Hospital & Hall
29. Gold Belt (dance hall)
30. Silver Bell Saloon
31. Senate Saloon
32. Vincent St. John residence (rooming house, 2nd floor)
33. Clipper Saloon
34. Brunswick Saloon
35. Andrew Floaten residence
36. Colorado House
37. St. Patrick's Catholic Church
38. City Jail
39. Masonic Temple/Hall
40. Betts/Haines/Roth Livery & Carriage House
41. Fred Pheasey residence
42. Pheasey Livery
43. Stubbs and Jakway lumberyards
44. F. D. Work Hardware & Colorado National Guard Armory (2nd floor)

A NOTE ABOUT THE PHOTOS AND ILLUSTRATIONS

All portraits in this book with the label "Restored Portrait" have been created by artist MaryJoy Martin from images in microfilmed historical newspapers. In the sample below, the image on the left is as it appeared on the microfilm printout; the image on the right is the completed portrait by the artist. Illustrations labeled "Restored Photo" have been created similarly, but the microfilmed images were clearer, thus have undergone less than thirty percent restoration by the artist.

Portrait of H. M. Hogg: from microfilm printout (left) and finished artwork (right).

CONTENTS

PREFACE

I have heard the Western Federation of Miners called the "Western Federation of Dynamiters." I have heard this union denounced as "worse than Highbinders and Black Hand societies" and the members labeled as the "scum of Europe," "rabid anarchists," "rank socialist agitators," and "blood-thirsty assassins" with a "huge slush fund" for the legal defense of their hired killers. At Telluride, Colorado, the WFM Local Number 63 was accused of "inaugurating a Reign of Terror" in which murders and assassinations were commonplace and those they disliked were said to vanish under the "most vile and mysterious circumstances." By numerous accounts the Miners' Union men, particularly the officers, were degenerate, corrupt, menacing, frenetic, bomb-throwing, felonious, godless, indefensible, illegitimate sons of soulless hags.

These were my kind of lads.

As an investigative journalist, I have been driven to seek what lurks beneath the capstone of any creature or organization with a reputation so malignant as to make it a monster. Was the Western Federation of Miners a monster? Were the officers of Telluride's Local 63 monsters? Had the members committed all the crimes laid at their feet? Were murders so commonplace in San Miguel County as to create fear in the citizenry?

History seemed to whisper it was so. The first time I heard of it was in 1976 when an old miner in Denver told me a startling ghost tale about a headless shift boss named William Barney who haunted the Smuggler-Union Mine. I was told Local 63 had killed Barney and buried his dismembered corpse in the forest. Barney topped a list of Telluride men murdered or kidnapped by Local 63, a list that had outlived the union itself, a list that was signed, sealed, and published for posterity by the Mine Owners Association and confirmed by those models of veracity and courage, the Pinkertons. The Pinkerton National Detective Agency, with its motto, "We Never Sleep," had impressed me in my childhood as an organization determined to end lawlessness in the West. Thus I was drawn to the victims the Pinkertons confirmed as murdered, especially to Barney, who appeared as a ghost in the very same year union men had shot him. He showed up in Robert Livermore's memoirs (Gressley, *Bostonians and Bullion,* 1968) as a grinning skull with the red hair still attached. He popped up in San Miguel County histories and journals. He exemplified all the victims of Local 63. He had a dramatic story to tell.

I rushed out to investigate this mysterious victim whose bones had been found on Boomerang Hill six years after his disappearance. Along the way I met his fellow victims, those who had been butchered by the members of Local 63, those who had been run off, those who had been shot in cold blood, and those who were brutalized and abused. I wanted to tell their forgotten stories and let justice speak on their behalf.

Usually what emerges from dusty archives and government documents is simply a clearer, more detailed picture of historical events, but in the case of the murder victims of Local 63 there was a startling about-face. The paper trail proved the officers and members of that long-silenced union were not monsters of any sort. Sinister whispers on the winds of history were only echoes of a time when hatred had its day in Telluride.

To some, the labor history of Telluride was and is a taboo subject. To others, it is an embarrassment. To yet others, especially those who are descendants of the union men, it is still a ragged wound, for the men and women who were driven from their homes were never shown justice or even offered a chance to tell their side of the story.

Telluride's labor history is a complicated chronicle, one without monsters, but one with men and women who did what they believed they should to achieve their goals, whether right or wrong. It has heroes who stood against powerful corporations to end child labor and create safe working conditions and fair wages for future generations. It has mine owners who were dedicated to their employees, and those who were dedicated to the profit of the company, employees be damned. It has journalists who were motivated by truth and those who were motivated by hatred, thriving on antagonism and discord. It has idealists who believed the law would protect them, and egotists who believed the law was for their own personal use. It has arrogant leaders who flaunted their power and humble leaders who were committed to service. Fear and loathing, compromise and forgiveness, stubbornness and strength of character, violence and gentleness played on both sides, yet the greatest injustice was perpetrated by the State, the Mine Owners Association, the Citizens' Alliance, and those sleepless Pinkertons.

Telluride's story buried me beneath National Guard reports, trial transcripts, Pinkerton reports, mine company records, prison documents, union documents, memoirs, letters, old confessions, cockroach-stained county ledgers, crime reenactments, and forensic investigations and consultations. It took me from the dusty archives of Colorado to the hallowed halls of Wayne State University Labor Archives, from Idaho State Historical Society to New Mexico's state library, from pathology departments to rifle ranges. More than two decades of turning stones from New York to San Francisco, and the dedicated assistance of researchers in Pennsylvania, Nevada, Colorado, and beyond, as well as interviews and correspondence with the descendents of the players in this story, have produced solid evidence to topple accepted but erroneous beliefs of the guilt of Local 63 and its members.

As documents pulled the skin and sinew of hatred away from the hidden facts, the raw, bloody anatomy of conspiracy was revealed, a conspiracy so unmistakable on the part of certain mine owners and Pinkertons that it could not have been confined exclusively to Telluride. The documents proving Telluride's plot reach beyond these mountains to the entire West, wherever miners' unions clashed with mine owners in those early days of the twentieth century. J. Anthony Lukas discovered such disturbing plotting while researching *Big Trouble* (Touchstone, 1997). In the Telluride cases, actual documents finally prove the conspiracy; prove some Pinkertons indeed manufactured certain "confessions." The documents exonerate those accused of murder and destroy the dark lies forever.

I have footnoted each toppled assumption and cross-referenced every event from primary sources. I reconstructed incidents from as many viewpoints as possible. This book speaks for those who struggled for justice in an era when lives were "cheaper than timber." But it is not just a compilation of dry documents, names, dates, and events. History shouldn't be so hard that you get saddle sores riding through it, for above all, it is made by the men and women who lived it. This is their story, the story of those who lived through the labor strife, their tears of triumph, their long nights of terror, and their boisterous companionship; their broken dreams, seething hatred, and shining hope.

For one brief moment may they live again in these pages.

MaryJoy Martin
Lone Tree Cemetery, 3 July 2003

CHAPTER

1

RATTLING THE BONES

"The only completely consistent people are the dead."

— Aldous Huxley

Like threads made of shadow and wind, the mystery of William Julius Barney has tangled the serenity and startling beauty of San Miguel County for more than a century. Barney had vanished from Telluride in 1901 and the Mine Operators' Association declared union officers had brutally slain him. The governor of Colorado and the sheriff of San Miguel County declared the same. The county commissioners, attorneys, editors, the adjutant general of the Colorado National Guard, and the man known as the "Great Detective," James McParland of the Pinkertons — all declared the homicidal Telluride Miners' Union, a local of the Western Federation of Miners, had murdered William Julius Barney.

High on the mountainside where Will Barney worked, a last bit of wall of the Bullion Tunnel tram house still clings to the cliff.

Photo by the author.

Barney's death was important to the Mine Operators' Association (MOA) and the Pinkertons, not because they loved and honored the man, but because they needed murders to prove a "Reign of Terror" had gripped Telluride — a reign bred by the Western Federation of Miners, a reign that McParland said made the infamous Molly Maguire killings look like child's play. Since one murder failed to create the reign, the MOA tucked in a few dead companions beneath Barney: John Mahoney, murdered by the union; Sam Servis, murdered by the union; Wesley Smith, murdered by the union. Four murdered men were enough to constitute this Reign of Terror, particularly when their names were mentioned so frequently that it seemed the four were dozens or had been murdered several times over. Little did it matter that none of the dead men left a body behind for the coroner. Once they were declared murdered by the union, their names were strenuously rattled like magic bones before the public: a lie repeated often enough, especially by the press, eventually becomes the truth. Pure magic.

The press was a crucial weapon in the war the MOA waged on the Western Federation of Miners (WFM), and Barney's murder was indispensable to the war. The MOA had begun its war as soon as the WFM organized in Butte, Montana, on May 15, 1893. Prior to this, mine owners and operators had little to fear from the few unaffiliated miners' protective associations and unions in the western states. Individual miners and singular local unions, such as the unaffiliated Knights of Labor, had no real power to negotiate with operators. If miners requested safer working conditions, compensatory wage increases, or protested wage reductions, they simply were fired. When local unions made such demands, mine operators forced them out of the mines. In the 1892 strike in the Coeur d'Alene district of Idaho, mine owners merely replaced union men with strikebreakers, bringing them in from out of state under armed guard, although it violated state law. The mines and mills hired a private squadron of armed guards who protected the strikebreakers behind barricades, menacing any union man who approached the "dead line." Federal troops, sent in to control the situation, forcibly shut down the few mines whose owners had readily consented to union requests and continued to operate with union men. Backed by military presence, citizen organizations compelled the legally elected sheriff and marshal to resign. The mobs herded union men and their sympathizers into bullpens and brought charges against hundreds of them. The courts quashed the charges.[1] The MOA merely used indictments as an object lesson.

Profit was the bottom line for mine operators, yet many knew that treating employees with fairness and respect was good business. Some truly believed the most important thing to come out of the mine was the miner; others believed the most important thing was the ore. Typical of the more callous mine operators was the attitude of John F. Geisel, the Terrible Mine foreman in Ouray County in 1887 who allegedly took no interest in recovering the body of a miner killed on duty. "What's the difference?" he told an employee, "Men are cheaper than timber."[2]

Mining was a perilous occupation, with numerous miners dying from accidents, explosions, or mine-related illnesses before the age of forty. If miners were crippled in accidents, they were let go. Many were forced to live in company boarding houses where the barest minimum of their needs was met. Yoked to long hours, wages halved by boarding and equipment fees paid to the company, dismal or nonexistent medical care, and no compensation to his family should he meet with accident or death, the miner's hope for a better future was in solidarity, a brotherhood of skilled men who could stand shoulder to shoulder across the West. Out of this hope the Western Federation of Miners was born.

That first WFM convention in 1893 consisted of forty delegates representing fifteen unions in Colorado, Montana, Idaho, South Dakota, and Utah. Their purpose was to organize for protection against such situations as the 1892 Coeur d'Alene tragedy and for advancement of their profession. The preamble to the WFM constitution began:

> *We hold that all men are created to be free and should have equal access and opportunity to the enjoyment of all benefits to be derived from*

> *their exertions in dealing with the natural resources of the earth, and that free access and equal opportunity thereto are absolutely necessary to man's existence and the upward progress of the human race...it is highly fitting and proper that the men who are engaged in the hazardous and unhealthy occupation of mining, milling, smelting, and the reduction of ores should receive a just compensation for their labors....*[3]

Some mine owners and operators believed their employees deserved a just compensation and dealt with the new union accordingly. Yet, with mining investments nothing more than a gamble, an increasing number of mine owners opposed organized labor because labor's demands threatened profit. Thus, as soon as miners organized, mine operators also organized. While the operators' associations assured employees that miners' unions were acceptable, at the same time the MOA embedded spies in the ranks to subvert the unions from within, a course of action taken against unaffiliated locals before the WFM existed. Spies, such as Edward L. Zimmerman of the Thiel Detective Service who got himself elected as a union financial secretary,[4] and Charles A. Siringo of the Pinkertons who sidled into the union as C. Leon Allison in 1891,[5] became commonplace in the WFM's locals.

A posed work scene in the Tidal Wave Mine at Ophir only hints at the difficult and dangerous occupation of the miner.

© Main Street Photography, Walker Studio Collection; used with permission (23T).

The financial and political power of the opposition quickly surpassed the negotiation power of the WFM and union men were subjected to the same scenario as before: they made their demands to mine owners in a businesslike manner, were refused, went out on strike, were replaced by strikebreakers imported from other states under armed escort, were harassed until confrontations ended in deaths and mine operators called for military assistance from the state.

The nightmare of 1892 recurred in the Coeur d'Alene district in the strike of 1899, this time far bloodier on both sides.[6] Federal troops were again the solution, followed by mass arrests without warrants and union men incarcerated in crowded warehouses and boxcars, they and their families subjected to assault and abuse. A proclamation by Idaho's Governor Frank Steunenberg prohibited the employment of any member of the WFM, declaring it a criminal organization and blaming it for numerous unsolved crimes. The MOA implemented the card system, barring known union men from securing work in the district and issuing cards that approved only nonunion men for employment.

Idaho condemned the WFM without investigation, evidence, or indictment, and attainted its members without due process. The state sanctioned the mine operators' lawlessness with the power of military might to drive the point home, clearly demonstrating to the miner he was worth less than timber.

Such glaring injustices drove miners into the arms of the WFM, swelling the ranks from the first fifteen locals and a membership of 2,000 to over 200 locals and a membership of 48,000 by 1902, embracing the West from British Columbia to Texas.[7] Despite the violent opposition to it, its purpose remained to unite miners, millmen, smeltermen, and engineers "into one central body, to practice those virtues that adorn society and remind man of his duty to his fellow man, the elevation of his position, and the maintenance of the rights of the workers." The organization sought to "secure compensation fully commensurate with the dangers of" the mining profession, to end the use of scrip as payment and end the dictation of employers as to where employees were to spend their earnings; to "strive to procure...the use of any and all suitable, efficient appliances for the preservation of life...and to labor for the enactment of suitable laws...to provide for the education of our children and to prohibit the employment of all children until they have reached at least the age of 16 years...to prevent by law any mine owner, mining company, or corporation...from employing detectives or armed forces, and to provide that only lawfully elected or appointed officers...shall act in any capacity in the enforcement of the law...*to use all honorable means to maintain and promote friendly relations between ourselves and our employers and endeavor, by arbitration and conciliation or other pacific means, to settle any difficulties which may arise between us, and thus strive to make contention and strikes unnecessary* [emphasis added]."[8]

Scrip, or payment issued by the mining company in lieu of cash or bank notes, was good only at the company stores. This was a huge source of conflict at many mines.
Author's collection.

The Western Federation of Miners, legally organized, clearly began as a body with noble purpose and not one of militant anarchy with the destruction of corporations or government as its goal. Yet it was in the latter colors that mine operators painted it. Colorado claimed nearly thirty-two percent of the total WFM membership, giving the state's mine operators a collective shudder. The power of its escalating membership meant mine owners could no longer arbitrarily impose wage reductions, long hours, or boarding constraints on their employees without waking the specter of the strike. Strikes were always costly. The WFM had to be stopped.

During strikes in Colorado, the MOA waged battles against the WFM as they had in Idaho. Confrontations invariably flared. It took men of conviction and courage to stand strong in the face of extreme frustration, abuse, threats, and tyranny, while watching nonunion men take their jobs, and all too often at the *very wage* they themselves had asked. When pushed too far, some Federation members became defiant, attacking the strikebreakers and the deputized guards, although the organization itself decried such violence and turned lawbreakers over to the authorities. Individual troublemakers or small renegade bands within the union bent on intimidating, assaulting, or shooting their perceived enemies only fed the animosity of the mine operators, contributing more blood-red paint to color the WFM as lawless or criminal.

Lawless and criminal, however, were labels that wrapped themselves equally well around the mine operators and their associates. If the WFM was militant, so too was the MOA; so too the armed citizens' alliances. While the mine owners accused the Federation of beatings and kidnappings, their own thugs beat and kidnapped union men and supporters. One brutal example was the castration of a miner named Joe Raiz who later died from his wounds;[9] another was the kidnap of Adjutant General T. J. Tarsney, an attorney representing arrested union men from Cripple Creek. During the time of the miners' trial in June 1894, Tarsney was struck over the head at midnight and taken from his Colorado Springs hotel by fifteen masked union-hating men. Dragged to a remote spot on the prairie, he was stripped and beaten, his flesh seared with tar and feathers.[10] The injuries contributed to his death in 1902.[11]

Many historians who were quick to label the WFM as the most militant union in the West failed to note those in opposition to the union surpassed it in instigating violence, intimidation, discrimination, deportations, and retribution, and outstripped it in militancy, having the power to purchase the state militia to drive unarmed, law-abiding union men from mining districts and to raid their homes and union headquarters at will. The WFM never hired Pinkerton infiltrators to disrupt business, manipulate funds, provoke discord, and plot crimes within the corporations, as did the MOA. The case of Pinkerton Operative No. 42, A. W. Gratias, illustrated this practice clearly: Gratias had become financial secretary in a Cripple Creek WFM local and was given instructions by the Pinkerton supervisor to cut strike relief funds "as much as possible, so as to cause dissatisfaction, and get the men against the union." He was to put the blame for the cut on the WFM's executive secretary, William Haywood, in order to turn union members against Haywood.[12] Many MOA members believed such guerrilla warfare with agents provocateurs entrenched in the WFM's locals was their right, no matter how criminal the activity. The WFM never had an army bristling with bayonets, Krag-Jorgensons, and Gatling guns. Nor did it have the state militia at its disposal, even when it urgently petitioned the governor for such protection.

During the Cripple Creek district strike of 1894, mine operators hired so large a force of armed guards, who were then deputized by anti-union Sheriff Bowers and "divided into infantry, cavalry, and artillery" units, that to any observer it was an imported private army. Violent confrontations between these approximately 1,200 deputies and the strikers who armed themselves against them ended with dead and wounded participants on both sides.[13]

According to the 1905 Senate Report read before the Fifty-eighth Congress, even the Governor of Colorado believed the mine operators and their backers were militant and lawless:

> *On May 26 [1894] Governor Waite issued a proclamation, in which he called upon the strikers to lay down their arms... He also declared that the assembly of a great number of deputy sheriffs, many of them recruited from other counties, was illegal, and he demanded that they immediately disperse. Under Section 856 of Mills' Annotated Statutes of Colorado, Governor Waite held that the sheriff of any county was authorized to deputize only citizens of that particular county. He regarded the act of Sheriff Bowers in swearing in deputies by hundreds and equipping them with stands of newly purchased arms as really the formation of an army. He held that the right to appoint deputies did not give the right to assemble an army and that the act of Sheriff Bowers was a usurpation of the power of the chief executive. Though the deputies were under the command of the sheriff, the governor considered the mine owners as backing the movement [the mine owners provided the funds], and that the object of enlisting and arming such a body of men was to force the miners to give up their struggle.*[14]

This same disastrous and tragic drama had played from Coeur d'Alene to Cripple Creek, British Columbia to Leadville, demonstrating that militant violence was a two-sided coin. When the Irish-born Edward Boyce, who had become president of the WFM, spoke at the convention in 1897, he advised each of the locals to have a rifle club and "to provide every member with the latest improved rifle."[15] His words sprung directly from the Coeur d'Alene and Cripple Creek tragedies, where men lost their jobs, their homes, and some, their lives, because they simply demanded a fare wage. Neither violence nor anarchy was his goal. He believed union men had to protect themselves.

Boyce was a dynamic leader, spurred by a deep passion that rose like flame toward revolution. Under his leadership he edged the WFM toward socialism, increasing the nightmares of employers. At the ninth annual convention in 1901 he told the delegates:

> *Advise strikes as the weapon to be used by labor to obtain its rights, and you will be branded as criminals who aim to ruin the business interests of the country. Change from the policy of simple trades unionism that is fast waning, and you will be told that your action is premature, as this is not the time. Pursue the methods adopted by capitalists and you will be sent to prison for robbery or executed for murder. Demand, and your demands will be construed into threats of violence against the rights of private property calculated to scare capital. Avail yourself of your constitutional rights and propose to take political action, and you will be charged with selling out the organization to some political party. Counsel arbitration, and you will be told there is nothing to arbitrate. Be conservative, and your tameness will*

> *be construed as an appreciation of the conditions imposed upon you by trusts and syndicates. Take what action you will in the interests of labor, the trained beagles in the employ of capital from behind their loathsome fortress of disguised patriotism will howl their tirade of condemnation.*[16]

Heavy with truth, Boyce's rally cries nonetheless failed to inspire the rank and file to embrace an actual labor revolution. Many miners believed demand, strike, arbitration, boycott, and political action were viable solutions. Socialism was beyond them. Constitutional rights were not. They did believe they had a legal right to protect themselves and their families and a right to arm themselves against private armies hired by the mine operators.

Ed Boyce, first President of the Western Federation of Miners.
Author's collection.

Despite the absence of any mass revolutionary action or the flame of socialism igniting, Boyce's words were reason enough for the mine owners to hurl further condemnation upon the WFM and seek the organization's destruction by whatever means they could, legal or otherwise. Some believed words and ideas were as dangerous as dynamite. Some feared the union because its numbers increased daily. Some hated it because it dared arm itself against their illegal armies. Some despised it because it gained political power and the support of such leaders in Colorado as United States Senator Thomas M. Patterson, State Senator William S. Buckley, and Representative F. A. Moore, the latter two introducing legislation to restrict the work day to eight hours. The WFM had organized the movement to gain passage of such a law, which the mining, milling, and smelting corporations bitterly opposed. Again, loss of profit was the bottom line. Although the public supported the WFM in its campaign for an eight-hour law, the MOA did everything possible to squash such legislation. Indignation rumbled through the citizenry, giving the WFM greater public support. The MOA realized the power to extinguish the Federation was linked to public opinion. They had to destroy its public support, which was no small task.

Whenever a mine disaster claimed multiple lives, particularly due to negligence on the part of a mine operator, it amplified public sympathy for the WFM. The public readily believed miners had the right to demand the full power of the law in such instances as when nearly 300 were killed at Scofield, Utah, in an explosion caused by the company's failure to comply with ventilation laws; or when 137 died in a Fernie, British Columbia, mine due to the company's negligence. Before all the bodies were removed from the latter mine, the manager reduced the survivors' wages. The public was outraged.

Senator Patterson, addressing the Fifty-eighth Congress, said the Mine Operators' Association and mining corporations "tell us these men are over-paid — that their demands for decent wages and safe conditions imperil the prosperity of the country. Merciful God! The prosperity that demands a constant repetition of these horrors is not fit to survive an hour."[17]

As owner of the *Rocky Mountain News,* Senator Patterson used his newspaper to fan the flames of public outrage and to promote the miners' fight for justice, instantly gaining the wrath of corporations, syndicates, and the MOA, and the disaffection of their allies, the Pinkertons. The MOA quickly joined the battle of words, a battle for the very mind and heart of the public.

Newspapers suited the MOA's war plans far better than actual battles on the ground, far better than scores of labor spies, and far better than legal harassment. Newspapers formed public opinion; they were the mass media of their day. They claimed to present the facts and the truth. Readers trusted them to do so. The melodramatic yet revealing statement of John Swinton, a New York editor in the late 1800s, illustrated that such trust was misplaced: "There is no such thing...as an independent press.... The business of the journalist is to destroy truth, to lie outright, to pervert, to vilify, to fawn at the feet of mammon, and to sell his country and his race for his daily bread.... We are the tools and vassals for rich men behind the scenes.... Our talents, our possibilities, and our lives are all the property of other men. We are intellectual prostitutes."[18]

Although the exploitation of the journalist was not as transparent as Swinton expressed, too often powerful people behind the scenes indeed manipulated public opinion with the subtlety and skill of surgeons. Propaganda was as old as sin, and perhaps the father of it. Yet the quintessential weapon in the MOA's public opinion war against the WFM was the publisher or editor who despised the Federation more than the MOA did, for seething animosity always failed to recognize truth or fact. In Telluride the local MOA found such a chambermaid in Francis Edward Curry, editor of the *Daily Journal*. Curry was a man whose festering bitterness assured he would willingly discredit, disgrace, or dishonor the WFM, spilling his twisted news across the state via the *Associated Press*. Guy E. Miller, president of Local 63 in 1904, pointed out, "It is only crimes in which they think there is a possibility of implicating the union or a member of it that are heralded abroad."[19]

Curry pronounced the WFM and particularly its Telluride Local 63 guilty of whatever crime and outrage he could. If a man who had contact with union men went missing in the district, Curry let it be known the union was most likely responsible for the victim's *death*; whether the man was an actual victim or dead was immaterial. If the victim turned up living, Curry refused to print the fact, allowing the accusation of murder to hang over the union. "The disappearance of men," said Miller, "has been charged to the union persistently months after the alleged victim was seen" alive.[20] In news stories Curry derided union men and their families. He advocated violence against them. He labeled them anarchists and the "disturbing element."[21] He refused to apologize for errors. If one disreputable union man actually committed a crime, Curry made it appear a common practice among the membership, creating the illusion the union indulged freely in criminal activity. If one belligerent union man uttered a threat, Curry reported many members as uttering the same threat. If those he slandered no longer lived in the region, Curry's scorn broke into print like a black flood, since his targets were unable to protest or defend themselves.

F. E. "Eddie" Curry knew the power of the press. He understood the power of repetition. He knew he played a pivotal role in the community: the *Daily Journal* and the weekly *San Miguel Examiner* were the main sources of news in this remote mining camp in western Colorado. He knew a local population that felt cut off from the world devoured news the instant it was off the press. Born in New York in 1856, he had come to the San Juan region at the age of nineteen with his older brother, John R. Curry. They settled in Silverton in 1875. That year John established the *La Plata*

Miner in that camp, and in July 1881, founded the *San Miguel Journal* at San Miguel City, moving it that same month to Telluride, with Eddie in charge. John sold the paper to Charles Painter and it was renamed the *Telluride Journal,* with A. M. Reed as editor. Although Eddie left the area briefly, he returned to work for Painter as editor. The *Journal* became his life. Had he kept his personal contempt from sticking to the printing press, had he not bedded with the MOA, had he promoted settlement and harmony instead of strife and enmity, Telluride's history might have played out with far less tragedy.

Even long after Telluride Local 63 had wearied under the suffocating persecution, Curry continued to condemn it. His only reason for the condemnation appeared to be the claim that it was a murderous organization, a claim grounded in false accusations, a claim he could not relinquish. By repeating lies in print, he assured their place in history as "facts." The pinnacle of this journalistic magic was William Julius Barney, whose "murder" had become fact.

To sustain the fact and keep it newsworthy, Curry made the murder more brutal and detailed with the passage of time, changing the original single shot to Barney's head to a vicious eleven bullets perforating the man. He boldly delivered the murderers' names to posterity. When Curry figuratively rattled Barney's bones with a triumphant see-I-told-you-so headline in 1907, it echoed over the San Juan Mountains for decades, and his fact crept into pioneer memoirs and history books, accepted as unalterable truth: *beyond the shadow of a doubt, Telluride Local 63 had murdered William Julius Barney.*

If that was truth, F. E. Curry stored nuts in his nose.

CHAPTER

2

A CORPSE NICELY DISMANTLED

"A person doesn't die when he should but when he can."

— Gabriel García Márquez

His brain reeled under the pressure. Sweat dribbled into his eyes. The bloated Pinkerton detective in the chair opposite him wanted details. *Details* would save him from the hangman's rope. The detective had stroked him with that promise since he first met him, but Steve Adams was short on details and faulty in memory. He sucked at the cigar the detective had given him as if to suck the dreadful details out of it.

"One of the hands was missing like animals had chewed it off," Adams said.

The Pinkerton seemed pleased, gently pressing him to continue.

"And the maggots had got their work in in the intestines," Adams recalled, "and were just in black balls about the size of eggs."[1]

The white-haired, bespectacled Pinkerton offered Adams another cigar. The man seemed so reassuring, as kindly as a grandfather. Adams knew he was James McParland, manager of the Western Division of Pinkerton's National Detective Agency. He knew his reputation was as big as his belly: this was the "Great Detective," the "Star Witness," the man responsible for the hanging of the Molly Maguires in Pennsylvania in the 1870s. At that time, McParland had worked as a Pinkerton Secret Operative among the Irish miners in the Pennsylvania coalfields. The coal company wanted him to ferret out an alleged clandestine group of murderers and thugs. Using the alias James McKenna, he joined the Ancient Order of Hibernians (AOH), an organization he believed was the same as a secret society in Ireland known as the Molly Maguires. He took part in beatings and murder plots, only to reveal himself later as a detective, his testimony condemning fellow AOH members to death.[2] In those days McParland signed his name "McParlan" on Pinkerton reports, since he was the son of Ewan McParlan and Mary Laughlin. He added the "D" to the end in the 1880s yet dropped it again in 1900, only to restore it in 1901. Perhaps undercover work caused him to forget who he really was. Since revealing himself on the stand in 1876, this agent provocateur was saluted as a hero, a savior, and a great detective — at least by the Pinkertons and their clients — and became known as the "terror of the Mollies."[3] Others saw him as a liar, a killer, a manipulator, and a goon, this hard-drinking man who was called a tiger in a brawl and McKenna the Tough while under-cover. Yet to Steve Adams, McParland's voice was soothing, his cigars a sign of friendship, and he was Adams's only hope of escaping the musty prison cell and a hangman's noose in the Idaho penitentiary.

Steve Adams had been arrested February 20, 1906, in connection with the assassination of ex-Governor Steunenberg of Idaho. His old friend, Harry Orchard, had led the State to Steve's door. Orchard, who had a pocketful of aliases[4] and more wives than the legal limit, had personally confessed to McParland that he blew the ex-governor to ribbons on behalf of the WFM.[5] That was exactly what McParland wanted to hear.

Not any sort of officer of the law, McParland was hired by the State of Idaho to investigate Steunenberg's murder. Having extracted Orchard's confession within days, McParland had informed Idaho's Governor F. R. Gooding on February 5, 1906: "Orchard said that there was a comrade of his named Steve Adams with whom he had an understanding that if ever he got into trouble Adams would assist him as a witness, and if Adams got into trouble he (Orchard) would do the same... we want Adams here in the worst way."[6]

Pinkerton James McParland.
Restored portrait © M. J. Martin 2003.

If arrested, Adams was supposed to be kept in a separate cell away from Orchard as McParland explicitly spelled out to the warden. To prime a man for confessing, McParland required solitary confinement, a penetrating silence, the watchful presence of a stony guard, and as few contacts as possible. Adams was to be denied access to an attorney, knowledge of his wife and children, and no information on the charges against him. McParland had left instructions with the warden at Boise and returned to his Denver office.

When Adams was arrested he immediately was placed in with the confessed bigamist serial bomber, giving Orchard opportunity to go over details with Steve and causing near apoplexy in McParland, who was unable to return to Idaho for a week. While together, Harry reminded Steve they had made a promise to help each other out should either get into difficulties. "Assist him as a witness" was a phrase of many dark colors. For nearly a week they shared a cell, sharing details, with Orchard explaining precisely what McParland wanted.

Both McParland and Orchard needed Adams to corroborate Orchard's own mind-boggling book-length confession. By law the confession would not stand on its own. McParland also needed Adams to confess to crimes committed in Telluride, Colorado, in order to satisfy his most important client there: Bulkeley Wells.

Long before the State of Idaho had retained the services of Pinkerton's National Detective Agency, Pinkerton's Great Detective, James McParland, had solicited and won Colorado Mine Operator's Association members as clients. Among these was Bulkeley Wells, president and manager of the Smuggler-Union Mining Company that operated the Smuggler-Union, Sheridan, Mendota, and Contention properties at Telluride. For the high fees paid to McParland, Wells wanted bodies hanging on gallows. A confession from Adams had potential to satisfy Wells's desired body count.

Just as Harry Orchard's confession conveniently solved dozens of murders, dynamitings, and assaults throughout the West, McParland wanted Steve Adams to clear up the unsolved crimes at Telluride, specifically those that could be blamed on the Telluride Miners' Union Local 63: the assassination of Arthur L. Collins; the disappearances (which translated to "murder" in McParland's mind) of William J. Barney, John Mahoney, Wesley E. Smith, Sam Servis; the murder of Ben Burnham, and the attempted murder of James P. Phillips. The convenience of having one confessor per region kept the paperwork simple. McParland assured Adams he was safe from the gallows himself, as long as he confessed according to McParland's instructions; that is, as long as he claimed the crimes were committed under the order of the officers of the WFM. McParland assured Steve he was "just a poor tool." Later, at one of several trials, Adams told the court McParland "wanted the names of the officers of the Federation used as much as possible all through the confession."[7]

As with Orchard, McParland guided Adams, supplying details, names, and dates in a simple question and answer session prior to obtaining a typed and signed document. No part of Orchard's confession revealed any details that weren't already published in newspapers or available to McParland. Similarly, no part of Adams's confession revealed any new details, but unlike Orchard, who keenly memorized details until the confession could be recited without mistakes, Adams had no memory for dates or names.

The thirty-nine-year-old Adams failed to recall what year he arrived in Colorado, what year he joined the WFM, what years he was arrested for petty crimes in Denver and Utah, where he was which year for what purpose, even though the years under discussion were only seven. McParland had to gently correct him when he switched names of WFM officers or their victims. Adding to Adams's confusion, McParland seemed to have massive memory lapses as well, using names from thirty-year-old Molly Maguire cases in reference to participants in Adams's crimes.

At the end of two days, Steve Adams delivered a twenty-four-page confession that satisfied McParland and gave Bulkeley Wells hope for at least one hanging. That hope came in the form of a decomposing corpse, the very one that Wells had sought for four years: the corpse of William J. Barney. It would give Wells the proof he needed to hang the man he most wanted on the gallows: Vincent St. John, once president of Telluride's Local 63.

In conversation Adams referred to Barney as "Barnum," indicating he actually didn't know who the victim was supposed to be, but McParland had his stenographer correct the name on paper. McParland's reports of his conversations with Adams also illustrated Adams had no idea when or where Barney was supposed to have died, if he was buried at one time and reburied two weeks or two years later, or never buried at all. Adams even failed to keep straight the name of the man who helped find the body in order to bury it when it came Adams's turn to do the burying. According to several conversations reported by McParland, a man named Kelly assisted Adams in locating the corpse. Since Kelly was not an officer of the Telluride union, his role was taken over by Oscar Carpenter, the union secretary, by the time the confession was on paper:

> *Two men were murdered in 1902 in the Telluride District of Colorado, one named Smith and the other Barney. I helped to move the body of Barney from the place where he was first killed. Assisted O. N.* [sic]

> *Carpenter to do this. Mr. Carpenter seemed to know I had killed Collins and that I was a safe man. He told me that St. John had once moved this body and had buried it but wanted it moved again. I did so and can take you to the place where the body is now buried. Smith was killed by St. John at Telluride, after the murder of Collins. All I know about it is by hearsay. Carpenter told me all about it.*[8]

The confession's erroneous dates and worthless hearsay were mere trifles for McParland and Bulkeley Wells. The prize was the name of the cold-blooded killer: Vincent St. John. All Wells needed was Barney's corpse.

Wells, who divided his time between the corporate offices in Denver and the mines at Telluride, had been organizing endless digging excursions in search of Barney's bones for several years, relying on jailhouse tips, solid rumors, and vindictive accusations from the enemies of St. John. Now he had a Pinkerton-procured confession. He demanded that McParland supply details immediately so he could exhume the elusive bones.

The superintendent of McParland's Denver office wired McParland in Idaho: "Wells suggests get sketch from Adams location Barney's body, mail it Telluride. Wells goes there Sunday night, motive, exhume body before others interested dispose same. Wells has notified Rutan to wire Nichols. Rutan leaves tonight or in morning for Denver for papers. Wells instructs locate Baston."[9]

Rutan was San Miguel County's Sheriff J. Calvin Rutan, a faithful servant to Wells and the Mine Operators Association. Baston was an alleged accomplice of Adams. Wells was conducting the investigation of the Telluride "Crimes Committed by Local 63" personally, giving instructions to both Rutan and McParland whom he paid monthly. McParland sent reports to Wells's corporate attorney in Denver, Jacob Fillius, and to Wells's legal arm in Telluride, Edward C. Howe, attorney for the MOA.

On March 14, 1906, McParland reported:

> *I took up the matter of the location of Barney's body, and [Adams] again informed me that it would be utterly impossible for him to describe where they put this body, and except by the best of luck it would take him personally a couple or three hours to locate this body. He said, however, he was satisfied he could locate it, but it was impossible for him to describe just where they put it. He said the time he was out riding with Kelly-the-bum when he got the latter to indicate where the body of Barney was located that he and Kelly met General Bulkeley Wells going into Telluride, and he would not be surprised but Mr. Wells might be able to remember that fact, but he might not be able to do so.*[10]

McParland's mention of Kelly the Bum was a peculiar addition to the Steve Adams saga. Kelly the Bum was a character from McParland's Molly Maguire case thirty years earlier. The Bum was a vile confessed murderer named Daniel Kelly, alias Manus Cull. McParland told Kelly's story to both Adams and Orchard as an illustration of how horrific murderers could be set free, thanked, and even paid a full year's wages if they turned state's evidence. Kelly was more than a cold-blooded killer. He actually enjoyed murder. Yet McParland used him and his confessions to hang other Molly Maguires and Kelly was set free.[11] Although freed three decades before, it was unlikely this Kelly the Bum riding with Adams in search of Barney's bones was the same man from Pennsylvania. Either the Great Detective's mind was slipping or he was fabricating evidence. His report was dictated to a stenographer, then typed in triplicate, read for errors, and signed by McParland. Thus the repeated reference was hardly a typographical error.

McParland continued to press Adams for the location of Barney's bones. As weeks passed, Adams seemed to enjoy the prospect of getting out for an excursion. He discussed the details with Orchard and insisted he needed to go to Telluride to find the site.

McParland took the proposal and both Orchard's and Adams's confessions to Denver with him, allowing his clients, Wells and Fillius, to study the documents.

"I am pleased to state that Mr. Fillius has grasped the idea of what is wanted and is working hard on the case," McParland reported to Governor Gooding.[12] Through Fillius and Wells, the MOA was pouring money into the prosecution of the WFM in Idaho in an undeniable, organized effort to destroy them.[13]

On the Colorado front, all Wells wanted was a noose for Vincent St. John. He became more absorbed in his need to find Barney's body. Without consulting state officials in Idaho, McParland boiled up a plan to have Wells sneak Adams out of Idaho for a few days. McParland was confident he was the one in control, not the state of Idaho, not Colorado.

He spelled it out to Gooding:

> *During our conference the matter came up as to how Steve Adams could be taken from Boise to Telluride in order to locate the place where Barney's body was buried. We readily agreed on a plan whereby this could be done without any person knowing that Steve Adams was in Colorado...it is all important that we should recover the body of Barney before Carpenter might take a notion to have the same removed, and while there is no risk in taking Adams into Colorado so far as anybody recognizing him is concerned, still there is one risk, that is in case of an accident and Adams getting killed we lost the most important witness we have got.*[14]

McParland's machinations aroused Wells with the idea that O. M. Carpenter, one-time secretary of Telluride Local 63, might beat a path to Barney's corpse before anyone else did. That was unlikely: Carpenter knew nothing of the corpse and was nowhere near Colorado, and no one at this point knew Adams had confessed to moving a body. Nevertheless, McParland wanted to give Wells something for his money, although the snow was so heavy in the mountains above Telluride at this time that searching for a grave in the timber at 11,100 feet was preposterous.

By May McParland had Wells worked into a fever. Wells convinced himself "the snow is about cleared from the hills."[15] The secret plot to sneak Steve Adams out of the state was put into motion. With a deputy sheriff's commission and in company with Telluride Town Marshal Willard Runnels, Wells traveled by train to Idaho. He hired Runnels as a bodyguard or gunman for the trip. Neither Wells nor Runnels had extradition papers, a warrant, or any other legal document giving them authority to take Adams out of Idaho, yet they were fully armed and wore their lawmen's badges.[16]

On May 27 Adams was brought from Boise to Orchard, Idaho, and put into the custody of Wells, Runnels, and a Pinkerton detective. They took him via rail to Telluride, secretly lodging him under guard at Wells's residence at the Smuggler-Union Mill complex. During the next two days they escorted Adams up Boomerang Road (also called the Alta road, which went up Boomerang Hill) to the site of an old sawmill, a distance of nearly eight miles from Telluride, one mile from the Gold King Mill, and less than a half mile from Alta. Barney's body was supposed to be buried

Extensive deforestation is evident around the Gold King Mill near Alta, the area where Will Barney was supposedly buried.

© Main Street Photography, Walker Studio Collection; used with permission (62T).

somewhere near the sawmill. Due to heavy snowdrifts still lingering at that altitude, Adams failed to find the gravesite.

Deeply disappointed, Wells returned Adams to the Idaho penitentiary. At this point Adams had not been charged with anything, especially not with conspiracy in the death of ex-Governor Steunenberg, for which he had been arrested. He still had no legal representation. His wife, Annie, and their small children had also been locked in the penitentiary shortly after his arrest for "their own protection," McParland had assured him, hinting that something dreadful might befall them.

Back in cell-side consultation with Orchard, Adams hashed out more details of the gravesite location. Curiously, Orchard knew more about the grave than Adams did, a fact confirmed by McParland's report of June 4, 1906: "Orchard informed me that in talking with Adams since the latter's return from Telluride he is satisfied that Adams has located the exact place where the body of Barney is buried, and he is very sorry that the snow was so deep that they were unable to exhume this body."[17] Did Adams's promise to be Orchard's witness mean Adams was taking responsibility for some of Orchard's dark deeds?

McParland went on to say that Adams was equally disappointed he failed to find the grave. The Pinkerton wrote, "He says that if General Wells is not able to discover this body after the snow has gone he could go there at night just as he did at the time that Carpenter and he re-buried the body, and could uncover it, as there is no doubt about it providing that it has not been removed, and he does not believe it has."[18]

Unconcerned how one could locate a remote gravesite better by night than by daylight, McParland continued to squeeze Adams for details of the burial with disguised promises of immunity and chilling hints of death by hanging or at the hands of a mob, furnishing handfuls of cigars to the confused and nervous prisoner.

Steve Adams was a simple man; he had no defense against the psychological warfare McParland so skillfully practiced. A moderately built miner of a little over five-foot-ten, Adams had a long nose set in an elfin face with a wide smile, light brown hair, and a complexion that took the sun poorly. His watery eyes were so pale blue that a Telluride jailer listed them as "pink" in 1907. McParland described them as shifty, but every member of the WFM described by McParland had shifty eyes, a trait more in the Pinkerton's mind than in reality.

After more than a hundred days at the game, Adams began to believe this rotund detective was his friend. He wanted to please him. He began spinning out details of beatings and other crimes, of which McParland could find no evidence. As long as Adams said the WFM officers ordered the crimes, this was useful information. Adams had gradually become more relaxed with the prospect of confessing to indict others, feeling secure enough to "lie down" on the "trap of the gallows" in the prison yard and "take a nap."[19] He also wanted another chance to breathe the fresh mountain air of Telluride. Thus he gave strenuous efforts to the Barney story. He begged McParland to tell Wells not to go digging for Barney's body for he was certain he could go "straight to the place where the body and clothes are buried."[20] At night, of course.

By June 9, 1906, Adams churned up the details of how he finally had found the body in September or October 1903 on behalf of St. John and in company with O. M. Carpenter. Why it took him so long to recall such vivid details only McParland could explain. Barney had one hand missing as if predators had gnawed it, and a few finger bones scattered nearby. Although murdered in June 1901, Barney was in astoundingly good shape for a corpse that clearly was never buried. Adams "suddenly came upon the body" where it was "sticking up...behind a log, fifteen feet from the road"[21] and only about one hundred yards from the old sawmill. McParland's reports confirmed the body was on the surface. Adams did not do any digging to remove it.

According to Adams's confession, his instructions from St. John were to find the body, remove the clothing, and bury the body separate from the clothing "so no one could identify the remains by the clothing." Adams told McParland, "The body had on a coat, vest and trousers, also a pair of hob-nailed boots." He knelt beside the body and stripped it down, dumping the clothes and boots in a sack. He said Carpenter "couldn't stand the stench and got deadly sick," leaving the job to Adams.[22]

Growing comfortable with the grisly details of his story, Adams continued, "I just pulled the feet out of the sockets at the knees and put them in the [second] sack, then pulled the thigh bones out of the sockets at the hips and put them in the sack." Although he told McParland, "The maggots had got their work in in the intestines," he later informed Marshal Runnels that he dumped the "entrails in with the clothing."

Adams pointed out to McParland, "The hide on the back was a little tough." Thus, after he had "squeezed the ribs into the sack," he rolled the hide and the head up...in a round ball and shoved them into the sack."[23]

Carpenter, still reeling from spewing his supper, regained enough control to carry the putrid sack of clothes, boots, and entrails down the hill. Adams carried the sack of rotting limbs, torso, and head. By now night had fallen. They went somewhere between thirty to a hundred yards from the body's original location to bury their bundles. Since the ground was rocky, with a covering of heavy spruce needle duff, they dug only shallow holes, burying the sack of bones separate from the boots and clothes. Oddly, Runnels later claimed Adams told him they "threw the shovel away and the sack with the entrails and clothing was chucked into the river,"[24] which meant they carried it several miles. Either Adams had yet another lapse of memory or Runnels was spinning what cowboys called "windies."

"The ground was level where they dug the grave," McParland reported, "and as they had an axe with them they cut some roots and brush and when they buried the body in the grave they put a lot of this brush on top of it so that if anybody was walking around there they would not notice the fresh earth."[25] If no one had noticed the rotting corpse for more than two years when it was lying only fifteen feet off the road, then why did they worry someone might notice fresh dirt a hundred yards off the road?

Adams said the holes were so shallow that "the boots kind of stuck up" and he "had to dig the hole a little deeper so as to cover up the boots." Carpenter hurried off ahead of Adams, his stomach still churning. Before Adams left he "cut a nick in an evergreen tree with the axe. This was not a very large tree."[26] The sawmill company was active at this location in the decades of the 1880s and 1890s, taking most of the large spruce from the Turkey Basin area, which was less than a mile north of the village of Alta and the Gold King Mill. The terrain was littered with logs helter-skelter and dotted by regrowth and the ragged younger trees rejected by the logging crews. A few stands of larger spruce clung to slopes and ravines more difficult to access. A nick in a small spruce in a logged area would be difficult to find after three years, particularly since many smaller trees had scars from large trees crashing down on them. But McParland valued details.

Trying to better guide Bulkeley Wells to the burial site, McParland stressed the axe mark in the tree: Adams "said he had thought this matter over thoroughly since he got back to the jail and he knew very well that he cut a little nick in the tree, and the body is not located over thirty feet from that nick and is on level ground; and possibly some of the remains of that brush that they put over the grave might be there yet. But of course that was a good while ago, and the brush may have all decomposed and turned into soil by this time."[27]

A few weeks after this report, Wells, still faithfully digging for bones, hit pay dirt. At the end of June he and Runnels were back at the site where Adams said he buried Barney and they discovered "boots and a few pieces of bone." Wells was one inch closer to hanging St. John. "[Have Adams] draw a description showing the location of the body to the place where the boots and clothes were buried," he requested of McParland, "as [I] think there must be more of Barney's body than the couple of bones that [we] found at the place where the boots were buried."[28]

Adams drew a rough sketch of one hole marked "body" and another hole marked "clothes," which was too general to be of use. He wanted to return to Telluride to point out the spot instead. In July McParland put the idea to the prosecutor in the Idaho case. Adams was rebuffed. The Idaho case against the WFM's executive officers, Charles Moyer and William Haywood, and another WFM man, George Pettibone, was heating up. The State didn't want to risk losing Orchard's only corroborating witness.

Although pleased with himself enough to boast, "I worked single-handed in the Molly Maguire case, and single-handed in this, and I know the ins and outs of both gangs,"[29] McParland was still insecure about his case against the WFM. He concocted an elaborate plot and carried out a plan with Idaho officials, Colorado officials, Bulkeley Wells, and others to kidnap Moyer, Haywood, and Pettibone from Colorado so they would have to stand trial in Idaho.[30] These men were arrested and charged on the strength of Harry Orchard's confession. This alias-laden killer had pointed his stubby finger at the officers of the WFM and they were immediately denounced as guilty by Idaho's Governor Gooding and McParland. McParland was so convinced of their guilt that he proclaimed to the press, "The officers of the Western Federation of Miners [at this point that included Vincent St. John, who was arrested as well]...will never leave Idaho alive.... Although they will not leave Idaho, I have information and proof of their connection with a dozen atrocious murders in Colorado that would hang them if they did."[31]

Bulkeley Wells had to wait through the winter before he could gain the proof needed to fasten the noose around St. John's neck. Once the snow cleared in summer 1907, he resumed digging for Barney's bones and persistence rewarded him.

On August 8, 1907, the Telluride *Daily Journal* bellowed in a headline so huge that it nearly ran off the page: BARNEY'S BONES ARE FOUND. The subheadings read: "Skeleton of Bullion Shift Boss is Recovered; Seven Years Search Bears Results; Remains of W. J. Barney, Brutally Murdered by St. John and his Fellows June 22, 1901, are Recovered...."

Nothing available to Wells could prove these bones were the remains of William J. Barney, nor did it matter that the sensationalized subheadings suggested Wells was searching for Barney a year before he disappeared and four years before Wells took up his avocation as a pit digger. All that mattered was that Wells had the bones and he was free to announce to the world Vincent St. John was a brutal murderer.

Not as intoxicated with the discovery as the *Journal*, the *San Miguel Examiner*, Telluride's weekly, devoted only the width of a column to its headline on August 10: Murder Mystery Solved. Nor was the *Examiner* as willing to denounce St. John as the murderer, although Wells insisted this was a fact.

Both newspapers spun an intriguing tale, their sources being Wells and Marshal Runnels. On Wednesday, August 7, Wells took his crew from the idle cyanide plant on a digging mission up Boomerang Road. With more than a dozen men wielding picks and shovels on his behalf, he was certain to find the body. They dug all morning without results. By noon, Runnels decided to build a fire to make coffee and, in his search for wood, according to the *Examiner*, noticed "a rib bone protruding from the earth under an old log." The pick and shovel crew quickly unearthed a complete human skeleton. The *Examiner* said an undershirt was with the bones.

The bones were taken to Coroner Hadley, who held no inquest, but arranged the bones on a table and declared they compared "in height to a man of Barney's size." (*Examiner*) This was what Wells needed to hear, although it was by no means accurate or conclusive. Hadley was a family doctor, with little experience in the forensic sciences of his day. Many things, such as cartilage, living tissue, tendons, and overall individual body structure, had to be factored in. Bones that matched the general size of Barney's six-foot-two-inch frame could easily belong to a man five-foot-ten-inches tall. Possibility was not verification.

Without consideration of weathering, decay, shovel or other tool marks, such as Adams's axe during dismemberment and burial, Hadley determined, "There is a bullet hole through two of the ribs, one through an arm bone and a bullet mark on the collar bone." (*Examiner*)

Wells gathered the bones and skull — *which had no fractures or bullet holes through it* — and placed them "in a safe receptacle." (*Journal*) Proper burial or an attempt to locate relatives wasn't even contemplated. These bones were Wells's property.

Despite having no known witnesses to Barney's disappearance or murder, and despite Adams having asserted all he knew about this body was from hearsay, the *Journal* and *Examiner* printed their own versions of the "facts" of Barney's harrowing last hours. Indulging in far more details than the *Examiner*, the *Journal* promoted Barney to shift boss and made him a deputy sheriff, although this was the newspaper that first pointed out in 1901 that Barney "was not a deputy" as some supposed.[32]

Both newspapers agreed William J. Barney became "involved in trouble" with union men his last night in Telluride, June 22, 1901. The *Journal* proclaimed, "Sufficient facts were available at the time of his disappearance to establish beyond a doubt that he was made away with by the desperate element managing the strike on the Smuggler-Union mine prevailing at the time."

The *Journal* (August 8, 1907) said Barney went to a livery stable to fetch a horse at 11:00 p.m. (*Examiner* said 10:30) and finding the stable closed for the night, aroused the man in charge, Sam Servis. (*Examiner* made no mention of Servis; this was the *Journal*'s scoop.) While Servis was saddling the horse in the stall, "suddenly half a dozen men entered the barn and assaulted Barney, beating him over the head with a revolver and pushing him toward the back door. In the melee a shot was fired when Vincent St. John, president of the Miners' union, and leader of the mob, excitedly cried out, 'Don't kill him here.' Barney was pushed out the back door and hurried away. As the mob left the vicinity two or three shots were heard."

The gang hustled the beaten, staggering Barney on foot down the railroad track. He was bleeding so profusely that his blood left a trail of clots for nearly a mile — from the alley near South Alder and Pacific Avenue to the railroad bridge a quarter mile below town. The cashier for the Rio Grande Southern, who slept in the depot, noticed the rowdy pack of men on the track, taking them for drunks. By now the mob had doubled in size.

At the railroad bridge Barney finally dropped his gun. Why his attackers hadn't disarmed him at the beginning was unexplained. They continued shoving and dragging Barney for another mile, until they reached the Boomerang Road to Alta. Here "he was given in charge of two men," the *Journal* (August 8, 1907) stated, "the rest of the party scattering and returning to town. They walked him up the road a distance of between eight and nine miles," until they reached the site of the old sawmill and bunkhouse ruins in Turkey Basin.

From the bottom of the road at San Miguel to the old sawmill was closer to six miles, not the *Journal*'s eight or nine; nevertheless, those six were up a mountain, from 8,745 feet to 11,100 feet. For a man who had been bashed over the head with a gun and beaten so badly that he left a trail of blood for a mile, it was amazing he could walk so far up a mountain, supported by only two men.

Up to this point, both newspapers presented similar stories, with only minor discrepancies. The *Examiner* (August 10, 1907) softened somewhat and said the union men merely intended to chase Barney from the district. He got in a fight with these last two, but they whacked him over the head with a gun again, and shot him "three or four times."

LOCATION OF CORPSE ON BOOMERANG ROAD

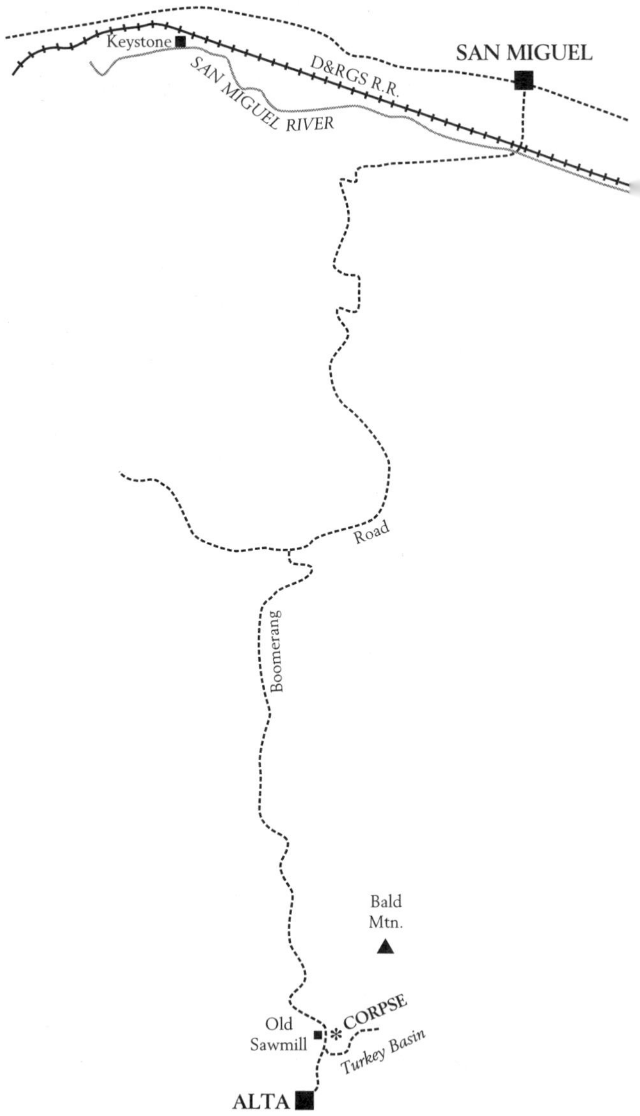

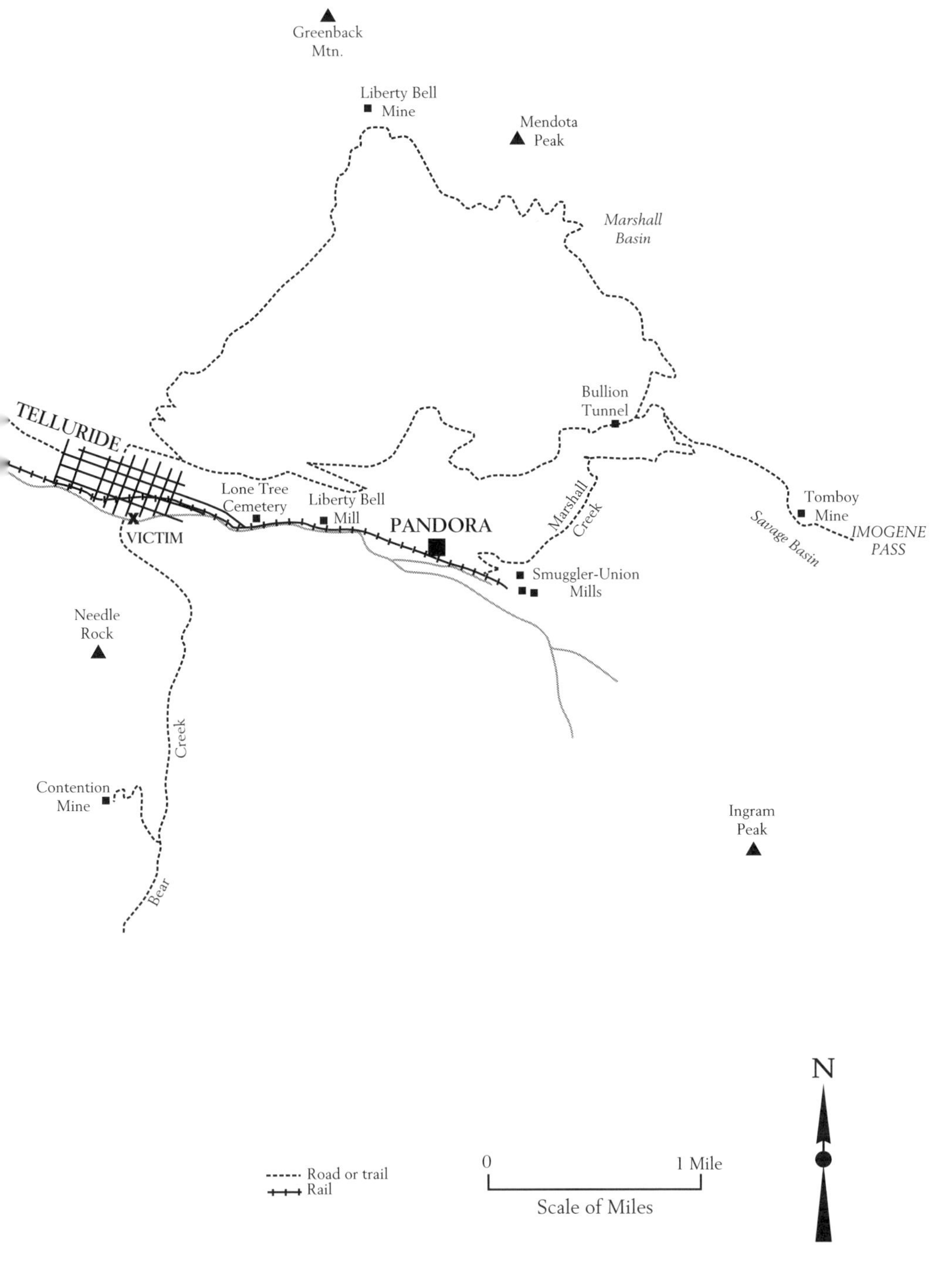
Greenback Mtn.
Liberty Bell Mine
Mendota Peak
Marshall Basin
Bullion Tunnel
TELLURIDE
Lone Tree Cemetery
Liberty Bell Mill
PANDORA
VICTIM
Marshall Creek
Tomboy Mine
Savage Basin
IMOGENE PASS
Smuggler-Union Mills
Needle Rock
Creek
Contention Mine
Bear
Ingram Peak
Road or trail
Rail
0
1 Mile
Scale of Miles
N

The *Journal* made them more brutal, showing their sole purpose was to kill him. "Just this side of the old saw mill Barney was led off to one side of the road, about 100 feet, and shot. Eleven bullets were put into his body.... It is said that as he was being marched down the track he had said, 'Men, you will be hanged for this,' when one responded, 'Well, you won't be here to give evidence against us.'" (August 8, 1907)

Whoever was on the scene to report this last conversation between Barney and his killers must have worked for the *Journal*; the *Examiner* made no mention of it. How the *Journal* decided Barney was shot eleven times was another mystery. At least the *Examiner's* conservative "three or four times" was consistent with Coroner Hadley's assumptions.

The last two men with Barney, the two brutal killers who put eleven bullets into a harmless man they hardly knew, were none other than Vincent St. John and Oscar Carpenter, president and financial secretary of Local 63. The *Journal* had declared it so in the headline and pointed toward them throughout the article.

Barney "fell near a log, or fallen tree, over which the body was rolled, and he was left lying as he had fallen," the *Journal* said. The *Examiner* disagreed: "The body was dragged to one side of the road and down over the bank about 30 feet where it was hid under a fallen tree."

The kidnapping, beating, and dragging of Barney on foot up the mountain — a total of about eight miles from the livery — and the return — a round trip of sixteen miles — took only a short time, according to the *Journal*, for St. John and Carpenter were back at the livery in "a couple of hours." They woke Sam Servis and "warned him at the peril of his life to never breathe a word of what he had seen and heard." Need they worry about a man who could yawn at the mobbing of Barney? Despite the discharge of a gun several times, the yelling, the beating, the blood spatter, and St. John's "Don't kill him here," Servis neither informed the police, nor anyone else. He simply crawled back in bed and went soundly to sleep.

In the meantime, Barney's body was left where it had fallen, remaining in the position for "about two years until the spring of 1903," according to the *Examiner* (August 10, 1907), "when just prior to the strike of that year the parties to the murder became alarmed lest the body be found, and hired a man living at Ophir to bury it. This party whose name is not given, after doing the work, came to Telluride and went on a protracted drunk. Again the murderer became alarmed that this man had not done his work properly or would give something away and here Adams comes into the story."

Adams was "hired to go strip the body of its clothes and bury each in a separate place to prevent identification." He went in search of the body "accompanied by O. M. Carpenter... and the man from Ophir," who "became suspicious that his companions intended to make way with him after they had learned the hiding place of the body and did not stop at the right place but continued on for some distance and gave them the slip running away. Coming up the road they had been talking about landmarks and the man from Ophir had remarked with a drunken lear [*sic*], 'I've got a landmark right around here some place.' Guided by this remark, Carpenter and Adams went back and found the body with the feet and shoulder sticking up out of the ground." (*Examiner*, August 10, 1907)

The *Journal* (August 8) reported a different set of facts, claiming Boomerang Road saw few travelers the summer Barney was killed (although this was the main route the Gold King/Alta miners took to Telluride):

> *...but some two weeks later, someone coming to town over the road, remarked upon a sickening stench prevailing at the point just below the old saw mill. This evidently reached the union leaders' ears [but no one else's] and they no doubt decided that the stench must be removed before it excited an investigation. Accordingly two men were sent up to bury the body. Later, through fear of the result of further investigations, Steve Adams and O. M. Carpenter, secretary of the union, were delegated to go up by night and move and rebury the body. They did so, but Adams says he had to do all the work, as whenever Carpenter got within range of the stench from the putrefying body he was seized with violent vomiting.*
>
> *The garments and shoes were stripped from the body, the idea being to bury them in separate spots so that if by any accident the skeleton was discovered, there would be no means of identifying it by the clothes. Adams said that with the clothes would be a pair of congress shoes, cut through down the sides and along the sole so they could be hurriedly taken from the decaying, stench disseminating feet.*

The *Examiner* (August 10, 1907) also made a point to mention the "congress shoes," claiming, "Adams accurately described the manner in which Barney's congress shoes were cut from the body."

Bulkeley Wells used the skull from the corpse on Boomerang Road in shop windows as a declaration against the Miners' Union and Vincent St. John, although the union no longer flourished and St. John was long gone from the area.

Reenactment photo by the author.

This was an unusual detail, markedly missing from McParland's reports. McParland was painfully fond of peculiar details. He wrote his reports on the very days that he interviewed Steve Adams about the Barney burial and described the footwear as "hobnailed boots." He used the word "boots" consistently, never "shoes." When Wells first found the boots at the end of June 1906, he referred to them as such, with no mention of them being cut through. Hobnailed boots were heavy working boots; congress shoes were lightweight high-top shoes, with distinct elastic panels in the sides and no laces. If

Adams was in a hurry to remove them from the decaying feet, why go through all the trouble of cutting through the leather soles, when that would take longer than ripping them off? Had Orchard cut these shoes and Adams knew nothing about them? This detail also conflicted with the report from McParland where he quoted Adams as saying, "I just pulled the feet out of the sockets at the knees..." Nonetheless, for the *Journal* it was the clincher: "This proves beyond question that Steve Adams was telling the truth in this matter, and leaves no doubt but that he told the truth in the other details of his confession." (August 8, 1907)

The *Journal* was triumphant. McParland was triumphant. Wells was triumphant. He exhibited Barney's skull in a Telluride shop window,[33] with a sign proclaiming, "See for yourself! The Grewsome Work of the Telluride Miners Union and Vincent St. John."[34] St. John was a brutal murderer and now they could prove their case against him and the WFM's Telluride Local 63. "The finding of the skeleton is important as it absolutely establishes the murder of Barney," the *Journal* intoned (August 8, 1907), "and disposes effectually of the efforts constantly made to convey the impression that he left the country either voluntarily, or through fear of violence."

Only one slight annoying inconsistency muddled their absolutely established facts: William Julius Barney was not dead.

CHAPTER

3

WILLIAM JULIUS BARNEY

"The traces we leave behind can only hint at the true heights of our joys and the despairing depths of our hurts."

— Anonymous

Raised glasses sparkled with good cheer. Every eye, every face was dazzling and blithe. Soft evening light embraced the party in the warmth of friendship as they cheered Will and Sarah Barney on their tenth wedding anniversary on February 14, 1899. The blazing fire and joyful laughter drove away all thoughts of the deep chill outside. Will was king of hearts for the night and his courtiers were the brightest diamonds of Ely, Nevada: George Brandt and his wife, H. A. Comins and his wife, Steve Williams and his, Mrs. W. N. McGill, John Webber, William Blair, and a handful of others who had surprised the couple with an impromptu celebration at their home on Murry Street.

Will gazed over their heads in a moment of sublime accomplishment. At age thirty-one he had his own successful transfer business in partnership with E. H. Decker, with a convoy of heavy wagons and horse teams. He had ice harvest contracts and hay and timber contracts that would keep him in the black for months. He had his own house. He had a beautiful daughter. His wife entertained the Ladies Social Club and gossiped with the wives of the movers and shakers of White Pine County.[1] Sarah was his queen of hearts, at least on this night for lovers.

Quiet, sometimes brooding, Will rarely let anyone know the inner workings of his mind. Not given to boisterous demonstrations or animated speech, he kept the wounds of his past wrapped in silence. He was a big man, standing a head above the average, built lean and powerful. His red hair, bushy red mustache, and bright blue eyes advertised his Scottish heritage, a birthright his father had always recited to him with pride.

Will grew up in the vainglorious shadow of his father, William Miles Barney, the first-born son of Henry Barney whose ancestors had fled Scotland. In wide-eyed wonder Will listened to his father's tales of the family pioneers who traveled from Connecticut to Ohio, where William Miles was born in 1823, and later to Michigan, where they took up farmlands. William Miles recounted his own adventures as a "blood brother to the son of [a] Potawatomi chief," who taught him bareback riding and survival skills.[2] He captivated his children with tales of life in the goldfields of California and pioneering feats in Colorado Territory. He had bought into placer gold claims in Gregory Gulch (Colorado), which proved less than worthless, then moved a few miles to Boulder Creek. He told how he and his partners, George Savory and Henry Ludlow, built the first cabin in Boulder County at Whiterock[3] and prospected for gold in the hills. He had unnerving tales of "Indian raids" and his enlistment as an "Indian fighter" in the Third Colorado Cavalry. Known as a "superb horseman and a crack shot,"[4] at age forty-one he joined Boulder County's Company D, serving as sergeant under the command of Captain David H. Nichols. His friends Savory and Ludlow enlisted as privates. This volunteer U.S. Cavalry regiment became known as the "Bloody Third" after they slaughtered a peaceful encampment of Cheyenne and Arapaho villagers at Sand Creek. The Third Colorado Cavalry's service lasted only a hundred days, but Sergeant William Miles Barney suffered the privations of an ill-equipped unit in the frosty months on the prairie, making an adventure of it in the telling, his Star's and Sharp's carbine his best friend.[5] His chilling stories would be forever stained by the depraved murder and mutilation of men, women, and children, who many of his time dismissed as "marauding redskins," giving themselves an excuse to kill. After mustering out December 30, 1864, William Miles became a homesteader and a valued member of the Masons, being appointed as treasurer of Valmont Lodge in 1867. His tales overflowed with bravado, wit, electricity; he had no equal. Will worshipped him.

Sturdy, energetic, often lacking in humility, Will's father knew what he wanted from life and grabbed it with both hands. By 1870 he had a thriving homestead in Boulder County, two hired laborers, cows, chickens, horses, and high-yield hay pastures. In January 1865 he had married Millison J. "Melissa" Rannells, who owned 160 acres of pasture ground two miles north of Longmont and seven miles east of Lyons. Thirteen years younger than her groom, Melissa was the daughter of homesteader Samuel F. Rannells, one of the pioneer organizers of the Boulder Valley Presbyterian Church (September 1863). Together William Miles and Melissa produced a family as sturdy and energetic as William, their firstborn arriving in October 1865, a girl with bright blue eyes and red-gold hair named Ella Cordelia. On her heels followed their only son, William Julius, born August 14, 1867. Named in honor of his father and his father's youngest brother, he was called "Willie" from the first day he howled into life. He soon had two more sisters, Katherine, called "Kate," and Emma Sarah, a small quiet creature with soulful eyes.[6]

Willie had a special gift with horses, a gift his father cultivated early on. The boy's quiet, imperturbable nature and fearlessness wove a blanket of harmony around the animals, allowing him to become "one with them." His heart belonged to these hay-eating beasts, but his soul belonged to his mother, Melissa. When she died February 11, 1879, the soul of eleven-year-old Willie was buried with her.

Devastated by the loss of Melissa, William Miles Barney entrusted Ella, Willie, Kate, and Emma into the care of Melissa's parents, Sam and Sarah Rannells. Although in their late sixties, the grandparents still ran their Longmont farm with the help of their adult bachelor sons, Newton and John (J.R.), along with hired laborers. Their married daughters, Clarissa, Sarah, and Grace, encouraged the motherless children whenever they could in those dark years, endearing themselves to Willie. Although Grandmother Sarah was the sun in the boy's galaxy, no woman could replace his mother.

Willie, who by the age of thirteen had become "Will," kept to himself, telling his secrets only to the horses. He attended school dutifully, helped on the Rannells farm willingly, his size and strength the pride of his kinfolk, his ability to handle horses a source of respect and even awe.

Will's father was no idle man in those years when he left the children in Sam and Sarah's care. He had inherited half of Melissa's acreage, the remainder being divided equally between her four children. At the age of sixty, William Miles Barney, still vigorous and possessed of lusty appetites, wooed and won thirty-five-year-old

Helen, daughter of Joseph B. and Jane E. Barclay, esteemed and landed citizens of Boulder County. Helen came with acreage attached. The couple and that acreage were joined until death on December 27, 1883.

Helen's eighty acres, hers outright, was two miles due north of the land William Miles inherited from Melissa. Helen also had the promise of a portion of her father's ground, upon his death, which would increase the Barney farm substantially. Barney purchased 160 acres adjoining Melissa's and Joseph Barclay's and additional pasturelands near Lyons. By 1885 Helen and William Miles had gathered his children onto a thriving dairy farm worth $4,800. They raised potatoes, hay, oats, wheat, chickens, swine, and horses.[7] Along with hired hands, seventeen-year-old Will, who had just graduated from high school that spring, provided unfailing labor to the farm.

Helen gave birth to June "Jennie" Barney in February 1885, but the child died before she reached her first birthday. Helen's vigor slipped. She required assistance from her sister Carrie Barclay, who moved into the Barney house in March. In 1886 Helen gave William Miles a son whom they named William Henry, in honor of his father and grandfather. William Miles was a proud man and his firstborn son with this new wife would bear his name like a standard proclaiming he was son of this pioneer. Although the baby lived only a few days, perhaps his brief presence splintered young Will's shadowed heart, having a half-brother with his own name, for Will became more distant.

Perhaps it was the stumbling sweep of death all around, or this woman replacing his mother, or some unseen ghosts at his back, but something pulled Will away from his family, his beloved horses, and the land he was to inherit from Melissa. The rift between him and his father had widened, and in the spring of 1886 he abandoned his father's farm for his grandparents. His oldest sister, Ella, two years his senior, invited him to work with her and her husband on their farm twenty miles north in Weld County. Ella had married a Berthoud man named Andrew L. Johnson on February 20, 1885, in a big family wedding at the Barney homestead. Soon afterward they settled at Windsor. Their new farm required intensive labor, and Will moved in with them as a farmhand in spring 1887. On April 13 of that year, Helen gave Will's sixty-three-year-old father another son, Ralph Barclay Barney, a jubilant flame for the aging pioneer.

At Windsor Will met a teamster named George W. Rutledge who had recently moved his family from Central City, Colorado. The combination of Rutledge's fine horses and his pretty daughters snagged Will into a job with George and tumbled him into the sticky maze of love. He and Sarah E. Rutledge married on February 14, 1889. Will was twenty-one. Sarah was seventeen.

A romantic dreamer longing to escape into the society of diamond-studded teatime gossipers, Sarah saw Will as her castle-builder. Yet this phlegmatic man crowned with sandy-red thatch was no king. He had a deep connection to the land and to horses, satisfied by a good day's labor and a slow walk across twilight-shrouded fields. He took up a 160-acre homestead in Weld County east of Wellington and Sarah simply became the wife of a big quiet farmer.

Called "Sadie" by her father and Will's father, she preferred Sarah, a name better suited to stardust and daydreams. She sought society, but in the middle of prairie farmland the only society was Will's horses. Her firstborn son withered and died on that prairie, and Sarah felt she would do the same. She convinced Will to turn his dreams to the west, to escape wind and grass. The prairie was too big for her.

Unlike his father, Will never boiled with the magma of adventure. He had heart ties to the land, to his sister Ella and her growing family, and to his beloved grandmother Sarah Rannells. The frail bond with his father hung on the threads of a land trust signed in 1893. William Miles Barney had gambled nearly all his land on a sheep venture that went over the edge when marauders drove several thousand head of sheep off the cliffs above Lyons. Unable to pay his loans, Barney lost most of his ground to the bank. He borrowed from his daughters, Kate and Emma, trying to get back on his feet. Later he borrowed from Will. In January 1898 he was shattered by the death of Helen. Eleven-year-old Ralph was heir to her estate, but the land immediately became tangled in dispute with the Barclay clan. Grasping the fragments of his pioneer triumphs, William Miles Barney and his son Ralph moved into a house in Longmont while attorneys vultured the remains of the land.

Will also had lost in his father's ventures and in 1894 he migrated to Nevada, leaving Sarah with her parents and family on the Rutledge farm near Timnath, a few miles northwest of Windsor. Always willing to work, Will took various jobs on the vast ranches in White Pine, Nye, and Lincoln Counties. By 1896 he had become ranch foreman of the White River Ranch and Sarah joined him in May.[8]

In July Will was hired as foreman of the Murry Creek Ranch near Ely, Nevada. Sarah hosted an open house, fluttering among the wives of the wealthy ranchmen who called on her and Will. A taste of society kept her hungry for more, but Sarah, frightened by the sharp edges of winter, had Will take her to Salt Lake City in October to pass the season in comfort. He returned to Ely without her. He and E. H. Decker formed a freighting partnership, hauling hay from White River to DeLamar. He was once again one with the horses he loved so well. The freight business was booming, with more than a hundred outfits running supplies in the region. Will moved ten tons of hay in that first month.

By June 1897 he had a house on Murry Street in Ely, a house Sarah filled with the glitter and gossips of the Ladies Social Club. Sarah appealed to Will to work closer to home, so he took a job in town as a stonemason. He engaged in work as a teamster only in the local area, but ventured off to California in autumn on behalf of his father. He returned to Ely in November of that year. Sarah's longings for bigger things still left her lonely in this far away corner. She convinced her younger sister Minnie May Rutledge to stay with her the following summer. May had become a schoolteacher at Fort Collins, Colorado, and Sarah's connections in the White River district landed May a teaching position in Preston, south of Ely.

After eight anxious years trying to have another baby, Sarah finally gave Will a beautiful daughter August 4, 1898, ten days short of his birthday. "He bears his new responsibilities like a man," congratulated the *White Pine News* (August 6, 1898), and William Julius Barney believed baby Florence would fill up all the holes in Sarah's soul. With partner F. Murphy, he expanded his work as a teamster, hauling hay, timber, and supplies in the booming region. Sarah and Florence spent weeks at a time as guests of various wealthy ranch wives. For the Christmas holidays she gathered friends and sister May, holding tight to her hopes, still unable to capture the light of her dreams.

May fell ill in the summer of 1900. Will bundled her, Sarah, and Florence off to the nearest train and sent them home to Timnath, Colorado. Will stayed on at Ely, taking on more freighting contracts from Ely to Pioche. Although May died in August at Timnath, Sarah refused to return to Nevada, surrounding herself in the warmth and safety of her parents, grandparents, and brothers. She found a secret comfort in the family's twenty-six-year-old farmhand, Winthrop D. Sylveste… young man from Maine had been a devoted assistant to George Rutledge for se… eral years.

Will carried on through the winter months in Ely, fulfilling contracts for James Bray and hauling lumber for James Stoddard's sawmill in Lexington Canyon, just over the Utah state line.[9] In February 1901 he pulled up stakes and sold out, all but a number of horses and one wagon. Returning to Colorado, he pastured his horses at Ella's farm at Windsor and stayed on briefly at the Rutledge farm with Sarah and Florence. Although usually content as a teamster or at ease with farming or raising livestock, and always capable of finding work where he wanted it, Will turned elsewhere for his livelihood. He left his wife and daughter at Timnath and rode 390 miles southwest to the remote mining camp of Telluride. He hired on with the Smuggler-Union Mining Company as a stonemason in late March 1901, specifically to construct stone ore chutes to replace the battered timber chutes. He focused on his work in silence, a solitary man disinterested in fellowship with other employees. "He was a queer sort of chap," said O. B. Kemp, the mine superintendent in 1902 (when Barney hired on, Kemp was the company's chief clerk and bookkeeper). "He attended to his business and never seemed to have any trouble with anyone."[10]

Perhaps shadowed by the glances he saw between Winthrop Sylvester and pretty little daydreaming Sarah, Will fled for remote places. Perhaps he was disillusioned by more blatant affection between them or bewildered by the abyss his father had spun into with unending legal battles for Helen's land on behalf of Ralph Barclay Barney. Or perhaps the lingering sting of his beloved grandmother Rannells's death

Colorado Avenue, Telluride's main street, is full of mule teams used to transport supplies to the mines. This was taken around 1902. The popular Cosmopolitan Restaurant is at the far right. Next to it is the barbershop and bathhouse run by union supporter, Charles Miller. Charles Rose, another union friend, had his jewelry shop in the ornate white building. The First National Bank, Sheridan Hotel, and county courthouse give the town an air of elegance.

© Main Street Photography, Walker Studio Collection; used with permission (41T).

and the ensuing legal tangle over the portion of land Will was to inherit from her, or all of these things thrust the thirty-three-year-old William Julius Barney deep into the San Juan Mountains.

These transcendent peaks separated Telluride from the rest of the world. She was a boisterous town, loud with the booming vigor of 1,300 miners, intoxicating in the noise of saloons, gambling dens, and hurdy-gurdy dance halls whose energy splashed into the streets late in the night. The atmosphere vibrated with the braying of hundreds of mules and the music of scores of accents. The flags of Finland, Sweden, Tyrol, Italy, Ireland, England, and elsewhere painted the sky in the ethnic quarters of town. Opera houses, churches, and fraternal organizations goaded the wild edges toward civilization, but men and women flowed in and out of the camp like a steady surging tide. They came for riches or anonymity, burrowing deep inside the majestic peaks, disappearing deep inside the distance of a place on the edge of nowhere. Telluride was a place where a man could lose himself, change his name, or shed his past. It was a place where a man could ponder, where he had space to think; a place where he could forget.

CHAPTER

4

TELLURIDE STARTS A WAR

"Their game of greed but hastens their own doom."

— Eugene V. Debs

A thundering rush of rock roared through the bowels of the Smuggler-Union Mine, spewing dust into the nearby stopes and drifts. No one screamed.

Before the dust settled, shift boss R. M. "Bob" Wright plunged into the stope, discovering the body of E. J. Oakland, his head pulverized by a mass of rock. Life had been snuffed out instantaneously and the men who dug out the body whispered a prayer, the prayer their comrade had no chance to utter. His wife collapsed in horror and grief when they brought Oakland's mutilated body down to Telluride.

Oakland was a contractor in the Smuggler-Union, working not for a daily wage, but for a set monthly fee on contract. This meant he had to mine a certain amount of ore per month in order to receive payment. This ore was measured in fathoms, a fathom being six feet high, six feet long, and whatever the width of the vein might be. Management fixed the price paid per fathom and a miner had to accept it or go without work. If he was working a narrow vein, he generally made customary wages averaging three dollars per day, but his expenses incident to his work, such as powder, tools, and board, were deducted from his pay at the end of the month. If he was working a wide vein of ore, he still had to mine so many fathoms daily, which was impossible to fulfill unless he worked more than eight hours per day. Men pushed themselves beyond the limits of safety and to the edge of fatigue in the deadly stopes in order to simply make enough money to survive. Yet oftentimes, at the end of the month, the contractor not only failed to receive any pay, but he also owed the company for working expenses. If he was a family man like Oakland, his efforts to fulfill the contract in order to be able to feed his family became more desperate, putting his life at greater risk in an already dangerous profession.

Previously Telluride's miners had never heard of the contract system. Most mines in the state paid a daily wage for work done per day, not work done per fathom. The contract or fathom system had been introduced at the Smuggler in 1899, strictly as a means to squeeze more profit from the mine.

The Smuggler-Union Mining and Milling Company, managed by a board of seven directors, was incorporated May 18, 1891, with capital stock of five million dollars divided into fifty thousand shares of $100 each. Its principal office was in Denver.[1] At that time the Smuggler-Union properties at Telluride were put under the management of Nate T. Mansfield, a well-loved local man and practical miner. He resided in the building that housed the company's local offices near the mills, just a little over a mile east of Telluride, near Pandora. During his eight years of management, underground miners were paid a straight daily wage of three dollars. Boarding at the mine was not mandatory. The main access to the workings was the Bullion

The Smuggler Mill and flotation plant had a grand view of Bridal Veil Falls. Author's collection.

Tunnel, at the mouth of which stood a complex of boarding and bunkhouses, tram house, powderhouse, office, livery, and the Smuggler post office and store, clinging to the cliff's edge at 10,949 feet in view of the company's mills far below. Mansfield gave members of the local miners' union permission to hold meetings in the Smuggler store's reading room. Union men were readily hired and often promoted to supervisory positions, since they were more experienced and more reliable workers. Although not yet passed as law, the eight-hour day was the norm.

This amicable marriage between management and labor was derailed when the Boston-based New England Exploration Company purchased the controlling interest in the Smuggler-Union Company on April 28, 1899, for $3,000,000.[2] The Boston company was affiliated with the vast Calumet and Hecla Copper Company in Michigan, composed primarily of the Livermore, Shaw, and Agassiz families of Boston and backed by English investors in London. Colonel Thomas L. Livermore's New England Exploration Company sent his twenty-seven-year-old son-in-law, Bulkeley Wells, the company vice president, to Telluride to oversee the transfer of the Smuggler-Union property.

The elegant Wells charmed Telluride's businessmen the few days he stayed in town, despite their trepidation at having another of the largest local mines controlled by outsiders. New England Exploration was an aggressive company. Extensive expansion of the properties at Telluride and dramatic management changes were to be implemented as quickly as possible. Although the Smuggler had paid its shareholders steady dividends under Mansfield's management, and the property had been expanded with new mills, tramways, and miles of ore exposed even during the Silver Panic of 1893, Mansfield was forced to resign.[3] An interim manager, Charles W. Miller, replaced him. The wage per day remained the same until New England Exploration installed Arthur Lancelot Collins as general manager on September 1, 1899.

Named for the legendary British king and his wife-stealing knight, Collins had a reputation as a successful director of immense mining enterprises from Britain to Africa, Norway to Australia. He was born July 8, 1868, in Truro, England, son of a prominent London-born mining expert, Joseph Henry Collins. J. H. Collins of

Salisbury House, London, was author of countless articles and books on mineralogy and accumulated a large mineral collection that eventually ended up in the British Museum. Arthur and his brothers were members of the father's business, J. H. Collins & Sons, Mining and Metallurgical Engineers. Arthur was sent to the United States in 1893 to manage a large mining company at Central City. With his brother, George, he invested in several mines in the region. The following year he married Marguerite Morton Becker, daughter of Gilpin County Judge Clayton F. Becker. At thirty-one he began his transformation of the Smuggler-Union Mining Company, a change that would eventually rip open the entire town of Telluride.

The five-foot-six-inch Collins wore his reputation like a crown. He knew mine owners "regarded and respected his opinion whenever it was given," allowing him to demand a high salary for his judgment and advice.[4] In addition to being a shareholder, as general manager of the Smuggler he received $10,000 per year, whether or not the mine was productive. His salary was $834 per month, while the best of his mine foremen drew $110 and the average underground miner drew $81.

Arthur Lancelot Collins.
Denver Public Library, Western History & Genealogy (#X-22182).

When he first arrived in Telluride, Collins granted an interview with a *Journal* reporter, stipulating certain rules: his name was not to be used in the paper, company business would not be discussed, and he was not to be quoted. Taken aback, the reporter quickly recovered when Collins informed him the Smuggler had the potential to become one of the best mines in the world. Collins went on to predict a "brilliant future" for San Miguel County.[5] Arthur Lancelot was at center stage.

San Miguel County had never suffered the extreme fluctuations of prosperity and population of a typical boom camp. As early as 1833 Europeans had been in the area, gradually increasing after 1875 when gold placers were discovered along the San Miguel River. Development of mineral veins brought investors, but because Telluride was predominantly a silver camp with lower grade ores, there was no great influx of prospectors. The silver claims that eventually consolidated as the Smuggler-Union were discovered in 1875 and slowly but steadily developed.[6] The town established in 1878 as "Columbia" soon changed to "Telluride," and the general assortment of banks, churches, saloons, outfitters, newspapers, and hotels sprung out of the mud. The railroad arrived in 1890, allowing for a quiet increase of activity in the decade of the 1890s, yet not in the frenzied beehive manner of Cripple Creek or other gold camps.[7] If Telluride had a brilliant future, Arthur Lancelot Collins would have to create it while on the throne.

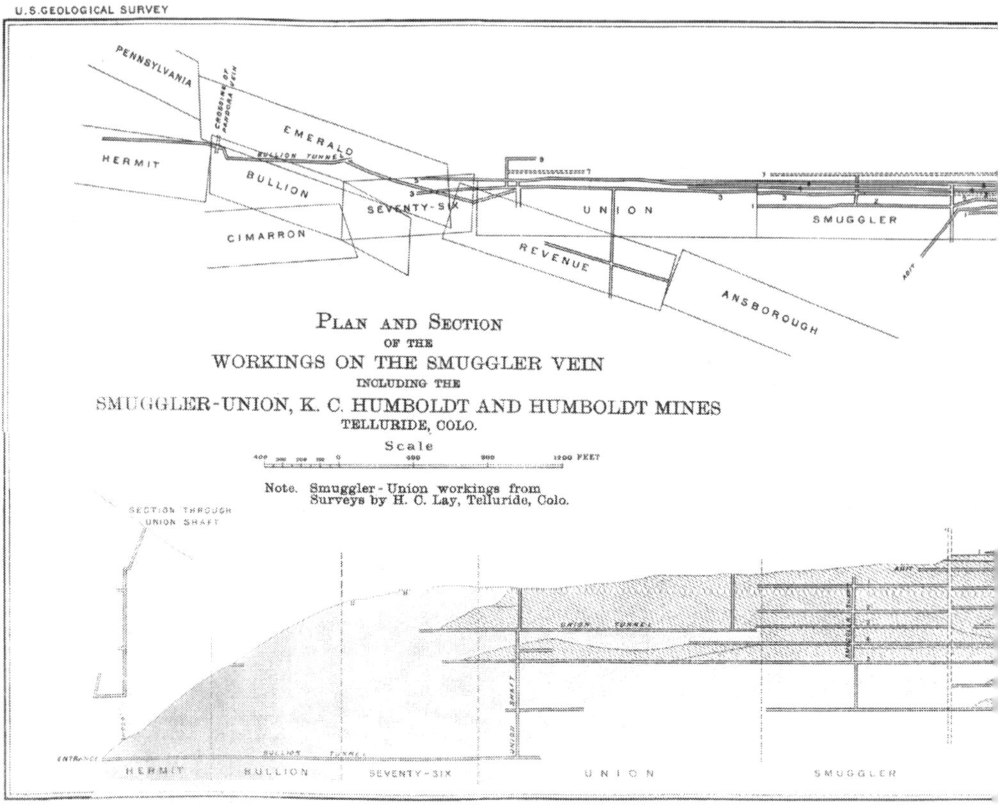

When Collins assumed the Smuggler management, silver was worth sixty-four cents per ounce. The mine consisted mostly of low-grade ores. This meant more rock had to be hauled out. More rock meant more miners. But at three dollars per man per day, this meant less profit, especially if the mine had three eight-hour shifts instead of two ten- or twelve-hour shifts. Collins needed a method to increase production and profit while decreasing production costs. He was particularly fond of what he termed the "old Cornish system" of mining, in use in his native Cornwall. This was the "contract system," which in reality made piecework of mining.

Collins implemented the contract system the month he arrived and at first many miners accepted it, some making more than the average three dollars per day because they were in "soft" rock, which "caved" more readily, making work less difficult. Gradually, as harder ground was encountered, the men found themselves working more than eight hours a day in order to fulfill their contracts. Determined to extort a greater profit from the mine, Collins decreased the price per fathom from twenty-six dollars to twenty-two and even as low as fourteen dollars. He increased the cost to the miner for tools and powder. To decrease the loss of time, boarding at the Bullion Tunnel high on the mountain's shoulder became mandatory, at a charge of ninety cents per day, or nearly one third of the daily wage. A man was free to board elsewhere, Collins noted; yet those who did were discharged.

At the inception of the contract system, the miner was simply required to produce the ore. Muckers on daily wages were employed to break the ore and shovel it into the mill holes. Ever diligent to give the company's absentee stockholders

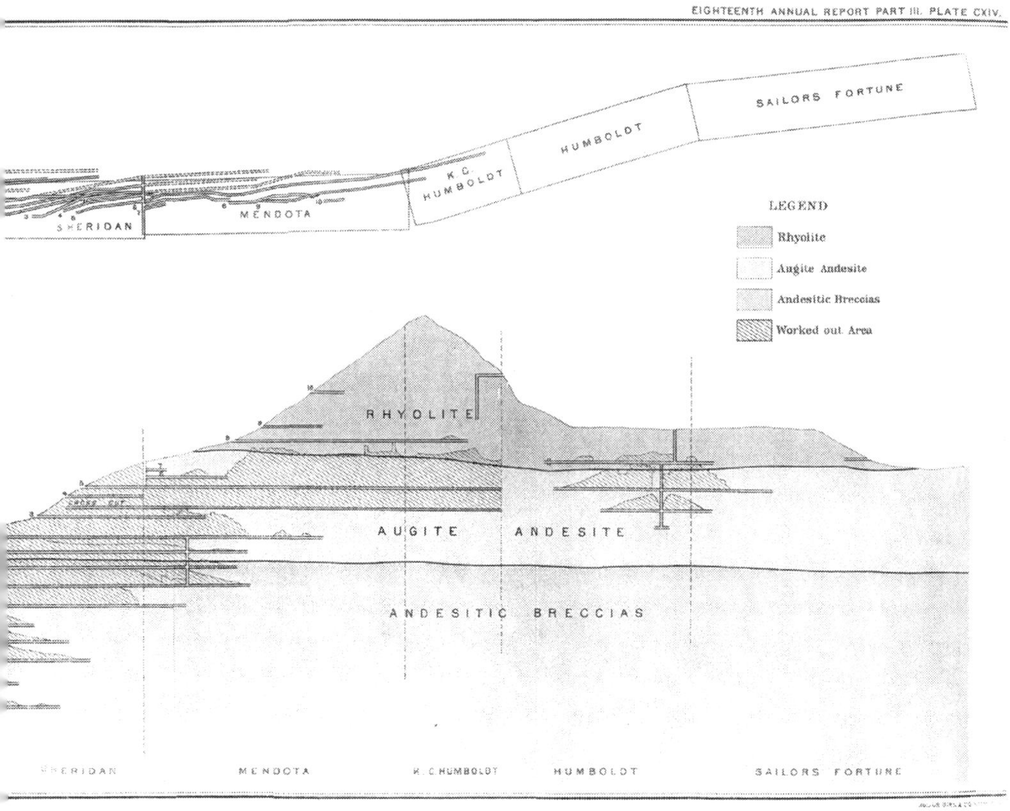

dividends, Collins reduced the need for daily laborers by requiring the contractor to also break the ore and deliver it to the mill holes, thereby tripling his workload. If the contractor disliked the new terms, he was free to seek employment elsewhere. To the miners this was tyranny. To Collins it was good management. Many miners found themselves in debt to the company and were forced to quit. In the meantime, while professing to support unions in general, Collins told Local 63 it was no longer to hold meetings on Smuggler property. John E. Conn, a Finn who owned the Jupiter Mine in Marshall Basin, offered the second floor of his building for a meeting hall.

From his ascension to general manager in 1899 until late 1900, Collins had little organized opposition from local union men. Addressing individual grievances only, he ignored the union as a collective body, thwarting any organized efforts. Union shift bosses had been gradually replaced with nonunion men. Long-time foreman Hugh "Hughie" J. O'Neil was cut to shift boss. Collins brought a number of nonunion bosses, such as Thorvald Torkleson, from the Gilpin County mines. To solidify his position, he had appointed his wife's older brother, Charles M. Becker, as mine superintendent, and his own younger brother, Edgar, as assistant manager, with Thomas Livermore, the Colonel's eldest son, as engineer. Union men had no voice in management and no strong leadership.

Local 63 of the WFM had been chartered August 3, 1896, as the "Sixteen to One" Telluride Miners' Union. The first president was Sherman Parker, who later moved to Cripple Creek, and the secretary-treasurer was Ed Olson. At that time Smuggler manager Mansfield willingly granted the union the reading room for its weekly meeting, encouraging organization.[8] John Carmichael, followed by John McGinley, then A. W. March, succeeded President Parker, with Ed Olson retaining his office throughout this time. March especially was limp in confronting Collins, although the district union called for direct action and even radical action, to prevent the further erosion of standard wages. The Sixteen to One Telluride Miners' Union was desperate for a man of strength and clear vision to lead them. In September 1900 they found their man in the baby-faced, twenty-four-year-old Vincent St. John. His comrades affectionately called him "Vint" or "Saint" — the latter nickname having nothing to do with piety or Vatican proclamations, but merely derived from his surname.

Vincent St. John didn't seek leadership. His friends Oscar M. Carpenter, Harry Jardine, Hughie O'Neil, Ed Thomas, John Lewis, and others shoved him into it. Despite his youth, even the older men, such as O'Neil, Thomas, and Lewis, saw in him their promise for a better future.[9] The union was composed of ethnic factions with conflicting ideas, but St. John had the ability to unite them. When Collins implemented the contract system, he created competition and division between these groups, by suggesting Cornish and American miners were better than the "lazy" Italians and Tyroleans, and Swedes were stronger than the "superstitious" Finns and other such malignant generalities.[10] Collins made it known he disliked Italians, yet readily hired them because they accepted lower wages in their desperation to work. The Italians banded together, safe in their language and customs as a group. Likewise did the other immigrants. St. John, whose mother was from Ireland, saw no difference between American-born and foreign-born miners. All were his brothers. All were equal. His was a genuine impartiality and Ed Thomas knew the young man could bring solidarity to the union.

Born in Newport, Kentucky, July 6, 1876, Vincent St. John was the only son of New York native Silas St. John and Irish immigrant Marian "Mary" Cecilia Magee. He had a sister two years younger named Helen. The family moved frequently, Silas going wherever he could to find employment as a clerk or bookkeeper. The children bounced from school to school, from Ohio to New Jersey, from Pueblo, Colorado, to San Jose, California, gaining a broader view of the disparity and poverty around them.

Vincent grew up working hard to supplement the family's income, eventually supporting his widowed mother and sister as a miner when he was seventeen in California and Arizona. At eighteen he found work in Cripple Creek's mines. At twenty he moved to Telluride and was hired at the Smuggler-Union in early 1897. With nearly a hundred other miners, he lived near the Bullion Tunnel in a drafty bunkhouse. His free hours were spent with books, music, chess, and camaraderie. Hughie O'Neil practically adopted him. His fellow miners loved him. His deep sense of purpose, passion for justice, and unflagging energy were contagious to them. They were captivated by his naïve idealism and the Irish wit and sarcasm inherited from his mother, now honed to an art form.

St. John spoke to the hearts of young and old alike. He believed in political action and direct action to bring labor out of the pit of wage slavery, but abhorred violence, always stressing the principles of the WFM constitution, "to use all honorable means to maintain and promote friendly relations between ourselves and our employers." His oft-repeated line was, "Violence never solves anything."[11] He had represented Local 63 at the district union meetings as a delegate and had served on auditing, funeral, and resolution committees. He did believe in Olson's resolutions that laborers had a constitutional right "to take up arms to defend" themselves for they found "the corporations and employers of labor [were] always ready to shoot down...the workingman if he does not submit to their tyranny."[12]

When he was nominated as a candidate for president of Local 63, he told them, "I am not experienced enough," and knew that in the "heat of passion I might say things I should not." They wanted him all the more for his humility and honesty. Despite his own fears, others judged him as "even-tempered and conciliatory."[13] Ed Thomas assured him he would stand beside him. At the general meeting of the Sixteen to One Local 63 in September 1900, the reluctant, boyish candidate was elected by an overwhelming majority. Edward D. Thomas became his vice president and Oscar M. Carpenter, secretary-treasurer.

The first order of business gave Arthur Collins sleepless nights. During January and February 1901, Local 63 sent letters to Collins requesting him to consider abolishing what they called the "corrupt contract system." In March a committee, consisting of President St. John, Secretary Carpenter, and Smuggler miners Louis Macari, Thomas Murley, and Dan Hoagland, called on him in person. They requested contract men be guaranteed a standard wage of three dollars per day if the contract failed to meet that wage. They assured Collins, "The Union is willing to submit the entire question to the State Board of Arbitration and abide by their decision."[14]

The *Journal* managed to overlook this last vital point when it presented Collins as the target of an unfair union attack and claimed none of the committee even worked at the Smuggler. The newspaper ardently defended the contract system, contending it allowed men to choose their own partners and to make better wages if they were diligent workers. Contract men were even given private quarters in the bunkhouses, which the union considered discriminatory. "The purpose of the contract system is not to reduce wages but to effect a saving in the consumption of supplies," the *Journal* crooned. The article (April 23, 1901) hinted that if miners weren't charged for their powder and tools, they would waste their supplies. At the first "intimation of trouble" Collins planned to "shut the entire property down tight and make no effort at operation." With a flourish of support for the company, the *Journal* summed up, "For fifteen years the Smuggler-Union has been the mainstay of Telluride. Without this property there would have been no Telluride." Even though the Smuggler employed less than a third of the miners in the area, the statement was clearly a threat that if the union didn't cease agitation, Collins would punish the town.

The union men called on the *Journal*'s editor F. E. Curry to give him a chance to correct his misrepresentations and present their side fairly. He politely told his readers the committee "was very moderate in their expressions and gentlemanly throughout." St. John and the committee pointed out the men who had called on Collins were indeed miners from the Smuggler (president and secretary excepted, due to their full-time positions as officers), and they asserted the contract system was not only corrupt, but the motive behind it was evasion of the eight-hour work day. In order to fulfill their contract, some men returned to work after supper or even took on another shift. Curry simply defended the company's position again, stating, "There are those who maintain that should a man find happiness in working ten, fifteen or twenty hours, it is his right to do so." (April 25, 1901) Collins himself denied contractors were forced to return to work after supper, and had they ever attempted to do so, "it would not be permitted." Collins declared "contracting is not compulsory,"[15] yet many men who asked for day wages were either given lower-paid outside work or none at all. Collins and Curry made it appear the union was lying and had no legitimate grievance.

This prompted the union to publish a statement to prevent any public misconception of its motives. Another union committee presented their grievances in a letter to Collins, publicly letting it be known that a strike would be called if nothing were done, closing with "We kindly ask you to counsel with the board. Respectfully yours."[16]

Collins replied in kind:

> *Gentlemen: I have your letter of March 30th and regret I cannot agree with the action which you tell me has been taken by Union No. 63. Less than half our employees are contractors, so those who do dislike the system have ample chance for other employment at the mine. But in some of the workings we find contracting best for all concerned, and we always have more applicants for contracts than we can employ.*
>
> *We have no desire or wish to force anyone to take contracts, but we cannot give up our clear right to make a contract with men who are willing and anxious to take it.*
>
> *Arthur L. Collins, Manager*[17]

Unlike Curry and the *Journal*, Charles G. Sumner, owner (in partnership with Charles Fluke) and editor of the *San Miguel Examiner*, endeavored to give equal treatment to both sides. He examined the union's proposition and Collins's reply. In an editorial (April 27, 1901) he concluded, "Possibly it is fair enough from the standpoint of the company and undoubtedly there is no desire to 'force anyone to take contracts,' but it looks as if the 'clear right to make a contract' is something of an elastic proposition...in the true sense of the word it is not a contract at all, but a proposition presented by the company, at its own price and figures, wherein the miner furnishes his own powder, steel, candles, and supplies and is ensured nothing." Simply put, the contract system meant that if the miner worked for a month but failed to complete his full number of fathoms agreed to by contract, then he was paid nothing, but still owed the company for his supplies. "It is true, perhaps," Sumner said, "that a man should know better than to take some of these contracts, but the fact that they do is sufficient evidence that they believe there is at least a day's wage in them."

Some contractors did make good wages, but these were generally in ground where work was easier. "Of course," Sumner continued, "if a man is losing money he retains the right and privilege to quit, but it so happens that he hardly ever quits until he has found out he has already lost money. Therefore, the proposition on the face of it is hardly fair. The Union has asked that men be guaranteed day's pay on these 'contracts' — that they be ensured a day's wages for an actual day's work, and not be required to lose their time and labor and reap no benefit there from."

In some contracts a clause allowed for the arbitrary recalculation of the price per fathom set by the company, subjecting the contractor to a reduction in price per fathom at the time payment was due. The cut could be as much as five dollars per fathom.

"It is not our mission to unjustly criticize the Smuggler-Union," Sumner wrote, "and we appreciate...that the company has millions invested and that its operations are of vast importance to the district. But at the same time, anyone is conscious...that the vast possessions could not operate twenty minutes if it were not for the miners, and they are entitled to fair treatment at all times. Under the contract system we cannot see how they receive it as a whole. The system is wrong."

In a general statement, aimed at the contract system itself and not Collins, Sumner wrote, "There is no law that grants a manager the right or privilege to include a bunco clause in his contracts."

Arthur Collins took immediate offense. "We consider the contract system better for both employer and workman on a low grade and narrow vein like the Smuggler-Union," Collins wrote, "For the employer, in that the men are encouraged to use their heads as well as their hands, and to break more ground with less powder. And for the men, because they are more nearly their own masters and can make higher wages...."[18]

Collins claimed that since the introduction of his fathom system, the miners on average had been making higher wages than the customary three dollars per day, and those who did not were simply poor miners or lazy. For his argument he used general averages, not the real wages of each. If one man made eight dollars on contract and another two, the average between them was five. It looked good on paper, but the majority of men would not have requested an end to the system if they were making five dollars per day. The company's concern over the quantity of powder used was a mere subterfuge since they charged the miner for all powder used, and charged him "a higher price than it can be purchased elsewhere" in the district (*Examiner*, April 27, 1901).

Sumner said he was "aware that the Smuggler is a low grade proposition, but efficient production has never yet been a condition of subjected labor. A man will invariably give better service if he is fairly paid for it."

Since Collins and the *Journal* claimed the contract system was fair and muttered against the union as the cause of discord, Sumner, who had been in the district since 1895 pointed out, "The Smuggler-Union mines have operated for years without friction of any kind. Labor has been well paid and the mines have made satisfactory returns. Then, we would ask Manager Collins, is not the inauguration of the 'contract system' prima facie evidence that it is not beneficial to labor? No trouble was ever experienced before its introduction."[19]

Vincent St. John and the union were anxious to avoid a strike and the hardships it entailed. They and most of the business community believed their request for termination of an unjust system of employment was reasonable. St. John made a conciliatory proposal that would allow Collins to assess the capability of a contractor for ten days, thereafter "all contractors who work longer than a period of ten days shall receive at least three dollars per day for the time employed."[20] As ten days was ample time for a miner to prove he was a qualified worker, the union believed there was no valid reason to refuse to pay standard wages after ten days. If he were unqualified, the work could be measured at the end of ten days, the man paid off, and discharged.

Still Collins refused, saying, "It is too expensive to measure more often than every thirty days." When it came to managing the mine, his bookkeeper explained later, Collins "had very pronounced views. He would not brook interference."[21] The bookkeeper suggested Collins would have done better to be more diplomatic. The *Denver Post* (July 8, 1901) said, "He is not easily approached.... The miners had several talks with Collins and his manner irritated them." The Ouray *Plaindealer* (July 12, 1901) said the manager had "many admirable traits, but is somewhat dictatorial in manner, and is endowed with what is commonly termed a 'mulish' disposition....

Manager Collins might learn a lesson from the management of the Camp Bird. This mine employs double the number of men in service at the Smuggler...the men are treated with the consideration due to humans...[they] would accept a reduction [at the Camp Bird] rather than work in the Smuggler at full pay."

As days passed and the union threatened to strike by May 1 unless its demands were met, Collins grew more adamant, finding excuses to discharge union men and nursing a personal animosity toward St. John.

"Our fight is against the contract system," St. John insisted, "not against Mr. Collins."[22] St. John refused to breed hatred and enmity when it was justice he sought. He asked Collins to meet him on the same grounds.

Ignoring the issue, Collins shot back that St. John was too young and too inexperienced and too irresponsible to be running the union. Collins was thirty-two. He had forgotten he was merely twenty-five when he was put in charge of the Central City properties. He had forgotten he named his own brother, Edgar, who was more than a year younger than St. John, as assistant manager of the Smuggler. Despite his own British citizenship, Collins insinuated the union was full of "foreign-born half-wits" who knew nothing about mining and even less about voting for a leader.[23] He suggested no one had even attended the general meeting but a handful of St. John's friends, thus his presidency was a sham. The union quickly informed him that every member had a vote in the election of officers and that the membership was made up of as many Americans as immigrants. Collins was none too pleased with Americans at that point and "boasted he was glad he wasn't one."[24] Several Colorado newspapers echoed his contention that the English, and particularly the Cornish, were by far the better mining men.

Insulted, frustrated, and losing patience, a group of eight to ten masked men, some said Finlanders, some said Tyroleans, decided to take matters into their own hands, in violation of union policy. They slunk out of the shadows around midnight April 26 and grabbed a nonunion trammer, a Finlander named John Stone at the Bullion Tunnel complex. They escorted him down the hill, ordering him to get his time check and not return. No one knew who the perpetrators were. St. John squarely addressed the entire union, his large gray eyes full of a fierce intensity. He severely condemned the act of intimidation and denounced the masked thugs as cowards, promising swift justice to anyone caught in such acts. Violence only damaged the union's cause he told them earnestly. There must be no violence whatsoever.[25]

The *Journal* enlarged the incident, the handful of men becoming a masked mob of thirty, armed for bear and bent on murder. Using dubious unnamed sources, Curry claimed, "It is said they led him down the hill with a rope around his neck."[26] He also claimed the mob left warning they were going to take another victim off the hill the next night. He sent the story out via wire, where it grew bigger and darker.

On May 11, Sumner chastised Curry and condemned the intimidators, defending the union: "The Union should not be thrown in disrepute for any petty act in which it did not participate.... We know that the officers of the Union are men of sense and discretion who will use their best efforts to put down any violence. They do not believe in a policy of that kind winning out when they simply ask for justice."

Collins capitalized on the episode. Enlisting the brass in Boston, he had Vice President Bulkeley Wells, who was a mere four years older than St. John, send a threatening letter to the newspapers, to impress upon the community the New England Exploration Company's importance in their midst. Wells reminded everyone the company had just swallowed up the Sheridan, Mendota, and Humboldt

Mines as well as the Newport placer claim and its water rights on the river. They were about to digest the large consolidated claims of the Contention, which would make them one of the largest employers in the district. They had also invested in several large mines at Silverton, just over the mountain to the east. Another Collins brother, George, was put in charge there.

Wells insisted under the contract system the average wage was three dollars per day and the system was in no way introduced to get around the eight-hour day. He also insisted management "never discriminated against the employment of Union labor. If it is the sense of the Telluride community that the employer and employee may not voluntarily enter into contracts, and that an armed force of masked men may drive from a company's properties men who desire to work under the existing conditions, the New England Exploration Company certainly proposes to make no additional investments and is forced to regret the investments already made in the Telluride district, and will not carry out its plans for enlarging the operations of the Smuggler-Union mines. It is the belief of the present management that the Smuggler-Union properties can be operated at a profit only under the present working system, and it will close the mines rather than abandon this system."[27]

The threat was clear. Wells disliked Telluride's support of the union. Only a few Telluride businessmen championed Collins, among them F. E. Curry, the *Journal*'s editor; Charles Painter, owner of the *Journal* and an insurance company; L. L. Nunn, a mine owner and president of the First National Bank of Telluride; A. M. Wrench, vice president of the same bank; Edward N. Lavender, a freighter and owner of the Smuggler store near the Bullion Tunnel; W. B. Van Atta, a merchant; and two or three others. The rest supported the union, believing the contract system was unjust, including Senator William S. Buckley, who owned the extensive Ophir Consolidated Mines on Yellow Mountain. Even the managers of other large mines in the district, the Liberty Bell, Tomboy, Nellie, Camp Bird, and Butterfly, all on standard day wages, backed the union. The general community response to Wells's threat was summed up by St. John: "...this is a company...composed of English and eastern capitalists and the total amount paid out to the employees of the company represents the sum and substance of the benefit this community realizes...."[28] The absentee owners and shareholders had no interest in the town.

With the force of Wells and Livermore behind him, Collins told St. John, "I am the manager of the mine and no one else, especially unions, shall interfere or have anything to say in the operation of the same."[29]

St. John made a final offer. Again Collins refused to discuss anything with the union. As a body the union had voted to strike if management rejected its offer. St. John informed the executive board of the WFM, which approved the strike, and offered his apologies to the public for the inconvenience the strike would cause, explaining, "We have used all honorable means to bring about a settlement of this matter, and have failed. And in drawing your attention to the statement of Mr. Wells...'that in the limits of the total force employed, men can secure work on day's pay if contract work is not acceptable,' we can say that we have given Mr. Collins to understand that we are ready to go to work at any time the company will say it will pay the going wages of three dollars per day of eight hours."[30]

On May 1, 1901, at the beginning of the Smuggler-Union's morning shift, 350 men walked off the job. Local 63 requested miners from other districts to stay away from Telluride until a "fair and just settlement" was arranged. The union assured the public the strike would be conducted in a peaceable manner and offered Collins "all assistance in the suppression of any lawlessness" that might arise, including an offer to post responsible armed guards at the mine at union expense, which was rejected. Instead, Collins and the company "purchased a full supply of arms" and ammunition.[31] This alarmed many union men, but St. John calmed their fears and cooled their resentments; they "must see the strike through without violence."[32]

Charles Sumner invited the union and management to present their concerns to the public via his newspaper in the hope he could promote settlement, but indignation made Collins more resolute. Driving aside the issues, Collins again focused on St. John.

Outraged by St. John's unwavering dedication to the union's cause and an ever-increasing public support of him, and alarmed by the sudden influx of the remaining employees joining the union and rallying behind St. John, Collins resorted to blatant attempts to discredit the young man. Union pressman, Harry Jardine, suggested Collins, who wore a full beard, believed brains were lodged in facial hair since Collins continued to declare the "downy-cheeked Kentuckian" was incompetent in his presidency and too dimwitted to know fact from fiction. Editor Curry jumped in by calling St. John an "illiterate agitator."

St. John ignored the name-calling and went straight to the issues. When Collins stated, "We have nothing against the Miners' Union," — just against the "boy" running it — St. John asked, "then why did your superintendent [Charles Becker] order us off the property and tell the deputy sheriff if we did not go, to drive us off? Why have you discharged every man that took an active interest in the Union? Why did the boss of the tramway [Orrin Sackett] tell his men that if any of them joined the Union they could come and get their time?"[33]

Collins had no answer. The Smuggler's production had stalled. A panic had tumbled Wall Street on May 9 and some Colorado businesses crashed. Money was tight. Collins was losing ground. This agitation left him no leisure time with his pet Belgium hares or excursions with his wife and babies when they came out from Denver. He had a mine to run and other mining interests to look after in other districts, and St. John continued to pulverize him with counterpoints and an insistence on arbitration. Descending to outright lies, even in print, Collins tried to regain his aloof English superiority and end the annoying affair by sheer power of ascendancy and facial foliation.

In exasperation he defended the contracts once again, saying, "It is not true that we ever 'cut' or alter this price before the month is up...contracts are often adjusted according to the greater hardness or softness of the rock, and men constantly exercise their judgment against that of ourselves, refusing this change or accepting another."[34]

In other words, yes, contracts were "altered" but not "cut" before the month was up. Semantics.

"There is no general attempt to cut down prices, as shown by the average price per fathom, keeping the same for nearly two years." In other words, although the price per fathom had started at twenty-six dollars and affidavits showed it was now as low as fourteen dollars, there had been no change. Collins figured averages on a seriously defective scale.

He clinched his argument by claiming the contract system was "more or less in use in every mine in the country." He mentioned the Calumet and Hecla copper mines in Michigan and the mines in Central City, all of which were owned by the company that employed him. He had introduced the system himself at Central City, and the unions had protested. Using as examples only mines that employed him was certain to convince the miners the entire country used the fathom system. In reality, very few mines did: a day's pay for a day's work was the standard.

Like Wells, Collins pointed out the importance of the Smuggler to the community: records showed that 140 of the 170 cars of ore and concentrates shipped from Telluride in March were from the Smuggler properties. St. John was unconvinced that the amount of ore shipped gave the company a right to take advantage of its employees. He refused to be derailed or intimidated, assuring Collins the men would remain on strike until the contract issue was resolved.

Collins angrily replied, "We are not spoiling for a fight, nor do we intend to be dragged into one over an issue which does not exist."[35]

With numerous signed affidavits recorded by the county clerk, Harry L. Servis, and published in the *Examiner* on May 4, St. John publicly proved the issue not only existed but also the Smuggler management was clearly taking advantage of its employees and discriminating against union members.

Several affidavits showed when the men requested work on day's pay, they were told if they wanted to work for the Smuggler they had to take contracts. George Mowat, an experienced miner, had discovered he was losing money under contract and applied for day work, only to be refused. Louis Macari, another experienced miner and leader among the Tyroleans, was told by Collins himself "any Italian or Austrian who would not work under that system could quit at once."[36]

Alex Ample said he had contracted for fourteen dollars per fathom, but when settlement was made at the end of the month, the company paid him only thirteen. Others said when the company measured up the work done, the pay was for less fathoms than the men had actually stoped. When Mike Musso and his partners found themselves cheated out of their pay in this way, they politely brought the deficit to the attention of Superintendent Charles Becker at the Bullion Tunnel office. At the time Deputy Sheriff William R. Beattie and deputized mine foreman Bill Jordan were in the office. Musso was attempting to discuss the problem but Jordan horned in, "Shall we run these fellows off?"

"No," replied Becker. "Talk to them like gentlemen, and if they don't go, then run them off."[37]

Musso and his partners failed to get the pay due them. Seeking legal redress would cost them too dearly.

With the *Examiner* publishing such affidavits, Collins accused Sumner of partiality to the union and opposition to the company. "By giving the union this publicity," Collins rebuked him, "you are turning away investors. You have already spoiled one mining transaction. This controversy will do the community no good."[38]

Sumner laughed. He suggested Collins had no real concern for the community or he would have sought settlement. A practical and objective man, Sumner assured Collins he didn't oppose the company, only the contract system, and invited Collins to submit any documents he wished for publication to prove his case. There was no reason to keep the public uninformed, he said, for the strike situation was no danger to investors.

Believing he had the advantage with the company's current records, Collins gave Sumner a complete list of the stoping contracts for the month of April showing the name of the contractor, the number of hours he put in, his average wage by day, and the amount paid after deductions. The total number of men on contract was 146 (about one third of the men employed) and the average pay appeared to be $4.06 per day. Men should have been flocking to this mine to work for that high pay.

St. John was not deceived. "Mr. Collins is evading the issue, as the 'average wages' have no place whatever in this dispute," he said.[39]

Of the 146 men on contract in April, seventy-nine appeared to make over three dollars per day. The remaining sixty-seven made less than three, some making as low as a paltry sixty-three cents per day. "That is good wages!" St. John grinned. The glaring item missing in Collins's list was a deduction for tools and materials, meaning the average wage was even less.

Additionally, the contracts were made after the union had presented its protest to the company, thus they were slightly better than those provoking the dispute. Contractors for April numbered only 146, which was fifty-four less than the previous month. Also, most of the contractors were in the upper stopes in softer ground. Collins claimed a contractor who made less than three dollars per day was "not a miner." St. John said Collins was mistaken. Nonunion Cal Rutan had mined in the area since 1881 yet the list showed he and his two partners made ninety-four cents per day after board. These same men contracted in a different stope and made $3.23 per day. Nothing was wrong with the miners; something was radically wrong with the contracts.

The union was unconcerned with miners who made more than average wages (if indeed any did). The concern was for the miner working difficult ground or large veins whose time was consumed "in throwing the ore down the mill hole, for which time he is not paid a red cent by the company."[40] (The miner on contract was paid only for fathoms stoped, not how many hours he worked.) Sixty-seven men who received anywhere from sixty-three cents to $2.76 per day were sixty-seven men on starvation wages, and that was sixty-seven too many. For these and any in the same situation, the union wanted the company to guarantee a three-dollar daily wage. It seemed a painless proposition.

"If there was not something wrong, the men would never enter an objection," said Sumner (*Examiner*, May 18, 1901). "A man does not throw up a job for the mere pleasure of getting out to hear himself talk."

Even miners on Collins's list who appeared to make well above average wages protested the contracts. Louis Macari, one of the affidavit men, and his partners made $5.51 per day in April, yet condemned the system.[41] If the contract system was paying above average wages, then why was management refusing to pay standard wages? Insisting on paying a higher daily wage while refusing the union's request to guarantee what would be a lower daily wage "is not a case of magnanimity on the part of the company," said Sumner (May 18). "Oh no. There is something rotten in Denmark."

The general manager was irate. Sumner and St. John both needed to be caged. "If the men were guaranteed three dollars many of them would not attempt to earn it," Collins snapped. He insisted certain ethnic groups tended toward a "natural laziness." He detested the idea of men being paid for standing around in the mine.[42]

"You know the remedy to that," replied St. John. Collins's excuses and mulish disposition were getting tiresome. The issue was clear to St. John and simple to settle, but Collins refused to even consider anything the union offered. One last time St. John asked, "Let us end this dispute by presenting it before the State Board of Arbitration. Even if they find in the company's favor we will abide by their decision."[43]

"There is nothing to arbitrate," Collins insisted. He walked away from the discussion.

The mine had been shut down to a skeleton crew. Both the old Pandora Mill and the new Smuggler Mill continued to operate until the last of the ore was processed; then the stamps were hung up. Collins used the down time to order improvements on the plants. He declared no strike actually existed, no differences existed; the company simply had shut down for timbering, remodeling, and repair. Curry eagerly promoted this line of thought and slung a resounding affront in the face of the union on May 4: "When the property resumes operation some very undesirable elements will not be re-employed in the mine." The strike was no more than a temporary inconvenience.

Temporary or otherwise, the presence of pickets on the road at the company's property line annoyed Arthur Lancelot Collins. If they crossed the property line he might be able to do something about them, but thus far the union men conducted themselves peaceably. He kept nonunion employees on as guards, particularly those men who were impressive in size. Will Barney, still constructing stone ore chutes, was asked to act as guard, which meant in his case he simply worked with a revolver strapped to his hip. Shift boss Melson "Mel" Robbins, thoroughly anti-union and willing to take one or two union men to the ground, was kept on as a guard. Foreman William J. "Shadagee[44] Bill" Jordan; trammer Ben Burnham; foreman Bill Thorn; shift bosses Bob Wright, Thorvald "Tork" Torkleson, Carey M. "Red" Barclay, Lawrence Fields, and Will McKenzie; timekeeper Isaac C. "Ike" Carruthers; watchmen John Shay and Samuel S. Servis; blacksmith John Wearing; steam engineer Matt Allen; D. E. Carroll, and a few others were asked to act as armed guards.

Some carried out their additional duties as guards discreetly, not wishing to offend anyone. McKenzie, a native of Scotland and long-time employee of the Smuggler, was esteemed by both union and nonunion miners and wished to keep it that way. Others, such as Sam Servis, beat a track elsewhere. Servis, a small man of twenty-three, was once a steeplechase jockey in New York. He had been employed for a year as boarding house watchman at the Bullion Tunnel. He moved back in with his cousin, Harry Servis, the county clerk, taking employment at the old George Betts Livery and Carriage House now run by John L. Haines.[45]

Collins planned to reopen the mine with nonunion men from out of the district and wanted his guards to have more authority to act against any union man who stepped out of line. He preferred an impressive patrol backed by the law, men who could legally carry concealed weapons. He requested Sheriff Oscar D. Downtain to deputize as many guards as would accept a commission. Collins assured him this was at company expense and the county would not be responsible for payment to these men as deputies. He simply wanted his guards to have the power to arrest offenders without having to wait for the sheriff to arrive on the scene — the main workings of the mine were at the Bullion Tunnel entrance, more than two miles of hazardously steep road above town.

Downtain was reluctant. He feared allowing Collins a force of deputies would provoke trouble, not prevent it. He had previously deputized shift bosses Fields and Robbins and foreman Jordan on April 27 on Collins's request because of the "mob" that had taken John Stone off the hill. Although cocky and tough, thirty-one-year-old Shadagee Bill Jordan from Chateaugay, New York, was a popular "man's man" and proved competent in his duties thus far, despite his reputation for "getting soaked." Mel Robbins had performed similarly. Collins insisted on having his force of

This group of Smuggler miners was taken prior to the death of night foreman William McKenzie, who is standing immediately to the left of the porch post on the right end of the bunkhouse.

deputies, "for the protection of property and safety of any additional employees" he might hire. He made it clear that if the sheriff was unable to reach the Bullion complex in time to stop trouble, the consequences would be on Downtain's head.

Downtain was a mild family man, unwilling to ruffle the power and might of New England Exploration. To keep peace he deputized reluctant Red Barclay, a good-humored man as big as Will Barney but broader. He had similar features, with sandy-red hair and a big mustache. Barney himself refused a deputy commission: he was not interested in tangling with the union in any way, particularly if Collins brought strikebreakers in to take union miners' jobs. Barney had enough troubles of his own and stayed clear of anyone else's. The only others to be deputized were Shay, Thorn, and Burnham.

Collins needed more men with the enthusiasm of Jordan and Robbins. By mid-May he was determined to reopen the mine as soon as he could secure a workforce and enough men willing to ride shotgun for the company. No one was going to tell him how to run his mine. Disgusted with the lack of local support, he put in a discreet call for assistance to his few friends in town. Within days, volunteers more eager than Robbins offered their services. These were cowboys from the west end of the county and surrounding areas, specific rowdy characters other working cowboys said were "hell-bent to cut their wolves loose." They were a breed of cowboy who hated miners just because they believed miners, especially the foreign-born, were an "inferior grubbing animal;" cowboys who wanted to shoot something just for the bragging rights; cowboys who promised Collins that as soon as he gave the word, they would "clean out the foreign sons-of-bitches in a holy minute."[46]

Among them was Jack Hyde, a man of "gigantic stature" with a reputation as a "notorious gunfighter."[47] His size alone could put the fear of God into the union. Hyde had a heavy brow, sour face, lantern jaw, and unruly mustache. His eyes were close-set like two rats hiding behind his burly long nose. With his hat thrown back and his blue jeans tight enough to advertise his stock of goods, he swaggered into Arthur Collins's office, a six-shooter strapped to his hip. He bragged how he could draw a gun before a man could spit. He boasted he "never yet missed any human target"[48] he had fired at and how much pleasure it would be to "do up some of these useless sons-of-bitches for Mr. Collins."[49] He was willing to take on a deputy commission, but Sheriff Downtain refused, citing Colorado Statute 856, that non-residents of a county could not be deputies. Hyde was from Ridgway, just over the mountain in Ouray County. Downtain refused to bend; the law was on his side. He disliked having hot-headed cowpunchers hiding behind the star so they could shoot up his county. Hyde was undeterred, offering his services as a mine guard. "If there's any serious trouble," Hyde promised the mine manager, "I'll just give you another mound in your local cemetery. Laws be blowed."[50]

Arthur L. Collins had found his man.

Jack Hyde, cowboy and Smuggler guard.

Restored portrait © M. J. Martin 2003.

CHAPTER

5

DEPUTIES, SCABS, AND A DASH OF CURRY

"And the next instant he was one of the deadest men that ever lived."

— Mark Twain

Cracking skulls and planting hot lead near the feet of union pickets was high entertainment for Jack Hyde and the other cowboys guarding the Bullion Tunnel and Sheridan dump. Shadagee Bill Jordan had been joining the fun, provoking a number of men to report his gunplay to the sheriff. Jordan was currently in charge of fourteen men working the Sheridan dump, just above the Bullion Tunnel, a part of the Smuggler-Union's extensive properties. One union committee approached the guards to make an inquiry about the number of men working the dump, calling to the workers to join the union's cause. Jordan ordered the miners to "get off this ground! And the next time you think to come here, you better bring your coffins."[1] He fired a few rounds at them to make his point. As the committee started away, Hyde and the cowboys fired a volley over the men's heads, bellowing insults and guffaws.

While Collins's guards and deputies tried to inflame the union men on the mountain, F. E. Curry built fires with discord in town. Hiding behind his frequent profession of "fairness" and statement that "it is vitally important that in handling this subject nothing shall be said that is unfair or that will tend to excite either side and aggravate the situation,"[2] he jumped in straightaway to aggravate the situation. He reprinted an inflammatory story from the *Ouray Herald* without sending a reporter to check the facts and interview the named parties. One John Barthell was said to be a nonunion Smuggler employee who had supposedly wandered over to Ouray and blathered about the "trouble" being nothing more than retaliation. The story claimed contractors, who had been "highgrading," or stealing ore, became angry with management for having them arrested and called the strike in revenge. Curry supplemented the tale by claiming another Smuggler miner (conveniently unnamed) read the story and said, "Barthell has the situation sized up pretty nearly correct."[3] The article not only maligned the union, it discredited the legitimate grievances against the contract system and hinted that Italians were responsible for the strike. Everyone knew a few Italians recently had been arrested for ore stealing at the Smuggler, their cases pending in district court.

Charles Sumner quickly responded by emphasizing ore stealing had nothing whatsoever to do with the strike, but Curry had done his damage. The only man from the Smuggler whose name was John Barthell was actually a union man, faithfully at his picket post daily, a twenty-seven-year-old native of Finland who hadn't been over to Ouray blowing wind. His cousins, Edvard Jarvi, Erik Bartin, and John Eila, also union men, defended him, but Curry brushed them off.[4] He was merely printing the news from Ouray. Quite likely, enemies of the union had used Barthell's name to generate distrust and hostility.

The following week a stranger calling himself Patrick Quinn sidled into Telluride looking for work at the Smuggler-Union. On Wednesday morning, May 22, union men pointed him toward Vincent St. John, just as Vint was coming out of the rooming house next to Frederick Pheasey and Kenneth A. McLean's Clipper Saloon on Colorado Avenue. St. John and Quinn stood on the sidewalk, discussing the strike situation, the Smuggler, and Quinn's ambitions, paying no heed to the pedestrians around them. Quinn said he "wanted work, but had no intention of mixing up in anybody's troubles."[5] St. John suggested he apply to the Liberty Bell or Tomboy instead of the Smuggler.

Shadagee Bill Jordan, having had a liquid breakfast potent enough for him to see saints and demons, ambled by the pair, and "for no reason, or probably from some fancy originating in his badly-muddled head," turned on St. John.

"Are you fellows roasting me?" Jordan demanded offensively.

"No one has mentioned you," St. John replied. "No one has as much as given you a thought."

Jordan huffed, muttering something about "fighting men," and snarled, "If anybody here is looking for trouble, they'll get it."

"You seem to be the only one looking for trouble," St. John told him. "Why don't you move on and attend to your own affairs."

Feeling uneasy, Quinn started to move away. Jordan stepped in front of him, compelling the stranger to go around him. As Quinn stepped to the side, Jordan pulled his gun, cocked the hammer, and leveled it on the man, at which moment St. John thrust Jordan's arm upward. A bullet blasted into the sky. St. John grabbed Jordan's arm, attempting to wrench the gun from his hand, while Jordan tried to cock the hammer again. St. John forced Jordan's thumb from the hammer, thrusting the gun downward and Jordan snarled, "I'll kill you, too, you little bastard!" Quinn leapt into the scuffle just as Marshal Mac McIntosh arrived on the scene.[6]

Both the *Journal* and *Examiner* reported the same story, as told by six or seven men witnessing the confrontation, which clearly was instigated by Jordan for no other reason than "to display a little cheap authority." Shadagee Bill was arrested and lodged in jail, where he "began to whimper" that "the gun was discharged by accident." Had he pulled it by accident and cocked the hammer by accident as well? This was not his first offense. He had by this time "achieved considerable notoriety from the frequency of his gun plays."[7] Despite testimony from Jordan's friend Lawrence Fields who claimed Quinn instigated the fracas, and a few words from Will Barney who said Jordan was merely out of control because of drunkenness, Police Justice Holmes bound him over for trial.[8] Friday morning Smuggler Superintendent Charles Becker and Fields paid Jordan's $800 bond and he was released to find something else to shoot before his trial. Later, bond was reduced to $500.[9]

District court convened the following week, with a full docket, including the Italian ore-stealing cases. The jury freed some of these men, but District Attorney Samuel G. McMullin freed Shadagee Bill. McMullin, who was from Grand Junction, was outrageously liberal in dismissing several cases, inspiring Sumner to write a lengthy editorial (*Examiner*, June 8) on the "rank farces" and "carelessness and incompetency" of the prosecutor, stating the district court had gone into "the burlesque business."

San Miguel County Courthouse, where District Court was held.

Courtesy William "Senior" Mahoney.

Reflecting the outrage of many in the community, Sumner wrote, "The district attorney announced, in several cases, that he did not believe conviction could be secured on the count, notwithstanding the substantial part of the testimony remained to be taken."

The dismissal of the Jordan assault case was a particular "surprise to all concerned," because the witnesses all presented the same facts. The district attorney claimed the "evidence presented on both sides was so conflicting that he did not believe it was wise to proceed with the case, and dismissed it." The only conflicting evidence was that of Fields.

McMullin stated, "It might do the Union no good if it were brought to trial."

"Rot!" Sumner protested. "That was the first instance cited that the Union had anything to do with it. It was an individual affair. McMullin merely took the brief testimony of the preliminary hearing, deciding on it, and prevented it from coming before the jury, though it cannot be presumed it was done without consultation."[10] Dark rumors flew in the streets that Collins might have been the one who "consulted" with the district attorney to turn Jordan loose.

Defending McMullin, F. E. Curry wove some hazy tale about Quinn not being a suitable witness, that he was now working at the Liberty Bell under the name Valark, and had been in jail, state reformatories, and would not be "respected in court."[11]

Yet Patrick Quinn was merely the complaining witness. Several others were parties to the incident, including Vincent St. John. St. John had gone to Denver to attend the WFM convention on behalf of Local 63 and to visit his sister, Helen, but he had returned on Wednesday, June 5, in time for the trial. McMullin made the trial evaporate.

"It was a case that demanded a thorough hearing and involved indisputable rights," Sumner thundered. "It was thrown out without any cause whatever, unless it was for some reason that is not public property."

Sumner suggested the district attorney had "conducted the vaudeville" show simply for the fees he collected. "If this method of legal procedure is to obtain there is little use in trying to seek redress in court." Something was amiss and Sumner knew it. He angrily stated, "The result of such proceedings of this term and the disposition of the cases, rather puts a premium on lawlessness."[12]

He was right. Shadagee Bill Jordan, the "man's man," was back at the Sheridan playing the menacing game.

On Tuesday, Charley Carlson strolled down the trail from the Liberty Bell Mine, across the shoulder of Mendota Peak to the Marshall Basin trail. One of the larger mines in the district, the Liberty Bell, clung precariously to the ragged heights above Telluride at 11,300 feet, just below St. Sophia Ridge and in the drainage of Cornet Creek. Although a trail went directly down the steep and deadly mountain to town, many Liberty Bell miners preferred the gentler descent of the Marshall Basin route. Carlson worked at the mine for the standard three dollars per day. He had just come off shift. In the crisp, sunny afternoon he headed down to the Smuggler post office at the Bullion Tunnel to collect his mail. Less than a mile from the post office he spotted Deputy Bill Jordan and the detested cowboy Jack Hyde on the road near the Sheridan Mine. Carlson hesitated, aware the cowboy had been pushing miners around since his arrival in the district, and apprehensive of Shadagee Bill's itchy trigger finger. He spotted a few other men on the Sheridan dump and felt confident to proceed.

Hyde suddenly loomed in front of Carlson, demanding to know his business on the road. It was a public road, Carlson told him, and he was merely on his way to get his mail. Hyde accused him of being there for some other purpose, giving no clue what that might be, and Carlson repeated he was there only after his mail.

Hyde's big stone fist slammed into the miner's face. Staggering backward from the blow, Carlson tasted the blood rushing into his mouth. The huge cursing cowboy closed in on him again. Another stunning blow sent Carlson reeling and the cowboy kicked him to the ground. Deputy Jordan laughed as Hyde kicked Carlson again and again and Carlson scrambled backward in terror and pain. Hyde bellowed his amusement and, pulling his revolver, he drove two slugs into the ground on each side of the choking miner's head.

"Now you move on!" Jordan told Carlson. "And tell your union friends to keep their feet off the ground."[13]

Carlson tottered down to Telluride, determined to lodge a complaint with the sheriff's office and swear out a warrant. He had done nothing to provoke a beating and was in no way trespassing on Smuggler property. The Liberty Bell was not on strike and he had never served as a union picket at the Smuggler's property line. Upon reaching town, his face and clothes stained with blood, he realized he had no witness to the attack. He knew Deputy Jordan was not on the side of the law. Other miners who had been assaulted and shot at had made complaint to the sheriff but nothing was done. Even those who had witnesses were left without assistance from the court. Thus Charley Carlson turned to the *Examiner* for help. Charles Sumner was fast becoming the union's champion.

Sumner called for the law to carry out its duties to protect all citizens and to rein in the rogue deputies. Pickets had been shot at ever since Collins hired his cowpuncher guards, which numbered more than two dozen. Hyde was slamming heads every chance he found a miner alone. Other guards were doing the same. The guards hurled insults at the pickets near the Bullion Tunnel, shooting over their heads for amusement, trying to provoke a fight.

"There is entirely too much of that sort of thing up there," said Sumner. "These gun plays are getting too numerous and it seems there is no official protection. Authorities should at least endeavor to put a stop to it. A large funeral is likely to result from it some of these days if they don't."[14]

Harried by Sumner and union supporters on the one side and Collins, Wrench, Painter, and Curry on the other, Sheriff Downtain grew more indecisive and apprehensive. On the one hand he dreaded alienating the powerful and on the other, he feared riling the masses. The camp had 1,300 miners and more of them were joining the union daily. Collins's guards and deputies were clearly trying to provoke a fight, but Downtain supposed if it stayed "on the mountain" it could be ignored. Ever since the guards had started taking potshots at the miners, union pickets were gradually arming for self-protection. Several pickets had arrived with rifles at the Smith store, carrying them while on duty. Smith's was a miners' supply store on the Smuggler to Tomboy road, east of the Smuggler-Union Company's Bullion Tunnel property line. Charles A. "Broken Nose" Smith operated it with his partner, John E. Conn, a union member, who owned the building. Smith had increased his stock of ammunition. The two-story building was provisional headquarters for the union men at the Bullion complex. Armed pickets and armed guards were becoming Downtain's nightmare. Something might explode.

Pickets near the Smuggler's mills were not armed on duty. Guards there seemed less inclined to provoke trouble since Collins's residence and office were in view. But once off duty, guards were deliberate in their efforts to inflame the miners, some going as far as to frequent known union haunts in town where they slung their contempt in union miners' faces.

On Friday evening, June 14, the gunplay returned to town. Mel Robbins, the deputized thirty-three-year-old Smuggler shift boss, boldly sauntered into the St. Louis,[15] a favorite saloon among the union miners. With him were Jack Hyde, John Shay, and Shay's friend, Sam Servis. Robbins, being a privileged Collins deputy, wore a concealed revolver under his coat. Hyde had to check his at the door. Robbins capitalized on the current gossip that Will Barney had been talking to St. John (which was possible since St. John made an effort to speak with anyone who worked at the Smuggler). Some claimed the stonemason was interested in joining the union, and Robbins was there to set the record straight; no Smuggler guard was a traitor. A word scrap ensued with Hyde fanning the flames. One of those involved was the St. Louis's owner William Hudson. He tossed off the entire clatter as hot air from Robbins and went below to tap a fresh keg. Robbins yelled after him and, pulling his gun, shot through the floor as if shooting at vermin in the cellar. Hudson was unhurt. When he returned, he informed Robbins he had sent for the night marshal, Joseph D. Corey. Robbins bellowed, "The marshal can't arrest me! I am a deputy and these men are the trouble."[16]

Corey not only arrested him, but Robbins had to contribute a fifty-four dollar fine to the town's treasury.[17] "It doesn't behoove any man to make the crack that the marshal cannot arrest him, as he will have to pay the freight for it," Sumner chortled (June 15, 1901). It was Robbins's first offense, thus he went straight back to his duties as armed deputy at the Smuggler by the following Monday with a mere warning from Sheriff Downtain.

Downtain had several private meetings with Vincent St. John concerning the growing tensions between guards and pickets. St. John appeared to be controlling the union men and Downtain believed the union president had some uncanny influence over them. The rumor was afloat that the unpretentious St. John "ruled the union" with absolute power and "his word was law."[18] Anyone who took on big Shadagee Bill and disarmed him with the swiftness and strength of a cougar, gained more than the union's admiration and respect. It didn't matter that St. John's hand was badly cut in that altercation or that he was left somewhat shaken. The ardor the men had for him only increased.

Nevertheless, the notion he ruled this body of miners was erroneous. St. John might have wished his word was law, but it took everything in his power to keep the peace. Factions within the union were enough of a strain with their heated ethnic squabbles. Louis Macari, leader of a contingent of Austrians, was becoming more outspoken and confrontational. Some of the Finns muttered against the Italians, claiming their highgrading habits had brought disrepute on the union. But the daily tension of the strike, with mine guards hurling insults, rocks, and hot lead at the pickets, and union men chafing under the frustration as weeks turned into months with no sign Collins would budge, sucked up all St. John's energy. Yet to many he appeared tireless. After an interview with him, Denver reporter Paul Thieman described him as having "catlike strength and pantherish courage," a young man with "indefatigable determination."[19]

St. John spent long hours talking to the men at the picket line, on the streets, and in meetings. He believed if he spoke to each, he might prevent tempers from flaring. He went among them to keep up their spirits, riding up to the Bullion complex and back to town several times daily, a hand on their shoulder to steady them, a word in their ear to encourage them. He ran patrols over the Marshall and Savage Basin roads in an attempt to prevent further beatings and warned Collins to keep his cowpunchers in control. He had dependable men with clear heads and calm dispositions keeping an eye out for trouble in the saloons and discouraged the union men from patronizing gambling houses. His word wasn't so much law as it was hope. He took his presidency as service to the members. He constantly reminded them that they could win against tyranny if they conducted themselves with honor. No violence. Stand firm in the face of insult. No retaliation. No intimidation. Have courage. The men took their strength from this small, wiry, fearless miner who had moved them to move mountains.

Since his election as president in the fall of 1900, St. John had relocated from the Bullion Tunnel boarding house to town, taking accommodation at Fred Pheasey's rooming house at 130 East Colorado Avenue. Pheasey, a thirty-three-year-old Englishman, was his closest friend and confidant; a businessman who owned several properties, including a livery and the Clipper Saloon and bowling alley, and who built the brick Stellmacher-Pheasey block two years earlier. Pheasey also had a pretty little sister named Alice. He hoped she would trip into St. John's heart, but St. John's obligations to the union men left no time for romance. As president he worked more hours than he had as a miner — often twelve and fourteen a day — for only a few more dollars in pay, an upgrade from eighty-six dollars per month to ninety.[20] St. John and the other union officers spent grueling hours seeking support and assistance in resolving the dispute with Collins, knowing that the longer the men were on strike, the possibility increased for anger to ignite into violence.

The region's eleven locals comprised the San Juan District Union, which put "full and ardent" support, "financial and otherwise," behind Local 63. All the WFM locals of the region had published resolutions condemning Collins's contract system as "corrupt" and tossing in his face that "the going scale of wages of the [entire] district is, and has been for several years, $3 per day."[21] On May 17 at the San Juan District Union meeting in Ouray, the delegates had enthusiastically elected Vincent St. John as district president, with Oscar Carpenter as district secretary-treasurer.

The additional duties had been no deterrent for the two dedicated men. They continuously sent committees to attempt discussion with Collins and conferred with WFM president Edward Boyce and members of the executive board. St. John solicited assistance from various respected businessmen in town, among them Mayor C. F. "Fred" Hilgenhaus; Gustaf Brickson, owner of the Sheridan Hotel; A. L. "Abe" Nathan, owner of a large dry goods store; Andrew H. Floaten, a respected lecturer and president of the Peoples' Supply Company; Dr. Anna Brown, county physician and union supporter; and Fred Pheasey, whose "Englishness" might be useful in bringing the English-born Collins back into discussion. Collins would not be moved.

Editor Curry supported Collins in his stubborn refusal to negotiate with the union, painting Collins as the one ready and willing to arbitrate, and always fair, while attacking the union as the source of discord and trouble in the camp. The local Federal Union 104 of the Western Labor Union (unaffiliated with the WFM) published resolutions condemning Curry for his continued "false misrepresentation of facts...calculated to prejudice the minds of the general public against" the striking miners. The Federal Union, comprised of clerks, barbers, tailors, musicians, cooks, waiters, and all branches of labor not directly employed in the mining industry and dubbed "flunkies," wrote, "The *Journal*'s columns have been used to defend the mining corporation in the retention of the contract system and to eulogize the management, a non-American, a non-citizen of this community and republic. We...appeal to the good sense of our fellow-citizens for their moral support in this contest now going on in our midst."[22] They proposed to boycott the *Journal.*

The WFM's executive board member, Thomas J. Sullivan of Leadville, made a personal trip to Telluride, offering compromises to the benefit of the company, but Collins dismissed him, saying he "would soon have all the workmen he needed and refused to concede anything."[23] Sullivan found no way to get Collins to reconsider, even by pointing out that if the mine were re-opened under scab labor, it would jeopardize the safety and peace of all. If trouble occurred, Collins replied, he would call on Governor Orman for state militia. He intended to reopen with nonunion miners and no one was going to dictate how he ran the mine.

On Monday, June 17, the mine whistle blew for the first shift since May 1 and fifty scab miners went to work at the Bullion Tunnel under heavy guard. Collins let it be known they were not working on the fathom system,

> *...but by the day, receiving the regular wages of the district; in short, the mine resumed operations with nonunion men upon exactly the same terms upon which the union miners were willing to declare the strike off and return to work. This action of the manager greatly exasperated the union men, it appearing to them to be a deliberate purpose upon his part to destroy their organization and work the mine with nonunion men; and after the union had been destroyed he would restore the fathom system.*[24]

Red Barclay, Mel Robbins, Norwegian Tork Torkleson, and Scotsman Will McKenzie, were in charge as shift bosses. Collins had employed Torkleson since the Norwegian arrived in the country in 1896. Will Barney, his stonework finished, was hired on as a full-time guard, not a shift boss as later maintained. Barney knew nothing of mining, having never worked as a miner, but Collins wanted him for his impressive size as a guard. Because he kept to himself, Superintendent Becker considered him one of the most conservative guards, one who never indulged in confrontations with the strikers or tried to provoke fights. The union men viewed him as no threat at all.

The union used "no violent means... to prevent the resumption of work,"[25] much to the chagrin of those who anticipated bloodshed. Because the union appeared non-reactive, Collins wired several Colorado newspapers, stating the strike was over and the mine would increase operations. The union quickly informed the *Denver Post* (June 18): "The strike is still on and the company has made no effort to settle." Collins's intention was to make it appear nothing troubled the district and those wanting employment were welcome to apply.

The union committees warned miners to stay away from the area until settlement was made; they were determined to maintain their rights, despite management's deceptions. Keeping the now-incensed striking miners from acting in frustration became a full-time task, with St. John and other levelheaded officers encouraging them to preserve their honor and the honor of the WFM. The officers patrolled late in the night and constantly reminded the men violence would only backfire. During that first week of scab operation at the mine, St. John, Edward Thomas, Oscar Carpenter, and union committees approached the nonunion miners when they came off shift, asking them to quit work and join the union for the good of all brother miners everywhere. Most were not persuaded, but two or three did quit.

By the end of the week, despite tension digging its talons deep into Telluride's back, the miners remained peaceful. The *Denver Post* (June 21) reported, almost as if it were amazed, "So far, no influence aside from argument or persuasion has been used to deter men from going to work, and it is believed that no other means will be invoked." Similar circumstances had sparked full-scale combat during many other strikes in Colorado. If the union resorted to violence in any form, Collins and his allies believed it would be to the advantage of the company, since the state militia could be called to break the strike. Rumor was rife that Collins was paying a union man or two to keep him informed of St. John's plans. Such rumors were common, a form of psychological warfare on the part of management to create enmity within the union, although many companies hired labor spies as well. The Thiel Detective Service had been infiltrating Colorado unions since 1898 and their rivals, the Pinkertons, were quick to chew into their territory.[26] Perhaps the union had learned the same game to play on the guards, for gossips spread word that Smuggler guard Will Barney had been seen speaking with union pickets or St. John on Friday, June 21.

By Saturday, June 22, all was still quiet when Will Barney informed Superintendent Charles Becker he was heading down to town. Was he on his way to meet St. John? No logical reason existed for the stonemason to join a miners' union, yet this rumor was so persistent that, even four years later townsfolk claimed Barney had been asked to join.[27] Unlike Jack Hyde, Barney had no hostility toward the miners and no known record of confrontation. One source of gossip claimed the shooting incident at the St. Louis involved "shift boss" Barney, but the miners all knew Barney was a stonemason and had nothing to do with the Smuggler's underground workings. They also knew Mel Robbins, who was indeed a shift boss, was guilty of using his deputy status as an excuse to shoot off his gun and his mouth. Robbins was fined; Barney was not. When Barney went into town that Saturday he was swallowed by the din and whirl of the hurdy-gurdy, topsy-turvy, rollicking and clucking, tiger-bucking Telluride throng.

On most Saturday nights nearly all the miners from the surrounding hills stampeded into town. This gave an average of fifty more men to every hotspot, each already bulging with patrons from the town's 3,700 citizens. Telluride had more than thirty-one saloons, gambling houses, and dance halls. East Pacific Avenue, South Pine, and South Spruce Street were infested with prostitutes. The profession was acknowledged as beneficial to the town's treasury. Individual girls lived in shacks and cribs on The Row where parasitic pimps dragged in "toms" to caterwaul. The Row was encircled by the big houses, such as the White House Palace, Gold Belt, Pick and Gad, Monte Carlo, and the hands-down favorite, the Big Swede, all within a block of each other. Run by Gussie Richardson, a fleshy fifty-nine-year-old Swedish widow, the Big Swede dance hall had something for everyone: a saloon, gambling, billiards, and girls of every stripe and flavor, with ragtime pianist Frankie Stillwell knocking the house down each weekend. Twenty of the town's saloons were strung for four blocks along Colorado Avenue, the main business thoroughfare, and one reliably raucous one, at South Townsend and West Pacific Avenue, flew under the flag of a temperance hall. This left the entire business district wide open to roving revelers who wandered from one end to the other, howling, singing, bellowing, brawling, often in gangs four or five abreast along the sidewalks.

Editor Curry squawked against the "outrageous antics of" those who "get drunk and parade the walks and streets in squads, making day and night hideous," until they had "passed the limit." This had been going on for many long months.[28] The "squads" forced men, women, and children into the street "to escape pollution by contact." Curry was afraid they would sully Telluride's reputation since they staggered into view near the New Sheridan, where diamond-encrusted society took tea. One crazed man, barefoot and wearing only his underwear, had gone flying into the Sheridan, waving a pistol and ranting about a mob chasing him. Others dared to raise their fists in the marshal's face. Curry was mortified. He suggested the only way to settle them down was "with a club." Such men deserved "a good beating." He especially was disgusted with "the class" from the temperance hall down in "Finn Town" near the railroad track. The Finns, or Finlanders as they signed themselves on a protest to Curry, had their own temperance hall a few buildings east of the Swede-Finns. They pointed out that the Swede hall "should be more properly termed a saloon." The Finlanders wanted it known they were not the culpable "class" deserving Curry's club.[29]

While everyone tried to figure out which wild, bellowing covey was Finn, Finlander, Swede or Swede-Finn, the Italians caroused at the Columbus and Torino Saloons while the Austrians, known as Tyroleans and speaking an Italian dialect, knocked each other down in the Silver Bell dance hall and Senate Saloon; the Germans pounded out their oompahs at the Telluride Beer Hall, while the Irish took whoever however wherever it suited them before stumbling into St. Patrick's Catholic Church to worship with the Italians and Austrians...and the tumultuous gases of life boiled on and on late into the night.

By Monday morning, June 24, the bits and pieces of civilization were reassembled and men were back on the mountain at their picket posts and scab labors, some nursing bruises and sore heads. One was missing. Superintendent Charles Becker discovered Will Barney had failed to return to his guard station at the Bullion Tunnel. Becker telephoned Bill Jordan, who was in town, informing him "Barney has not got back; you better look him up and bring him with you."[30]

St. Patrick's Catholic Church. Seventy-one percent of county residents belonged to this church.

Photo by the author.

Using his official status as deputy, Jordan thumped around in the tenderloin district, making inquiries at saloons and dance halls where Barney "was in the habit of resorting." He heard rumors of threats against "that deputy," but found no trace of Barney. By evening Jordan reported to the sheriff's office: William J. Barney was missing.

Sheriff Oscar Downtain was out of town, thus his payroll deputies, Spellman and Beattie, made immediate inquiries, following the leads Jordan provided. Barney, or at least a man similar in description, was last seen about ten o'clock Saturday night in one of the tenderloin's dance halls. Witnesses claimed a "considerable crowd" was outside the place and men in the crowd were "heard to make threats that they were 'laying for that deputy and would do him.'"[31]

Spellman and Beattie also found two witnesses who saw a party of about a dozen men half dragging a man down the railroad track late Saturday night near South Willow Street. The witnesses took the party for drunken comrades staggering home. Another witness, who lived near the railroad in Finn Town, saw a man running down the track with a crowd of men chasing him late Saturday night. Yet another witness claimed he saw a fight of about a dozen men pounding on one man about eleven o'clock Sunday night. This fracas was in the flat behind the old Pheasey livery barn, which squatted at the corner of Willow and Colorado Avenue. East of this was Stubbs and Jakway's lumberyard at the edge of the track. The witness said the mob's victim broke away and one shot was fired. Having run uptown to fetch a friend, the witness found no one in the area upon returning. Night Marshal Joe Corey was elsewhere and never heard the shot. Another party heard the shot, but discovered the sound was from a "canon firecracker." With the July Fourth holiday less than two weeks away, firecrackers were frequently sizzling and popping.

Deputies Spellman and Beattie investigated the area on Tuesday morning, discovering what appeared to be blood on the railroad ties near the Stubbs and Jakway lumberyard. They traced the blood from this point on South Alder Street, down the tracks, to the footbridge across the San Miguel River at the bottom of South Spruce, a total of three blocks. As if Barney were the only man with blood in him, they deduced the blood was his and he was murdered, his body tossed into the raging river.

On Thursday and Friday, Spellman and Beattie and two volunteers searched along the San Miguel River between South Alder and South Davis for Barney's corpse. Roily and muddy from spring runoff, the river was out of its banks, bogging the willow thickets, and refusing to turn loose a corpse. The search was abandoned. Perhaps Barney had been tossed in the river farther downstream.

At the village of San Miguel, a mile west of Telluride, Beattie found a man who had noted a party of men on foot with another man on a horse leading several saddle horses. This group was seen Monday morning, June 24. They followed the river down, then crossed on the road at San Miguel, mounted the horses, and rode back toward Telluride.

Thursday evening, June 27, the *Denver Post* had published a brief paragraph under the headline, "Foul Play Suspected," declaring "W. J. Barney, a trusted employee of the Smuggler-Union, has been missing...evidences have been found that strongly point to foul play." Telluride's *Journal* had mentioned nothing about the disappearance. By Saturday, June 29, the town buzzed with the Barney mystery and multiple tales of assault. Numerous witnesses suddenly found it their civic duty to report brawls and altercations from the previous weekend, brawls so common that they thought little of them at the time. Eager to be heroes, boys delivered a revolver to Sheriff Downtain, claiming they had found the gun five days earlier (June 24) while playing under the second railroad culvert one half mile below the depot. The gun was said to be similar to the six shooter Barney carried as a guard and since it had one spent cartridge, great minds in the sheriff's office deduced it to be Barney's and the one shot fired was the very one heard by witnesses on Sunday night.

City Marshal Lawrence N. "Mac" McIntosh unofficially asked around and concluded Will Barney had simply left town for a few days, or until "the trouble had been settled."[32] He doubted the man had been assaulted or induced to leave town by anyone. F. E. Curry believed otherwise.

Curry inflated the simple story of a missing man that had absolutely no logical or tangible evidence to point toward foul play into a front-page event on June 29. He subtly hinted the union had something to do with Barney's disappearance. Carefully refraining from actually naming the union or union officers, Curry used motive as his pointer. No one had reason to kill Barney, except the union.

The editor first presented his own reliability by explaining the *Journal* had not printed the story earlier, at the "request of friends of the missing man and the officers at work on the matter." Such requests never stopped him before. He was unaware Barney had no known comrades in town, only acquaintances from the Smuggler. The company's timekeeper said Barney "kept to himself," and the bookkeeper said Barney "was a queer sort of chap." Management barely knew him — he had worked only three months and as a full-time mine guard for only one week — and neither Becker, Collins, nor Jordan knew the man was married, had a child, a well-connected father and relatives all over northern Colorado, or even knew his age. Nevertheless, Curry not only gave Will Barney friends, he had the friends keeping tabs on Barney's habits:

He never indulged in "discharges of hot air" or insults or threats; he "was not given to over indulgence in drink...he never took more than a few glasses of beer. Whenever he came to town and left word that he would be back at a certain hour, he was always there."[33] This was meant to illustrate Barney was so utterly dependable that nothing but death would keep him from returning to his post. If death caused his absence, then it was most unnatural.

Francis Edward Curry, Editor, Telluride Daily Journal.

Courtesy © E. N. Wright; used with permission.

Curry cleverly pointed out that since the strike Barney was employed as a guard and was "not a deputy sheriff but was supposed to be by those who, as is usual in such instances, entertain a bitter feeling against sheriffs and their deputies, or supposed deputies." Curry knew that, outside of criminals, the only men in the district who disliked deputies in such instances as strikes were the union men. They had no complaint whatsoever against Sheriff Downtain and his legitimate payroll men but against the Collins deputies, particularly Jordan, Robbins, and Barclay. McKenzie and a few others still had the respect of the miners, even though by this time (June 29) they were at work with an additional forty-five nonunion men who had taken striking miners' jobs, making a total of ninety-five underground miners and another sixty millworkers. The union men knew Barney was a stonemason and simply a guard, not a deputy, but Curry had to give them a motive in order to make them killers. Otherwise there was no reason at all for union miners to kill, beat, intimidate, or drive Barney from the district. "Supposing" a man was a deputy was a far-fetched motive, since those shift bosses who were indeed deputized had not been murdered, beaten, or threatened in any way by union men, even when the deputies continuously bullied or assaulted them. Even the most detested guard, Jack Hyde, had been left unharmed.

Gathering the sum of disconnected information gleaned by Deputies Spellman and Beattie, Curry made a grander leap of logic than the officers, concluding the man in the dance hall Saturday night whom the crowd was rumbling to "do in" was the same man — Will Barney — who was dragged down the railroad track dribbling blood, seen running from a crowd, and who dropped his revolver under the railroad culvert below town: "all leads up to the conclusion that violence has been done," said Curry. Barney had pay due him, "a considerable amount," and he would not have left town without it, unless he had been forced to leave, or killed. After publication of this sensational tale, more witnesses came forward with suspicious sightings. A railroad employee who slept at the depot reported he had been uptown Sunday night, June 23, returning to the depot about 11:30. The *Journal* reported:

> *Just as he was about to open the depot door he noticed a crowd of men coming down the track towards town and he stopped and stood on the platform until the party had passed. It was quite dark, the moon having gone down, but he could plainly see two men leading and supporting a man between them who was apparently unable to support himself. Right behind these two were probably a dozen men. The man says he judged from their walk that they were foreigners and he supposed they were helping a drunken companion home, but not a word was uttered by anyone, which is unusual for a party of drunken men. The railroad man, after the crowd had passed, entered the depot and went to bed, thinking no more about the circumstances until he saw the* Journal *of Saturday.*[34]

Without asking, "Is it likely all these various scenes are separate and unrelated incidents, typical of any given weekend in the tenderloin?" Curry was determined to link all the initial information with the additional, even if it caused a cranial rupture. He failed to realize some events occurred on Saturday night and some on Sunday night. He decided the crowd threatening Barney at the dance hall Saturday night

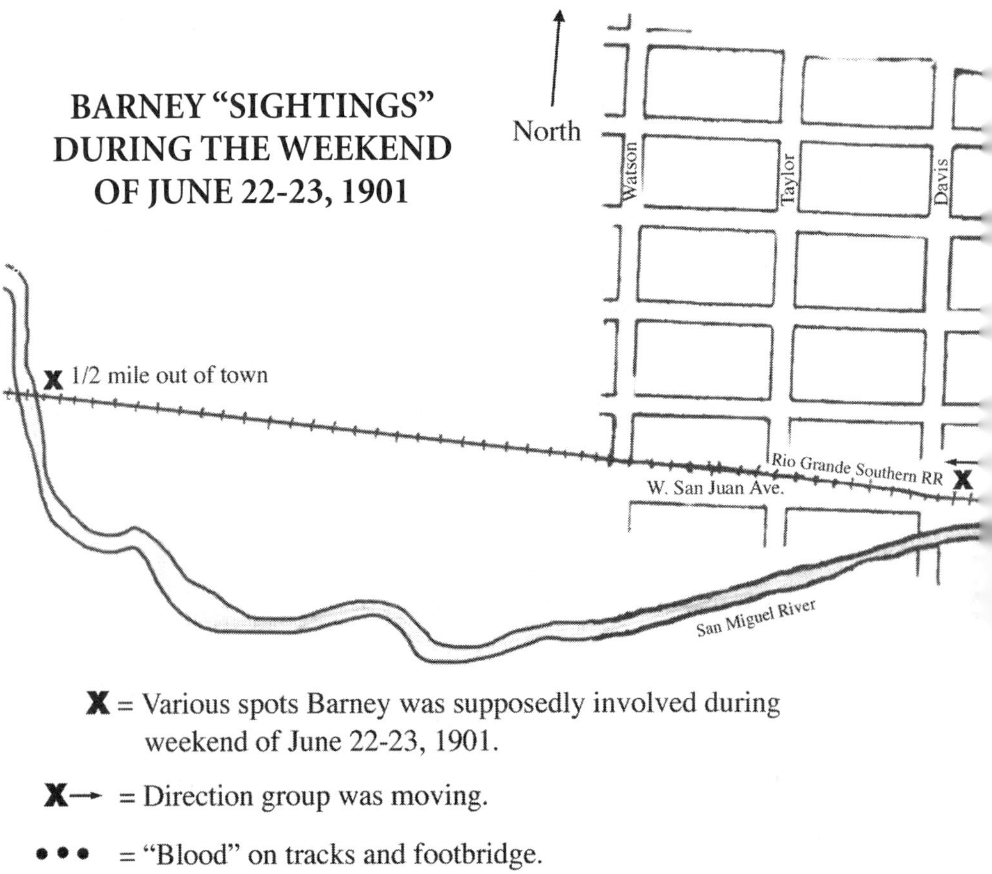

seized the man because they thought he was a deputy. He ignored the probability that the crowd, if indeed any crowd was making threats, was most likely after someone else. Since Barney was not a deputy and *the miners knew this,* the crowd may have been rumbling against Robbins, Jordan, or especially Barclay, since he was a big man with sandy-red hair similar to Barney and he even had a similar name. Barney was only a few months in the district and relatively unknown and reserved. Twenty-seven-year-old Barclay was from Delta. He had been in the Telluride area for a few years, was more gregarious and jovial, more likely to be noticed. A number of townsfolk thought the entire missing man story was indeed about Red Barclay, because he was involved in some quarrels that same weekend, showing up at the Bullion complex on Tuesday or Wednesday with "sticking plaster" and a few dents.[35] Curry failed to mention this.

Curry was certain that, after seizing Barney, the gang roughed him up, first on Saturday night, then again on Sunday night, this time in the flats near the Stubbs and Jakway lumberyard. He broke away, firing his six-shooter, and ran down the tracks. The mob of twelve pursued him back to *Saturday night,* where he was seen running west (out of town) on the track near Finn Town, about one half mile away from where he was being dragged, spilling his blood from Alder Street to the Spruce Street footbridge. Meanwhile, in the moonless black depths of Sunday night, he was seen by the depot man in custody of men who walked like foreigners, who, amazingly, were taking him toward town. Perhaps this walk of foreigners caused them to go in circles,

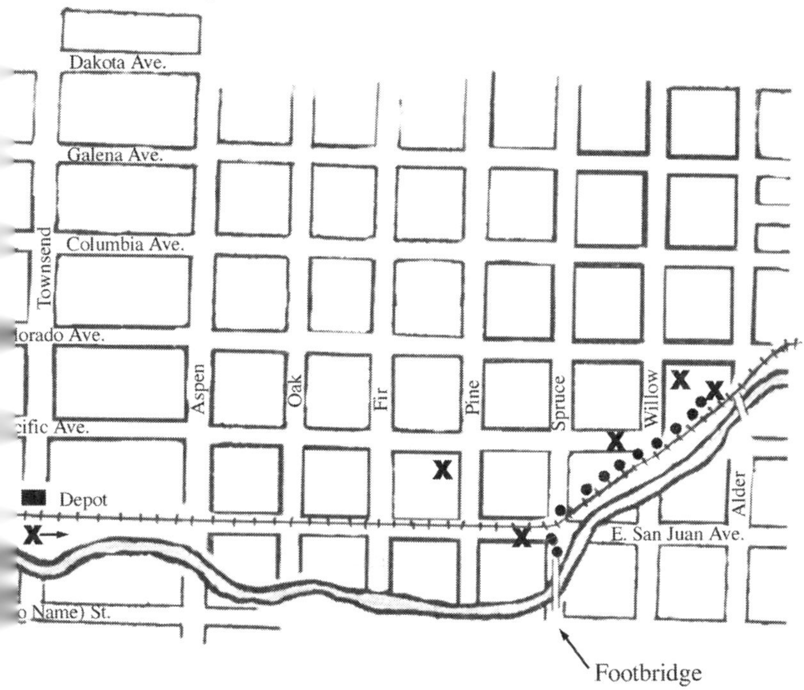

for, Curry concluded (July 1, 1901), "Barney was taken to the railroad bridge [this was the culvert another half mile out of town where the gun was found] where he was either killed, stunned by a blow or in an unconscious condition from some dope previously administered, was thrown into the river." Adding the dope was a nice touch to suggest premeditated murder.

If one was to concur with Curry's conclusions, one had to believe that Will Barney, a big, powerful, armed stonemason, spent Saturday and Sunday night in brawls that covered a space of more than a mile, with a short detour down to the river to bleed on the footbridge at the bottom of Spruce Street. Had the events all occurred on one night, in some sort of logical sequence, glaring facts still remained:

> *1. No witness to any altercation or gang of men dragging their companions actually identified Barney as a participant or as much as mentioned a conspicuously big man.*
>
> *2. The blood found on the tracks on Tuesday morning might have been there as late as Monday night or as early as a week or two before.*
>
> *3. The blood might have come from any other person in town, especially since brawls were common in this area; also a violent robbery was reported in this exact area*[36] *behind Pheasey's barn on June 6 and other similar robberies had been reported in the neighborhood. The blood might have been an animal's (stray dogs plagued the town*[37] *and several people shot at them; and the Great Eastern Circus had been in town on June 13, complete with tigers and maniacal baboons, the train stopping here).*
>
> *4. The six-shooter found at the culvert a mile from the Sunday night fight scene could have been there days or weeks; the gun was identified as one similar to Barney's, not positively his. Curry said the gun was "too valuable for the owner to throw it away,"*[38] *thus he reasoned it had to be dropped when the victim was pursued too closely to recover it. Yet any staggering drunk could have dropped it without realizing it.*

Although Curry's conclusions were glaringly illogical — having a man brawling two nights (did he rest in the daylight hours on Sunday or spend the day praying at St. Patrick's with his attackers?) and covering a space of more than a mile — he declared (July 1), "Today but one opinion is heard, and that is that Barney's body is somewhere in the river." That one opinion was Curry's, not the consensus of the citizenry.

Telluride had 3,700 citizens in 1901, with another 1,300 miners descending on the town on weekends. Curry himself whined about the continuous brawls and carousing that shattered the peace not only on weekends, but during the week. At other times he noted City Marshal McIntosh and Night Marshal Corey always spotted everything in the tenderloin, and if they didn't see all, the girls of the row did, eagerly informing the officers as an exchange for protection from cruel pimps. Drunks fell into the river, not to be discovered for weeks. Men shot themselves and barely gained notice. Pimps set prostitutes on fire and hardly a nod went their direction. Ex-City Marshal James Clark was shot in cold blood in the street a few years before and no one investigated. Men came and went without informing their employers and no one mentioned it in the newspapers. The only difference with William Julius Barney was opportunity.

Forty-five-year-old Francis Edward Curry saw Barney as an opportunity to hang the union out to dry. Standing firmly behind him were Charles Painter and bank vice president A. M. Wrench.

CHAPTER

6

DEATH STALKS THE HOLIDAY

"The ever memorable and blessed revolution, which swept a thousand years of villainy away in one swift tidal wave of blood."

— Mark Twain

Since conjecture and innuendo had more importance than facts in the *Daily Journal*, the miners christened it the *Daily Diaper* for the load it carried. Curry's sour attitude had slipped into rancor, inspiring him to discourage even the July Fourth celebrations with the assertion, "the conditions prevailing this year are not favorable" because of the depressing influence of the strike.[1] In past years Telluride businesses had raised funds for a grand slam Fourth, with purses for miners' drilling contests, horse races, parades, bands, ballgames, and dances. With Painter, Wrench, Van Atta, and Nunn in agreement with Curry, such plans were scuttled, seemingly as punishment for the union and its supporters.

Neighboring towns, including Sawpit, Ouray, and Silverton, invited the miners to celebrate with them. Baseball matches were scheduled with those towns. The Telluride firemen still planned their grand ball for the evening of July Fourth. Despite the men in the corner who muttered about agitators and anarchists, Telluride planned to make the best of things.

Behind the scenes, WFM executive board member T. J. Sullivan, still working with St. John and Local 63, held joint meetings with them and the Federal Union. They invited local businessmen into the discussions, seeking their input to bring a settlement to the dispute. The general community supported Sullivan and the union, believing the WFM and Local 63 were "conducting everything on a fair and honest basis." Sumner praised Sullivan as "a man of excellent judgment and conservatism."[2] Nevertheless, Collins had brushed away any further discussion and left for Silverton and Denver to close business deals on other mining properties. Sullivan devised a plan to approach Collins by appointing an official citizens' committee that included the ardently anti-union A. M. Wrench and his more moderate ally W. B. Van Atta, an outfitter and owner of a mercantile supply company. Other members of the committee were Gus Brickson, Senator Buckley, and Edward I. Field, general manager of the prosperous Nellie Mine, located up Bear Creek and owned by the North American Exploration Company.

The idea was to create a local mediation committee personally invested in the community. No one wanted strife and a prolonged strike. Sullivan's genius was in convincing Wrench to be a member. This not only stroked the banker's vanity, but also compelled him to perform on behalf of the entire community. It also forced him to bring his friend Arthur Collins into the negotiations, which he agreed to do as soon as Collins returned from his business trip. How could Collins refuse to discuss a resolution of differences when his most zealous and powerful supporter invited him to the table? Hope was in the air. Sumner had been a party to the meetings and congratulated Sullivan for his competence and St. John and the union for their willingness to include a man who, by the mildest definition, was an enemy of labor. "It may accomplish much in the way of bringing about an understanding," Sumner said (June 29, 1901).

To a handful of strikers the presence of Addison M. Wrench was acid in their wounds. The New York-born banker, although well educated and refined, spared no occasion to denounce union men as anarchists. He was a petite, sweet-faced, bespectacled thirty-two-year-old with striking "cerulean eyes and a beautiful complexion."[3] He wore his flaxen hair in a Kewpie doll curl at the crown. His ego belied his small delicate frame, for he was used to having things done his way. Second son of a textile factory foreman, he had been married nine years and had a five-year-old son named for him and a newborn baby. He made certain his wife, Minnie, lacked nothing, including a faultless German housekeeper and several servant girls. He had long served as an officer on the Telluride Board of Trade and had begun his career with L. L. Nunn as bank cashier and partner in building various business blocks in Telluride. His alliance with Arthur Collins was cemented in profit, and he believed unions had no business dictating to management.[4] St. John was magnanimous enough to accept him as an intermediary, but other union men felt the barbs of betrayal. St. John gave his word he would not take any action until the committee met with Collins.

Addison M. Wrench as he looked in 1895, age 28.
Restored portrait © M. J. Martin 2003

In the quiet hours before dawn on Wednesday, July 3, 1901, a delegation of strikers gathered at Smith's store on the mountain. The night before someone in town tipped off the Smuggler superintendent that strikers might be planning something and to keep an eye out for trouble. The strikers were determined to call out the nonunion men from the Smuggler properties and convince them to join their cause or find work elsewhere. By now 200 men were working for the company, eighteen of whom were established management, fifty-four were nonunion millmen, about twenty-four were strikebreakers working under Foreman Jordan at the Sheridan dump, and another eighty-five were at the Bullion Tunnel under Superintendent Becker.[5] The rest were guards. Some of the men were working the night shift; others were asleep in the bunkhouses, or just rising for their breakfast. Vincent St. John was still asleep in his rooms in the Pheasey block in town, unaware of the rumblings on the mountain.

St. John's work to unite the ethnic groups within the union was finally realized, for the delegation consisted of Italians led by James Roner and Vic Boggia, Austrians led by Louis Macari, Irish led by Jerry O'Rourke and William Brennan, and Finns led by John E. Conn and John Barthell.[6] They numbered about four dozen. At 5:00 a.m., just as the night shift was coming off work and the next shift was about to go on, a delegation of about a dozen strikers approached the men at the Sheridan, demanding the nonunion men quit work immediately. The union committee told them if they left peaceably, "there would be no trouble."[7] They preferred that the men would join the union's cause. The *Examiner* reported the delegates were unarmed, but the men in the ranks behind them were carrying weapons.

Jack Hyde and the guards yelled their usual insults and ordered the committee off the property. Trigger-happy Shadagee Bill Jordan came out of the office and John Barthell "did something which he had no authority to do and which seemed completely lacking in good sense."[8] He stood in front of the delegation, unarmed by all accounts, and shouted at Jordan, the guards, and the scabs, "You are under arrest." This was the best phrase he could manage in his broken English. Jordan and his men answered with a fusillade of hot lead, a bullet ripping through Barthell's neck. The young Finn dropped to the ground, his life instantly snuffed out. Incensed by the cold-blooded killing, his comrades scrambled for cover and returned fire.

For several minutes a hail of bullets came from both sides, with some nonunion men fleeing in terror. Shadagee Bill took a bullet in his right leg near the groin that went out the other side, tearing a ragged hole but not damaging bones. In a stumbling run he fled as strikers fired at him. He found safety in an adit, where he collapsed, bleeding profusely. A few moments later one of his men dove into the same excavation to hide with him. A number of union men stormed the bunkhouses, knocking down fleeing scabs as they rushed outside. In the clamor and confusion Joseph Lujan, a thirty-four-year-old native of New Mexico working as a mule packer and trammer, stepped into the line of fire with his hands raised. He tumbled down the tailings with a bullet through his chest. Deputized trammer Ben Burnham took a bullet in the abdomen and another in the right thigh, his companions escaping with him to the safety of the Bullion Tunnel boarding house three quarters of a mile below.[9]

A squad of strikers rounded up the nonunion men remaining at the Sheridan while another group went to the Bullion Tunnel to demand the surrender of those scabs on the premises. Whether shooting had broken out here during the clash at the Sheridan dump was uncertain, since eyewitness reports from the tunnel site were conflicting. Some said the deputies and guards opened fire on the union men as they came down the hill with John Barthell's body, which was shortly after Burnham's comrades dragged him to the boarding house half dead. Others said the union men had captured the Sheridan scabs and regrouped at Smith's store when the Bullion Tunnel guards and deputies opened fire on them. Some said that as soon as Jordan and his men had opened fire, the union men had scattered in several directions and general firing had erupted at both sites. An account in the *Durango Democrat* (July 5, 1901) as told by Thomas H. Ballard, a strikebreaker, said fifty or sixty union men took up position about 200 yards directly up the hill from the Bullion Tunnel and demanded the scabs surrender.

"Raise the white flag or we will open fire," the union men yelled, according to Ballard.

"A shift boss ordered us to get under cover, and turning to the strikers, yelled back his defiance, saying, 'Turn loose as soon as you are ready.'"

A hail of bullets blasted through the mine office, but no one was inside. The ensuing gunfight continued several hours, the scabs pinned inside the building at the tunnel's entrance and the strikers taking refuge in Smith's store.[10]

Another account said the sound of gunfire at the Sheridan brought an armed force of nonunion men up from the Bullion Tunnel, led by Superintendent Becker, Mel Robbins, Bob Wright, and Red Barclay. This squad of about two dozen joined the conflict, but being out-numbered at the Sheridan site or encountering the men dragging Ben Burnham to safety, returned to the Bullion complex, the shooting continuing on the run. This is probably closer to the truth, since Wright's and other nonunion testimony at a 1906 hearing support this account.[11] Having taken refuge in the building at the mouth of the tunnel, Becker's force continued firing at the strikers in Smith's store and on the hill nearby. Becker was shot through the left arm, the slug shattering the bone. A shift boss named John Nicholson was shot in the face, the bullet entering under the right eye, glancing along the cheek and jaw bone, and exiting the back of his neck.

Shortly after the din of gunfire began at the Sheridan, union miners at the Tomboy Mine (a mile east of the Bullion Tunnel) and Liberty Bell (a mile west of the Sheridan) abandoned work, grabbed their firearms, and charged down the mountain to assist their union comrades. This swelled the ranks to about 250 men within the next few hours.

During the first hour after Barthell was slain and Marshall Basin seemed engulfed in combat, a desperate telephone call went out from Smith's store to town, alerting the union president to come immediately to prevent a bloodbath. Vincent St. John at once notified Sheriff Downtain and Dr. J. G. Sheldon. Then grabbing a horse, he raced out to Pandora, where the Smuggler trail angled abruptly up the mountain. Downtain was a few minutes behind him. As St. John galloped the horse headlong up the precipitous trail, hot lead rained down on him from both combative parties, the union men unable to recognize their president. With bullets screaming by, he pressed on. Downtain failed to follow since mill guards advised him it was impossible to get up the hill. The sheriff raced back to town.[12] Rumor of mass slaughter shrieked through the streets as frantic scraps of news reached Telluride. First accounts had as many as fifteen slain and buildings blown up, pitching Downtain into a panic.

Dr. Sheldon had hurriedly packed his horse with medical gear and flew up to Pandora, where he met the Smuggler Company's tramway superintendent Orrin Sackett and assistant manager Edgar Collins, Arthur's brother. St. John had telephoned Collins and Sackett from Smith's store, advising them they should help him arrange an end to hostilities. Becker had also gotten word to them at the office, saying he and the nonunion men were "practically helpless up here," that his arm was severely damaged, and Burnham would bleed to death without medical care.[13] About this time phone lines between the Bullion Tunnel and town went down or were cut, leaving Telluride without any news from the mountain. Advising St. John not to "trot about unarmed," union men at Smith's store handed him a gun belt and revolver, and he strapped it on.[14]

As Sheldon and Sackett rode into view of the buildings, they were caught in the crossfire, a bullet grazing three fingers on the surgeon's right hand, and Sackett's horse catching the cutting edge of several bullets. They waved a truce flag and yelled until someone recognized them. Six miners armed with rifles acted as a bodyguard for Dr. Sheldon while two others carried his saddlebags, assisting him to locate the wounded. Sackett met St. John at the livery stable.

"What does this mean?" Sackett demanded.

"It means these fellows have got to get off the hill," St. John replied, meaning the nonunion strikebreakers. He hadn't fully sized up the situation, but he knew the combat would end only if the scabs departed and the mine shut down. The incensed union men could not allow John Barthell's death to be in vain.

"What do we have to do to stop the shooting?" said Sackett.

"All you have to do is get these fellows off the hill."

"It is not safe getting down with all the shooting going on," Sackett told him. "We must be protected. If we could arrange to have the shooting stopped until we could get the men together and get down off the hill." He added he had no authority to do anything on behalf of the company and Edgar Collins, being the one in charge in Arthur's absence, was the one through whom arrangements needed to be made.[15]

Sackett returned to the main Smuggler office near Pandora and conferred with Edgar Collins, who had sent an urgent telegram to his brother. En route from Silverton to Denver, Arthur received the telegram at Durango about noon, where he chartered a train straightaway for Telluride. Meanwhile Edward N. Lavender, the postmaster of the Smuggler post office, had just collected the mail in town about 9:30 a.m. He was determined to deliver it up the mountain despite the rumors of death and destruction. A twenty-nine-year-old wiry adventurer, Lavender telephoned

The steep trail to the Smuggler-Union's Bullion Tunnel complex zigzags up Marshall gulch (visible to right of central buildings), and connects to the Tomboy road via a switchback. By 1902 the Tomboy wagon road from town was completed (shown here from center left at the boarding house level). © Main Street Photography, Walker Studio Collection; used with permission (68T).

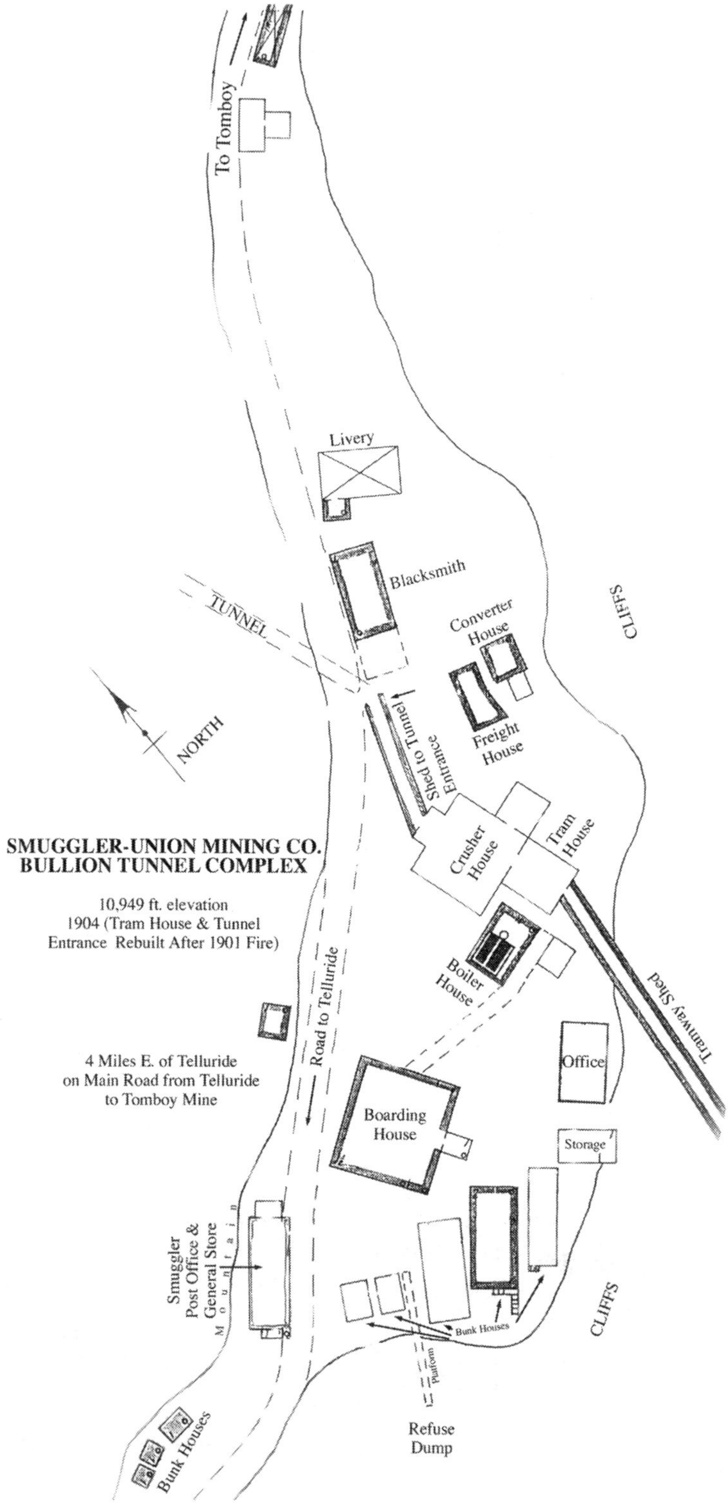
To Tomboy
Livery
Blacksmith
TUNNEL
Converter House
CLIFFS
Freight House
Shed to Tunnel Entrance
NORTH
Crusher House
Tram House
SMUGGLER-UNION MINING CO.
BULLION TUNNEL COMPLEX
10,949 ft. elevation
1904 (Tram House & Tunnel Entrance Rebuilt After 1901 Fire)
Road to Telluride
Boiler House
Tramway Shed
4 Miles E. of Telluride on Main Road from Telluride to Tomboy Mine
Office
Boarding House
Storage
Smuggler Post Office & General Store
Mountain
CLIFFS
Bunk Houses
Platform
Refuse Dump
Bunk Houses

Smith's store, warning them he was bringing up the United States mail and he was not to be molested. The trail went up Marshall Gulch from Pandora in steep switchbacks, connecting to the Tomboy road, where the Smith store, Bullion Tunnel complex, and Smuggler post office were strung along the edge of it. At the Smith store, also known as "the Joint" or "Miners' Rest," Lavender saw nearly 200 armed union men. They left him alone. He continued a short distance to the post office, without any hint of trouble. The shooting had become infrequent, and he found far less carnage and damage than rumor had painted in town.[16]

Using the rumor mill to his advantage, Addison Wrench hammered Sheriff Downtain to wire Governor Orman for Colorado National Guard troops. Downtain was reluctant, afraid the militia would inflame the situation and uncertain if circumstances warranted such measures. At the same time, he could not bring himself to go up the hill to investigate. St. John was waiting for him, expecting him to be at the scene shortly since Downtain had been only a few minutes behind. Wrench played on the sheriff's fears, pressing him to request troops. This and his later actions made it appear Wrench intended to sabotage the citizens' committee, making it impossible for the union and Collins to come to terms, despite his agreement to be a mediator. He knew if rioting could be incited, it would give the State a reason to send troops to crush the union. Wrench or his cohorts had contacted District Attorney Sam McMullin in Grand Junction, in the dawn hours before any news of rioting had reached town, instructing him to wire the governor.[17] Dutifully, without knowing any facts, McMullin sent Orman what he thought would remain a private message, not wanting to appear he was in collusion with anyone since he actually had no information of active hostilities at this point. McMullin's telegram stated:

> *Strike situation at Telluride becoming serious. Will go there first train and investigate and co-operate with sheriff. From advices it may be beyond the power of sheriff to control and I advise you now so that if sheriff calls upon you there may be no delay. Keep this private.*[18]

When Wrench finally prevailed on the apprehensive sheriff, Downtain had capitulated. Downtain wired the governor precisely as instructed:

> *Strike situation is beyond my control. Hundred and fifty armed men surround Smuggler properties. Have blown up two buildings. They guard all trails and allow no one to pass. One man has been killed and several more have been wounded. Mine office building of Smuggler Company is being fired on constantly. I hereby officially request you as commander in chief of the National Guard of Colorado to send 500 militia here with all possible speed to assist in quieting trouble. Judge Gabbert can explain to you the lay of the country around the Smuggler mine and building. I advise sending a portion of the troops by way of Ouray, balance directly here.*[19]

Wrench's allies fired off other telegrams demanding troops and presenting the situation as much worse than it was. Howlers claimed the union miners from the Camp Bird and Revenue Mines had come over Imogene Pass from Ouray County and joined the Smuggler, Liberty Bell, and Tomboy men, swelling the armed force in Marshall and Savage Basins to as many as eight hundred. Although the tale was untrue, this was the reason Downtain requested some soldiers be sent via Ouray.

Telegrams and telephone calls flooded Governor Orman all that day and long into the evening, most pressuring him to send troops without delay. The

Smuggler-Union Company's Denver attorneys and directors called on him, demanding he comply with Sheriff Downtain's request. Denver businessmen and prominent mine owners offered their advice, many becoming "loud in their demands for sending the military force."[20] Orman, a man not easily pushed into decisions or given to over-reacting, consulted with Adjutant General George Gardner of the Colorado National Guard. Gardner was enthusiastic to deploy troops. Nevertheless, the governor deemed it unwise to send a force of soldiers to the distant scene of trouble without first determining if such action were necessary. James Bradley Orman had been elected because he was a man of wisdom and substance, unwilling to bend to political manipulation. Democrats, Populists, and even some Republicans had put him in office. A Colorado pioneer of 1869, Orman had contributed much to the prosperity of the state and he refused to act exclusively on behalf of the corporations. Labor leaders urged him "to look carefully into the state of affairs before moving with the military force, as many mistakes had been made in sending the militia to settle labor difficulties when they could be better settled in other ways."[21] The governor agreed. His decision was confirmed when he received the following urgent telegram from State Senator William S. Buckley, a Telluride resident and highly respected mine owner:

> *Governor James B. Orman, Denver, Colo.:*
> *Before you send troops to this place kindly listen to my version of situation. Uncalled for now. Please answer.*
> *W. S. Buckley*[22]

Orman ordered General Gardner to mobilize one troop and four companies to ensure no delay in deployment if the circumstances justified it, but he resolved to first gain complete information from Telluride, wiring a reply to Senator Buckley. Again Buckley sent a message saying not to send troops; he would look into the situation personally and inform the governor as soon as possible.

The mood on the hill was improving, with St. John gaining some control of the strikers and Sullivan, Carpenter, and Senator Buckley taking charge in town. St. John had met with Edgar Collins while waiting for Downtain at Smith's store. He had telephoned Lavender and Lavender's English employee Louis C. Lomax, who were at the Smuggler post office. He asked Lomax to go to the bunkhouses to see if Sheriff Downtain was inside. St. John preferred to work with the authority of the sheriff at his side. He believed that if the sheriff and his payroll deputies were on the scene, the situation could be resolved.

"I'm not about to go over there with bullets flying about," Lomax told St. John.

"Wave a white flag and the shooting will cease," said St. John.

Lomax followed instructions but found Sheriff Downtain was not present.[23] He reported to St. John, who began negotiations with Edgar Collins, trying to get the shooting stopped altogether and the mine to terminate operations in order to give the citizens' committee, the union, and Arthur Collins a chance to negotiate. Edgar was willing to call a truce only if the scabs would be allowed to get their belongings and go to town. St. John assured him they would, demanding the agreement be put in writing. He also insisted the company must withdraw all employees with the exception of a few watchmen who met the approval of both the company and himself, and that they make no effort to resume work "until the strike was amicably settled and officially declared off." Collins balked, but agreed to write up a document. In addition to St. John and Collins, parties to this conference at Smith's were shift bosses

Will McKenzie and Bob Wright; Louis Lomax, Lavender, and union men John Conn and Harry Jardine, a Scottish Socialist who had been at the Sheridan dump when Barthell was slain. A crowd of union men stood outside, keeping a wary eye on the Bullion Tunnel buildings.[24]

Edgar Collins wrote the following document with terms for a mere three-day cessation of operations, instead of the terms St. John had agreed to verbally.[25] Anxious to prevent any further violence or deaths, St. John signed despite the change:

> *It is hereby agreed between the Miners' Union, by Vincent St. John, president, and the Smuggler-Union Mining Company, by Edgar A. Collins, that all work shall cease in said mine for the space of three days, ending Friday evening. Also that the said miners' union will refrain from violence, either to person or property, for the same period.*
>
> *The said Smuggler-Union mine is to have the right to keep four men as watchmen at the Bullion tunnel, one at Pennsylvania tunnel and one at Sheridan dump.*
>
> *Agreed to this 3d day of July 1901.*
>
> *Smuggler-Union Mining Company*
> *By Edgar Collins, Assistant Manager*
> *Vincent St. John*[26]

By 11:00 a.m. the gunfire had nearly stopped. Occasional blasts echoing on the hills made young Collins nervous and he suggested some sort of truce flag should be raised. A flour sack was run up a staff above the tunnel entrance to inform the men on the hills the fighting was over. Strikebreakers and nonunion employees gathered their belongings from the bunkhouses, many of them hastening down the hill before anyone could stop them. Some of the more quarrelsome men the strikers grabbed and disarmed, rounding them up with the cowboys and guards. At this time St. John went to the boarding house to look in on Ben Burnham, who Dr. Sheldon was tending. With a bullet in his gut, Burnham was writhing in agony and growing more delirious. At noon Dr. Sheldon sent a message to town that he needed to get the wounded to Dr. Hall's hospital. Superintendent Becker had a serious but not dangerous wound and Nicholson had a chance to survive, but Burnham was slipping closer to death.

Five of the men who had escaped from the Sheridan rode back up the hill to find Shadagee Bill. At about 11:30, they discovered him and the man who had taken cover with him, still hiding in the adit. Jordan was weak from loss of blood. The men got him up on a horse and headed for town. Jordan was terrified union men would try to kill him if they saw him, so he had his bodyguard take him via the old Sheridan trail that zigzagged down Royer Gulch.[27] When six men armed with Winchesters rode into town at 2:00 that afternoon escorting Jordan, it "caused intense excitement."[28] Rumor had Shadagee Bill murdered on the mountain. He was taken to Dr. Hall. Once patched up and safe in the hospital, he spun a tale of complete innocence in the initial gunplay that ended John Barthell's life. He claimed he was in the office when the shooting began and, having stepped out, was immediately shot.

Jack Hyde and the nonunion miners and guards, who were rounded up from the Sheridan in the dawn hours, had been marched to the top of Imogene Pass, two miles southeast of the mine dump. The union men had pointed Hyde and his associates toward Ouray with instructions to "keep moving and don't come back!"[29] These union men knew nothing of the agreement St. John and the company signed at 11:00 a.m. Upon their return to the basin around noon, they saw the flag of truce above the Bullion Tunnel and swooped down on the scabs who were now gathering their belongings. They surrounded the scabs, shoving them into a line, preparing to march them over the pass as well. St. John angrily protested, telling them he had given the men his word they would be allowed to go to town. Lomax joined St. John in his efforts to convince the strikers to let the men go, but the crowd grew more intense, the leaders snarling at St. John, "You have nothing to say in the matter!"[30] About a hundred union men took charge of sixty-eight strikebreakers and guards and started them up the road toward the Tomboy. St. John pulled one of the strikebreakers, William Merrifield, from the mob.[31] Although many union men hated Merrifield and he them, St. John had known and worked with the man for years. The old but hearty miner lived here, had family here, and St. John had promised safe conduct. He tried to intercede for other resident miners, but the union men yelled that no guards, no deputies, and no cowboys would be allowed to stay, and they ordered the scabs to move.

Smuggler postmaster, Ed Lavender, met St. John near the company store, telling him, "You ought to stop them!"

"I can do nothing with them," St. John replied in anger and exasperation.[32] If only he had the "absolute control" over them as so many claimed. He was one man against a volcanic mob. The sheriff and his deputies still had not arrived. Other enraged strikers still ranged the hills above the mine. If he pushed the mob too far, wholesale combat was likely to erupt again, this time in a bloodbath, since the strikers were armed and the nonunion men no longer had weapons. St. John's only motivation was to keep men alive. He had to find some middle ground in order to move closer to reconciliation. If the cowboys and scabs were sent over the mountain, they could always return when the flames had died down. Red Mountain Town and Guston were only a five mile walk from the Bullion Tunnel; Ouray was sixteen and Silverton about the same. The day was sunny. St. John knew the best choice was to allow the strikers to send the men over the range. After the mob pushed the column of strikebreakers and guards onward, St. John assigned a few reliable union men to go with the armed strikers to see that no one was beaten or molested in any way.

A large threatening gang of volatile Italians halted the captives near the Tomboy Mill, three quarters of a mile east of the Bullion Tunnel. The Italians accosted them with rocks and jeers, some screaming, "Shoot the scabs!" St. John's assigned guard had reached the group and told the Italians to push on. William Brennan, William Banks, Bob McGinnis, and a handful of others kept the enraged gang from harming anyone, instructing the nonunion men to move along quickly.[33]

They continued up the road, but as they neared the Tomboy's upper workings, the Italian gang came up behind them. Thomas Ballard, a scab, was near the rear of the column with Red Barclay. Although Barclay was a Telluride resident, St. John was unable to intervene for him because he had been deputized and used his gun and his fists on the pickets a few times too many. Now he was unarmed, a captive in line with the rest, at the mercy of the enraged gang of Italian strikers. A few dozen yards behind Barclay and Ballard one screaming assailant discharged a load of buckshot into the column before the bodyguard could stop him. Several pellets ripped into the back of Ballard's left arm and three hit a man in front of him in the leg. The tip of Barclay's finger was gashed, but the injured men remained calm.

The once great Tomboy Mine was the site where angry Italians accosted strikebreakers.
Photo by the author.

Brennan, Banks, and the bodyguard warned the Italians to back off or they would be shot. The column moved on again, but the gang followed, shouting "the vilest abuse." Ballard said the Italian mob was "simply frantic."[34]

"They had evidently been drinking," he told a reporter for the *Durango Democrat* (July 5, 1901), "and were worked up to a point of frenzy. Every one of them had guns, which they flourished threateningly. Some of them wanted to massacre us all on the spot. Seven or eight men, I believe they were Americans or Irish, quieted down the turmoil and said there should be no more shooting. Several noticed the blood trickling down my sleeve and suggested that I be allowed to return to Telluride. This plan was altered, however, because they were afraid I would be shot on the road."

Several squads of strikers were still uninformed of the truce and needed to be brought under control.

Ballard said the "men who assumed command would not allow the Italians to accompany us to the top of the range, and we were marched up the steep zigzag trail which crossed right at the Tomboy, without further incident. At the top we were told to strike out in any direction we wanted to, but never come back to the basin."[35]

Most of the nonunion men beat a trail to Ouray and Red Mountain Town, some heading out to the ranches in Ridgway and Montrose. Ballard continued walking until he reached a friend's house in Silverton by 9:20 that evening. The friend called a doctor who dressed his wounds. Three buckshot pellets were in his elbow, another in his shoulder, and a fifth was in his upper arm. He also had a flesh wound in his hip from the same blast.

"The truce arranged for and granted," Ballard said, "was most shamefully broken."[36]

The truce was not so much broken as it was still in the process of implementation. Communication across the high rocky expanse was slow at best, with no means to inform everyone the truce had been arranged and the mine was shutting down. Downtain never came up the hill. St. John was relatively on his own; only a few men from the dawn delegation were composed enough to assist him. Thus, after the mob drove away the scabs, St. John took to the hills to personally disband and bring in the various squads of armed strikers. One by one he spoke to the men, giving each the good news tempered by a severe warning that they must cease all combative posturing at once; anyone violating his order would be expelled from the union and turned over to the authorities. He told them they had a right to defend themselves, but the hostilities were ended and they must now conduct themselves with "utmost propriety because what you do now will make a difference for everyone else in the San Juan and in the future."[37] He put his trust in them. He placed a guard of twenty union men armed with Winchesters at Smith's store to ensure no one entered the Smuggler property. He organized search parties to check the Sheridan and mountainside for any injured men. He found Joseph Lujan's body stiffening on the Sheridan dump. St. John was sick at heart: during his years as a Smuggler miner he had boarded with Lujan and ate with the man at the same table. Now the trammer was left like carrion for the crows. It didn't matter if he was a nonunion man. St. John ordered a group of men to bring Lujan's body to town. Earlier he had sent men down with Barthell's body.

Dr. Sheldon had removed the injured men from the mountain, getting Becker and Nicholson to the Hall Hospital around the same time men arrived in town with Barthell's body in the early evening. Ben Burnham's injury was so critical that Sheldon took great care in bringing him down the steep trail. Sheldon, C. P. Gulberson, W. Hamor, and Lawrence Fields finally arrived at the hospital with Burnham by 8:30 that evening. The only other victim of the riot who sustained injuries serious enough to warrant hospitalization was striker J. A. Pierson, accidentally shot near the Tomboy around 3:00 that afternoon.

Each injured person or armed party arriving from the mountain caused tremors of excitement in town. Gossips howled that a full-scale war was still raging in the basins and the governor was sending the state militia to subdue the anarchists. Rumor that dynamiters blew up two buildings became established fact — the *Ouray Herald* (July 4, 1901) claimed the strikers blew up the Smuggler post office. The *Journal*'s interview with Jordan, who claimed total innocence, implied the union men attacked without provocation. Although by early afternoon many nonunion Smuggler employees had arrived in town, the tongue-waggers insisted all the company's men had been driven at gunpoint out of the district, many severely beaten, some murdered in cold blood, their bodies left in the hills. Some gushed that Jack Hyde's perforated corpse was coyote bait at the top of the mountain and that Manager Collins would be shot dead as soon as he stepped off the train.

Outside of the yowling and yapping tongues, town was actually quiet, with only knots of gossipers on the street and no sign of armed throngs. Mayor Hilgenhaus had ordered all saloons closed from 6:00 that evening until 6:00 the next morning. Marshal McIntosh and his night policemen joined forces with the payroll deputies to keep order in the streets. The Smuggler-Union mills were shut down and guards posted. Late into the evening, union vice president Ed Thomas and secretary Oscar Carpenter were in consultation with T. J. Sullivan, Senator

Buckley, Edward Field, Van Atta, and Brickson, while St. John went among the members, pouring hope into their ripped open, turbulent souls, telling them "a bird with broken wings can fly again."[38] They mustn't give up. They must trust him. They must do as he directed them. John Barthell died a martyr, but his death was not for nothing. They must believe a settlement would be made because justice spoke in clear tones from Barthell's silenced form; justice was on their side.

At 8:02 that night Arthur Collins's chartered train steamed into Telluride. He stepped off, unarmed, without a bodyguard, unafraid of the frenzied mobs or lurking assassins of whom his advisers had warned him. None existed. Not even one union man, no hecklers, no raised fists, appeared at the depot. His brother Edgar met him, along with several Smuggler officers, Addison Wrench, and a few reporters. They were grim. Yet Collins was certain he had won. Blood stained the union's hands. He was certain the annoying Local 63 and its puerile president were damned and the governor would send him troops shortly. He had no fear of reprisal. The union's goose was cooked.

Collins was informed the governor had not given the order to deploy troops. He was also informed Senator Buckley and the mediation committee were that very moment trying to arrange terms suitable to both the union and management and Collins was requested to meet with them at his earliest convenience.

Collins went livid. Under no circumstances was he meeting with the union. "We will adhere firmly to our resolution not to submit to their demands,"[39] he declared and ordered Sheriff Downtain to telegram the governor again. Orman obviously was uninformed of the facts, Collins asserted.

At 10:25 that night Downtain fired off another telegram:

> *Smuggler property has been forcibly taken possession of by strikers while a truce was supposed to be in effect. They ran all the Smuggler employees over the range and still have possession of property. There is no active violence at the present moment. Two men have been killed and four wounded.*[40]

Downtain had succumbed to the extreme pressure of Collins and Wrench. He knew the strikers never had possession of the property. He knew Edgar Collins had shut down the mine and posted guards and was at present still firmly in possession. But Collins and his allies wanted troops. The sheriff was their best means of getting Governor Orman to execute their demand. Although the town was quiet, Downtain had to convince Orman things were otherwise, his desperation leaping from every fabrication.

Not quite fifty-three, Oscar D. Downtain was a family man, with a quiet wife named Mary and two children, Clarence Edward, thirteen, and Florence, eleven. Born in Kentucky in 1848, he had come to Colorado at the age of twenty-one, working as a hostler at Ft. Lyon in 1870, as a stage driver for the Barlow and Sanderson Stage Company at Lake City in 1875, being later promoted to division superintendent. He was elected as Hinsdale County Clerk and Recorder two terms and by 1883 he had settled in Telluride as stage and express agent. Here he served as alderman, city clerk, mayor, and two terms as sheriff. Having faced stage robbers and brought in gunmen and thieves, he was a hero to the children of Telluride. Despite a painful limp caused by a draft horse stepping on him, an injury that never healed properly, he was constantly in action with his job and his horses. His friend, ex-Marshal James Clark, a man he had met in Ft. Lyon when Clark was a cowboy, was gunned down in the streets in Telluride in 1895, and Downtain had been helpless against the town council to bring the killer to justice. This night the temperate and laudable pioneer was helpless again, torn by lies and crippled by the power and might of Arthur Lancelot Collins. Downtain could do no more than he was asked.

CHAPTER

7

A BIRD WITH BROKEN WINGS

"The communication of the dead is tongued with fire beyond the language of the living."

— T. S. Eliot

After a rough night, Governor Orman returned to his office at 9:00 Thursday morning. It was the Fourth of July and he had cancelled most engagements. "You must send Manager Collins his troops!" was the dominating noise of the day. Assistant Adjutant General J. M. Johnson handed him the last telegram from Sheriff Downtain. Orman's irritation rose. Senator Buckley had told him nothing of a mob of strikers forcibly taking possession of the Smuggler property. Information from various Telluride sources was too conflicting to know what was actually happening.

General Gardner and the National Guardsmen in Denver were eagerly mobilizing at the Curtis Street and Glenarm armories. Cavalry, artillery, and infantry were in readiness, and "all the boys displayed great enthusiasm." Several other National Guard companies of infantry and cavalry were put on alert in other parts of the state. Gardner, a man who despised labor unions, pushed for immediate deployment, advising not just a portion, but all one thousand soldiers be sent out. "I cannot see how we can take less than that number," he declared emphatically.[1]

Orman was vexed by the warlike atmosphere and lack of clear information. At 12:45 he received a response from Senator Buckley: "Conditions unchanged; miners in peaceable possession of mine. Manager Collins is here. Attempts toward settlement being made. Life is not endangered."[2]

Buckley's telegram sent up roars of outrage. Union opponents interpreted it as meaning the strikers were in peaceable possession of the Smuggler. What effrontery of Senator Buckley to act as if such possession was tolerable or even acceptable. The governor was told his only choice was to send the National Guard immediately. Orman was unconvinced the esteemed senator meant the strikers had taken the mine; that made little sense. He believed the "miners" were simply those who operated the property, for Buckley's statement, "life is not endangered," indicated the situation was not critical. The governor needed someone on the ground at Telluride.

At 6:00 that evening Orman received a telegram from A. L. Collins, who was incensed that no troops had been sent. Although the mine was concretely in Collins's possession and never had been seized by the strikers, and the district had been quiet for more than twenty-four hours, Collins deliberately continued the deception. He even made the strikers a "masked" mob when none had cloaked their identity in any manner.

Telluride, Colo., July 4.
Governor Orman, Denver, Colo.:
Unprovoked attack made upon our property Wednesday morning by several hundred armed masked men. Valuable property destroyed. Two

men killed. Several severely wounded. Rest driven across range. Desperadoes now have forcible possession. Sheriff says he is powerless to give us possession of our property or protect lives of employees and has called for troops, which were denied. Without this protection, property must be indefinitely abandoned. Will you telegraph me advice and furnish protection?
The Smuggler Mining Company,
Arthur L. Collins, Manager[3]

Governor Orman's anger was tangible. Collins seemed the sort of man who would not stoop to lies just to get his way. A reply was sent immediately:

Unlawful possession of property will not be tolerated in this state and if property is not immediately surrendered to rightful owners, immediate action will be taken by state authority and all parties implicated severely dealt with. THE SHERIFF HAS NOT BEEN DENIED TROOPS.
James B. Orman, Governor[4]

Although many opponents of organized labor suggested Orman was withholding troops because the Smuggler-Union Mining Company was delinquent in paying assessed taxes, the statehouse said the idea was absurd. Orman's telegram proved he "stood firmly for the enforcement of the state laws."[5] He simply refused to act impulsively.

Lieutenant Governor David C. Coates, a Pueblo businessman, offered Orman his services to personally investigate the situation. He had been in contact with several Telluride businessmen and union officers.

"The whole trouble was started by a fool deputy sheriff," said Coates. "He fired a shot and wounded a miner, and that naturally caused the other miners to assert themselves."[6]

Eyewitnesses had told Coates Bill Jordan had shot John Barthell. If Barthell hadn't been killed, the miners would not have returned fire. The attack was not unprovoked. To Coates it was self-defense.

"If these deputies and their ready guns could be kept out of the road in such emergencies," he said, "these little difficulties between labor and capital would invariably end peaceably. I fully believe that, unless this man Collins, the mine manager, continues to act arrogantly, things will settle down and there will be no further cause for alarm. Collins, I understand, is a hard man to deal with. He has refused so far to treat with the union. I do hope the local arbitration committee will do something. It seems to be well organized, and if Collins and the company will do their share, I think the strike and the trouble will soon be over. If the troops are required down there the cost to the state will be enormous, and I sincerely trust they will not go."[7]

The use of troops in the Cripple Creek strike of 1894 had cost the state $67,174. The cost of Leadville disturbances two years later was $214,185.[8] Troops again were used during the Lake City strike of 1899. The state was forced to issue bonds to pay its bills and was still paying interest on those bonds. If troops were sent to Telluride, which was 422 miles by rail, the cost of transportation alone would be exorbitant. The statehouse estimated a full campaign could run up to $300,000. Orman refused to sink the state into debt and endanger lives if other means of settlement were available.

Shortly after the last dispatch from Collins, Orman appointed a commission to go to Telluride to investigate. Lieutenant Governor Coates, Judge Theron Stevens of Ouray, and John H. Murphy of Denver made up the commission. The men were presently in Denver conferring with the governor. Coates was also secretary of the

State Federation of Labor. Stevens, judge of the Seventh Judicial District (which included Telluride), was familiar with most of the businessmen and officials of that town. He personally knew Sheriff Downtain, Wrench, Painter, and numerous others. Murphy was the WFM's general counsel, a well-respected Denver attorney who had once been master mechanic for the Union Pacific Railroad until he was blacklisted for joining Eugene Debs's American Railway Union.[9] Although suffering from tuberculosis, Murphy was deeply dedicated to the rights of labor. Governor Orman announced to the press these men would not only investigate the situation for him and "endeavor to restore peace and prevent further violence," but they would also be "armed with the authority of the state."[10] He wired Collins that he expected the mine manager to give the commission his full cooperation. The other contending parties were given the same order.

Anti-labor howlers took to Orman like a piranha pack, verbally shredding him for sending a commission dominated by members of organized labor. "How can these men be considered impartial," said mine owner A. E. Reynolds in Denver, "with their obvious union ties?" The notion that the commission was unbiased gagged F. E. Curry who called Coates the "patron saint of anarchy."[11] J. A. Davis, a former Smuggler-Union Mining Company president, castigated the governor and praised Collins and the New England Exploration Company. "I consider that they have been treated with the greatest injustice by the union in Telluride, which has denied them the right to work by contract in the mine."[12] He directed the governor to send troops at once.

"If money is to be invested in this state for working our mines," Davis cautioned, "such investments should be protected by law. In my opinion this strike, to say nothing of the violence, is entirely uncalled for, and will not be upheld by any fair-minded person."Davis claimed Manager Collins was always "ready and willing" to meet the union's representatives and willing to have the conflict arbitrated, pointing out that the trouble resided in the union president, who, he was informed, "was a mere boy, just about twenty-one years of age."[13]

The president of the State Federation of Labor, H. E. Garman, countered Davis's accusations. "The men in the Telluride union have always seemed to me a particularly levelheaded lot. I know that the union, as a union has nothing to do with the disturbances of yesterday, and if the men individually got into this trouble, it is safe to say that they were crowded into a position where they had to fight."[14]

That evening at 9:30 Coates, Stevens, and Murphy left on the train for Telluride. A shoal of newsmen engulfed them at the Denver depot, gulping down every word they uttered for the late editions. The Telluride drama was selling newspapers by the thousands. The *Denver Post*, *Denver Republican*, and *News* were sending reporters with the governor's commission.

Lieutenant Governor Coates fed the reporters pure optimism, convinced a settlement acceptable to all would soon be made. Judge Stevens explained, "Governor Orman wants us to go there and see what we can do to settle this trouble without calling out the troops. Lieutenant Governor Coates and Mr. Murphy represent the labor interests, and I — " the judge gave a jolly smile, "I represent everybody."[15]

"Good nature and good fellowship sit upon the broad countenance of the judge like a halo," the *Denver Post*'s Paul Thieman wrote of him. "His spectacles shimmer with the glint of the humorist who loves his fellow men — the only genuine humorist."

Thieman was smitten by this judge who began his career as a blacksmith. Elected as justice of the peace in Ouray years ago, Stevens had enjoyed the law so much that he "studied it with sledge-hammer force and became a noted lawyer." He was now judge of Colorado's largest judicial district, which took in most of the western portion of the state.

"He is a short, broad-shouldered man, whose big head sits down close to his shoulders," Thieman wrote. "He speaks with a peculiar whimsicalness. The great man talks to those about him with a laugh in every word."[16] But when it came to the law, Theron Stevens conducted himself with utmost seriousness and carried out his mission on behalf of the statehouse with gravity and respect. He would not be trifled with.

In Telluride the July Fourth holiday was a quiet one, "though there was no lack of excitement on the third," Sumner quipped (July 6, 1901). A cloudless azure sky domed the warm day while the band performed concerts in the park. The only bangs and blasts were a few firecrackers lit by small boys. Several scores of citizens went to Silverton to take in the baseball games, Telluride winning eight to four. A number of others, including Sumner and his wife and several friends, went picnicking at Trout Lake. The day concluded with the annual Firemen's' Ball interrupted briefly by smoke billowing from a bakery. The firemen rushed to the scene in their spike tails and shirtwaists, only to find a smoking oven. The dance continued until 1:00 in the morning to the strains of Dray's Orchestra, with a large crowd of elegantly gowned ladies and dazzling gentlemen. Meanwhile, ongoing dialogue between the citizens' committee and the union seemed more hopeful, prompting the district attorney to wire the governor at midnight:

> *I have just seen Buckley. He says chance of settlement in morning. If not settled by noon he will join with me in appropriate suggestions.*
> *S. G. McMullin, District Attorney*[17]

In the morning Coroner I. L. "Ike" Glenn impaneled a jury and held an inquest over the remains of John Barthell and Joseph Lujan. Since Barthell's funeral was scheduled for that afternoon, this was merely a preliminary investigation, with few witnesses summoned. Andrew H. Floaten, D. N. McLeod, J. J. McPhee, George Cline, Kinney McLean, and W. M. Gillett returned a verdict of death "from the effects of gunshot wounds at the hands of parties unknown."[18] Ben Burnham had died from his wounds as well and a preliminary inquest was scheduled for 5:00, but the coroner planned to adjourn after that until July 12, giving the authorities time to gather witnesses and secure testimony.

Telluride Miners' Union Local 63 took charge of Barthell's body. Word of his funeral brought hundreds of miners out of the mountains and from neighboring towns, causing the Tomboy, Liberty Bell, and other big mines to close at noon. At 2:30 they gathered at Ike Glenn's undertaking parlor and marched with the casket to the First Congregational Church on Columbia Avenue, near Aspen Street. Vincent St. John and his fellow union officers walked directly behind the casket, leading an imposing 600 to 700 miners and mourners in procession. The *Journal* reported this was the largest funeral ever held in Telluride.

"It was really an impressive spectacle when President St. John called the men to halt," wrote Guy Lacoste of the *Denver Post* (July 5), "and for seven blocks they stood facing the point where their dead brother lay." A foreboding silence hung over the crowd while the band played a solemn dirge. "Men of hard lines are these. Upon each face is written in deep, scarred lines the story of ceaseless toil. There is no spirit of vengeance among them, but there is a manner, a determination, in those faces that bodes no good for the man who crosses their purpose."

Although vicious chatter sliced through town that the 700 men were armed and ready to attack Telluride, no violence of any sort reared to break the peace. They had come to bid farewell to a young brother miner who "always had proved himself to be a true and faithful worker in the ranks of organized labor and was always willing to assist his fellow man in time of need; was always willing to render his services when called upon and at all times ready to lend his support to adjust the wrong."[19]

First Congregational Church at 428 W. Columbia Avenue.

Courtesy William "Senior" Mahoney

Twenty-seven-year-old John Barthell was born Johan Bertills in Kovjoki, Vörå, Finland. He had joined his cousins in Telluride in 1900, Americanizing his name and joining the union. Every union man considered him a martyr, a brother who had sacrificed his life for his fellow miners and his dedication to justice. The church was too small to contain all who came to give him their last respects and the doors were held open to the enormous crowd of mourners outside, while the choir sang and Reverend W. S. Hunt delivered a somber message.

"Violence never advances a just cause," said Hunt. "It may seem to, but I am sure it does not. The violence, which resulted in this death and in other deaths, is as foolish as it was wrong and as wrong as it was foolish."[20]

Hunt spoke of righting wrongs via the ballot and gave hope to the mourners that John Barthell's life was not in vain and that they would follow his body to the cemetery as surely as they would "follow the light of truth."[21]

His coffin was opened and hundreds of men and women walked past, gazing on his lifeless face for the last time. As a mournful march rose from the assembled bands, Barthell's casket was draped with a black velvet pall and hoisted onto the shoulders of six stalwart union miners. In a silent column, two by two, 700 people walked from the church to Colorado Avenue and out to Lone Tree Cemetery, escorting the body to the grave.

"The funeral was one of the most impressive ever occurring in Colorado," wrote Guy Lacoste. "The procession was over a mile long and its purpose evidently was to impress the citizens' committee and all others with the strength and power of the union and the futility of trying to force them to go beyond the proposition they have made to the mine owners."[22]

On the sun-bright hilltop overlooking town and in view of the mighty Smuggler-Union properties, John Barthell's body was lowered into the earth to the slow beat of the drum. T. J. Sullivan and Vincent St. John each addressed the throng, asking the men to observe forbearance and maintain peace.

"Our brother John Barthell's death should never have been," said St. John. "But it is not for us to seek blood for blood. Such violence cannot be justified, can never be. No, my brothers, we must let his lifeless body speak for us, for indeed his death amplifies the truth, and only justice can come from truth."[23]

Perhaps it was in that hallowed moment that those 700 miners took courage from their young president, for from that point forward they observed his every word, and not one hint of hostility disturbed the district. Many miners flocked to join the union because of him. "Probably the most wonderful demonstration of the love of the miners for him and his control over them was on the day of the funeral," wrote Paul Thieman, almost mesmerized by him. These were men of every nationality, some seething with hatred, some bent by rage, each man believing "he was grievously wronged, that a Boston corporation was forcing his family to possible starvation." Just a mere "spark would have ignited such a flame of indignation that not a Smuggler-Union building would have remained standing. St. John went among the men all that afternoon and evening and there was no disturbance of any sort, nothing but the most perfect order."[24]

To ensure this perfect order, St. John personally had asked the saloonkeepers to close their businesses for the day and that night. Amazingly, they complied more readily than when the mayor had ordered the same on July 3. Somehow Vincent St. John, the small quiet miner with soulful gray eyes and a gentle boyish face, had won the hearts, the respect, and trust of most of the town. Many people believed he had single-handedly stopped the riot and combat on the mountain and for that reason he had become a champion for all of Telluride.

The miners gave him their absolute trust, consigning him "full power to make any terms he believed advisable with the Smuggler management."[25] He and his fellow union officers continued to make concessions to Arthur Collins, but Collins declined to settle.

At noon that day Collins had flatly rejected the union's final offer, announcing his orders from London were to "bulkhead the mines" if the union refused his terms. He was willing to allow union officers on the mine property and would not discriminate against union miners, but he held his contract system sacred, refusing to pay men standard wages should their contract price fall short. He issued a written ultimatum that the company would permanently shut down the property unless his demands were "accepted forthwith."[26]

"That ends negotiations," said St. John. "Let him shut down. The men have gone farther than they should in an effort to compromise. It is impossible for them to go a step further."[27]

Few believed the New England Exploration Company and its London partners planned to abandon a property in which they had more than $3,000,000 currently invested. Collins fully intended to operate the mine with nonunion labor if he could get the governor to furnish troops. Wrench had his full support and promised the support of other local businessmen.

When a reporter told Vincent St. John of the unconcealed scheme, which was already suspected, the union president lost his patience, angrily stating, "Collins will never operate the Smuggler until he comes to our terms. We will protect the mine from injury, but let him put scabs to work and the union men will drive them out as they did Wednesday."[28]

In an unparalleled move the union wired a message to the major newspapers in the state, which the *Denver Post* published on the front page:

> *To the People of the State of Colorado: All is quiet about the Smuggler mine and there is no danger of violence unless the Smuggler management provokes it by attempting to start with scab men under armed deputies. The troops are not needed in Telluride and if sent for it will be at the instance of the Smuggler people for the purpose of working their mine with scabs under state protection and to shoot down union men.*
>
> *V. St. John, President No. 63*
> *O. M. Carpenter, Secretary No. 63*
> *T. J. Sullivan, Member Executive Board,*
> *Western Federation of Miners*[29]

The announcement minced no words; Arthur Collins would be forced back into negotiations if he wished to salvage the company's public image and his own reputation. He had just "consummated one of the most important deals in the history of Silverton,"[30] gaining the North Star properties there. The company already had control of the Sunnyside group in that district, and the majority of Silverton's miners were union men, fully supportive of the Telluride strike. If Collins failed to compromise with Local 63, the future of all the company's holdings in the region would be endangered by similar strikes — unless he could persuade the state to send troops instead of commissioners. Collins reluctantly went back into dialogue with the union and citizens' committee.

In the meantime, the governor requested further information from the sheriff. Still anxious to obtain troops, Downtain replied, "There has been no violence or property destroyed since the 3d. I don't think I can protect life or property in case of riot. It is usually done before one knows it."[31]

The governor's commissioners arrived on the train at 6:00 that evening. Feverish nattering greeted them, but nothing more. Immediately they wired Orman:

> *Conditions here are quiet and seem favorable for settlement. Conflicting elements are in conference, and there is absolutely no immediate danger to either life or property. We are just going to scene of trouble at mines with Manager Collins, and will send further information as speedily as possible.*
>
> *Theron Stevens*
> *D. C. Coates*[32]

About 7:30 that evening Stevens, Coates, and Murphy rode up to the Smuggler mills with Collins, T. J. Sullivan, Charles A. Smith, Ed Thomas, and O. M. Carpenter to view for themselves the present state of affairs. Although a frantic man warned them armed men were coming after them, this proved to be nonsense. Collins declined to accompany them to the Bullion Tunnel. At the Bullion complex they found everything quiet, with four guards and Edgar Collins on the premises.

At Smith's store, a few hundred yards from the Bullion Tunnel, they found a dozen union men on guard, but no disturbing force anywhere. Collins admitted the strikers never had possession of the property and no violence had occurred since Wednesday after the eighty-eight scabs and cowboys were chased over the pass. Even the sad tale that the thoroughly hated Jack Hyde was coyote bait proved false, for Hyde had actually turned up Friday in Telluride unscathed. Despite his having beaten, threatened, and shot at unarmed union men for weeks, when he was captured by them and escorted out of the district, they did him no harm. The commissioners found no proof of blown up buildings despite so-called eyewitness tales of such reaching town and being printed in the *Journal* and other papers. The only evidence of property damage was hundreds of bullet holes in the Bullion Tunnel office and through the walls of a few other buildings. Smith's store "looked like a sieve"[33] from the bullets fired by deputies and scabs. The telephone line to town had been repaired.

The party reported to the governor via a midnight telegram, July 5:

> *Have just returned from Smuggler mills and mines; everything peaceful and no armed force anywhere about. Manager Collins and brother say they have had peaceful possession of mine since Wednesday, and there has been no indication of trouble since that time and no reason to believe there will be any more. No need of troops whatever from present indications. Further conference of management and men tomorrow.*
> *D. C. Coates*
> *Theron Stevens*[34]

Governor Orman was still far from assured the trouble was over, due to having received other telegrams at extreme variance with those sent by his commissioners. Wrench and his associates continued demanding troops, although these men "had been no closer to the seat of hostilities than two miles and were depending upon rumor for their information."[35] In the morning Orman again consulted General Gardner who advised to hold the troops in a state of readiness in case the rioting recurred.

The governor's commissioners sent him a telegram Saturday afternoon, indicating all requests for troops were exaggerated:

> *Telluride, Colo., July 6.*
> *Hon. James B. Orman, Denver, Colo.:*
> *Have had long talk with Sheriff Downtain. He agrees with our estimate of conditions, and that there is no necessity for troops unless further trouble should arise. Conferences are now going on, and indications are for a peaceful and early settlement.*
> *D. C. Coates*
> *Theron Stevens*[36]

Orman replied, "Advise me as soon as conference of management and men has terminated, keeping at all times in mind that the laws must be enforced and all citizens have their rights protected."[37]

Arthur Collins had arranged the day before to meet once again with the union officers and T. J. Sullivan Saturday morning at 11:00 at the Sheridan Hotel. Sullivan, St. John, Thomas, and Carpenter arrived but Collins failed to appear. When they telephoned him, he told them he was too busy and put them off until 2:00. He informed a reporter, "I will not further concede a single material thing. I see no chance for an agreement."[38] The *Denver Post* reporter, Guy Lacoste, gave this information to Sullivan.

"If that is Mr. Collins's position you may say positively there will be no agreement," Sullivan replied. "Further, you may say we will refuse to hold any more conferences at a later date. The matter must be settled today."

Energized by the slight flair of animosity, bank vice president Addison Wrench told Lacoste, "Unless an agreement is reached today you may say most emphatically that the citizens' committee will urgently request the governor to send troops and protect the operation of the mine, and that nearly every businessman in town will join in the request. If the Smuggler mine is closed down irreparable injury will be done Telluride. Half of our income will be taken away. Manager Collins wants troops and unless the union will accept his fair propositions, we should have them, and we are going to demand they be sent. Of course, Coates will advise against it, but he is an anarchist."[39]

Lieutenant Governor Coates, an intelligent, poised man fond of law and order, had never exhibited signs of anarchy. Wrench labeled him thus merely because he was a member of organized labor. Anarchists sought to overthrow the government; organized labor simply sought justice and equality. To Wrench anyone acting on behalf of the worker was an anarchist, the worst of the breed in Telluride being Vincent St. John. Unlike Collins, who made certain his public conduct was above reproach, Wrench intentionally goaded the union officers, causing St. John to step out of the conference on occasion to keep himself from saying things he might regret. Both Coates and Judge Stevens deflected the hostility with unflagging effort, the judge often injecting humor to move business forward.

At midday Coates, Stevens, WFM attorney John Murphy, and reporter Guy Lacoste rode up to the Bullion Tunnel to see the scene again, this time in full daylight. Once more they found no evidence of damage other than bullet holes. At Smith's store twenty union guards were on duty, composed and vigilant. These are "intelligent men," said Lacoste, impressed by their determination and courage. Numerous Winchester and Mauser M1898 rifles were in clear view inside the store and the men constantly swept the mountainside with field glasses. They allowed no one to approach the Smuggler properties, either to damage the mine or "permit its operation until matters are settled." From their position they held command over the narrow, steep, zigzagging trail to the mine.

Lacoste mentioned the possibility of soldiers coming up that trail and the union men laughed. One matter-of-factly pointed out, "It would be massacre to send soldiers up here." A hundred miners with rifles could hold the trail "against a thousand soldiers" since the soldiers would have to march up the trail in single file. Although they intended no such violence, they were prepared for the worst.[40]

Collins was likewise prepared, believing he could force the union out while holding them at bay with promises of dialogue. He made a last-ditch effort to convince the Tomboy and Liberty Bell managers to shut down those mines in support of him, which would throw several hundred men out of work, rendering the situation critical. They refused. The Tomboy's general manager, John Herron, was born in British India and the company's corporate headquarters was in London, yet Herron was well liked by his employees and he intended to keep it that way. He was nicknamed "Genial John." His most trusted foreman, William "Hutch" Hutchinson, was a steadfast union man and friend of St. John. Herron especially saw no reason to put company profit at risk. The Tomboy force had returned to work that morning.

Utterly obstinate and aiming to dodge the 2:00 conference, Arthur Collins conferred with Wrench, but Lieutenant Governor Coates and Judge Stevens confronted him.

"Under no circumstances," Coates sternly assured Collins, "will Governor Orman send you troops for the purpose of working the mine with nonunion men. You have only two choices left."[41]

Bulkheading the Smuggler never had been a choice, merely a threat. Backed in a corner, Collins joined the union officers in a final conference.

After a preliminary discussion with the citizens' committee, Collins, Sullivan, St. John, Thomas, and Carpenter were left alone to arrive at terms of settlement.

An hour later the citizen's committee and governor's commission were summoned. The agreement was fine-tuned and finally signed by the representatives of the company and the union, Arthur Lancelot Collins underlining his name with a superior flourish. Judge Theron Stevens and David C. Coates signed as witnesses.[42]

> *This memorandum of agreement, on this 6th day of July, A.D. 1901 entered into between the Smuggler-Union Mining Company and Local Union No. 63, of the Western Federation of Miners, of the County of San Miguel and State of Colorado, WITNESSETH:*
>
> *That the differences, heretofore existing between the parties hereto have been amicably adjusted on the following basis, to wit:*
>
> *First. The Union expresses its entire disapproval of the recent outrages.*
>
> *Second. The Company agrees not to discriminate against the Union, nor the members thereof, and the Union, and the members thereof, undertakes to not interfere with nor molest nonunion men.*
>
> *Third. The Company agrees to let the Secretary or President of the Local Union have full access to its surface property at all reasonable hours, provided that the work of the men is not interfered with.*
>
> *Fourth. The Union agrees to use all its influence to stop the illicit selling of liquor in Marshal Basin or around the mine.*
>
> *Fifth. The Company is to have the right to let contracts to any men who wish to take them. All such contracts to be on printed forms, which are hereafter to be drafted by J. H. Murphy, representing the Union, and Jacob Fillius representing the Company. Such contracts are to be for the period of one month, at the end of which time the work done under the terms thereof shall be measured up, and if it shall appear that the contractor has not earned Three ($3) dollars or more per shift, the Company undertakes or agrees to pay off the contractor at the contract rate or raise the contract price pro rata; that is to say, the price for the work will be raised so as to make it equivalent to Three ($3) dollars per shift for future work of the same contractor.*
>
> *Sixth. If the President or Secretary of the Union shall at any time represent to the management of the mine that any contractor is inefficient or dangerous to the safety of other workmen, or unable to carry out the requirements of the contract or agreement, the management agrees to at once measure up the amount of work already done by such contractor, and in the event of such amount of work being less than at the rate of Three ($3) dollars per shift, the Company will pay him as is provided in paragraph five of this memorandum of agreement.*

> *In witness whereof, the Smuggler Union Mining Company has caused its name to be fixed by its general manager, under the seal thereof, and the Local Union Number 63 has authorized its President and Secretary to execute this agreement on its behalf and to affix hereto the seal of the Union.*
>
> *Done in the City of Telluride, San Miguel County, State of Colorado, this 6th day of July, A.D. 1901.*[43]

It was Vincent St. John's twenty-fifth birthday, and despite having slept very little since July 3, he was exultant; strikes so rarely ended favorably for the unions. In a sincere gesture of reconciliation and goodwill, he firmly took Arthur Collins by the hand.

"Thank you, Mr. Collins," said St. John, his simple phrase speaking a thousand words. Both parties had made concessions, yet St. John had gone more than half way to achieve a settlement, giving in to Collins's demand for a thirty day measurement instead of fifteen. Collins kept his fathom system, but at least now attorneys on both sides would legally draw up the stoping contracts, leaving no room for bogus clauses and additional charges, making it a system less likely to be used by management. St. John won the union men their standard wage. He won the allowance to conduct union business on the property. He won respect for Local 63.

Collins congratulated him and shook the hand of each man present. He, Stevens, and Coates immediately telegraphed the governor:

> *Conference just closed, and agreement made which ends entire trouble satisfactorily. Strike will officially be declared off and operation of mine at once resumed.*
>
> *Theron Stevens*
> *D. C. Coates*
> *Arthur L. Collins*[44]

Senator Buckley also wired Governor Orman with the good news, stating, "Strike amicably settled, indicating your wisdom in not sending troops. Accept my congratulations."[45]

Orman was ecstatic. "I tell you, I feel better," he grinned and heaved a huge sigh of relief, commensurate with the elation he felt. "But I have had hope of this all along."[46] He telegrammed his thanks and congratulations to Collins, Buckley, and all those involved in the settlement.

"Telluride strike settled!" flashed across the wires from one end of the state to the other. Smuggler miners who had been leaving the district in droves to avoid impending military occupation heard the unexpected news and stampeded back to town, meeting jubilation with celebration. The brass band erupted into "A Hot Time in the Old Town Tonight" and marched up the main street tailed by joyful miners, giggling children, and barking dogs; the echoes of ragtime music bouncing off the glorious summer peaks that embraced the town. Saloons filled with buoyant crowds and foaming tankards were raised to the heroes of the day, who were many. Arthur Collins made an appearance, congratulating the miners and "greeting them good-naturedly."[47] Energetic singing and dancing overflowed into the streets. This was Telluride's overdue Fourth of July and the entire town turned out as only a mining town could, boisterous celebrations tumbling into the night. The spirited multitude gave Marshal McIntosh remarkably little worry, except for an attempted robbery in one saloon.

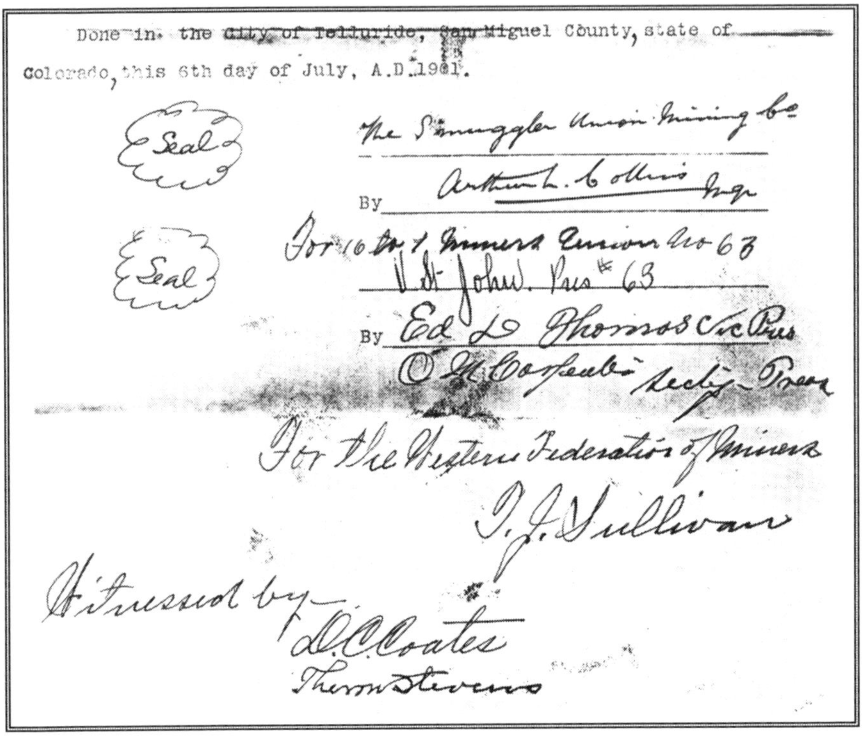
Done in the City of Telluride, San Miguel County, state of Colorado, this 6th day of July, A.D. 1901.

Seal

The Smuggler Union Mining Co

By Arthur L. Collins Mgr

Seal

For No. 1 Miners Union No 63

V. St John. Pres #63

By Ed L Thomas Vice Pres

O M Carpenter Secty Treas

For the Western Federation of Miners

J. J. Sullivan

Witnessed by

D. C. Coates

Thom Stevens

The signatures on the agreement between the mine company and the miners' union as they appear on the original document.

Courtesy Idarado Mining Company.

At the Clipper Saloon, Fred Pheasey and Kinney McLean raised a glass to Vincent St. John, honoring his birthday, his mother, and declaring him, in frolicking fun, "patron saint of Telluride." Both Fred and Vint were members of the local Catholic parish, and many of their closest friends and fellow parishioners were present, San Miguel County being seventy-one percent Catholic[48] and Pheasey being generous with free beer. A mob of miners hoisted their president on their shoulders in exaltation and, singing and shouting, "Saint! Saint! Saint!" they hauled him into the street. He protested their adoration, telling them "credit for the settlement belongs to everyone." They were deaf to him, even Ed Thomas and Oscar Carpenter acknowledging him as the "lion at the gate."[49]

"This St. John is the most picturesque and remarkable personage in the difficulty," wrote Guy Lacoste (*Denver Post*, July 8, 1901). "He has a beardless face and youthful expression. But powerful and magnetic, he absolutely controls every miner in the union. They worship him."

And they did that night, wherever he went, pressing him for details of the riot, the shooting, the gore, the deaths, and the triumph. He had won them justice through determination and truth. He never wavered. Even bullets failed to drive him back. In the end, the bird with broken wings flew so high it touched the sun.

CHAPTER

8

WALTZING ON THIN ICE

"I regard you with an indifference closely bordering on aversion."

— Robert Louis Stevenson

The signed agreement between Collins and Local 63 was presented to the entire union body at Red Men's Hall Sunday evening for the purpose of ratification. Nearly 400 members were present. The document was endorsed and the vote to officially declare the strike over was unanimous.

Arthur Collins immediately announced the Smuggler would start operations on Monday, July 8. Newspapers published his notice: "All men forcibly driven from the Smuggler property on July 3rd will be at once reinstated on their return. If wounded or otherwise incapacitated, their positions will be reserved for them, but all other vacancies will be filled without any discrimination against union men, preference being given to old Smuggler-Union employees and local residents."[1]

Some of the eighty-eight men who had been driven over the pass returned that week, including Charles Stoltz, W. S. Peterson, D. E. Carroll, Edwin Thomas, and Red Barclay. Since nearly a quarter of the men had been guards and not miners, the cowboys had no reason to return. Only a few of the eighty-eight men were locals, thus less than a third returned. No one menaced or assaulted the returnees, disproving Curry's tale that these men were warned never to return at the peril of their lives.

When the whistle blew Monday morning, about 150 miners went back to work at the Smuggler. Since the mills were still under remodeling, the full workforce was not yet required. Numerous nonunion employees soon after joined the union. Sulfurous Harry Jardine, the Socialist union campaigner and irreverent "Press Committee" for Local 63, gleefully wrote, they have "seen the error of their ways and have joined our ranks."[2] St. John's reputation for accomplishing impossible tasks drew in more members daily, swelling the ranks to nearly 900, making it one of the strongest locals in the WFM. Before St. John's presidency Local 63 had a weak and spiritless membership. His dynamic presence invigorated it, and the victory of the Telluride Miners' Union affected the entire WFM throughout the West, increasing the membership to nearly 48,000, with more than thirty percent of that in Colorado.[3] Although the *Miners' Magazine* heaped praise on St. John and Carpenter, declaring them men of "sterling character and efficiency," the two refused to accept credit alone. "To gain such a signal victory," they wrote in a joint letter, "required the earnest cooperation of all concerned, and this was granted so heartily that we feel justified in saying that an equal amount of credit is due all, and that by no particular man or set of men can the honor be claimed." In conclusion, they applauded each member of Local 63: "When redeemed humanity shall emblazon on the pages of history the names of its honored heroes we feel assured that the brave boys of the San Juan will hold a conspicuous place on the records."[4]

The Pandora Mill (left) and Smuggler Mill (right) and cyanide plant (center) after 1904. The photo shows the renovated office/manager's house. Bulkeley Wells enclosed the porch and had the second-story "porch" bedroom added above it. A dormer extended the attic story. When Collins lived here, no dormer or enclosed porches were on the house.

© Main Street Photography, Walker Studio Collection; used with permission (65T).

Telluride businessmen nevertheless praised the president and secretary, and Charles Sumner commended them and T. J. Sullivan in the *Examiner*, particularly Vincent St. John. Once an unknown miner, St. John now was recognized on the street, men and women flocking to greet him, to shake his hand, to thank him, to hear his story.

The district's unabashed adoration of St. John galled Wrench and chafed Collins and caused Curry to nearly foam at the mouth. Publicly Collins played the role of the refined conciliator. Privately he seethed. Despite the agreement, Collins felt he had been forced into settlement by the threat of violence. In letters to the *Denver Post* he criticized the "unwarranted proceedings by the union's younger and inexperienced officers" who, in his estimation, were the "fruitful cause of trouble."[5] He meant only one: Vincent St. John. Oscar Carpenter was a married man older than Collins. Likewise was Ed Thomas.

The editor of the *Examiner* finally had enough of this ceaseless denunciation that St. John lacked brains or wisdom due to his age. Sumner, who had become acquainted with St. John only in the months prior to the strike, had developed an unshakable admiration for the union president. The more he knew him, the more he liked him. Sumner was a Republican, St. John a Democrat, ideologically at opposite ends, yet Sumner found the union president more than just idealistic, warmhearted,

Vincent St. John (left) and Oscar M. Carpenter were photographed for the Miners' Magazine *as the "sterling" leaders of Telluride Local 63, summer 1901.*

Courtesy Linda Paulsen.

and "genuinely honest."[6] The twenty-five-year-old miner spoke of going beyond settlement to forgiveness, believing justice "put things to rights," but only "forgiveness heals."[7] To him each individual was important, each life priceless. "St. John may be 'nothing but a kid,'" Sumner wrote in the *Examiner* (July 13, 1901), "but we incline to the belief that when it comes to ability, energy, determination, common sense and cool-headedness, there are many older heads who could sit at his feet with profit. Old men are entitled to wise judgment and discretion, but it is often found they slip up on the proposition."

Collins chafed all the more. He was expert at giving the impression of benevolence while doing his best to provoke St. John and drag him from the lofty place where the miners and most of the citizens had raised him. Using the *Journal* for his mouthpiece, Collins gave all the credit for the settlement to Vice President Edward D. Thomas's "good judgment, sound and conservative reasoning," as if St. John hadn't existed.[8]

Unruffled and ever willing to make peace, St. John and the union published a gracious notice to all, and sent similar cards to each individual involved, including Wrench:

> *We hereby extend our sincere thanks to Governor James B. Orman, Lieutenant Governor D. C. Coates, District Judge Theron Stevens, Attorney John H. Murphy, Senator Buckley, the Citizens' Committee, and the editor of the* San Miguel Examiner, *for their untiring efforts to settle the difficulties between the Miners' Union and the Smuggler-Union Mining Company.*
>
> *We heartily appreciate Governor Orman's good judgment in not sending the troops, and Senator W. S. Buckley's efforts in keeping him informed of the true state of affairs.*[9]

Despite the settlement and the union's efforts, Collins, Wrench, Painter, Curry, and their circle of associates festered with bitterness. Collins was unable to stomach what he viewed as an affront to his authority and the failure of the State to protect his rights. Wrench and Curry rebuked Telluride businessmen for lack of support of Collins. They wanted to affix blame. Wrench even went so far as to eliminate any business dealings with the *Examiner* and advised Collins to do the same.

Although Charles Sumner had never personally opposed or attacked him, Collins took exception to him for supporting St. John and the union. Sumner praised the manager as a "courteous and agreeable gentleman" and made it clear the fight never had been against Collins. "That he bears us antipathy for antagonizing his interests is his legitimate right and it is perfectly right to withdraw his patronage from this office should he choose to do so." Sumner had no interest in prolonging ill feeling. He had merely supported the union because he believed the "existence of the town is dependent upon the wage-earner and upon no one else. There is due consideration for all capital invested here, but the money expended by capital circulates in local channels through no other medium than the laboring man, and if the laborer does not elect to spend his money, there is no town."[10]

Addison Wrench held the opinion that labor would not exist without capital. He insisted the future prosperity of Telluride depended on outside capital and when investors developed properties in the district, that property "must be assured protection and their right to conduct their business without interference from the manipulations of violent organizations must be guaranteed," or the investor would look elsewhere.[11] He alleged the union was thoroughly criminal in its actions and Collins was its victim.

Rossiter W. Raymond, a nationally known mining engineer writing for the *Mining and Engineering Journal*, echoed Wrench's sentiments. He asserted that the most "serious obstacle to profitable development" was organized labor. "It is within my personal knowledge...that the loyal performance of a fair day's work has been habitually discouraged and forbidden by the miners' unions."[12]

The statement was preposterous. Most mine companies sought union labor, as union miners were generally more experienced, dependable, and conscious of safety. The WFM deplored strikes, using them only as a last resort, since the cost was dear. "There is a popular misapprehension that labor organizations favor and advocate strikes," said John Mitchell, national president of the United Mine Workers of America.[13] The previous general manager of the Smuggler-Union, Nate T. Mansfield, defended the union men as "excellent employees," and said the "strike was an unfortunate result of failure of management to treat with the union's business propositions."[14]

In agreement with Mansfield, Senator Buckley shed his usual restraint with reporters and informed Guy Lacoste that Arthur Collins "was to blame for the whole affair. His unfortunate manner of dealing with the miners was the chief cause of the strike."[15] The Ouray *Plaindealer* said Collins was a good businessman but had no talent for handling "a large body of men of different nationalities."[16] If harmony was to be restored, Arthur Collins needed a touch of humility and honest grace. The miners were already thawing toward him, despite their dislike of him, since he showed the nerve to walk among them totally unarmed during the height of the dispute. This won him some admiration. The miners were willing to forgive if Collins was willing to respect them. "He will more than live up to the agreement," his friends assured them.[17]

Perhaps. He seemed more interested in holding union men liable. Along with the county commissioners, he offered a joint reward of $500 per each case "to anyone furnishing evidence which will lead to the conviction of persons who shot into the buildings of the Smuggler-Union Mining Company, or at its employees, on July 3rd."[18] The reward equaled nearly half a year's wages and was certain to bring witnesses out of the woodwork. Collins refused to consider that most witnesses already said the deputies and guards initiated the gunfight. Many actually named Bill Jordan as the first to pull the trigger and as the man who shot John Barthell.

"The conflict was not sought by the men," Sumner pointed out (July 13, 1901). "They maintained a peaceable stand all along." The union officers had no foreknowledge of events that led to gunfire and death. None was near the mine property that morning and St. John risked his own life to stop the riot.

The *Plaindealer* (July 12, 1901) minced no words: "Again there must be taken into account that ever present institution of the devil, 'the deputy;' always a prominent factor in 'perfecting' strikes. If a strike lags, the deputy may be relied upon to bring it to the boiling point, and seven times out of ten he is a rogue in the law's livery, who, if he had his deserts, would be wearing a ball and chain rather than a badge of authority."

Mel Robbins paid only a fine for one incident of gunplay out of many. Red Barclay and Jack Hyde were never prosecuted for assault. Nor were any of the other guards. Shadagee Bill, with his reputation for drunkenness, threats, and violence well established, was untouched by the law. The investigation into Barthell's death was terminated, as were the Burnham and Lujan inquests, allegedly for "fear of arousing excitement."[19] District Attorney McMullin stated, "I don't see how convictions would be possible."[20] Under the circumstances of a riot it was unlikely to determine who shot whom, but with twenty or more witnesses claiming Jordan shot the unarmed Barthell, a conviction in that case was possible. Rumor had it Wrench and Collins influenced McMullin's decision to adjourn the coroner's jury in order to protect Bill Jordan. The WFM attorney, John Murphy, said these men had "demanded of me to state whether or not I was going to defend persons whom they designated as murderers." He told them the Federation planned no defense of any man who committed wrong, but was determined to protect any union man from false charges and that if any investigation occurred, he would be certain it included inquiries into who had supplied "the deadly weapons to the deputies and the men employed by the company." The investigation would not be "a one-sided affair."[21] After much discussion, McMullin concluded, "It was in the best interests of the community and all concerned that no prosecution should be had...and the matter was considered forever closed."[22]

The union published a "Resolution of Respect" (*Examiner*, July 13) in honor of Barthell, neither accusing Jordan nor demanding justice for the young Finn, yet they made it clear nonunion miners ended Barthell's life only because he asked them to join the union's cause. This was for the record.

Truth was immensely important to St. John, whether anyone acted on it or not. He determined to at least put before the public what he viewed as correct and accurate information. In the wake of the strike and settlement, the *Denver Post*, calling strikes repugnant and unnecessary, attempted to absolve the current Smuggler management by intimating that Local 63 called the strike because it disliked Collins's more strict and efficient management and that in the past the "details of the mine were conducted without proper supervision."[23] While offering the *Post* the benefit of the doubt that the misinformation was unintentional, St. John and his officers defended Nate Mansfield, the former general manager: "Having been treated with the utmost fairness by the old management in all matters, we desire to place ourselves on record as saying that our pleasant relations in the past have come from the fair and businesslike manner in which the old management conducted the affairs of the company."[24] Mansfield had been a conscientious manager; the safety of his men came first. He even had fire guards who patrolled against the ever-present danger of fire; Collins eliminated them. Profit came first for Collins.

By the end of the second week after the settlement, the Smuggler-Union had a force of 200 men at work. The *Post* had made a gloomy prediction that it would take years for the company to recover from the conflict, but Collins announced that by September he would have 700 men employed, far more than before the strike. The new cyanide plant would be complete by then and both the Pandora Mill or "old mill" and the Smuggler Mill or "new mill" would be working full capacity. The tram from the Contention group of mines up Bear Creek would be running to the Smuggler Mill and the Smuggler-Union Company's output would be more than half the entire production of the county.

Despite the optimistic projections, Collins was determined to make the union pay for interrupting profit. Most of his employees were now union men. He still failed to promote union miners to supervisory positions, all foremen and the majority of shift bosses being nonunion. He raised the boarding house rate from ninety cents to one dollar per day, gaining a straight profit of nearly $2,000 per month above cost.[25] The fees for powder and equipment also were gradually raised. St. John and Carpenter attempted negotiations with him, generally conceding in order to keep the peace. When it came to any infractions in the new stoping contracts, St. John immediately brought them to Collins's attention, keeping ever vigilant on behalf of the union men. He required Collins to make "rebates to the miners" when the contracts fell short, and insisted the miners' grievances be rectified without delay, much to the manager's extreme irritation.[26] Since the other large mines operated without stoping contracts, but on a daily wage, the Miners' Union proposed a uniform "scale of wages" for all mining employees in the district, starting with underground miners at three dollars per eight-hour shift, machine men at four dollars per eight hours, and including every employee from engineers and hoisters to blacksmiths and tool sharpeners.[27] The managers of the Liberty Bell and Tomboy, Arthur Winslow and John Herron respectively, were willing to accept the scale, but wanted to negotiate longer hours for the millmen. The millmen were not organized under the Miners' Union, thus St. John was unable to negotiate for them. He started a campaign to organize them.

Cautiously, Collins sought every means to crush the union's growing power. He wanted Local 63 indicted for more than the bullet holes in the Smuggler buildings, but no witnesses were forthcoming. The injured company men had no clue who shot them during the riot. Superintendent Charles Becker, Collins's brother-in-law, had been in danger of losing his arm due to suppuration of the wound until a specialist from Denver tended to him. His arm sustained permanent disability. Nicholson surprised everyone by recovering from what seemed a likely fatal head wound, and Jordan was back on his feet within a week. Jordan remained noticeably subdued, making no further public statements concerning the events of July 3. Since the district attorney adjourned the inquest, Collins knew no union man was going to be prosecuted for shooting Burnham or Lujan — that was the cost of protecting Jordan. Few legal avenues for retribution remained, unless the rewards brought witnesses.

To rely on the avarice of a disgruntled union member was a waste of time. Collins needed a means to discredit Local 63 in the public's view. Even the lowly hay hands in the agricultural regions of the county had been inspired by St. John, refusing to cut hay at the critical time unless their wages were increased. They won a fifty-cent raise.[28] Organized labor was becoming a deadly threat to profit. Wrench, Painter, and Curry were all too willing to assist Collins in curtailing union power, particularly that of Local 63.

They took up with Curry's earlier innuendos that the union was responsible for the absence of Will Barney. Although Collins had made no attempt to offer a reward for Barney's whereabouts before the riot and settlement, he and the company's board of directors in Denver, along with the San Miguel County Commissioners, now offered a reward "of $100 for the recovery of the body of W. J. Barney, if dead, who disappeared from Telluride, Colorado, on Sunday June 23rd, 1901; and offer a reward of the further sum of $500 to anyone furnishing evidence which will lead to the conviction of the party or parties who caused the death of said W. J. Barney. Said W. J. Barney was about six feet two inches in height, weighed about 210 pounds; had sandy complexion; had sandy mustache, and was about 35 years old."[29]

This advertisement, paired with the riot reward notice, was placed in newspapers throughout the state, including the *Examiner* — Collins was all too happy to patronize Sumner in this instance. Sumner thought the wording was rather curious, since no evidence indicated thirty-three-year-old Will Barney was dead, let alone dead at the hands of other "parties." The notice so carefully asking for his *body* hinted they knew he was alive and had no intention of paying, since the usual wording in such a case as a disappearance was a reward for the whereabouts. Why they chose Sunday and not Saturday, when Barney was supposedly last seen, was another curiosity, leaving an entire day between brawls unaccounted for. Nevertheless, the company had struck the first blow in the campaign to attach infamy to Local 63. This was no real offer of reward; rather, it was subtle and cunning warfare, a message condemning the Telluride Miners' Union as murderers.

Sumner saw right through it, boldly suggesting, "If Mr. Collins and the Commissioners would offer a reward for Mr. Barney alive, he might be easily located."[30]

CHAPTER

9

SEND IN THE SPIES

"To the man who has only a hammer in the toolkit, every problem looks like a nail."

— Abraham Maslow

Vincent St. John was a problem. The admiration the ordinary citizens of Telluride had for the quiet young man had turned to sincere affection. As the union increased in membership and the members readily paid their dues, he inspired the entire body toward great enterprises, one of which was a funding campaign to build a brick hospital to provide the best care for miners and their families. Many miners had died from accidents and pneumonia because of lack of proper care. St. John had presided over too many heart-breaking funerals since he became president; not a month went by without several miners being injured or killed. He solicited the expertise of Dr. E. D. Allen and surgeon J. G. Sheldon to plan the most efficient and modern medical institution possible. The hospital building would also house a permanent miners' union hall, as well as meeting rooms for other unions. Arthur Collins and Addison Wrench made an effort to "block the progress" of the Miners' Hospital and "to hinder the good work in every way," for no other reason than personal loathing. "There is a strong undercurrent in a few quarters," Sumner wrote, "to put a crimp in the Union at every opportunity, but it will not work out now nor hereafter, and the sooner this practice is abandoned, the sooner will harmony prevail."[1]

Paying no heed to Collins and the bitterness of his associates, St. John worked selflessly for the union. He believed education was the best means to free minds from slavery and initiated plans for a union library and loafing room. He wanted to create a program to assist the foreign-born in learning English, for their own advancement and the safety of fellow miners, appointing Louis Macari and three others to coordinate the project. He specified union literature was to be available in several languages in order to reach all the miners in the district. He cemented cooperation with the Federal Labor Union and the Carpenters' Union, inviting them to assist with funding for the hospital and hall. Tirelessly at his side every day was his right-arm man, Oscar Carpenter.

Born in Troy, Alabama, September 16, 1867, Oscar M. "Carp" Carpenter was a quiet man with blue-gray eyes, thin blonde hair, and a comically bushy mustache. He was an "exemplary father and husband," with exceptional financial skills.[2] He managed the budgets of Local 63 and the district union without as much as a cent unaccounted for.[3] The WFM's practice of having committees from other unions audit each local's books assured financial responsibility. Carpenter's honesty and total lack of avarice secured that in Local 63. After leaving his father's cattle business in Texas at nineteen, the five-foot-six-inch, stocky-built secretary had begun his career as a practical miner in Bellevue, Idaho. In 1895 he, his young wife Ada, and their two babies moved to Telluride, where he worked as a miner at the Liberty Bell until he was elected financial secretary of Local 63 in September 1900.[4] He and Ada bought a small house on the south side of the tracks on San Juan Avenue to raise their son, Carl, and daughter, Bertie Esther. His skills as a fundraiser were instrumental in bringing the vision of a hospital and union hall to reality. He quickly gained the backing of many local businesses, from Obert's Vienna Bakery to the Floaten brothers' People's Supply.

Oscar M. Carpenter, Secretary-Treasurer Local 63.
Restored portrait © M. J. Martin 2003.

Both Carpenter's and St. John's energy and dedication to their work increased their popularity among the general population of the county, and within only a few weeks after the settlement, citizens proposed them as candidates for public office. The very idea was more than repulsive to Collins, Wrench, Painter, and Curry. Carpenter soon made it clear he would not run, suggesting former union secretary Ed Olson seek the nomination for county treasurer, but he and his fellow Democrats rallied behind St. John to run for county sheriff. The proposal caused Curry to convulse in a rage against the "caucus manipulations" that made it possible "to secure the nomination of most unworthy men."[5] The local caucuses were scheduled for October and Curry warned the citizenry more than a month in advance to see to it St. John was not nominated.

Sheriff Downtain, a Democrat, had been humiliated and insulted by Wrench for failing to convince the governor to send troops. He was also embarrassed by rumors that he had acted cowardly during the riot and was disappointed in his inability to overcome his fears. Some accused him of sulking. He announced he would not seek reelection as sheriff another time and gave his support to the union president. St. John was unsure of taking on such a position, but when Senator Buckley, Marshal McIntosh, Fred Pheasey, and Kinney McLean, all Democrats, urged him to seek the nomination, he consented.

The only other Democratic aspirant for sheriff was John R. Hogan, a longtime resident of the county and Smuggler foreman. Curry promptly designated him as the best choice for the Democrats (Curry was a Republican) because Hogan was "competent" and "impartial" and without "the suspicion of a blot...on his honorable upright life."[6] Curry applauded all aspirants for all the offices, Democrat and Republican, except St. John and Charles Sumner. Sumner was seeking the Republican nomination for county clerk. Incumbent County Judge John Wardlaw was given a slight rebuff for his support of the union, but Curry expected him to keep his seat. In the end, the most important contest was the office of sheriff and Curry considered both Republican aspirants, John Calvin "Cal" Rutan and James Real, men of character and worthy of the job. Cal Rutan, a big robust man with blue eyes and a slow smile, had previously served as sheriff from 1891 to 1895. "No man need blush when he votes" for Rutan, Real, or Hogan, Curry pointed out, leaving St. John entirely out of the picture.[7]

Driven by apprehension and repugnance, Curry rummaged for any legitimate condemnation to make certain St. John's nomination crumbled. He couldn't blame the recent highway robberies or rash of burglaries, including a break-in at Senator Buckley's house, on the union president. Nor could he blame him for the continued drunken brawls among the Swedes and Finns, even if the majority were union men. Curry needed an obvious union infraction to prove St. John was a disturbing element in the community and to promote the image of him as an anarchist.

John Calvin Rutan, Sheriff of San Miguel County.
Restored portrait © M. J. Martin 2003.

Opportunity presented itself as an incident of intimidation at Ophir, and Curry seized it like a cougar on cornered mutton. He lashed out at Judge Wardlaw, the local accredited correspondent for the Denver newspapers, for refusing to send the story across the state via wire. He accused Wardlaw of suppressing the "high-handed, outrageous proceedings" on behalf of the union, just because Wardlaw wanted to secure the labor vote in his reelection bid.[8]

The incident was indeed outrageous, but had nothing to do with Local 63, since the men involved belonged to the Ophir Miners' Union "Bryan" Local 64. A Pennsylvania man named Forest H. Wogan was hired to teach school in Ophir, but since he had worked one day (July 1) at the Smuggler-Union as a roustabout during the strike, his name was allegedly put on a scab list. Three irate union men in this mining village thirteen miles south of Telluride tried to force Wogan to resign. He took the matter to the school board, who upheld him; they had a contract with him. He said the union men forced him out of the hotel at Ophir by threatening the proprietor with a boycott. Wogan and his young wife rented a house and, on September 2, he opened the school. That night the three union men and their cohorts — Wogan said twenty to twenty-five — forced him to sign a resignation. In the group was Ophir school board member Dr. G. W. Compton, who advised Wogan to resign and leave. Defiant and obviously not threatened with death as rumor cried, Wogan filed a complaint with the county commissioners and Sheriff Downtain. The sheriff said he would send deputies with Wogan to protect him if necessary.[9]

Transforming one incident into universal practice, Curry elatedly condemned the entire Local 63, claiming it was making it impossible for decent men to earn a living in the region and that the "law respecting element...may have to abandon their holdings and migrate elsewhere."[10] He asserted Ophir would have no teacher for the entire school year.

St. John knew nothing of the intimidation until it screamed from the *Journal*'s front page on September 4. He at once vigorously condemned the action. On September 5 he personally went to Ophir to investigate as president of the district union, and to reprimand the perpetrators. He found only six men responsible, not twenty, and recommended Ophir expel two from membership. He offered his sincere apologies to Wogan and the matter was courteously settled. Ophir would have its teacher. Curry slipped a quiet notice in the *Journal* that "this is the proper course and will win the commendation of all classes and creeds."[11] Wogan turned around and took other employment in the village, forfeiting his contract with the school board. Another teacher had to be hired.[12]

Instead of destroying St. John's prospective nomination for sheriff, the incident advanced it. Citizens admired his ability to take control of bad situations and resolve them for the good. He seemed to have a natural talent for it. That week both he and Carpenter were unanimously reelected as union president and secretary-treasurer respectively. They were also reelected to their same offices in the San Juan District Union. Around this time St. John began receiving death threats via anonymous letters, which he laughed off as nonsense. He tossed them out, only mentioning them to Sumner and Pheasey. Someone badly wanted him off the stage.

On September 6, 1901, a different political figure was violently taken off the stage, turning local events in Telluride upside down. While holding a reception in the Temple of Music at the Pan-American Exposition in Buffalo, New York, President William McKinley was shot. The man behind the trigger was Leon Czolgosz, a crazed follower of Emma Goldman, the self-proclaimed anarchist. A copy of her speech, proposing "anarchism aims at a new and complete freedom," was found in his pocket.[13] "The assassin was immediately overpowered and taken to a police station on the Exposition grounds, but not before a number of the throng had tried to lynch him."[14]

According to James Creelmen, one of the news correspondents at the scene, McKinley asked the guards not to harm the gunman. The dying President said, "'It must have been some poor misguided fellow.' He seemed to be filled with amazement by the thought that any man in free America could have found a motive for seeking his death."

A gentle-souled man, McKinley had worked for peace, believing differences could be adjusted through arbitration. "Let us ever remember," he said in his last speech, "that our interest is in concord, not conflict; and that our real eminence rests in the victories of peace, not those of war."

On September 14 he died from the bullet in his stomach and the consequent botched surgery to remove it, his frail wife and numerous friends at his side. "Whatever else history may say of William McKinley, those who knew him will bear witness to the forgiveness that shone through his character. It was the crown of his life, the virtue that distinguished him among American statesmen. He died without an enemy, forgiving the hand that shed his blood."[15] His assassin was executed in the electric chair on October 29.

Since an anarchist brought about McKinley's death, Curry wove sinister links to those he labeled "anarchist" in the Miners' Union. It mattered little that labor came to like McKinley in his second term because he condemned the huge corporate trusts as "dangerous conspiracies against the public good."[16] No Miners' Union man was an anarchist or devotee of Emma Goldman. Some WFM leaders were Socialists, but Goldman scorned Socialists and they, her. In *Journal* issues following the assassination, Curry poured out innuendo as if to drown St. John in it before the Democratic caucus. He continued to harp on the recent strike as having seriously hindered progress and prosperity, blaming St. John for it. Yet at the same time he claimed the

Smuggler-Union would soon have a monthly payroll of $75,000. He demanded citizens support the interests of the Smuggler-Union and reject St. John.

The demand failed. At the Democratic primaries most delegates chosen for the caucus were union members, assuring St. John's nomination for sheriff. Curry shuddered. Something had to be done. Collins consulted the company's directors in Denver.

An alarming report suddenly appeared in the *Denver Post*, the source being a Telluride mining company with corporate offices in Denver. The report claimed the Smuggler-Union miners were threatening another strike, which would involve Silverton and Ouray as well. It warned that if St. John became sheriff, the mine owners would have no help from him in case of trouble. It also declared that the union was cruelly driving out the Chinese in Telluride.

Sumner jumped on the article as entirely false, but Curry used it to promote discord by reproducing it with the headline, "Miners Threaten to Strike Again; anti-Chinese movement begun."[17] Sumner headlined it for what it was: "A Campaign Lie; Malicious report calculated to discredit Telluride."[18] He pointed out the Smuggler-Union was the only Telluride mining company with a Denver office, and Collins and Bulkeley Wells (who worked at that office) were the likely sources.

Curry, on the other hand, piled more campaign lies on top of the *Post*'s, attempting to trigger a rift in Local 63 by stating St. John had a rival in the union for the office of sheriff and the possibility of St. John getting the nomination was very remote. Curry must have thought if he put it in print he could make it happen. No rival existed. Another union man, Mose Farragher, was considering the nomination if St. John backed out, since St. John wasn't sure he wanted it. Brazenly Curry went on to note St. John lacked the full support of the union and his "administration of union affairs" had "resulted in a considerable disaffection among the membership."[19] This was as far from the truth as honey was from sulfuric acid. St. John had been re-elected as president without opposition.

Both Curry and Sumner clarified there was some slight discord at the mines: the boarding house waiters' had requested an increase in salary at the Liberty Bell and Tomboy. No strike was threatened.

As for the Chinese, both newspapers reported they had not been driven from the camp. Recently many restaurants had affiliated with the Federal Labor Union, and since the Chinese refused to join, they were not hired as cooks or waiters. For years union men had avoided patronizing the Chinese laundries, but this had nothing to do with St. John. He did not run the Federal Union, Sumner pointed out. If he did, he would invite the Chinese to join and insist they be paid fair wages.

Telluride had only twenty-one Chinese residents in 1900, all men, most working in seven laundries and three small restaurants in the tenderloin district.[20] The laundries were simple two- or three-man operations. The Chinese population remained steady for several years, with one merchant/laundry business moving uptown to 207 East Colorado in 1903. Eventually they did depart, since by 1910 only two Chinese remained. The extensive bigotry from all quarters in town was the likely reason, yet the complete lack of Chinese women may also have been a factor in the men leaving Telluride. In Denver and other cities, many Chinese immigrant women were virtually owned by the men who brought them into the country, and farmed out to the highest bidder. Since April 1901, Denver detectives were investigating this criminal practice, and some Chinese freely admitted to it, explaining, "this was simply a custom that prevailed in their own country."[21] In Telluride the Chinese were a source of cheap services in their own niche in "Lower Town" and were left undisturbed for that reason, although not accepted as equals by anyone. Charley Fong Ding was a silk and porcelain merchant who also ran the largest laundry in 1901. Toy Moon operated a small Chinese restaurant and Chun Lee was proprietor of another laundry. They serviced mostly the girls of the row and the big dance halls. Many members of the Federal Union censured them for being little more than slave shops, the employees working for pittance. Labor unions abhorred reducing any human to slavery, and thus the Chinese oftentimes exasperated the unions, adding to the ugly prejudice prevailing throughout Colorado and the country in those years. Newsmen, including Sumner and Curry, used offensive terms when referring to them, more from blind stupidity than hatred. In Telluride that autumn the Chinese were quietly doing what they always had done. No one was chasing them out, especially not the Miners' Union. Collins and Wells had simply spoon-fed the *Post* and it swallowed the campaign lies in one gulp.

On Saturday, October 5, the Democrats opened their county convention at the courthouse. St. John's term as chairman of the Central Committee (since autumn of 1899) was up. Senator Buckley and Oscar Carpenter were nominated for the position, Carpenter elected by a majority of sixteen votes. Curry decried it as a defeat of the conservatives — Carp was no longer merely an anarchist, but also a radical and destructionist, like Vint.

The first nomination was for sheriff, and no other name except Vincent St. John's was presented for that office. He was chosen by unanimous acclamation and a standing ovation. F. E. Curry frothed.

On the following Tuesday the *Journal* (October 8) praised the Republican candidate for sheriff, Cal Rutan, as "upright, law-respecting" and standing "at all times for justice and right." Without making his attack on St. John appear personal, Curry claimed a very large number of citizens considered the union president's "possible election with serious apprehension. It is generally felt that he is chiefly responsible for the unfortunate and disastrous conditions prevailing last May and June, culminating the first week in July in riot, rapine, and murder; and that should he be elected to the chief office in the county it will seriously hinder the prosperity and development of the community."

Even the mine managers appeared to think Curry was smoking some sort of weed, for they readily offered a scale of wages that week for the entire local mining industry, an unprecedented move toward reconciliation, prosperity, and real peace. They submitted this to the Carpenters' Union, the Federal Labor Union, and the Miners' Union. To strengthen their own position, the mine managers officially organized as the Telluride Mining Association, which included Manager Arthur Winslow and Superintendent Charles Chase of the Liberty Bell, John Herron of the Tomboy, and Arthur Collins of the Smuggler-Union and Contention Mines. The scale of wages for the miners was nearly the same as St. John had proposed with the exception of longer hours and less pay for blacksmiths, machine men, and a few others; the millmen, still unorganized, were offered \$3.50 and \$4.00 for the various positions, most on a twelve-hour day; the waiters were offered a small increase, and the cooks, dishwashers, and meat cutters were all offered uniform wages throughout the district. All mines would charge one dollar per day for board. This was agreed to by the mine managers and was to be posted at the mines on October 10, when it was to go into practice. The managers even proposed that any employee receiving a higher wage at present would not have his wages reduced by the new scale. They fully expected the various unions to sign the three-year agreement. St. John and the Miners' Union thought the proposal needed further negotiation, and Winslow, Chase, and Herron willingly met in conference with him and his officers.

Collins balked. He thought the mine operators' proposal was more than fair. St. John insisted the hours for blacksmiths, engineers, and millmen were too long for safety. In a surprising turn, Curry wrote an editorial advocating the universal eight-hour day, urging Congress to enact such a law. Sumner suggested Curry's about-face was only to swing labor to vote the Republican ticket, and he may have been correct, since the disappearance of a man named John Mahoney on October 10 became an instant opportunity for Curry to blame the union.

Having worked at the Smuggler Mill for only a few weeks, Mahoney went to town that last Thursday evening with forty-two dollars, was seen at the National Club, and, according to Curry, then vanished into thin air under "mysterious circumstances." No railroad conductor had seen him leave town, nor had the marshal noticed him anywhere. The company still owed him twenty-seven dollars. Claiming Mahoney's name was on a scab list, Curry pointed a finger at the union: "The suspicion is gaining ground that he may have gone the way of W. J. Barney, who was no doubt murdered and his body done away with."[22]

Was it more malicious propaganda or was John Mahoney lying dead, his blood on union hands? This was the second mysterious disappearance laid at the Telluride Miners' Union's door in five months. Add the strike, the riot, and the union president's candidacy for sheriff, and Curry delivered a feast for James McParland, Chief Superintendent of the Denver branch of Pinkerton's National Detective Agency.

McParland solicited an appointment with Bulkeley Wells at the Denver office of the Smuggler-Union Mining Company. He explicitly instructed his assistant superintendent how to conduct the interview and sent him to see Wells. The assistant explained to Wells, Collins, and the company directors "that these outrages were undoubtedly the doings of the Inner Circle" of the WFM. He pointed out how strikes had flared around the country, that the WFM would be moving its headquarters from Butte to Denver's Mining Exchange building soon, its ever-growing power clearly a threat to mining corporations in Colorado. He made it clear Will Barney and John Mahoney were definitely victims of the WFM; such baffling murders were the labor organization's modus operandi, meant to instill terror in the remaining populace. He gravely assured Wells the WFM was worse than the Molly Maguires ever had been. He recited the heroic undercover work of James McParland against the evil Molly Maguires in Pennsylvania's coalfields; how McParland, "at the risk of his own life, ran down and broke up this gang of cutthroats and murderers. He gave the Telluride mine owners to understand that they would never be safe either in person or property, so long as the Federation had a foothold in San Miguel County, and that their only recourse was to hire a Pinkerton operative who would do to the Federation what James McParland had done to the Mollies."[23] He pointed out how Governor Orman was too receptive to labor. The mine owners must take matters into their own hands.

On offer were the services of a "Secret Operative," a special class of detective that was actually nothing more than a labor spy, with the express purpose of keeping unions in check, disrupting their affairs, and destroying their organization. These were not detectives gathering information or investigating crimes, but agents provocateurs meant to incite disorder and criminal activity if possible. McParland actually ordered his spies in several miners' unions in the West to sabotage relief funds, to promote violence among the hotheaded members, to create chaos, and to discredit union officers with lies.[24] The operative also was supposed to inform management of any employee "killing time or violating the firm's rules" or who was a "dangerous union agitator" in order to have the offender discharged on some pretext without violating union contracts of nondiscrimination. It was his duty to "meet with his fellow employees...treat them to drinks, and cultivate their friendship" so he might gain their trust, only to later betray it and any solemn union oath he might take.[25]

The operative would infiltrate Local 63, and if possible, get close to St. John and the officers of the union and prove their crimes. He would learn their every plan and report it to McParland, who would produce weekly reports for the Smuggler-Union Company and send appropriate instructions to the spy. The operative's identity would remain unknown to the company, unless he failed to get hired on his own. In that case the agency would give the operative's name to the manager only, so he could hire the spy without further delay. The operative was to take rooms separate from other miners in order to be free to write his reports to the agency without detection. The reports were to be written in pencil. This was a strict agency rule, since typewritten reports or typewriters were likely to attract attention.[26] This multi-layered web of secrecy was the direct product of McParland's massive paranoia, a complete distrust of humanity. He alone was the appointed savior of the West, through "an act of Devine [*sic*] Providence,"[27] in his own words.

This was the perfect hammer for the Smuggler-Union's problem, and the company signed with the Pinkertons that fall. Operative Number 36, a man with actual experience as a quartz miner and presently (rather conveniently) working at the Camp Bird over the mountain from the Smuggler, would be briefed and sent "inside" within the next several weeks. His name was George W. Riddell, a twenty-four-year-old native of New Orleans. Tall, slightly stooped in the shoulders, with black hair and a black mustache,[28] Riddell was a top-quality spy, the Smuggler management never knowing who he was until some years later.

In company with Bulkeley Wells, Collins returned to Telluride on October 17,[29] Wells inspecting the mines and progress of the new cyanide mill and sniffing around for signs of anarchy among the Smuggler employees. After all, if they were going to pay several thousand dollars for a Pinkerton Secret Operative, the investment had better be needed. Wells ascertained St. John was indeed a deadly menace, and his candidacy for county sheriff a danger to the entire community. If he succeeded in winning the office, "measures" were to be taken. Meanwhile, everything possible would be done to prevent his success. Both Wells and Collins were intensely satisfied with McParland's allegation that St. John or his fellows murdered Will Barney and John Mahoney, especially since Curry affirmed it.

Just as Curry was preparing to publish his theory of the concealment of John Mahoney's remains, those remains walked into the city clerk's office at Salida and signed an affidavit.

CHAPTER

10

THE LITTLE NAPOLEON OF LOCAL POLITICS

"Politicians are like diapers. They should both be changed frequently and for the same reason."

— Anonymous

John Mahoney was never missing; his name was never on a scab list. His friends knew where he was and where he went from the moment he left Telluride to visit his family in Salida. He had requested the pay due him from the Smuggler on October 15. He had written several letters to his friends in Telluride, including one to S. L. Starks, asking him to collect money from a friend who owed Mahoney. Collins had never reported the man missing, nor had the company made any effort to locate him or offer a reward for his whereabouts...or his corpse. After the sensational report of Mahoney's mysterious disappearance yowled from the *Journal*, several railroad men said they *had* seen Mahoney on the train between Telluride and Salida.[1] Mahoney was expected to return to town in time to vote — and probably Republican, Sumner quipped. The entire story was fabricated by Curry and for one purpose alone: political humbug.

Curry refused to issue a correction or follow-up story. He knew Mahoney was alive; S. L. Starks had confirmed it and showed him the affidavit and letters. The Smuggler bookkeeper, O. B. Kemp, confirmed it — the millman had collected the pay due him. Curry's refusal to publicly rectify the erroneous information brought Charles Sumner's wrath down on the *Journal*, thus beginning one of the most entertaining press wars in the history of Colorado.

Sumner rebuked Curry and his "anti-labor crowd" and called the Mahoney story the "idiotic supposition of trouble breeders." He tagged Curry as one of the "henchmen of the powers behind the throne," maintaining, "there has been nothing small, cheap or contemptible that has not been utilized to throw suspicion and discredit upon the labor organizations of the camp." Curry used "every opportunity to denounce honest workingmen as 'agitators, murderers, assassins' and the like" and had "harpooned the Miners' Union" at every turn.[2] The *Journal*, which was an evening publication, became the "twilight glimmer" and its editor, the "bewhiskered marionette" or "hired puppet" who "never assumed startling bankable" judgment. Curry's "flatulent" and "puerile prattlings" were "ridiculously amusing." Lampooning the *Journal*'s column "Pertinent Political Paragraphs" as "Pungent Political Propaganda," Sumner pointed out that the newspaper "indulged in a deceptive game," and failed to "explain to its readers why it opposes St. John save that he is the head of the Miners' Union."[3] The point was legitimate: Curry's hostility toward St. John appeared to be unqualified personal loathing.

The office of the San Miguel Examiner *as it looked some years after Charles Sumner sold out.* Courtesy E. N. Wright.

In an unexpected editorial, the *Denver Post* endorsed Vincent St. John as an excellent candidate for sheriff, calling him "a remarkable young man" and recounting his fearless efforts to end the riot of July 3 and advance a settlement.[4] The *Examiner* published letters supporting St. John, including one in Italian from an Austrian and one in Finnish. The Finns and Tyroleans affirmed they were entirely behind St. John. Sumner was as well, asserting that a visiting attorney for one of the large mining corporations said, "The best thing the mining companies of this county can do is elect St. John sheriff. It will promote harmony and better feeling and the benefit will be mutual."[5] It was "foolish prejudice" that claimed, if elected, St. John would run the office of sheriff in the "interest of one element," since such behavior would make him or any officer "subject to removal from office." St. John "has an executive head and ability equal to any man in the county," said Sumner, and he would "make a careful and conservative officer."[6]

Senator Buckley flatly told reporters, "Vincent St. John is the only man in Telluride who did not lie to me during the strike."[7] The union president's honesty was a matter of record, for even Arthur Collins, who told a few bold lies during the strike, placed "implicit faith in his word."[8] The president of the Smuggler-Union Company, Benjamin B. Lawrence, had been in town and when he met St. John he was impressed by the young man's intelligence and integrity. They didn't fear St. John as a man; they feared his power to unite large bodies of laboring men.

The more Curry and Wrench lashed out at St. John, the more businessmen stood behind him. Many were Republicans — the GOP was the majority party in town. Sumner echoed them in the columns of his weekly, pointing out, "There is not an incompetent streak in his makeup. He has been accused and abused and vilified...when the fact of the matter is, no one is deserving of greater credit for the discretion and judgment used during the affair at the Smuggler." Anyone who knew St. John, knew he was a "man of quiet disposition...a man who would instantly discourage any violence or lawlessness, and who would exert every measure to prevent it...."[9]

Sumner used as much ink to praise most of the Republican candidates as well, supporting the Republican ticket, yet projected by name four Republicans and five Democrats as winners. He had a quarrel with candidate County Judge Wardlaw due to an "unwarranted proceeding" at the Republican convention, but predicted Wardlaw as a winner. Having had charge of the convention, Wrench and his party had unseated the legal delegates from Ophir because they were pro-labor. Wrench then proceeded to hassle Wardlaw, a documented vacillator, to withdraw his support from Republican labor nominees, and Wardlaw finally fell in with the banker, telling Sumner, "I am against any labor candidate and against the Union and I will stand with the balance of the delegates." Wrench and his group spouted about labor's "evils" and their own "principles" so much that Sumner christened them "The Saints" and "the pure and holy gang," and Curry their "purchasable angel." He scoffed at Wrench as an "ardent advocate of 'principle' — usually spelled the other way, with interest at two percent."[10] He then commended one of the Democrats' resolutions, which was another assault on Wrench. The Democrats recommended funds from the county treasury be deposited in both banks, instead of just Wrench and Nunn's First National Bank. The other was the Bank of Telluride, whose president was Nate Mansfield, a friend of the Miners' Union. Sumner was by no means anything but a stanch Republican; he said he simply couldn't "digest" the crowd behind the Republican candidates, which meant Wrench, Painter, Curry, and associates.[11]

Curry countered that if Sumner could "perform that gastronomical feat" he would have "more brains in his" belly than in his cranium. He jumped him for wearing "the flimsy mask of hypocritical pretense," calling Sumner a Democrat in Republican clothing.[12]

With a laugh, Sumner replied that if he did swallow the crowd in question, it "would produce the same purgative effect it had on" Curry of the *Daily Diaper*. To illustrate his reason for not stomaching the crowd, he added that Brother Wrench, the honorable banker and "little Napoleon of local politics"[13] who seemed to be "running for everything,"[14] had admitted that during the Republican caucus he and his generals "did stretch our authority to some extent in throwing out a delegate who was elected by printed ballots."[15] What was done was done, and Wrench had no intention of correcting it, prompting Sumner to accuse him of manipulating the Ophir precinct for his own gain. Wrench was incensed.

Sumner added to the petite banker's ire with the statement that the "working people are greater than the banks or any party" and that the GOP needed the labor vote and needed to support the workingman.[16] Wrench openly avowed he had "no

View inside the Bank of Telluride.
Courtesy William "Senior" Mahoney.

use for labor or labor candidates" and insisted the Republican delegates take the same stand, causing a rift at the convention. "Is Mr. Wrench the whole Republican Party?" Sumner growled. "It is not my purpose to tolerate labor unions or labor movements," Wrench snarled back.[17] He bitterly opposed Sumner in his bid for the county clerk nomination and convinced several merchants to withdraw their advertising from the *Examiner*. Nevertheless, Sumner considered the well-manicured "cerulean-eyed pug" the manliest of the anti-labor crowd since Wrench was the only one to castigate him face to face over his preference for several Democrats, especially for sheriff.[18] "Supporting this species of creature is intolerable," Wrench hissed in a superior how-dare-you tone. Sumner gaily predicted the "species" would win the sheriff's seat by several hundred votes.[19]

If St. John won, it would be by fraud and intimidation, the *Journal* bleated. "Having no hope of success if a fair election be held, they [union members] determine to stuff the ballot boxes"[20] with votes cast by dead men and non-citizens. Having "dead" men cast votes was a possibility if Will Barney and John Mahoney returned. "Marshall Basin precinct will bear watching on Election Day," the *Journal* warned. "It is reported that the registration there is over three hundred, while it is doubtful if there are more than two hundred legally qualified voters."[21] To drive the point home, the penalty for fraudulent voting was printed several times: one to fourteen years in the state prison.

While the candidates performed civilly and collected votes via their congenial personalities, the *Examiner* slammed the *Journal* back, saying it was written by an "irresponsible hireling...a demented atom." Sumner concluded, "The convulsions of the lydiaepinkham[22] trash department in claiming illegal registration in the basin precincts are ridiculous. The outfit knows that those precincts will go dead against them and they will promote every disreputable scheme to have them thrown out." Mostly union miners populated the Marshall and Savage Basin precincts.

A week before the election Wrench, Curry, and "The Saints" appealed to the miners to vote the Republican ticket, since it was the only choice that "guaranteed the observance of law, the maintenance of order," without which, capital would refuse to invest in the district, putting miners out of work. They also appealed to women to "get out and vote" and offered carriages to the polling places.[23] This was an effort unique to the district, since the majority of miners had no wives and the Republican businessmen had the notion their own Republican wives would vote the way they were told, thus doubling votes against St. John. To his fellow businessmen Wrench hissed about the continued mysterious disappearances of Smuggler employees at the hands of union anarchists, as if their murders were a fact and men were vanishing in quantity. He also complained St. John had neither money nor property, which somehow made the young man more vile and incompetent.

Sumner sneered at the shallow solicitation toward the labor vote by Wrench's "spasmodic puppet" and roundly slammed the "holier than thou outfit" for the "pusillanimous rot" against St. John that had become "vicious in the extreme." Sumner was sick of the unwarranted attacks. No other Democratic candidate was being so thoroughly assaulted by the "gang of angels." For the editor of the *Examiner*, the "angels" had gone too far. "St. John does not represent all the crime and disruption on earth," he growled, "nor is he an angel like some of his calumniators. But he is a man of honesty and ability, with a sense of equity and justice that would honor official capacity."[24] St. John had avoided responding to any of Wrench's venom, and had gone out to briefly campaign in the west end of the county, to "satisfy the farmers" he didn't "have horns and dragon's teeth."[25] And as far as mysterious disappearances went, Sumner pointed out, John Mahoney wasn't missing, and it was most likely Will Barney wasn't either. But the Republicans "might go a little farther back and inform the public of the 'mysterious disappearance' of certain funds" from the county clerk's office.[26] That was a mystery indeed, with far more secrecy surrounding it than Barney and Mahoney. The Republicans had completely hushed the abrupt resignation of county clerk Harry Servis, a friend of Painter and Curry. Servis had worked for the *Journal* as a printer. No one could get information on his resignation and the missing funds. It was a Republican cover-up, and Sumner was ready to sling it back at the "purity gang."

Harry Jardine of the Miners' Union wrote a letter to the *Examiner* to counter Wrench's charges as well. He said the notion that St. John's election was a "danger to themselves and the money they represent" was pretense. They knew St. John was a good citizen. Jardine also pointed out that "the possession of wealth is not a gauge of honesty nor a qualification for office."[27] The Republican candidate for sheriff, Cal Rutan, owned little more than St. John did, having only a small cabin and a few mining claims. Jardine jabbed Wrench with the point that in San Miguel County, workers outnumbered bankers.

Harry L. Servis, state penitentiary photo.
Author's collection.

Had it continued much longer, the scorching ink war might have caused nuclear fission, but Election Day was fast approaching. On the first day of November Vincent St. John and the union broke ground for the hospital and union hall building at the corner of Columbia Avenue and North Pine, ground that was purchased a few weeks earlier. That evening a French-Canadian named Norbert Savignac was assassinated on Wilson Mesa, ten miles west of Telluride, and, incredibly, Curry didn't blame St. John or the union, despite the genuinely mysterious action of the killer who removed the bullet from his victim's breast. The ranching colony of French immigrants was maligned, and a nephew of the victim was arrested but later released.

Tuesday, November 5, brought a record number of voters to the polling places, none of them dead. Republicans dominated the city of Telluride, yet they elected Democrats for the offices of county treasurer, assessor, superintendent of schools, and coroner. Coroner Ike Glenn, a popular undertaker (if undertakers can be popular), retained his usual 300-vote majority, the highest of any candidate. "People are afraid to vote against Ike," said Sumner. "He will have the last say."[28]

Five of the nine offices went to Republicans, including the highly contested office of county sheriff, Cal Rutan, squeaking in by a mere thirty-six votes. In the Ophir, Marshall Basin, Savage Basin, Liberty Bell, Turkey Creek, and south Telluride precincts, Rutan barely made a showing. But in north Telluride, where the bankers and businessmen lived, and the west end of the county, where the cowboys lived, Rutan was king.[29] Sumner had predicted St. John as winner, and blamed his narrow defeat on the miners who neglected to go to the polls. Records showed 200 of the registered voters in the labor precincts had failed to vote. Sumner concluded they had already conceded St. John's election and "this anticipation resulted in a measure of carelessness."[30]

"The tremendous uprising against St. John failed to materialize," Sumner pointed out on November 9, and even a large number of the town's Republicans had voted for him, more than any Democrat had received in the past. The only precinct where the voters "conceived a violent feeling against him — just why, it is hard to fathom," was in the west end of the county. Many cowboys held miners in contempt. The greater portion of the business community favored him and "public sentiment is still with him," Sumner said, because "St. John was a clean, honest, manly candidate." He was gracious in accepting defeat.

The winner, Cal Rutan, a forty-seven-year-old unmarried Ohioan, had arrived in Telluride in 1881 and currently lived at the village of San Miguel. His small mining claims coughed up few profits and occasionally he took leases on the Smuggler, working them with his partners. Having been sheriff before, he was well acquainted with most of the county's long-time residents. St. John congratulated him, believing he would be fair in administering the office, and kept the channels of friendship open.

Even the *Journal* stooped toward peace, taking an uncharacteristic about-face and admitting on November 6, "The candidates conducted themselves in a dignified and honorable manner without exception; there was no recourse to personalities and vilification...there was not a suspicion of so much as a quarrel at any polling place in the county. If there was any foundation for rumors, or fears, that attempts would be made at illegal voting, it was not manifest at the polls."

Meeting the charitable admission, but suspecting someone penned it other than Curry, Sumner replied (November 9), "The animosities and disruptions engendered during the heat of the campaign are off so far as we are concerned."

The cooler weather and hint of snow in the air had chilled the editors for now, leaving a quiet poetry of longing and sleep over the barren hills. Stripped of their blazing September gold, the aspens stood naked against the shadowed green of spruce and fir, and life in San Miguel County seemed to be drifting back to normal...the "normal" being barroom brawls, injuries and fatal accidents in the mines, the digging of the foundation for the Miners' Union Hospital under the direction of T. G. Hall, new telephone service to Durango and other points, and extravagant performances at Red Men's Hall and the Stubbs and Jakway Opera House.

In the frosty hours of Thursday morning after the election, two Austrians danced forehead to forehead at the Silver Bell dance hall, clinging to each other's collar in a drunken embrace. Both were members of the Miners' Union, Joseph Mengoni[31] and Angelo Maconi, partners and friends. They had been arguing for days, reconciled, and broke into heated words again. This night they had been dancing until 3:00 a.m., their rapid dialect not understood by anyone in the room and generally ignored. The dance hall girls were half asleep, the barkeeper lazily sopping up suds. Suddenly Maconi tossed leftover beer into his friend's face and lightly slapped him. Mengoni sprang back, drew a revolver and pulled the trigger, the bullet cutting clean through Maconi's chest and out the window, grazing a passerby. Patrons and the barkeeper pounced on the shooter while Maconi grabbed a chair and struck Mengoni over the head. Confused and overtaken by fear, Maconi then staggered behind the bar and slipped to the floor, his life wheezing out as blood drowned his lungs. Night Marshal Joe Corey rushed into the hall in time to break up a mob from the street that was beating Mengoni into senselessness. The dazed man was locked in jail. On Saturday, a battered but sober Joseph Mengoni, desolate with the realization of his senseless act, pled guilty to murder before the police magistrate and was bound over for the next term of district court.[32]

"It but illustrates the disastrous results attendant upon carrying a gun," Sumner wrote (November 9). The law forbade carrying concealed weapons, yet all too often it was impossible to enforce. Joseph Mengoni admitted he had purchased the gun to use against his friend after one of their recent arguments, but having changed his mind, had confessed his shame to Angelo Maconi while they were dancing and asked forgiveness.

Angelo's brother, Louis, and their cousin, Florence (Florenzo) Ziller, came over from Ouray for the funeral, which was conducted under the auspices of the Miners' Union, St. John presiding over the union burial ritual. After the funeral, Louis and Ziller riled their fellow Austrians, claiming the murder was premeditated, and threatened to mob the jail and drag out Mengoni to hang him on the nearest telegraph pole. Sheriff Downtain sternly warned them not to try it, and the newspapers echoed him. "Telluride is the only San Juan town that has never had a lynching and she must never have one," said Curry (November 9). Downtain was going to be happy to hand over the office to Cal Rutan in January. He posted extra deputies at the jail and conferred with St. John to see if he could influence the Austrians to keep the peace. Although this was not a union matter, St. John again quietly pulled through, the Austrians listening to his plea. Mengoni was left unharmed.

Satisfied with the outcome, Curry refrained from condemning the union and held his truce with Sumner. Likewise satisfied things were in order since St. John was defeated and the wage scale negotiations were still on the shelf, Arthur L. Collins departed with a bevy of mining engineers for an extended business trip to Mexico — the company was interested in purchasing mining property there. He left his younger brother, Edgar, in charge of the Smuggler-Union.

Never satisfied with anything to do with St. John and labor, Addison Wrench proposed to Painter and Curry the idea of establishing a local association of businessmen. Curry presented this as a means for businessmen to get together and share ideas, shortening the irksome winter evenings, but for Wrench it was to be more than a social club. He wanted to influence the local professionals and businessmen in his way of managing the world, building unity against the power of Local 63. Yet on Wednesday morning, November 20, tragedy struck, sweeping away all differences and smothering all enmity, if only for a moment.

CHAPTER

11

HEROES IN THE SMOKE

"There is no remedy for the ravages of death."

— Eugene V. Debs

In the frigid hour before dawn, Robert Dempsey tossed a few chunks of coal into the stove, closed the drafts, and dropped into bed. His cot was in a corner of the tram house at the Bullion Tunnel. He shared the small partitioned space with a man named Hall. Suddenly Hall was in Dempsey's face, gasping about fire and Dempsey flew off the cot, pulled on his trousers, and grabbed a fire extinguisher. He turned it on the flames that were devouring the wall near the stovepipe. When it was empty he thought the fire was snuffed out, but flames exploded from the joists between the ceiling and the floor above him. He emptied another extinguisher to no avail, the flames crackling through the wood of the ceiling, growing in intensity. Smoke drove him and Hall outside, where he grabbed the small fire hose that was attached to a water tank, broke the tram house window, and turned the fire hose on the blaze. The heat forced him back and he and Hall ran in stocking feet toward the office.

"The tram house is on fire!" Dempsey screamed to the first man he saw, Scottish shift boss Will McKenzie. McKenzie was just coming out of the boiler and engine house. He sounded the alarm and ran to the tram house with a bucket, meeting Dan Morrison as Dan fled from the burning building. Morrison and several other men helped McKenzie with the small, almost useless fire hose and got another fire hose attached to a pipeline, but the pipe seemed frozen and little water came out. The dry wood of the building gave the fire intense vigor, with huge flames reaching out toward nearby buildings. Most of the structures were clustered around the mouth of the tunnel, a corrugated tin snow shed connecting them to the boarding house. The tram house was about 200 feet from the tunnel.

Having arisen just before the morning shift went on at 7:15, Edgar Collins was getting dressed. Now interim mine superintendent until Charles Becker was fully recovered, he was in the office, which was about fifty feet from the boarding house. It was now 7:45 and the fire was devouring one full wall of the tram house. If it leapt to the boiler house, nothing would stop it and the powder magazine would be endangered. Collins instructed McKenzie and timekeeper Ike Carruthers to remove the powder from the magazine. He told Norwegian shift boss Tork Torkleson, "You had better go in and get the men out of the mine." He felt assured the miners were not in danger, but if the fire went out of control, the situation could become deadly.[1]

Less assured than young Collins, Torkleson rushed into the Bullion Tunnel (level nine) as if time were an enemy. The first man he came upon was Jack Thomas, a trammer loading an ore cart hitched to a big black horse. Another trammer with a mule was just behind Thomas. "Get out as quick as you can!" Torkleson ordered breathlessly as he ran by the men. "Everything is on fire at the mouth of the tunnel!"

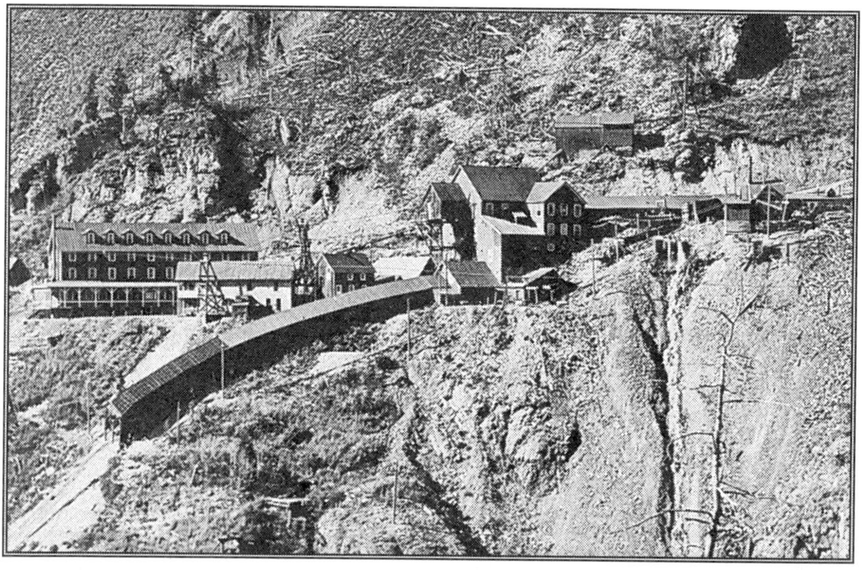

The Smuggler-Union's Bullion Tunnel complex clings to the cliff's edge at nearly 11,000 feet. The large building on the left is the boarding house, with one of the bunkhouses visible in front of it and the office immediately to the right of the tower. The large building in the center is the tramway and crusher house, with the tram shed curved down the cliff face. The long building connected to the right of the tram house is another tram shed, leading to the Bullion Tunnel entrance.

© Main Street Photography, Walker Studio Collection; used with permission (67T).

Thomas unhitched the cart from his horse and, holding the animal's tail, whipped it into a frenzy toward the tunnel's mouth. The nearer he got to the entrance the faster his horse went, and suddenly his mind reeled and blackness engulfed him. He clung almost helplessly to the horse as it dragged him onward through billowing smoke, bursting through the burning tunnel house. Most of the buildings were now engulfed in flames and acrid, dense, stifling smoke was sucking into the mine tunnel. Carbon monoxide — silent, unscented, and deadly — rushed ahead of it.

Minutes before, Collins had attempted to blow up the tram house to prevent the fire's spread. McKenzie and John Nicholson (who had recovered from the bullet in his face) and a few other men assisted him, the effort failing. Drums of heavy grease, a lubricant for the cable, were now burning in the tram house, pouring out a smoke so sickening that men were unable to breathe. Boxes of ammunition, purchased to use against the strikers and now stored in the building, were exploding and zinging in every direction. Two men were hit by flying bullets. Fire had spread to the machine house, adding to the poisonous smoke, and flames now billowed from the hay stored near the tunnel building.

Miners, waiters, blacksmiths, and bakers tumbled out of the bunkhouses and boarding house, many attempting to fight the ravaging flames. With frosty air outside, the mine sucked in the dense smoke and men screamed for the tunnel doors to be shut or all miners inside would perish. Carruthers fought against intense heat and smoke to get to the tunnel doors, but it was impossible. He grabbed some miners from the crowd, Martin Machina and his father Joseph, asking them to "bunch powder and fix it with fuse" so McKenzie might attempt to blow the tunnel mouth shut to save the men inside. More than a hundred men were on the ninth and seventh levels.

McKenzie hurled a bundle of dynamite into the tunnel's mouth through the clouds of smoke, the explosion shaking the ground beneath him. Unable to see if he had done any good, he prepared another. He was using a "shot of thirty pounds," fearing a heavier charge would explode the powder magazine. The second charge failed to shut the entrance, for the tunnel continued sucking in smoke. In desperation, Carruthers ran up the hill to the Little Mary Mine, calling for assistance.

Deep inside on the ninth level, Torkleson ran into shift boss Red Barclay, and the two men split the level between them, one running north, the other south, to warn the men to get out. Their lungs burning, their hearts pounding, the two raced against time, shouting and searching, refusing to turn back until the last man was found. Unable to get out the Bullion Tunnel's entrance, miners rushed into the manways, climbing the ladders 265 feet to the seventh level, some men choking from the acrid smoke, others dropping instantly from the carbon monoxide, falling onto those below them. Then suddenly the entire mine went inky black, leaving the miners in a stumbling panic.

Staggering, choking, nauseated men started coming up into the seventh level from the ninth and from there to the fifth level, seeking a way out of the mine. They met Mel Robbins, seventh level shift boss, at the fifth level cage station and told him the mine was full of poisonous gas, and he must get his men out. At that point Collins and George Wearing came down on the cage, intending to get the men out of the seventh, informing Robbins the buildings outside were on fire. Many of the men had escaped via the Sheridan Tunnel, but many were still believed to be on the seventh level. Collins and Wearing took the cage down from the fifth to the seventh, although Robbins warned them not to go because of the gas. They shouted to let him know they were all right, but as they reached the seventh level, their candles went out and they "felt themselves succumbing." Desperately Collins and Wearing signaled to be pulled up. They rushed outside, shaken, but safe.

Robbins pulled two nearly unconscious men out of the manway, carrying them to safety. He took nine men up to the third level on the cage, where they escaped through the Union crosscut. He, Claude Harvey, and William Richardson went back inside again, searching for others in the manways, the smoke burning their eyes, fear gripping their hearts, as they dragged helpless men into the fresh air. Fearing some were unable to climb the manway ladders and were waiting for help in the darkness below, Hughie O'Neil and a young man named Joe Nelson joined Robbins and his men, volunteering to go below to the seventh level. Robbins angrily advised against it. They told hoist engineer Dick Cogar not to raise the cage until they signaled. Robbins watched them as the cage disappeared into the deadly darkness. A few minutes passed without any sound. Cogar signaled down to them. No sound came back. Quickly he hoisted the cage back to the station, finding both men prostrate in the bottom of it. Nelson gave two short gasps.

On the edge of despair, Robbins yelled for help, trying to resuscitate the men. Nelson was dead, overcome so quickly that no spark of life was left in him. O'Neil was brought back from the brink and several men began the difficult trek out of the mine, stumbling through the dark with his limp body. Robbins started back inside, soaked in sweat, his face black with the greasy smoke, and his chest heaving, but Harvey and Richardson restrained him, and he sank to the ground overcome by heart-wrenching sobs.

As great dark plumes of foul smoke had curled over the ridge in view of Telluride, and a frantic telephone call had reached town that smoke had entered the Bullion Tunnel and "serious consequences were feared," Dr. John Q. Allen immediately summoned Drs. Ochsner, Barnes, and DeLannoy to accompany him to the mine. Reporters from both newspapers joined them, the group going up the narrow Pandora trail on horses laden with medical supplies. Having heard the news, Vincent St. John grabbed a horse and flew up the trail after them, passing them as he spurred his horse to the brink of peril, his only thought for his brothers trapped in the mine.

Arriving about the time chaos swept over the mountain, St. John met Billy "Hutch" Hutchinson, foreman of the Tomboy, and a crew of fifty men who had just come down with Hutchinson from that mine. Hutchinson took charge of the mouth of the tunnel while St. John took charge of the men fumbling with the worthless fire hose. Grabbing a sledgehammer, St. John plunged through the heat, the bitter smoke, and past the toppling buildings, to a second pipeline running along the hillside. He smashed open the pipe and fitted a larger hose to it. He and several other men got water down to the tunnel entrance, enough to cool the red-hot iron and stifle the flames to make way for Hutchinson and Charley Johnson to blow the tunnel.

With the precision that set him apart from many, Hutchinson completely caved in the tunnel mouth, sealing it against further smoke intake. They set charges to the two and three-story burning buildings to collapse them away from the tunnel and stop the fire's advance. Collins was relieved and deeply grateful that Hutchinson and St. John had taken charge, his own mind near panic, his face swelling from the intense heat. He didn't know how many men were inside the mine, or how many might have escaped, since so many had fled over the mountain in panic upon reaching the fresh air. He did say a number of the men were not at work, being sent back to the bunkhouses due to a main hoist not operating. That out-of-operation hoist had saved lives for that reason; yet at the same time, when men first fled the ninth level, the hoist was the reason they could not escape quickly and had to climb hundreds of feet of narrow ladder in the manways, a daunting task in the best of conditions. The previous night, boss plumber and steamfitter James P. Phillips had attempted to get the steam engine repaired without success.

While hope remained that men might still be alive inside, Hutchinson and St. John lost no time in seeking a means to get to them, but just as they arrived at the mouth of the Union crosscut, St. John's longtime friend and mentor Hughie O'Neil was brought out of the mine. St. John flew to the older man's side, gently taking his hand, his heart sinking, for the unconscious man's spirit fluttered in the shadows of death. St. John, Hutchinson, and two others worked O'Neil's limbs in an attempt to get him breathing. When he seemed improved enough to transport down to the boarding house, St. John told the men, "anything that can be done, must be done" to bring O'Neil back. He was put in the care of the doctors now arriving at the scene.

With smoke boiling from the mouth of the Union crosscut, entry had become impossible. Hutchinson and St. John took McKenzie and a few men from the Tomboy and went over the ridge to the Sheridan crosscut to see if they could gain access. After penetrating the black throat for a few hundred feet, greasy acrid smoke drove them out. Gulping fresh air, they returned; this time one of their party was nearly overcome, and they hurried him out.

Racing back down to the Bullion Tunnel, they saw the upper tram terminal crumple in flames, dropping the Sheridan cable to the ground. The Smuggler's huge cable swept over the rollers and down the mountain toward Pandora "like a streak of

The Tomboy Mine complex in Savage Basin was at 11,412 feet. Just over the top of the tram house roof (center right) one can see the Bullion Tunnel complex on the cliff.

Author's collection.

lightning," tearing away towers, hurling ore buckets over the slope, and snapping off trees in a deadly shrieking assault.

Intent on getting any possible survivors out of the mine, Hutchinson and St. John took Collins, Wearing, and four Finns to see if they could access the ninth level via the old abandoned Cimarron workings. The Cimarron was below the Bullion Tunnel, lethal with rotting and collapsed timbers and unstable ground. The men came across gaping shafts and had to go back for ladders to use as bridges across the chasms. The going was difficult and dangerous, the party finally crawling to the end of the workings, where they blasted into the ninth level. The bodies of William Merrifield, a young Finn, and two horses were slumped in the darkness. Merrifield was seventy-three and had been employed as a trackwalker. The bodies could not be removed via the Cimarron. Estimates of the number of victims stood between twenty-five and seventy-five, and by noon it was believed no survivors would be found inside the workings. The body of Joe Nelson was taken to town at 1:00, but Hugh O'Neil clung to life.

Afraid to risk moving O'Neil down the steep trail to town, the doctors put him in a cot in the boarding house, which had not been touched by the fire. He remained unconscious but seemed to be in extreme agony, his head tossing, his lungs heaving. St. John looked in on him whenever he could, his prayers written on his smoke-blackened face. He desperately wanted O'Neil to live and telephoned the man's friends in town to send up whatever help they could. Dr. Charles DeLannoy, a forty-five-year-old father of five small children, tended O'Neil with steadfast dedication, almost around the clock.

At 2:30 that afternoon, the fire had died down enough to reopen the tunnel to begin the search for victims. By this time Mose Farragher and Denny Burns had appeared, informing Collins, Hutchinson, and St. John that they had gotten the men out of the seventh level when men first came up from the ninth. They felt certain all but one was accounted for: a young Finnish timberman named John Nevala was missing. He knew very little English and it was possible he did not understand what was going on in the dark and smoke. The ninth level was a different story.

Hutchinson, St. John, and their crew blew a hole in the rock blocking the tunnel's mouth and cleared away enough rubble to enter. By 3:30 they gained entrance, finding timbers still on fire. The flames were snuffed out and the men started inside, the party consisting of Hutchinson, Johnson, St. John, McKenzie, Wearing, Collins, Lee Nylund, and a few Finns and Austrians. The first bodies they dragged out were Merrifield's and Karl Maki's, the Finn with him. The crew went back in, this time deeper into the ninth level, but heavy smoke lingered in the depths and deadly carbon monoxide subdued one of the rescuers, who had to be carried out.

Hutchinson and St. John, with a fresh crew of mostly Finns, Austrians, and Italians, reentered the mine. Since the electricity and steam engines had gone down, the mine was filling with water and the rescue team had to slog through muck and mud as they searched for their dead comrades. The work was dangerous, exhausting, and heartbreaking as their candles illuminated crumpled bodies in the stopes and manways. Some miners had dropped as if into sleep; others died in fear. One man had curled up gripping his hat in his teeth, his corpse frozen in that last moment of terror. Another miner had dropped so suddenly that his candle burned his hair and scalp. The bodies and tunnel were cloaked in thick, greasy, black sediment deposited by the smoke. Slowly, painfully, and silently the soaking wet, grime-covered men carried the bodies to the surface. By 5:00 that afternoon they had taken six out of the tunnel. A reporter cornered St. John as he emerged with the crew to ask how many victims they expected to find. He quietly told the newsman they had found fourteen more miners inside, and, without a moment's respite, he hurried over to the boarding house to check on Hugh O'Neil.

After a consultation with the other doctors, Drs. DeLannoy and Allen had taken sixteen ounces of blood from O'Neil in the hopes of relieving his system of the poisons he had inhaled. St. John joined the team of Charles Miller, Kinney McLean, and Hugh Hegarty under Dr. DeLannoy's instructions to continuously work O'Neil's limbs to help his heart and lungs function. The old miner began to breathe more normally, his ragged, tortured body finally relaxing in a peaceful unconsciousness. But Dr. DeLannoy shook his head; little hope of recovery existed.

St. John returned to the tunnel and, with another fresh team of men, he disappeared back into the realm of darkness and death, working straight through the night. By 2:00 Thursday morning his men had brought out fourteen more corpses. A reporter recorded the chilling, gut-wrenching scene: "Strong men worked in silence while tears coursed down their cheeks, and naught was said as the lifeless forms of their comrades were laid side by side on the floor of an improvised morgue."

Each body was taken down the hill by twelve men and placed in the care of Coroner Glenn. At dawn, only the bodies of Thorvald Torkleson and Red Barclay remained at the Bullion Tunnel. Torkleson had been found slumped on a ladder about thirty feet up a shaft. An Italian miner said Torkleson had saved his life, but while climbing in the manway, Torkleson gasped out a warning, "Look out below, I'll have to quit," and tumbled down the shaft. Like him, Barclay had tried to save his men before himself and was found near an ore chute where he had gone to warn the last miners.

With poisonous air still lingering in the depths of the mine, Hutchinson, St. John, and a crew went in to set charges to force the gases out. They made an attempt to get back to the breast of the ninth level, but one of Hutchinson's Tomboy men, Ashland Weston, dropped suddenly in near unconsciousness, and the party rushed him back to fresh air. He took several hours to recover.

While Hutchinson and St. John continued searching for bodies, Collins assigned a crew to begin clearing away the debris of the buildings and to repair the tram towers. He paid off several of the contract men and discharged others with their pay up to date. He had wired his brother in Mexico and Arthur was expected to arrive by Monday. The company's Denver attorney, Jacob Fillius, the state mine inspector, Fred Nye, and several Denver newsmen hurried to the scene. At noon D. N. McLeod brought up Jap Womer and Louis Evi to tend O'Neil as his nurses. Three wives of Smuggler employees also came up to look after him. They washed him, changed his grimy clothes and put him in a clean cot, but his condition was unchanged.

Crews brought out one of the dead horses, and dousing it in kerosene, set it on fire and tumbled it over the mine dump. The other horse and mule inside were in a difficult location and would take more effort to remove.

Carbon monoxide had penetrated to the upper workings of the mine, forcing the engine room crew out. The gases needed to be moved out before the mine would be safe. St. John, Ed Matthews, and Mose Farragher, along with Edgar Collins, Bob Wright, Ed Field, and Harry Moyle, went inside at 3:00 Thursday afternoon, searching as much of the workings as possible for several hours. They still had not located John Nevala, and it was suspected he was dead in the mine, although Mel Robbins and Edgar Collins insisted Nevala had escaped. Finding no more bodies, they set off charges to force the gas out, and part of the team returned to the surface. St. John, Wright, and Matthews continued searching every stope where men were known to have been. Within 300 feet of the breast of level nine, the silent killer engulfed them and they buckled under its power. Helping each other back out, they staggered to the surface at 6:30 in a state of severe nausea and near collapse.

A half hour later, someone brought news to St. John that O'Neil had taken a turn for the worse, and the young man raced to his friend's side, despite his own exhausted and ill condition. He stayed with O'Neil all through the night, but at dawn he went back in the mine with a Finnish and Austrian crew in search of Nevala. The Finns brought out Nevala's body from a stope on the seventh level at 11:00 Friday morning, raising the total number of dead to twenty-three. St. John and his crew continued searching every portion of the mine until they appeared at 2:30 Friday afternoon to report to Collins that they had covered the entire workings and no more bodies were found. At that point, St. John received news Hugh O'Neil had died at 2:10. The Miners' Union president bowed his head in wordless grief, his hope gone, his heart broken. He and O'Neil's comrades took the body down the mountain in silence, arriving at Glenn's makeshift morgue at 6:00. Arrangements were made for a mass funeral.

Having no sleep for two nights and no rest since Wednesday morning, Vincent St. John was escorted home by his close friend Kinney McLean late Friday night. Exhausted, sick from carbon monoxide, and his lungs burning and permanently damaged from the greasy smoke, St. John collapsed in his bed at Pheasey's rooming house. The depth of his grief was unknown, his tears unseen. He had given strength to his men, unflagging assistance to Edgar Collins, and poured his heart into Hugh O'Neil, but now it was over and he wept alone.

The mayor ordered all businesses to be closed Saturday in honor of the dead, and all of Telluride was draped in bleak mourning. Sixteen of the dead were union members; most were young immigrants and several had wives and children. Torkleson's wife, Ranghild, was in Norway, visiting family, when she received the cable of his death. Torkleson was thirty-three. Red Barclay's body was shipped to his parents' home in Delta. Merrifield's son took his to his home in Hotchkiss. The body of Iva Sundstrom, a Swedish miner, was shipped to his wife and brother in Silver Plume. Joe Nelson's remains were sent to his family in Spanish Forks, Utah, while William A. Graham's body was taken to his hometown of Sharon, Pennsylvania, by his brother Clyde. The bodies of Hugh O'Neil and Torkleson were to be held until relatives sent instructions for burial. The father of Antoine Anesi, a young man of twenty-one, requested his son not be buried until he arrived from Rock Springs, Wyoming. Sixteen men, union and nonunion, would be buried that Saturday.

In the chill hours before dawn, while wind moaned over the snow-swept peaks, Ike Glenn's assistants grimly placed body after body into simple pine boxes in a cramped warehouse. As they made macabre jests and avoided looking into the faces of the dead, they screwed down the lids on sixteen coffins in preparation for the funeral. The coffins were draped in black crepe and taken to Finn Hall. After Reverend Matt Takala said the rite over the remains of the six dead Finns, the sixteen coffins were taken to Lone Tree Cemetery in a procession of carriages, hearses, and 1,200 miners on foot, for a total of 2,500 citizens, the bands playing a somber march. Not since the nightmare avalanche of 1883 had taken thirteen lives at the Liberty Bell had Telluride been cut so deeply to the heart.

For that moment, wives and daughters, miners and merchants, bankers and laundrymen, saloonkeepers and mine managers came together as family, honoring the heroes who had given their lives to save others; honoring the youths who would never see their distant homes again; honoring the fathers who would never hold their children again. Every band in the county was present, every church choir, every labor union, every priest and minister. Every mine in the district was closed. The Miners' Union conducted the services at the cemetery and the priest or minister of each faith gave their burial blessings. St. John, as president of Local 63, conducted the solemn burial rite for the union men and Reverend Hunt conferred the benediction.

While the coffins were lowered into the frosty ground and the choir sang and the bands played a dirge, the sister of one dead miner quietly worked her way through the crowd. She searched the nameplates on the coffins until she found her brother's and fell upon it in uncontrollable sobs, unable to say goodbye. Her name was Hannah, and she had only three weeks before been reunited with her brother John Peterson after fifteen years without knowing where he was.

Those buried that day were Matt Stark, John Nevala, Alex Fellman, John Ahone, Karl Maki, August Kaanta, Billy Jones, Mark Zadra and his cousin Frank Zadra, Gus Sundborg, Emil Dahlstrom, John Peterson, Louis Borzaga, Allan Hendrickson, John Rals, and Oberto Rafati. Torkleson was buried on Sunday at Lone Tree; O'Neil's body was shipped to his boyhood home in Watkins, New York. When Antoine Anesi's father arrived, the boy was buried from the Catholic church with St. John and the Miners' Union conducting the same services as for the others.

With ravenous reporters combing Telluride and the scene of wreckage and tragedy for raw scraps of detail, St. John dodged out of sight. His heartache had plunged him into silence and his friends knew he was physically unwell from the smoke and exhaustion. They could do nothing for him. The young man was spent. Although he had endured Curry and Wrench's assault on his character for months in good humor and unruffled grace, for a moment his spirit was crippled by the loss of his brothers.

Although the *Journal* made no mention of St. John in the first report of the fire, when it became widely known the union president had acted selflessly at the risk of his own life on behalf of union and nonunion men, even Curry had to print the truth. In issue after issue, a *Journal* reporter spelled out details of St. John's nearly sixty hours of work without a moment's respite, proving to everyone this man's soul was deep-rooted in giving himself to save lives, hardly the instinctive action of Curry or Wrench's "murderous anarchist." Hutchinson and Bob Wright praised him as heroic. Edgar Collins was so grateful for Hutchinson and St. John's help that he spoke of them for days to anyone who would listen. He put it on record at the fire inquest that "I cannot say too much in praise of these two men." Collins's gratitude and high regard made him rashly promise that, in the future, union men would be promoted to bosses and outside positions, for union men proved they were more experienced and knew what to do in an emergency. Young Collins and others were haunted by the "if only" of hindsight: if only such experienced men had been on the ground, no lives would have been lost.

Billy Hutchinson had saved the day when he blew the tunnel's mouth, but many angry men wondered why Edgar Collins or his people had not done the same at the start, preventing death altogether. Why had it been a union foreman from the Tomboy who knew what to do? Why had none of the nonunion Smuggler bosses done the job? Enraged men wondered why the tunnel's heavy timber doors were not shut tight as soon as the fire was discovered, preventing smoke and gas from even entering. They wondered why Collins had been more concerned with saving the property, ordering the removal of materials, than saving the lives of those trapped inside. In their anger and loss they needed someone to blame.

Young Collins had given his all in an effort to save the men and the mine, but he had no prior experience with such events. He had worked tirelessly and had risked his life to warn men on the seventh level. He had done whatever he could do and whatever Hutchinson and St. John told him. His face and eyes were still painfully swollen from the intense heat he endured, his body fatigued from the endless hours at work. Yet some angry Italians flung blame toward him and the Smuggler, grinding down his weary heart. Some Austrians kindled blame on boss plumber Jim Phillips, believing he was at fault for not repairing the steam hoist or keeping the pipelines open for the fire hose.

On Sunday night, while memories of the dead were whispered through the town and stories of the heroes fell from numerous lips, a group of angry Austrians spotted Phillips at the Gold Belt dance hall. He spent a short time inside with friends and as he started for the door, a mob of men set on him. Someone thumped a rock into the back of his head, ripping a four-inch gash in his scalp. The mob closed round him, striking him over the head with a gun or pipe, screaming at him and shoving him. He broke away from them, stumbling only a few dozen yards when a rifle bullet slammed through his elbow, fracturing the bone. His friends hustled him off to Dr. Allen, who dressed his wounds and sent him home.[2]

Rankled by all the praise of the man he despised most, editor Curry used the Phillips story to attack the union and St. John. He wove innuendo with perfection, claiming the situation in the region he dubbed "Hell's Half Acre" was "becoming somewhat serious." Anyone visiting the tenderloin was in danger of losing his life, "especially those who incurred the ill will of that element during the troubles of the last few months. It is known beyond peradventure that one man has been made away with from that region, and it is clear that last night's assault was a premeditated...attempt to murder Jim Phillips."[3]

Everyone knew the "element" was St. John and the Miners' Union, and the one man who had been "made away with" was Will Barney. Knowing John Mahoney was quite alive, Curry refrained from adding his name to the list. The assault was not likely a planned event, and more likely exploded out of the seething Austrians who noticed Phillips in the Gold Belt and became enraged, attacking him because they needed someone to blame for the fire deaths.

Sumner heard the attack stemmed from the strike, but Phillips was out on a ranch during those weeks, and not involved at all. Nor was he on any scab list. Others thought it was caused by personal enmity with an Austrian over a dance hall girl. Several of the Austrians worked at the Smuggler, and they had blamed Phillips for not having the hoist working, thus this seemed more the reason for the attack. The Austrian who allegedly had shot Phillips was arrested a few days later, a man named Joseph Macari, cousin of union committeeman Louis Macari.[4]

Despite having no evidence of premeditation or union involvement, Curry had scrapped his truce, once again accusing the union of assault and murder. He called for the town authorities to take action against the dance halls to prevent further assaults on decent people and continued disappearances.

Inflicted with an "acute reform" spirit, the city council heard Curry's condemnation and took immediate action. Assaults, brawls, drunken encounters, and suicides were all too frequent in the tenderloin. The mayor, aldermen, and city council clerk drew up orders closing the dance halls and variety theaters, although their liquor licenses weren't revoked. In other words, they could no longer carry on the dancing end of their businesses — which translated to prostitution — but could still sell drinks. Rumors rumbled that the proprietors might retaliate with a crusade against the up-town saloons that engaged in illegal gambling. The council also decided to hire an additional night policeman, which many citizens said should have been done months ago. Acute moral reform by the city council only added angry proprietors to the growing list of angry men.

An undercurrent of unrest chewed at the edges of the coroner's inquiry into the fire deaths, while everyone awaited the jury's verdict. The fire inquest began on Monday morning before Arthur Collins returned, but the Smuggler company attorney, Jacob Fillius, was present, as were attorney L. E. Kenworthy of the WFM and Deputy District Attorney Edward Howe, on behalf of the people. The inquest continued through the afternoon, while much chatter of company negligence filled the streets. Dempsey recounted how the fire extinguishers were useless on the blaze when it first began and miners told of harrowing escapes. Carruthers and McKenzie explained their actions and fears. Mine Superintendent Collins testified, "I did not think there was any danger to the men inside" when the fire was first reported.[5] He praised the efforts of all his men, repeatedly giving highest praise to Billy Hutchinson and Vincent St. John, two men who were not in the company's employ, men with proficiency in instantly grasping the situation, organizing rescue parties, and handling explosives. Max Born, a kitchen hand, cast doubt on Collins by claiming he told Carruthers there was no need to notify the men in the mine. Collins said Born misunderstood; he had said there was no need because he already had sent Torkleson inside. St. John was subdued, answering questions in a matter-of-fact manner but crediting Hutchinson with "accomplishing the greatest good."[6] State Mine Inspector Nye testified that the company had complied with the legal requirements for fire safety two months previously. A heavy timber fire door was at the mouth of the tunnel, the minimum number of fire extinguishers were on the premises, and, despite inadequate water pressure, the minimum hoses and nozzles were installed. It was the minimum, but it fulfilled the law. Apparently the law didn't require anyone to know how to use the equipment or stipulate that the fire doors be shut immediately.

The official verdict concluded, "While we, the jury, do not charge criminal negligence on the part of anyone, we do find from the evidence that the Smuggler company had a very inexperienced management on the ground at the time of the accident."[7]

Such a verdict, absolving the company from a legal standpoint while casting a shadow of blame over Edgar Collins, the nonunion bosses, and over Arthur Collins for putting such inexperienced men in charge, failed to quell the rising anger against the Smuggler-Union Company. Defending Edgar, St. John said the young man must not be condemned, for he had risked his own life to do all he could and thought was right. But Arthur Collins he could not defend, for the manager had deliberately replaced experienced union foremen and bosses with men who couldn't even handle powder. He had complied with fire safety laws, but only the minimum, not in the interest of making the mine safer, but in the interest of keeping the state out of his hair. Nate Mansfield's "fire guards had been made redundant,"[8] men who would have had the job of shutting the fire doors immediately on first sound of the alarm.

Transporting huge timbers for the rebuilding of the Bullion Tunnel complex was a "thorny proposition" since the ragged wagon roads were little more than goat tracks. This team is crossing Royer (Royal) Creek on the Tomboy road, which was completed in 1902.

© Main Street Photography, Walker Studio Collection; used with permission (3T).

Death had no remedy, nor did the wrenching pain of loss, but management had to be held accountable for the lives of men who worked in extreme danger for a mere pittance. A miner's wage was "scarcely enough to support himself and his family," wrote Edward Boyce, president of the WFM and editor of the *Miners' Magazine*, "while the owners are traveling in a foreign country enjoying all the luxury that life can afford upon the product of his toil and at the risk of his life."[9] Too often mining companies were more concerned with property than the lives of employees, Boyce said, because, "as has often been stated, men are cheaper than mining material." For Boyce the fire at the Smuggler seemed to illustrate this, since at the outbreak of the fire the "man in charge" had "directed his efforts more to saving the property than warning the men underground." The Smuggler-Union Company would simply hire men to replace the dead and carry on in the name of profit. But families couldn't replace their loved ones. "This horrible disaster in the appalling loss of life is only one of thousands of accidents that occur in the mines every year." Boyce called upon the entire Western Federation of Miners to make greater efforts toward demanding safety in every mine in the West. "The loss of one man is of more importance than all the mining property in the state of Colorado," he said.[10]

The Smuggler-Union Company sustained a property loss of more than $50,000, most of which was covered by insurance. Rebuilding began straightaway, with plans for more substantial and more comfortable buildings. Yet a dark wind swept through the miners, especially those of the union, shadowing their hearts with a deeper dislike for Arthur Collins. Although the company had paid for all the burials, including the cost of shipping the bodies, the families were paid a miniscule compensation for their losses. Several later initiated civil suits. When Collins wrote a public letter of thanks to the mayor and citizens of Telluride, the only man he mentioned by name was his personal "trusted helper,"[11] Thorvald Torkleson, making the dark wind blow colder. None of the men had any complaint against Torkleson, but their old foreman Hugh O'Neil, beloved by all, deserved to be named as well. It was salt in open wounds.

Realizing he was aggravating bottomless grief and anger, Collins suddenly took a leap toward reconciliation. The evening of November 28 he agreed to meet with Miners' Union officers and other mine managers to settle the wage scale conflict. Shoving his English pride into his back pocket, he negotiated with St. John as if with an equal. The result was a three-year contract between the members of the newly formed Telluride Mining Association and the Sixteen to One Local 63 of the WFM. Since the millmen were still unorganized, the contract did not include them, but many of the ten-hour men were cut to nine-hour days, without wage reduction. The stipulation that boarding at the mine was no longer mandatory was added to the contract, although Collins refused to lower boarding rates. St. John had "conceded $17,000 a year in the price of board...simply to preserve the peace."[12] Stoping contracts were practically abolished.[13]

The entire district heaved a sigh of relief; the three-year contract meant Local 63 would have no reason to call another strike, unless management violated the agreement. Arthur Lancelot Collins put complete trust in the man he viewed as his adversary for he knew Vincent St. John was a man of integrity. Collins had nothing to fear. Trusting his enemy's word implicitly was a rare trust indeed. The contract was signed Thanksgiving evening and Telluride bowed her head in genuine gratitude...all except Addison M. Wrench.

CHAPTER

12

WINTER SPORTS

"If you haven't got anything good to say about anybody, come sit next to me."

— Alice Roosevelt Longworth

Addison Wrench wanted the Miners' Union to crawl into a hole and suffocate. If it refused, he would do the job himself or with the aid of his confederates. As Charles Sumner picturesquely put it, the "bantam banker" merely had to "jerk the strings" of his "flatulent puppet" and ink was spilled on his behalf, putting hatred into print every day but Sunday. The "gang of saints and angels" kept Sunday holy only because the *Journal* was not printed that day.[1] They didn't trouble themselves with holiday cheer either. Even before the Bullion Tunnel embers had cooled, and on through Thanksgiving, Christmas, and the New Year, Curry let no opportunity pass to publish slurs and innuendos against the union. The *Journal*'s truce was as Sumner suspected, only a hiccup in the machinery when F. E. Curry was out of town.

Since Joseph Macari was arrested in connection with the assault on Jim Phillips, and since he was a cousin of union committeeman Louis Macari, Curry used that as his proof to promote the assault as premeditated attempted murder carried out by the union. To him, the union was so evil that it even attempted to assassinate the deputy who took Macari into custody. Curry spread word Deputy Sheriff William R. Beattie, affectionately known around town as "Cheerful Bill," was shot at through an open window while he and Macari were in the jail office at the courthouse examining a pistol. Macari was not as much as handcuffed. Cheerful Bill was chatting with him as with a friend. A bullet supposedly went zinging through the window in broad daylight, just missing the deputy's head and lodging in the wall. The *Examiner*'s reporters investigated, only to find the incident was "an assassination josh" and a joke around the courthouse. The bullet had actually come from the gun in Deputy Beattie's possession, and he had accidentally fired it himself.[2] Another "Curried" fabrication crashed to the floor with a thud. To top it off, no evidence was found against Macari in the Phillips assault and he was released. This only served to rekindle the ink war between the newspapers and Curry began calling Sumner the "handmaiden of the union."[3]

Charles G. Sumner was handmaiden to no one. Although he was a loyal Republican, he was fair to Democrats, Socialists, labor, and capital, providing these were honest with him. He disliked liars of any stripe and particularly newsmen who published false stories as a means to discredit or humiliate those they opposed. He actually liked Charles Painter and had been friends with Curry and Wrench before the strike. They had attacked him for taking a stand on behalf of labor, and he insisted he bore no resentment toward them. His newspaper and printing business was next door to Painter and Curry's *Journal*, and he saw Curry nearly every day. Yet

Curry forgot the past five years of their friendship, allowing disagreement to wither into animosity.

Tall and slender, Sumner was one of the most amiable men in Telluride, unwilling to hold grudges against anyone. Known to friends as "Charley," he was called dashing and charming, his dark blue eyes, mop of wheaten curls, smooth-shaven cheeks, ready smile, and dimples making him the darling of the ladies. Somewhat of a dandy with a fondness for flamboyant suits and neckties, he was a cartoonist and a popular member of the local brass band. He always promoted music and first-rate theater in Telluride. His most serious failing was a latent prejudice against Chinese and Native Americans, an attitude that seemed to fade as he grew older.

Born in Colorado in 1874, Sumner grew up in Grand Junction where he married his high school sweetheart, Naomi Marksberry, on June 15, 1895. His first job was as printer's devil at the Grand Junction *News* when he was sixteen, and in a short time he became a reporter. He and Naomi moved to San Miguel County in spring 1896 where he went into partnership with Charles Fluke and established the *San Miguel Examiner*. His youngest brother, Ernest, came to work for him, and the entire staff esteemed the Sumner brothers and Fluke. Their shop paid union wages, one of the few newspapers doing so in western Colorado. Sumner's editorial policy was to take the underdog's side and to give all participants in any conflict fair representation, but never to be the organ of any man or organization. If unfair and unwarranted attacks were made on anyone, Sumner intrepidly defended the innocent and humorously reproached the aggressor. Colorful epithets were his rejoinder when other editors resorted to vulgarities and slander. He took his stand even when it meant losing powerful friends who threatened to damage his business. The stubborn refusal of men to seek compromise for the welfare of the community triggered a grave response from Sumner, yet he never published libel against anyone. He said people could get along, despite their differences, if only they met each other halfway. His defense of St. John and the Miners' Union made him some powerful enemies but he vowed to do the same for anyone unjustly accused, reviled, or attacked. Curry was all too willing to give him cause. Sumner personally advised Curry that if he persisted in attacking labor organizations, "no good will come of it."[4]

Charles G. Sumner, Editor/Owner
San Miguel Examiner.
Restored portrait © M. J. Martin 2003.

But the cantankerous, overweight F. E. Curry enjoyed gouging Local 63. It was more than a personal grievance. Curry was the mouthpiece of Wrench and his clique, and their motive was to malign the union in the public's eyes or better still, to provoke the union into unlawful retaliation.

If unwarranted accusations failed to produce the desired result from Local 63 and its president, perhaps thorny action might. Despite Edgar Collins's promises and

Arthur's claims that he "regarded a union card as the highest recommendation of skilled and competent labor,"[5] in the weeks after the Thanksgiving contract he promoted only nonunion men to positions left vacant by the fire deaths. This man who seemed to "speak out of both ends of his trumpet"[6] also promoted George Wearing, a man with a marked dislike toward organized labor, as foreman of the Sheridan in Bill Jordan's place. Jordan had moved to Delta in the belief that a lower altitude where miners were in short supply was better for his health. Manager John Herron of the Tomboy also subtly abraded the union, finding excuses to discharge union men who refused to board at the mine. This violated the November contract, and when St. John brought it to Herron's attention, he glibly replied, "All men find it necessary sometimes to make contracts they know they can't keep."[7] Pressured by Wrench, other managers began participating in the "Wrenching" game as well.

Almost as an affront to the "gang of angels," district court Judge Theron Stevens closed the fall term of court on December 9 by appointing a committee to examine the county books. This was a customary action on the part of the judge, but his choice in the appointments plunged Wrench and Curry to the depths of venomous disbelief. The three appointees were W. A. Nicodemus, Warrington Robinson, and Vincent St. John. Robinson had been a recent labor candidate for county judge on the Democratic ticket and was currently filling the position of county clerk after the resignation of Harry Servis. He also was a justice of the peace. These men had not sought the auditing position. Their honesty, fairness, and experience brought it on them and they simply accepted. St. John had served in a similar position more than once in the WFM. In May 1901 he had the unpleasant task of recommending action against a district union secretary, John M. Hogue, who had squandered union funds.[8]

Curry was so enraged at the appointments that he labeled the district judge a "friend of anarchy" and snarled to Wrench, "How dare Judge Stevens" appoint "this diabolical murderer" to "a position of honor and responsibility!"[9]

The "holy outfit" was doing their best to protect the fallen Harry Servis from court action. He had been charged with embezzlement, but his case was held over for the May session since the evidence had not been gathered.[10] The *Journal* mentioned not one word. Sumner attempted to gain the facts. He was thwarted at every turn. Even District Attorney McMullin, a friend of Wrench, seemed disinterested in prosecuting Servis. The matter of the missing county funds was supposed to vanish, and it would have quite nicely if Judge Stevens hadn't appointed Democrats and, worse of all, St. John, to audit the books. Nightmare! Insult! Outrage! Harry Servis, who drank too much, had bought his wife Nellie so many fine things — including a top-of-the-line piano — with county funds. Servis, who gambled too much, had begged $300 from city attorney Kinikin because Nellie "needed a major operation" in New York: Kinikin signed Harry's note at the bank. An operation wasn't the need, but Harry's gambling was, and Kinikin had to pay the note. He prepared attachment papers on Harry's house, which was purchased from Painter. Painter paid off Kinikin.[11] They had to keep the dirty laundry hidden. Things looked grim for the angels. Servis, a thirty-six-year-old native of New York with stringy black hair, an enormous floppy mustache, and gray eyes, made himself scarce, taking work at the Tomboy for the winter. His nerves were frayed; his wife's unraveled while she clutched five-year-old Freddie to her bosom. Harry's cousin Sam stayed close to the Haines Livery, the entire family keeping heads bowed and lips sealed while the auditors did their job.

St. John carried out his civic duties conscientiously and impartially. His naïve confidence in the decency of humanity blinded him to the danger of accepting the appointment. Had he known this would be the definitive blow to Curry's sense of law and order and Wrench's sense of principle, perhaps he would have refused.

Nevertheless, St. John was not so blinded that he failed to realize his enemies were legion. He simply didn't believe they would do him any personal harm. He took "enmity" as a thing aimed at the WFM, not himself. He knew the WFM was being assaulted from without and within. Pinkerton Secret Operatives were rife in the Cripple Creek district, where a spy known as Joseph Woodward was recently exposed in the Victor union and beaten so badly that he died from the injuries two days later. Before he died, he informed his wife in Denver that he was a Pinkerton and told her where to get important records so she might collect insurance from the agency and the Miners' Union. She was devastated. She thought he was a miner. The Pinkertons whisked his body to Denver and announced the cause of death was pneumonia. They did everything to get the documents from the wife, shut her up, and suppress the death. She was terrified they would take the papers by force. They flatly denied Woodward worked for them, going as far as emphatically declaring union miners hadn't killed him. His papers revealed at least five other Pinkertons were working as miners at the mine where he had been and as many as a hundred more Secret Operatives had infiltrated the Cripple Creek district unions.[12] The unions alerted the entire WFM, including Telluride's Local 63, where rumor already whispered that a spy was "inside." Woodward had been exposed when the Victor union hired a professional detective to discover how union business was being leaked at every turn. St. John and the union officers undertook to do the same, not so much to try to find the spy among them, but more importantly, to put an end to Curry's regurgitation of the Will Barney "murder," which was the *Journal*'s basis for discrediting organized labor. Constant unwarranted accusations against Local 63, if repeated often enough, could do more damage to labor relations in the district than an idiotic spy licking under the table. The union had nothing to hide; they were, however, sick of being called murderers and intended to find the truth about Barney.

Since even the Smuggler-Union Company knew little about Barney, finding information on him was challenging. The detective's preliminary report to St. John revealed Barney was married and was having difficulties with his wife. That was enough for the union to act. St. John and Carpenter took this information to Curry, politely asking him to retract the slanderous accusations of the past several months, and to refrain from attacking organized labor. Curry utterly refused, his boiling animosity over St. John's appointment as an auditor preventing him from even being civil toward them. He called them liars; they had no proof Barney was alive, married, or anything else. He had it on authority Pinkerton detectives had been searching for the man and found no trace of him. Barney was dead and they were the cold-blooded killers. This attitude left the union with only one choice: boycott.

They had proposed such action in July, but several businessmen, including Sumner and Fluke, discouraged them in the interest of allowing wounds to heal. They had followed that advice, giving Curry a fair chance, and had remained unresponsive to his slurs and innuendo. But Curry never ceased. The battering was damaging the Miners' Union. St. John found no humor in it. His usual sarcastic wit was gone. He was virtually sick of the unwarranted attacks. His mood was somber, even somewhat disheartened since the fire, and for several months afterward, perhaps due to physical damage from the carbon monoxide. But he knew any action taken against

Curry and the *Journal* had to be done prudently and in complete compliance with the law. While the December snows howled over the ridges and piled higher in the basins, the objectives and consequences of a legal boycott against the newspaper were presented to the union members throughout the district. No boycott would be called unless the majority of members voted in favor of it, which would take some weeks.

The heavy snow promised a long and dangerous winter in the high country. The war between the dance hall proprietors and the city council was heating up, but not enough to melt the snow. The proprietors pushed the city to shut down gambling halls in what seemed a sudden "moral wave" before Christmas. Business was slow for the dance halls, which were operating as nothing more than soft drink parlors — to skirt the law, those with dancing girls sold only soft drinks; those without girls sold liquor. Many of the girls had decamped until the holy fever subsided. Gambling was actually prohibited in the state, and saloons were supposed to close at midnight and on Sundays.[13] None of these statutes was observed in Telluride. Ten businessmen signed a petition on behalf of the dance halls, asking the city to reissue the dancing/variety theater licenses, but the moral crusaders on the council refused, threatening to publish the names of the "immoral" ten.[14] County Judge John Wardlaw, a city council member, advised against it. To further complicate matters, several businessmen took their grievance against the gambling halls to the district attorney. They wanted the city fathers to feel the crunch in the loss of revenue. Christmas was going to be the holiest on record. Sumner laughed at the obviously retaliatory move and said, "Very soon all will go their way and sin no more."[15]

Respectable music and dancing were still resident in town, with the Catholics offering their annual Christmas ball and many fraternal organizations holding masquerades. As a benefit for the new hospital, the Miners' Union put on a New Year's ball that was attended by hundreds of citizens, with Dray's Orchestra playing until 3:00 in the morning.

After everyone stumbled to bed on New Year's Day, Local 63 delivered a nasty baby on the *Journal*'s front step. St. John, Carpenter, and Kinney McLean distributed notices to all the town's businesses. The notice read:

> Whereas, *the Telluride* Daily Journal *has on all occasions taken a stand against the interests of organized labor in this section, and throughout the United States wherever corporations have opposed the demands of organized labor, thereby proving that the Telluride* Journal *is an enemy of the working class and a subservient tool of corporate wealth, and has, by the stand taken by the said paper, forfeited the respect and the right of the support of the workers of this county; therefore be it*
>
> Resolved, *by the 16 to 1 Miners' Union that the Telluride* Daily Journal *be declared unfair and a boycott be placed thereon; and all friends of organized labor be requested to withdraw all patronage from said paper.*[16]

The immediate reaction was support of the boycott, and Curry whimpered that the paper never "antagonized the cause of labor." He invited examination of the paper's files for any proof of such false accusations, claiming the *Journal* was always impartial. "It did not favor the election of St. John as sheriff," he wrote, "believing it to be inimical to the public welfare and the peace and prosperity of the community. And herein lies the occasion of this attack and the effort to ruin our business. It is personal spite, which the president seeks to avenge. And we have no doubt but that had a full membership of the Union been in attendance and voted on the proposition, it would have been hopelessly defeated. But the larger and better element of the membership being engaged in the mines, and out of town, the resolution passed." [17] Curry couldn't help himself; his bitterness produced wormholes in his memory.

The response from St. John, a frosty but businesslike explanation "To the Public," detailed the reasons for the boycott. It had nothing to do with the *Journal*'s opposition to his candidacy for sheriff, he said, and for the newspaper to use "deliberate falsehoods" as its only defense clearly showed "the justice of the Union's action." He explained that "The Miners' Union will never allow itself to become an instrument to satisfy the personal spite of anyone, the president or anyone else; and any official of this Union who would endeavor to use the organization for the purpose of satisfying personal spite would be deserving of the utmost contempt and scorn, and to be branded as a traitor to the cause and a greater enemy to organized labor than the Telluride *Journal*." [18]

He pointed out that the "better element" had indeed voted and the vote was "unanimously sustained at every meeting place." He then quoted several instances where the *Journal* labeled various labor organizations and officials, including Lieutenant Governor Coates, as a "mob of revolutionists and anarchists." St. John had a substantial number of quotes to prove his case, and could have included dozens more. No one would have doubted his word; everything he wrote was accurate. Some sort of committee must have clipped Curry's every lie, slur, and insinuation since even those the *Journal* had sandwiched between cheerful bits of local gossip were brought to view, including Curry's habit of never naming his "reliable sources" or "union members" who supposedly supplied him with information. His advice to corporations — to put labor leaders in "soft jobs with good pay" and thereby do away with strikes — Curry defended as sympathetic, but such low blows implying labor leaders could be bought off were proof of his hostility.[19]

St. John assured the public Local 63 was "not desirous of being a disturbing element, but insists on the rights of its members being fully guarded, and that without having to bear the insults of the *Journal* and in the case of this boycott, they are acting within their moral and legal rights."

He put his trust in the public, asking the community to support the union. He concluded that the "day is not far distant...when the workingmen and women will have freed themselves from the conditions under which they labor today, and that industrial peace, not oppression, will reign over all this land of ours." [20]

His appeal was heard. In the first week the *Journal* staggered under the loss as businesses withdrew patronage. Sumner disliked his fellow newspaper's predicament, and was not in favor of boycotts, but pointed out, "the fault lies entirely with our neighbor." He firmly believed "any person or organization is subject to fair and open criticism," and if it was "legitimately presented, can often accomplish great good. But to be perfectly candid...our neighbor has frequently let indiscretion get the better of its judgment and articles have been given publication that were quite unnecessary and only excited antipathy." Sumner praised the "little daily" as a credit to the district, and hoped it would quickly make a compromise with the union to restore its prosperity.[21]

Compromise wasn't in Curry's vocabulary. He maintained the *Journal* was correct in all its assertions and the Miners' Union was simply proving itself to be "the disturbing element in the community." Furthermore, the *Examiner* was aiding and abetting the union. Curry and the frothing angels were now certain Sumner was "responsible for the boycott." [22]

Amiable Charley Sumner went through the roof. He fired off an editorial with such sharp edges that he almost seemed poised for war. "To anyone with an atom of judgment it must appear that *Examiner* is in no wise connected with the boycott on the *Journal*. When we have to personally urge a measure of that kind on a competitor in business, this plant is for sale."[23]

Sumner had defended the *Journal*, had protected it against boycott in July and August, and had not been consulted in the present situation. Empathizing with a fellow newsman, Sumner said he could not believe Curry was responsible for the malicious reports accusing him of masterminding the boycott, but he personally knew the twelve businessmen "of a diminutive mental caliber" who were spreading the reports. He knew Wrench and his cohorts used threats to induce other businessmen to withdraw patronage from the *Examiner*. Wrench had done the same before. Sumner challenged anyone to bear out the claim he or Fluke were involved in the boycott, saying he in turn would prove them to be "deliberate liars or consummate fools." No reason existed for the *Examiner* to boycott the *Journal*, especially through the Miners' Union. "When we have a fight on our hands we will take care of it ourselves and ask no assistance," Sumner wrote. This fight was between the union and the *Journal*. He asked citizens to leave it that way.[24]

Addison Wrench relished the rumblings of war. He bundled off his wife and babies to Utica, New York, for the winter and fired up the war room. He warned the mine managers the union's boycott of the *Journal* would likely spread and threaten the prosperity of the mines. Over the next three weeks he, Arthur Collins, Charles Painter, F. E. Curry, and W. E. Wheeler organized the Telluride Businessmen's Association.[25] Although the articulated objective was to maintain the *Journal*, the specific purpose was to curtail union power. By painting the union as vile as possible, the association's organizers used fear to bring the merchants to their side. Twenty-six of the town's merchants had joined the association and on January 23 they held a meeting at the Telluride Club. Sumner, Fluke, and any known friends of St. John deliberately were not invited. Resolutions were adopted to prevent the further spread of the boycott and "thereby aid in the adjustment of any dispute and difference."[26] They asserted the *Journal* was entitled to free speech and, therefore, the citizens should refuse to comply with the union's boycott. Only half the businessmen signed the resolutions. Wheeler was elected president and a committee consisting of Arthur Collins, John Herron of the Tomboy, Charles Chase of the Liberty Bell, and W. B. Van Atta, was appointed to confer with the Miners' Union.

Willing to listen, St. John and the union officers met with the committee, submitting a resolution that said the boycott would end as soon as the *Journal* issued a public apology and agreed to refrain from publishing further unwarranted accusations against organized labor. It was a straightforward, simple proposition. The committee flatly rejected it. They demanded the union's immediate capitulation. St. John told them the "Miners' Union is nobody's monkey."[27]

In the wake of the stillborn conference, malevolent rumors were manufactured in a deliberate effort to undermine the union from within: dissension, resistance, and turmoil reigned, it was said, and the majority of the members wanted St. John ousted. St. John quickly countered the rumors with public statements that "there has been no dissension of any kind" and those who spread such reports "make a specialty of acute lying."[28] The underhanded strategy of Wrench and company served to bring even greater unity to the union members, and some took up picket posts at the *Journal* even in the harshest cold.

With twenty-six men in the new association, Curry was inspired to let loose, demolishing any doubts about who was the breeder of lies. While maintaining the Businessmen's Association was comprised of the best citizens in Telluride, Curry slammed the *Examiner* as "the boycotter's organ." He claimed Sumner and his kind didn't approve of the association and did not want "peace and prosperity. They thrive best on turmoil and unrest, and their positions are safest when adversity possesses their following."[29]

Sumner had enough. Fellow journalist or not, Curry had "the instincts of an untaxed mongrel" and Sumner blasted him in a sarcastic column that declared how businessmen had already expressed their "disgust at the publication of such offensive items and contempt for the ill-concealed attempt to charge this paper with the boycott." The *Examiner* refused to discuss the boycott's "merits or demerits" even though Wrench's undeclared retaliation boycott was affecting business. But if Curry, the "purchased instrument" and "organ of the labor-slugger," wanted a "newspaper scrap," Sumner was "quite ready to accommodate our small neighbor of the huge anatomy." The war was on.

Businessmen were indeed already disenchanted with the association and departing from it. Wrench and "a mighty half dozen" ran it as a tool opposed to Local 63. Saloon proprietors organized their own group, having no interest in Wrench's "principles" or purpose. They stood in support of the Miners' Union. Now Curry howled and snarled as if to ignite the planet, calling Sumner "the *personal* organ of the *chief* of the boycotters." The chief was St. John.

"Having discontinued its whimpering," Sumner replied, "the blackmailing organ is now endeavoring to shift the boycott on *Examiner*. Fine tribe of saints behind the scheme, and the editor stands as the shining moral divinity." Sumner slung dozens of new epithets at Curry, calling him "our treacherous neighbor," or "the huge corporation marionette next door," or "a confirmed trouble-breeder."

"Curry evidently desires the airing of a few little personal grievances," Sumner chortled. "He is the first we can prove to be a 'deliberate liar.'" If any friendship was left between the two editors, it likely died in February. On the first Saturday of that month, Sumner growled, "*Examiner* has naught to say reflecting on any organization and the businessman who will voluntarily endorse the contemptible slurs of Curry against this paper is short of judgment. The editor of that sheet is following the same disreputable methods that were responsible for the boycott on that paper, and in voicing the sentiment, 'What was done this week should have been done eight months ago,' he but admits that he 'knifed' the Miners' Union with equal enjoyment at that time."[30]

News of the boycott sidled off to other cities and returned to Telluride a beast of a different color. Outside newspapers claimed the miners were going to call a strike on A. L. Collins because he was importing nonunion labor. Collins telephoned both the *Journal* and *Examiner* to publicly and emphatically deny this. He was being beleaguered by phone calls requesting details and wished to inform everyone his relations with the union were harmonious. He "preferred" union employees; they were "more competent." *Examiner* gave him the benefit of the doubt, yet slipped in the phrase, "By deeds shall ye be known,"[31] since Collins was Wrench's twin in the boycott battle.

Sumner cheerfully gave as much free news space to Collins as he did to the union, proving the *Examiner* was no "organ." If anyone wanted more space, they were welcome to it at the going rate for a column inch. The Miners' Union said that was fair enough and on February 6 launched its own little four-page sheet, the *Telluride Miner*. It was not a newspaper, the editors said, explaining, "Not having any organ of our own we had to make one, and it must be understood that this is an organ, as it appears to be one of the requisites that a newspaper must be under the control of Mr. Anti Labor Collins." A. L. Collins's initials inspired the new union title for him. Written in a playful spirit, the paper poked fun at Collins, Wrench, Curry, and the Businessmen's Association, inquiring where Eddie Curry had his "little nighties" washed (since he had no wife) and asking if Mayor Hilgenhaus was "on the fence or merely trying to get under the bottom rail." It included what appeared to be want ads that spoofed all of the above. Real advertisers numbered seventy in the first issue.[32] The irreverent Scot, Harry Jardine, was the editor, yet the paper declared it had hundreds of editors because anyone in the union was free to submit an article.

Jardine detailed the reasons for the boycott against the *Journal*; it had nothing whatever to do with politics for it was "contemplated many months" earlier when the Federal Union boycotted the paper. The law regulating the right of trade unions to boycott was quoted; the boycott was legal and justified. The *Journal* had "started the fight," Jardine explained, and persisted until "our patience was exhausted and we took measures to protect ourselves against it."

Some merchants who had never advertised in the *Journal* had suddenly done so "simply to demonstrate that they were totally against" the union. "The effort of mining companies [Collins, Herron, and Chase] to coerce merchants to take a stand had no other motive than a desire to kill the Union, and they failed in that idea as they have before." Far more proprietors and professionals supported the union than the Businessmen's Association. The association continued losing members due to "some very obnoxious unwritten principles behind it." Wrench was the "Dictator" and he held court at the cheapest restaurant in town, the Saddle Rock. He "must be anticipating a run on the bank," laughed Jardine. "The members of the Miners' Union are desirous of living in peace and harmony with everybody, but do not feel like they can afford to let anybody tramp them under foot...workmen don't kick until they are trampled upon."

The *Journal* lost dozens of advertisers and, had Wrench and Collins not interfered, Charles Painter and Eddie Curry would have been forced to make peace with the union. Instead, Painter and Curry sought advertisers from Denver and national firms, signed on with the wire service, and continued misrepresenting labor organizations and hurling unjustifiable accusations against Local 63. Putting his own pride before the welfare of the community, Curry gleefully went into the persecution business and Telluride became a house divided. Because he had arrived in San Miguel County in the early days, Curry was considered a "trail blazer," and the union finally understood his mode of breaking trail. "The paths of some supposed snowslides are accounted for," Jardine chortled.[33] And Curry would never lower himself to apologize for the wreckage he inflicted.

"A house divided" deeply troubled St. John, yet he and the majority of the miners believed if they failed to take a stand against abuse, they would be easily driven out like dumb cattle. Whenever they made a step forward, they were beaten back by two. The Miners' Union settled in for a long boycott, thanking those who stopped patronizing the *Journal* and attempting to persuade those who did patronize it to cease. One of the latter was James J. Blazer, proprietor of the Saddle Rock Restaurant, who utterly refused the union's request. In anger Local 63 posted pickets at Blazer's. He, in turn, filed a complaint and Judge Wardlaw issued a temporary injunction against the union's pickets since Blazer was not specifically causing an injury to the labor organization.[34] Local 63 complied immediately: the union had no intention of breaking any laws. Instead of posting pickets, every union man wore a blue ribbon that said, "Look for the blue card." The union men devised a system where businesses supporting them displayed a blue placard with large black lettering proclaiming, "Fair House: this establishment can consistently be patronized by all union men." It bore the signatures of both the Miners' Union and the Federal Union officers. The system proved to be more effective than pickets and violated no law or injunction.[35] Yet it served to stoke the furnace of enmity among the anti-labor citizens.

Enmity was a highly infectious thing. It gobbled up Telluride's sewage and swarmed across the state like a wind-borne virus. St. John would know no respite wherever he went; whatever measures he took to safeguard miners' rights were condemned as immoral or malicious or uncompromising. While he and Carpenter were in Denver on union business they had to dodge rabid reporters who insisted Local 63 was wreaking vengeance on Telluride with its boycott.

The vengeance theme was popular that winter, used with aplomb by Adjutant General George F. Gardner of the Colorado National Guard. Although he himself had never met a single member, George "the Prophet" (as the Miners' Union called him) was certain even God hated the Telluride Miners' Union.

The fatal fire in November had been the first sign. In the first week of February 1902, an explosion at the Japan Mine killed two men and injured seven others. Never mind that they had placed four boxes of giant powder on a dirt-covered sheet of steel on the stove to thaw it out and five more on an electric heater. Miraculously the explosion failed to set off the powder house, which held enough dynamite to "turn a corner of the county around."[36] God must have been saving that for another time. Since it was customary to thaw powder this way in the freezing months, this was the

Pro-union merchants willingly placed the Blue Card in the windows of their businesses to show their support for the unions. Not the sole idea of the Miners' Union, this was initiated by the Federal Union and later the presidents and secretaries of both unions signed the placards.

Author's collection.

second sign. On February 27, two Tomboy miners were killed by "bad air." Several miners landed in Dr. Hall's Hospital with pneumonia. These were all signs, indeed.

The ultimate sign came at dawn on February 28. Snow had done the traditional San Juan thing and dumped several feet on the range in twenty-four hours, adding to the immense accumulation already in residence. The Liberty Bell Mine bordered a gulch high on the mountain above Telluride, with a tram running at angles down the steep slope to the Liberty Bell Mill halfway between Telluride and Pandora. The snow on the ridge cut loose, roaring through several of the mine's buildings, crushing men, mules, and machinery in a blinding instant and dragging everything into the gulch. The tram house man had his head flattened between the bull wheel and the chute. Another man escaped by grabbing the cable and riding to safety. A large rescue party from town that included several doctors, merchants, Vincent St. John, and Oscar Carpenter headed up to the scene in the still-driving snowstorm. While they were digging injured men out of the hardpack, another avalanche thundered over them, burying Dr. DeLannoy, Carpenter, and several others. They were rescued, but nine were killed. Crews continued the work, carrying the injured men down the slope. A third slide bombarded them and three more rescuers met white death. That evening another slide tumbled over the Sheridan Tunnel and as men were digging through it, more snow rushed over them, killing one and injuring eight others. By nightfall, nineteen avalanche victims were dead and seventeen more were in the hospital. Several bodies would not be found until spring thaw. No hope remained for those listed as missing. The mountains had become too dangerous to continue digging.

In the next several days, avalanches battered Alta, Silver Mountain, Marshall Basin, and Bear Creek. The Marshall Basin slide took out the Smuggler tram, dragging Henry Martin beneath tons of snow down Marshall Gulch. Two other men were injured. This was followed by a large slide in the same gulch and the Ajax slide thundered off the cliffs above the Smuggler Mill, demolishing the house of mill superintendent Van Law and dragging timber for a hundred yards across the railroad track. Collins ordered all the Sheridan men off the hill and laid off the 150 men at the Bullion Tunnel. On March 4, two more miners died beneath a slide up Bear Creek. Another slide took out a small mill. Several slides threatened to cut loose, thus to protect lives, many mines shut down.[37] In silence and tears, wretched grief wrapped itself around Telluride. "We are helpless," Curry wrote, at a loss for words.[38] As widows sank in anguish and fatherless children wept, Adjutant General Gardner proclaimed the snowslides were "God's vengeance."

Printed in the *Denver Times* on March 7, 1902, Gardner's words sliced across the state raw and bloody. He had stated in authoritative and measured tones, "I believe that the disastrous snowslides at Telluride was a judgment of God upon the miners of that section who have for some time past conducted a reign of terror there." God was peculiar in choosing victims, since most of the dead were not members of the Miners' Union. Providence also missed the opportunity — twice — to snuff out St. John and Carpenter when they were rescuing others. That mattered little to Gardner, for he droned on, "I cannot but believe that the recent series of disasters are the judgment of Almighty God upon the crimes that have been committed in that camp in the name of labor, and I am not a religious fanatic either. But I believe in the words of Scripture, 'Vengeance is mine, I will repay, saith the Lord.'"

Men in kilts were guilty of something in Savage Basin.

Photo by the author.

Since Local 63 had committed no crimes in Telluride, the general cited an imaginary incident about the union driving barefooted "old men...over the range," one dying "of want and exposure.... His blood is upon their skirts."

The statements were so preposterous that the miners' first reaction was loud guffaws. God knew only one or two among them wore "skirts, and those being plaid," thus the guilt for imaginary dead old men was heaped on Scots such as Harry Jardine who "should expect a snowslide momentarily."[39] The WFM local at Victor wrote, "In the past we have heard God Almighty and the Miners' Union accused of various things, among them being murder, arson and dynamiting, the war in South Africa and in the Philippines, but it remained for Adjutant General Gardner to clap the climax by accusing God Almighty and the Telluride Miners' Union of forming a corner on the snowslide business." They demanded the immediate resignation of the "blasphemer and slanderer."[40] A number of WFM locals across the state issued resolutions calling for Gardner's removal. The Federal Union followed suit, declaring the general's words a "profanity" that made the surviving family members "weep anew."[41]

The most scathing condemnation came from St. John and Carpenter who said the general "committed the grossest and vilest injustice...and desecrated all principles that teach men reverence for the Deity." Gardner's statements were "menacing to the promotion of equity; threatening to common decency...blasphemous to the dead; libelous to the living." They demanded Governor Orman remove "this odious individual" from office.[42]

Nearly every newsman denounced the utterance as obscene and monstrous, calling for Gardner's dismissal as well. Even Curry offered something of an objection, yet with that genuine talent he had for innuendo, he still managed to point a finger at St. John and Carpenter. He was certain nature alone was responsible for avalanches and God didn't "wreak his vengeance upon a whole community for the shortcomings of one or two individuals."[43] His entire readership knew which one or two he meant, especially since he followed this with a quote from the Grand Junction *Daily Sentinel* that commended George Gardner and nailed St. John to the cross. Gardner was articulating what "thousands of others may have thought," the *Sentinel* intoned. "The doctrine of visiting judgments is what is taught from the pulpit every Sunday.... General Gardner has simply expressed a conviction arising from a profound belief in Holy Writ." If James Orman removed him, it would prove the governor had an "intensely weak nature." The *Sentinel* was in deepest sympathy with the miners and their families, yet "the acts of Vincent St. John and his attitude...have engendered a hostile feeling against him among men of brains and conservatism throughout the state."[44] Curry let it stand without a squeal of comment and hurried off for a vacation in Aurora, Illinois.

This left Charley Sumner to straighten out the *Sentinel*, which he did with enthusiasm, calling it "dough-clawed and despicable. It hates organized labor and pays printers $25 a month. It approves Gardner's smallness. It is a small sheet."[45]

Sumner also gave a lengthy editorial lashing to the adjutant general, asking, "On what basis does this irrational military pygmy establish authority for such blasphemy? What right entitles this public pap-sucker to charge anything against organized labor...to offer rancorous opinions on affairs that do not concern his official graft?"[46] Sumner believed Gardner should be dismissed.

Governor Orman merely directed Gardner to apologize. Instead, the general flatly denied he ever said any such thing, although he had confirmed it to numerous newsmen the day it was published. The reporter who had initially interviewed him publicly called Gardner "a cheap liar and contemptible coward."[47] Labor organizations roared for his discharge. The San Juan District Union sent a resolution declaring him "unfit to be at the head of the National Guard because he has proved himself an ass for uttering the slanderous statement and a coward for repudiating it in the manner described."[48] Governor Orman called a court of inquiry.

As if General Gardner hadn't done sufficient damage, Judge Wardlaw heaped insult on the obscenity. Wardlaw, the official news correspondent for the *Denver Post*, wired a story presenting the Finnish miners as superstitious rank cowards during the Smuggler fire and the Liberty Bell avalanches. Under Wrench's anti-labor spell, Wardlaw went too far for his master, merely because the Finns as a group held St. John in such high regard as to show genuine love for the man. Creating dissension seemed to be the aim. But the Finns stood strong, refuting the charges with facts. In written statements, Arthur Collins and Charles Chase backed them up, extolling their bravery in both disasters and successfully putting a bung in Wardlaw's bunghole.

Telluride was ready for the sweet scent of spring.

CHAPTER

13

DEAD MEN DON'T DIVORCE

"A lie gets halfway around the world before the truth has a chance to get its pants on."

— Sir Winston Churchill

Judge Wardlaw had to back up and spill a few apologies. Some of his friends said his Adam's apple seemed to be more prominent than his brain during his alliance with Wrench. The lanky thirty-one-year-old judge was generally a dependable news correspondent. He had the job for more than a decade, servicing the *Rocky Mountain News, Denver Post,* and other papers. Sumner had been his friend and still had enough affection for him to take him aside to point out the absurdity of attacking the miners who were the mainstay of the county; what was the purpose in insulting them with false accusations?

"The Miners' Union has been accused of about everything under the sun and most of the things that happened after the sun went down, but it still lives," said Sumner. "A local element outside the organization has endeavored for some time to disrupt the Union, and once more their attention should be called to the fact that it is a foolish policy, an undertaking that only engenders hate and bitter feeling."[1]

County Judge John Wardlaw.
Restored portrait © M. J. Martin 2003.

No matter what devious or libelous measures Wrench and company employed, Local 63's membership increased monthly and not because of fear, coercion, threat, despotism, or militancy. Vincent St. John's selfless dedication and integrity brought them in. Even when he had wearied under the strain of unremitting personal attack and for weeks showed the fatigue of impaired health, he continued to buttress the miners and to rally them in a united brotherhood. This was what disturbed Addison Wrench and Arthur Collins the most. The vast majority of citizens, including Collins's own younger brother, *liked* St. John. Several of Wrench's allies said Vincent was "personally all right."[2]

St. John was no orator, no politician, no man of might or wealth. Simply a miner, he never sought position or power and actually was uncomfortable addressing crowds and despised mass meetings, preferring person to person encounters. He wasn't the dictator his enemies imagined. "There was no false pride in him," his contemporaries noted; "he was not flamboyant." His compassionate manner balanced his sarcastic wit and "intensity of purpose and quickness of mind." St. John "had a quality that inspired others," and a dependability of word and action.[3] In March 1902 he and Oscar Carpenter were again voted in as president and financial secretary of Local 63. The election told Wrench, Collins, and all the enemies of labor that Local 63's policy was "United we stand." Vincent St. John had become more than a leader; he was a symbol of unity. If Wrench, Collins, and their allies wanted to succeed in destroying the union, they would have to first destroy the man.

A simple bullet in St. John's head would only make him a martyr, giving even greater power to the symbol, and to what he stood for. Death threats in the form of anonymous letters had done nothing to disturb his service to the union. Venomous newsmen, such as David F. Day of the *Durango Democrat*, viciously attacked him to no avail. Perhaps St. John failed to see and understand the danger closing in on him and the union officers because he believed justice, right, and truth would prevail. He mistakenly put too much faith in the judicial system, in "right" not being subjugated by the power of individuals or corporations. He went about his business and the prosperity of the district did not falter as was predicted if St. John remained in the region.

Despite the strike, riot, and natural disasters of the past year, San Miguel County sent $1,165,263 in bullion to the Denver Mint, with the total output of the county just short of three million, an increase of $516,874 over the previous year. The Smuggler-Union produced $833,524, the Tomboy $819,800, the Liberty Bell $612,500, and many of the smaller mines between $40,000 and $200,000.[4] It was clearly a prosperous year, but the mine managers wouldn't blame St. John for prosperity. They and Curry insisted the president and the union had caused quite the opposite. If they could blame St. John for anything, they did, since an underhanded campaign against him was their only recourse.

Charley Sumner poked fun at the nonsense as if to lighten St. John's burden, demonstrating an ever-increasing admiration for the union leader. "It snowed again this week," Sumner wrote, "and it is not believed St. John was responsible for it this time, as he and Carpenter and McLean were busy finding out who the apostles were when the storm was tapped."[5]

For the most part the union leaders and members ignored the sewage spread by Wrench and his composters. The storm-causing abilities of Local 63 became a standing joke, and General Gardner's "reign of terror" was met with a "let them eat at the Saddle Rock!" To labor organizations the reign of terror phrase reeked of Pinkerton Agency propaganda, for it was a longtime favorite of James McParland, head of the Denver office, and the WFM knew mine owners had recruited an army of Pinkertons in recent months. Since Gardner's use of it, Curry harbored the phrase as the *Journal*'s flagship, claiming a reign of terror came upon the camp with the presidency of Vincent St. John.

Even if one lifted every rock in the county and poked inside every closet in town, no sign of this reign could be found. Outside of the legal and orderly boycott on the *Journal*, 1902 progressed normally. The first of the year, Cal Rutan had taken over the office of sheriff, keeping Beattie and Spellman as deputies and adding Gus

Roth to the payroll. Rutan complained he inherited Downtain's rheumatism with the office, but he didn't blame his troubles on St. John or the union. With rheumatism the worst of his concerns, he must have failed to notice any reign of terror. The city council, fully recovered from sanctity, reissued dance hall licenses and simply doubled the fee to pay for an extra policeman, allowing gambling to remain wide open. J. C. "Chris" Geyer was hired as an extra night policeman and put under Kinney McLean's charge. McLean, who had been elected constable of the first precinct the previous November, was a union man and friend of St. John and thus a "reign of terror" man. The city council apparently knew nothing of the reign either, despite McLean's signature on the boycott notice. A native of Nova Scotia whose father had come from the Isle of Uist, Scotland,[6] McLean had married Fred Pheasey's little sister, Alice, the previous April and was expecting his first child, his elation softening a number of the local bachelors for matrimony, including Kinney's own partner. Fred, a sworn bachelor of thirty-four, sneaked over to Rico to marry twenty-two-year-old Araminta Blanche Moran when no one was looking. Mel Robbins also had stumbled to the altar in a daze, forgetting his need to thump union men. His wife made a puppy of him. Peace dripped from the thawing snowdrifts all over town. Dr. J. Q. Allen bought out his partner Dr. Hall, to become sole owner of the Hall Hospital. Anti-labor Ed Lavender sold his Smuggler store to union-loving Billy Smith, who also ran the Tomboy store, and the city council sold more land in Lone Tree Cemetery to the Miners' Union...perhaps to accommodate all the victims of the reign. Dances, masquerades, parties, fraternity benefits, and church festivities all saw union men and women celebrating life with nonunion and even anti-labor folks, making one wonder if Telluride's brand of terror had something to do with the moldy nighties under Eddie Curry's bed. The reign was certainly very quiet while Curry was off in Illinois, and dropped no heads until he returned on April 10.

While Curry was out of town those weeks, a Liberty Bell shift boss went missing and the *Journal* failed to blame the Miners' Union. The disappearance was mysterious since the man, Wesley E. Smith, wrote to his wife in La Junta to tell her he was coming to see her but never arrived, his baggage still at the Victoria House on North Pine Street.

After the fatal avalanches wiped out the Liberty Bell buildings on February 28, the mine was shut down, and Wes Smith went to town, taking a room at the Victoria. Expecting to be out of work for more than a month, he wrote his wife on March 1 to inform her he'd be leaving Telluride no later than the fourth. He drew his paycheck for $85.50 and cashed it at Nathan's, the union clothing store that was less than a block from the hotel. On Tuesday, March 4, he went back to Nathan's and purchased new underclothes and shoes. Here he met a man named C. B. Wilson, who asked about a job. Liberty Bell Foreman Martin Anderson happened to be in Nathan's, and Smith pointed him out to Wilson and left. He returned to the Victoria, changed his clothes, packed his valise, put it in the keeping of the hotel clerk, and settled his bill. He asked the clerk to wake him in the morning for the train. He went back out about 7:30 that evening. He was never seen again.[7]

In La Junta Wes Smith's wife and two sons, age twelve and fourteen, became alarmed when the man failed to arrive. Mrs. Smith was living with her sister who was married to the La Junta town marshal, S. W. Zickefoose. On March 14, Marshal Zickefoose set out for Telluride with a petition signed by Otero County officials and businessmen requesting Telluride to cooperate with him while he investigated his brother-in-law's disappearance. The Telluride police had already initiated an investigation, but found no trace of Wes Smith. He carried a life insurance policy with the Ancient Order of United Workmen, a benefit payable to his dependents. Without a body, the wife and sons would receive no AOUW funds.

Interestingly, a man listed as missing for a week after the snowslides, didn't realize it and was enjoying life in town. No one noticed. Finally he discovered he was "missing" and made himself known.[8] Another man, State Representative W. A. Taylor, said he was once "killed" in an avalanche — at least the newspapers all listed him as dead in the Telluride slides of 1882, unknown to him. His family thought it was fact. Fourteen years later he went for a visit and his shocked family told him he was supposed to be dead. It took him that many years to discover the newspapers had demolished him.[9]

No one thought Smith had gone back up the mountain to the Liberty Bell and was caught in a slide or lost in the white expanse. No one thought he had gone off with a friend to some other snow-laden area that late in the evening. Both were possibilities, especially if he had business of some kind at the Liberty Bell or elsewhere. No one thought he had been a victim of robbery, even though he was carrying nearly a month's wages on his person. Marshal Zickefoose interviewed dozens of people in Telluride for a month, finding no evidence of anything, except nasty rumors. Rumor claimed some Italian miners, with whom he had allegedly clashed, had murdered Wes Smith. The motive was revenge. Smith had supposedly been a deputized shift boss at the Bunker Hill and Sullivan concentrator in Wardner, Idaho, during the labor disturbances there just prior to the Bunker Hill explosion in April 1899. Everyone liked to blame the WFM for that dynamiting; however, some historians said the huge trusts had more motive than the union.[10] In November 1901, Smith arrived in Telluride and worked without any conflict. He cashed his checks at Nathan's, a union stronghold. He dined at union restaurants and fraternized with union men at the saloons. Someone claimed his real name was "Merritt," although nothing substantiated this except a brief sentence in the La Junta *Tribune* calling him Wesley E. Merritt. He had no trouble with anyone. He simply vanished into the night, leaving his belongings behind.

Wesley E. Smith.
Restored portrait © M. J. Martin 2003.

On March 24, Charles Chase of the Liberty Bell published a description of the man, asking for information as to his whereabouts. Wes Smith was about forty years old, five-foot-ten-and-a-half inches tall, 165 pounds, and had dark brown hair and mustache, blue eyes, and a "small mouth with teeth badly decayed."[11] A few days later Zickefoose convinced Chase that Smith was dead, and the Liberty Bell Gold Mining Company officially offered a reward of $500 for information leading to the recovery of his body. The La Junta *Tribune* printed a sinister paragraph implicating union men, claiming they had threatened Wes Smith in Wardner, telling him that one day they'd kill him. If he had been threatened with death in Wardner, it would be sheer stupidity on his part to seek work in Telluride where nearly every miner was WFM and the county had a reputation for making nonunion men vanish, according to the *Journal.* Then again, it sounded like a Pinkerton plot: a "marked" man moves to Telluride only to suddenly "disappear" (whisked away by Pinkertons) so all the world would suspect the union of "foul play" — brilliant in its simplicity. Or perhaps Zickefoose invented the plot because he thought the mere rumor of death was enough to convince the AOUW to release the benefits for the destitute widow. How much truth was in any of the rumors will never be known, but that didn't matter. The gem was the implication.

The disappearance and allegation proved the reign of terror existed. After all, this was the second man who disappeared from Telluride during St. John's presidency — a few newsmen still counted John Mahoney, but locally they knew Mahoney was quite alive. If a dozen men had vanished from an agricultural town of the same size no one would claim a reign of terror was on hand. Mysterious disappearances were common enough in most towns in Colorado. All too often men disappeared because they were tired of their wives. One prominent man had left a wife and children in Indiana, came to Colorado, changed his name, and started an entirely new family. But in Telluride, if a man disappeared, the union was guilty. It was a simple solution.

The Miners' Union as an organization had no reason to kill Wes Smith. His death would serve the union or its officers in no possible way. Revenge for his part at Bunker Hill was a flimsy motive — otherwise, why not take out Bill Jordan down in Delta, since so many actually believed he was responsible for Barthell's death? An individual with a personal motive or a stranger with robbery as the motive seemed to be the only likely option, if Smith was murdered. Nothing indicated he was. The possibility that he was caught in a slide or lost on the trail to the mine was real. Except on March 3, when guards were posted to keep everyone out of the avalanche area just that day, men were still going up and down the mountain day and night, digging out bodies, clearing out the demolished buildings, preparing for rebuilding, and looking for their belongings. One man, who refused to give his name, was arrested at the mine on March 13 for stealing a canvas coat from the wreckage earlier that week. Everyone knew when the tram went down that highgrade ore was pitched all down the slope, and some men thought they might acquire it under cover of darkness. Wes Smith might have even decided to walk away from his wife and sons for his own motives, making it appear he intended to go to La Junta by making a point to leave his baggage with the hotel clerk. Many possibilities existed and no evidence of any sort pointed to murder, let alone premeditated homicide carried out by the union. But that failed to stop the rumors.

Soon Wes Smith's disappearance was linked to Will Barney's. Gossips hinted both were murdered, not just by any union man, but by St. John and Carpenter. Gaping holes drained that theory for anyone who paid attention to facts: Wes Smith was last seen alive on the evening of March 4 and St. John, Carpenter, Charles Finke, Kinney McLean, and S. R. Fitzgerald had gone to Denver on March 3 on business with the Democratic Party. St. John and Carpenter were in Denver while Wes Smith was talking to his foreman at Nathan's. They returned to Telluride on March 10.[12]

At the end of March, while rumors of union murders boiled beneath the surface, Arthur Collins and Charles Chase went to Denver to join with Arthur Winslow, the Liberty Bell general manager; A. D. Snodgrass, the chief clerk in the Smuggler-Union Company's Denver office; and several prominent mine operators to organize the Colorado Mine Operators' Association. The Telluride Businessmen's Association had little power and no persuasion to mount an offensive against the growing power of the WFM. If the state's mine operators united, they could work on a statewide strategy to combat the demands of labor organizations. Their initial order of business was to gather as many mine operators as possible. Twenty-seven members began the group, many of those from Idaho Springs,[13] where the WFM was strong. Their declared purpose was to protect their interests, especially against the engulfing power of the smelter trusts and the railroad syndicates. The *Mining and Scientific Press* seemed to suggest the "protection" reached farther than that: "It is full time that they should, for protection, avail themselves of the same weapon of organization that is used so effectively against them."[14] The *Press* suggested organized labor should align itself with the MOA in battling the trusts to ensure prosperity. General Frank Hall of Denver spelled out clearly that the MOA would not only be able to proceed against the syndicates but against the "evils connected with" labor. "The greatest danger with which the commonwealth has to deal is that of labor strikes.... With such an organization as the owners and managers have perfected here in this city, it will be possible to eliminate such deplorable affairs from our working system as far as the mines are concerned."[15]

The *Journal* applauded the new organization, anticipating the MOA would soon have the strength of membership to put the WFM in a coffin. The newspaper's policy was "Labor is good, as long as it is silent." Curry was back at the helm, and his latest method of slamming labor organizations and Local 63 came via the wire service. Instead of a frontal attack from his own pen, he printed any story with even a slight slant against labor from other newspapers, thus not being directly responsible for what was written. He continued his innuendos and attacks on Local 63, but sent them out on the wire for other newspapers to print rather than printing them locally. The union "murder" of Smith and Barney skillfully arrived in Durango's Dave Day's hands and fledged from there. The propaganda-mill around the district had so closely tied the Wes Smith disappearance with the Will Barney disappearance that people accepted as true that whoever had "done away with" Barney had done the same dastardly thing to Smith.

Combating the rumors was difficult, since several local government officials suddenly turned into clams, not uttering a word of the real facts. And they indeed possessed real facts: Judge John Wardlaw, Sheriff Cal Rutan, several local attorneys, and a jury of three Telluride men absolutely *knew* William Julius Barney was alive.

On April 4, 1902, Judge Wardlaw heard only one case in county court, a case filed just thirty days before by a dead man. William J. Barney was the plaintiff in a divorce suit against his pretty little wife, Sarah, on the grounds of adultery.[16]

In most divorce cases of the era, the plaintiff or complaining party was the wife. In this case it was the husband, for Sarah Barney had been in an adulterous relationship with her father's farmhand, Winthrop D. Sylvester, since late fall of 1900. Will Barney had witnessed their affection toward one another when he returned from Nevada in February 1901. He left her side so abruptly after not having seen her for months that it was likely he had decided to file for divorce immediately, but finding a few hurdles, had to wait. His first hurdle was residency: in Colorado, the law stated he could not file for divorce unless he was "a citizen of this state for one year prior to the commencement of the action."[17] He also had to file in the county in which he resided. Will wanted to spare Sarah and himself the shame of her offense by choosing distant San Miguel County for his purposes and had moved there for no other reason. Adultery was viewed as an appallingly immoral offense in those days, shaming the woman and humiliating the man, and was also a crime in Colorado, punishable by fine or up to six months in the county jail.[18] Barney didn't want his wife punished; he simply wanted a very quiet divorce. He had arrived in Telluride in March 1900 and filed for divorce in March 1901, fulfilling the law's requirement of a year's residency. The law also required that he produce "at least one credible witness" to prove he had resided in the county for that year.[19] This was done to the satisfaction of the court and Barney's attorney, Luther C. Kinikin, who worked as the city's attorney as well, asked for the "cause and the question of the guilt of the defendant, Sarah E. Barney," to be heard before a jury of three. Wardlaw granted this and Sheriff Cal Rutan personally selected the jurors.[20]

Sarah E. Barney had arrived on the train the day before trial, and with her was Winthrop Sylvester. As the defendant, Sarah could have let the trial go on without her, since the court was free to conclude the case without the defendant present, but she had a deep concern for gaining custody of her three-year-old daughter Florence. The plaintiff, on the other hand, could be neither absent nor dead, *under no circumstances*. The state statute demanded, "No trial of any case for divorce shall be had unless the plaintiff be personally present."[21] William Julius Barney was physically present in Judge Wardlaw's courtroom on that Friday in April 1902. If he hadn't been, the case would not have proceeded. Only in recent times a divorce case proceeded wherein the plaintiff was dead (after filing), but the case set precedent, and had such a thing occurred in 1902, it would have set precedent then. W. J. Barney versus Sarah E. Barney moved forward as an ordinary divorce suit, except for the secrecy surrounding it.

During the trial, Sarah didn't confess or plead guilty to the offense of adultery. She was found so by the evidence presented. Will Barney himself testified that he had immediately ceased cohabiting with his wife when he became aware of her infidelity. Winthrop Sylvester was named as the other guilty party. The decree of divorce was granted, with Sarah given sole custody of Florence. Judge Wardlaw forbade both parties from remarrying any other person for the period of one year as the law ordered. Had Will Barney been as dead as Curry and all his allies claimed, not only would no divorce have occurred, but Sarah, so deeply in love with Winthrop, would have married her lover immediately. Instead, Sarah and Winthrop abided by the law precisely, because they both knew Will was very much alive. They did not marry one another until exactly one year and ten days had lapsed, the event solemnized on April 15, 1903 by Reverend A. E. Chase at Timnath, Colorado, where the couple continued to reside on her father's farm.[22]

The divorce hearing and decree of W. J. Barney versus Sarah E. Barney, Case No. 975, was duly documented in the county court record in detail. It indisputably proved Will Barney was alive a full year after he was "murdered." It also proved that several men, who knew he was supposedly murdered by the Miners' Union, *knew* he was alive. Those the document established as present during the case, in addition to the divorcing parties, were Judge Wardlaw, Sheriff Cal Rutan, city attorney L. C. Kinikin, and an attorney from the local office of Hogg and Watson, as well as the three jurors and some unnamed local witness for Barney. Perhaps Barney's witness and the jurors knew nothing of his supposed murder, but Wardlaw, Rutan, Kinikin, Hogg, and Watson all knew a reward for the body of the plaintiff was still on the books. Since Wes Smith's disappearance, Telluride Mayor Hilgenhaus was out soliciting signatures to induce the city to offer a joint reward for both Barney and Smith. Wardlaw, secretary of the city council, and Kinikin, city attorney, would have been aware of this.

If St. John or Carpenter or any officer of the Miners' Union had known the document existed, they would have had the vital proof they needed to squash the false accusations against them. Around this time, the detective hired by the union gave St. John information that directly stated W. J. Barney was alive and had been involved in a divorce suit, and was "trying to avoid paying alimony."[23] St. John's mistake was in not demanding documentation, a mistake that would plague his future with accusations of Barney's murder. He accepted the divorce as fact without knowing where it took place. Having a clear conscience himself, he concentrated his service on other pressing business, the organization of the millmen, and increasing involvement in the Democratic Party. With the gubernatorial election approaching, an election critical to labor's cause across the state, St. John took frequent trips to Denver on behalf of the Democrats and the WFM. His naïve belief that truth was all he needed to sustain the Miners' Union against vicious attack demonstrated the dangers of his idealism and his youthful innocence. No matter how much he valued integrity, it didn't follow that those who opposed him held the same values. He failed to realize the depth of resentment in his adversaries, or the length they would go to stop him and the union. Keeping silent about Barney's divorce served them incalculably and was painless on their part.

The common practice of publishing divorces in the paper was not done in Will Barney's case. If the secrecy was due to threats against his person, he could have had the case removed elsewhere or had a bodyguard of deputies or even Pinkertons to protect him. Under protection, he could have easily come forward and from his own mouth let it be known the Miners' Union had threatened him with death or had run him out of town a year ago, if that were the case. Had he been run out, his testimony would have been crucial proof against the union and St. John, proof that Wrench and his allies would have gone out of their way to obtain and make public.

Secrecy did not originate with Barney; he was basically unknown in Telluride. Only a few people, such as Becker, Kemp, Carruthers, Collins, Robbins, and a few in town would have recognized him on the street. This allowed him the freedom to come into town the day before the divorce and leave the evening of the trial without notice. Nearly a year had gone by: even those who were acquainted with Barney at the Smuggler might not notice him; not one had been his friend. Unlike the reward notice for Wes Smith, no photograph of Will Barney was ever used, and the last published reward for his body was nearly a year past. He had no interest whatever in the union, or Collins, or anyone else. He clearly was taking advantage of the remoteness of the camp for his own purposes: avoiding the humiliation of his wife's disgrace. Had he no concern a reward was offered for the recovery of his corpse? From a later encounter with a man who knew him in Denver, it was clear Barney had no knowledge whatever of the reward or that he was listed as murdered.[24] If Wardlaw, Rutan, and Kinikin failed to inform him, he had no way of knowing since mass communication of the era was limited to weekly and a few daily newspapers. At the time of his divorce, Barney's murder was old news, mute and forgotten, except in Curry's nest of vipers. Had anyone informed Curry that Will Barney was in town divorcing his wife, Curry would have been unlikely to publish the fact; however, since a lean thread of honor still ran through a portion of the newsman's anatomy, Curry might have discontinued the murder theory as he had when John Mahoney turned up alive.

Kinikin's silence was troubling, since the man was an honest, businesslike attorney who had begun his Telluride career as a court reporter under Judge Gabbert in 1891.[25] He was neither anti-labor nor pro-union, but he may have been persuaded to remain silent if given a plausible reason. The simplest explanation was that Kinikin was probably informed this William Barney was not the same as the missing man — a few men with that last name resided in the county. Perhaps he was told this Will Barney had never worked at the Smuggler and was one of the cowboys from the western part of the county — it was possible he took work on a ranch in the interim. But for those who knew Barney, no doubt existed this was the murdered man who came to work for the Smuggler Company a year before. Will Barney's half brother, Ralph Barclay Barney, documented Will's disappearance from the Smuggler-Union, yet had no knowledge of Will's divorce, which indicated how deeply disturbing Sarah's adultery and the divorce had been for Will: he revealed it to none of his family, except his eldest sister, Ella Barney Johnson.[26]

Why did Wardlaw and Rutan refuse to divulge the crucial fact of Will Barney's divorce and thereby put an abrupt end to the poison and hatred in the town? At this time Wardlaw was under the anti-labor influence of Wrench. The judge was a notorious political vacillator, never quite certain what he believed in, yet he was loyal to those he chose during his time with them. Although strict in upholding the state's statutes, Wardlaw slid down the slope of slander on several occasions. Silence was an easier task to perform; yet it must have disturbed him, for after the Barney divorce he gradually unlinked himself with Wrench. His wife, Minnie, may have influenced him, for she was a very active Catholic and fellow parishioner of St. John and many other union men and their wives. Rutan was a man who had not, to date, shown any anti-labor actions, yet since he was strongly backed by Wrench in the election, it was likely he shared some of the banker's convictions.

If either the county judge or the sheriff announced Will Barney was alive, they would lose the power that sustained them, that of Wrench, Collins, and the Telluride Businessmen's Association, because they would appear to be vindicating the Miners' Union. Although such backing may seem minor from this point in history, the remoteness and size of Telluride at the time made Wrench, Collins, and their friends the defining power and authority in the county. The only thing with more political and visible power was the Telluride Miners' Union, simply because of its numbers. Collins and Wrench's power was so greatly threatened by the nightmare of labor's growing political strength as to induce these men and their allies to find any means to stop the union.

Exploiting the absence of Will Barney or any missing man as a union murder was ingenious and successful on so many levels that it became the standard operating procedure. The union had no means of disputing a disappearance unless they could find the missing person themselves. Rewards for the body would never have to be paid, but the reward notices served to condemn the union as murderous. Rumors of how the victim came to his death could be as monstrous and bloody as possible, without any physical evidence to back them up or dispute them. Since the only crimes or deaths applicable to the Miners' Union resulted from the riot of July 1901 and seemed to be in self-defense, the anti-labor associates avoided exploiting those. Most citizens believed the riot was instigated by hired guards and deputized Smuggler employees. Instead, to claim the union carried out a series of mysterious murders as cheap reprisals would gradually drape so evil a pall over the organization that no respectable citizen would defend it. Keeping Will Barney dead was of the utmost importance, especially since his sudden appearance in the flesh would not only discredit the liars who claimed he was dead, but would absolutely vindicate Local 63, giving the union a sweeping victory and strong public support. Even Arthur Collins seemed uniformed; the company's reward for Will's body remained intact. Collins had arrived from Denver the same day Barney came to town, but Collins was in a private coach with B. B. Lawrence and other company men, all too involved in the business of the state MOA to notice who was in town. None of the Smuggler employees knew of Barney's divorce. None of the city or county officials other than Wardlaw and Rutan apparently knew. Wrench might have been informed, but the least number of persons with the knowledge of Barney's divorce would lessen the chances of the Miners' Union finding the truth. Instead of a full-blown conspiracy, all that was necessary was the silence of the judge and the sheriff.

Were the Pinkertons involved in the cover-up and continued assertions of Barney's murder? Given James McParland's fondness for twisted plotting and his personal belief that he had a divine mandate to obliterate the WFM, the probability was very strong. On the strength of Barney's murder the Pinkertons had solicited the Smuggler-Union Company as a client. To retract that allegation would be to admit error. Yet, if they were truly investigating Barney's whereabouts for the past year, why hadn't they discovered him? A cut-rate detective for Local 63 found Barney alive, so why not the Pinkertons? The anti-labor faction in Telluride quoted on authority that the Pinkertons had searched every corner of the state for Will Barney and found no evidence he was alive. Was Barney a Pinkerton operative?

The possibility existed that Barney was an operative since he remained almost invisible for the next several years. A plot to send men to Telluride and secretly whisk them away in order to blame their disappearance on the union was neither farfetched or out of character for the Pinkertons or more specifically, James McParland, who had done similar undercover maneuvers in the past. But given the circumstances of Barney's divorce, his choosing to go to Telluride for only the year required by law for residency, and the excruciating humility of being sexually discarded by the woman he had been married to for ten years, it is unlikely he was a Pinkerton spy. More plausibly, McParland, like Wrench and Curry, simply took advantage of Barney's secretive, antisocial ways. Barney inadvertently created a perfect opportunity and they seized it. Yet McParland asserted his investigators searched for Barney and failed to find him, concluding he was dead. Was the agency grossly incompetent? Or had the investigation been just another agency lie?

Quite thorough in actual investigations, the agency showed no signs of incompetency in existing historical records. Under McParland, misinformation had become the Pinkerton modus operandi. If the Pinkertons, touted as the greatest detective agency in the country, said they investigated and found no trace of Barney, then everyone absolutely had to believe the man was dead. They merely had to report it so; they need not investigate. Will Barney the living man was incidental to their purpose; Barney as a murder victim was indispensable. To accuse the WFM or its leaders of murder, one needed murder victims, and the WFM just wasn't providing them, thus victims were assembled from the material of missing persons. Charley Sumner noticed this. In an effort to illustrate other mysterious disappearances ignored by the Pinkertons, Collins, Curry, and friends, Sumner pointed out how one Henry Thanning, a good nonunion Smuggler man, had recently vanished and no one knew where he was or even when he left. They did know "the personal belongings of some of the boys on the hill accompanied him."[27] After the Smuggler fire, he had been paid to accompany the body of Joe Nelson to the family home in Utah. While there, he took $175 from the parents, wrote to the Smuggler under false pretenses and received forty dollars due Joe, and hit the boys in the basin for another seventy on his return. He claimed he was tired of the petty thieving at the Smuggler bunkhouse and moved to town...the thieving stopped on the hill and went with him to town. The *Journal* made nothing of it.

McParland's own correspondence made it plain the Secret Operatives under his command weren't acting as investigators with an aim to solve crimes. Nor was it ever McParland's goal to clear the WFM's name; quite the contrary: he would do everything in his power to condemn the Federation. McParland believed the WFM was a veritable blight on the West and its officers "bloodthirsty assassins,"[28] although he had not one scrap of evidence against them, not even anything unearthed by his spies on the inside. Will Barney, Wes Smith, and John Mahoney were all he had to fasten on Local 63 and Vincent St. John. The glue was suspicion, yet even that had no foundation.

Two of the victims were undeniably living, but for McParland, facts were inconveniences requiring burial. By the authority McParland vested in himself, Will Barney was murdered and Vincent St. John was the murderer. McParland had never met St. John, yet he profiled the young man as a depraved killer, devoid of all humanity, godless, soulless, and unfit for a shred of mercy, merely because St. John had become the president of Local 63 and a rallying force in the WFM.

For the Great Detective, St. John was a man good for only one thing: the gallows. If the Saint failed to commit murders, then McParland would fasten any and all suspicion on him, using a string of unsolvable murders — unsolvable because no corpses existed.

CHAPTER

14

BUCKSHOT IN THE BACK

"Hell is truth seen too late."

— H. G. Adams

In spring of 1902, Vincent St. John was ignorant of the noose James McParland was carefully knotting for him. He was unaware union member George Riddell was McParland's worm in the Telluride Miners' Union. He knew a spy was supposed to be "inside," yet a spy of any kind was irrelevant since the union had nothing to hide. He was unconscious of the vast importance McParland and the MOA attached to Will Barney's "disappearance," allowing the matter to be dropped once he was satisfied Barney was alive. Being a logical man, St. John saw no threat in groundless assertions that a living man was "murdered" by the union. He had more important business to manage. Yet his not pursuing the Barney matter until he had Barney physically before the public gave McParland and the MOA a malignant advantage. Although the disappearance of John Mahoney had been quickly dropped in the Telluride district once he was found alive — Mahoney had recently run off again, this time to Cripple Creek, his wife and child left destitute in Telluride — McParland refused to acknowledge the fact. The latest disappearance, that of Wes Smith, was added as a "third brutal slaying" attributed to the union in McParland's files,[1] validating the existence of the reign of terror.

Still no terror actually existed, the general brawling and crimes committed in the district being less than the same per capita in Denver and unrelated to the Miners' Union. If miners used their fists in an argument the union could not be held liable for it, any more than ranch owners could be held liable when cowboys used their fists. Citizens walked the streets in Telluride at night without fear. Normal business continued, including the *Journal*'s, despite the ongoing boycott. Shadagee Bill was still quite alive in Delta, purportedly in the bootlegging business, since that county had gone "dry."[2] Neither the union nor the law seemed to make Bill shudder at lower altitude. The only thing causing fear in anyone was the Miners' Union's increasing membership and success in organizing the millmen and the fear was restricted to Arthur Collins, A. M. Wrench, and the five or six active members in the Telluride Businessmen's Association (TBA).

In April the San Juan District Union once again elected Vincent St. John and Oscar Carpenter to the posts of president and financial secretary, "a testimonial of regard for the boys," according to Sumner, "as the district union changes officers every session; usually selecting them from the next town where the Union is to meet."[3]

St. John and Carpenter energized the district locals into a dynamic organization of solidarity with the potential to act as one body throughout the region. Wrench, Collins, and the MOA saw this as potential to paralyze the entire district with a strike. They made it clear they had "a very profound dislike for St. John and Carpenter, as officials of the Miners' Union," letting it be known they would "feel better disposed toward the Union if it had different officers."[4]

Charley Sumner cringed. "That is a very shadowy statement for even a corporationist to make," he said. "The trouble is, there is not enough money in this county to buy St. John and Carpenter, and there is not enough influence to swerve the Union from honest principles."[5] All too frequently union officers were offered "all kinds of inducements" to "betray their trust."[6] He pointed out that the union members knew who qualified as officers. "They elect no men who can be bought, bribed or handled."[7]

Walt Kenley, a cowboy from the Delta Club Cattle Company, was illegally deputized and later joined Troop A, working as both a deputy for the mine owners and a private in the militia.

Courtesy © E. N. Wright; used with permission.

Wrench knew St. John was beyond his reach legally. It was only a matter of time before he and his cohorts began bending the law in order to accomplish their ends. The TBA and Telluride Mining Association (TMA), which amounted to the same thing, decided to assess the members a monthly fee toward hiring "man-killers" as deputies under Sheriff Cal Rutan. They told Rutan the county would benefit from the gunmen without putting out county funds. One of the first hired was Walter Kenley, a quarrelsome, rock-fisted, trigger-happy, big and brawny cowboy from the Club Cattle Company in Delta County.[8] The Colorado law stipulating sheriff's deputies were to be residents of the county was ignored. The main object of the deputies was to intimidate union miners, and Walt Kenley satisfied Wrench, Collins, and the TBA in that department completely. Union men kept an eye on their comrades when Kenley pushed his way into their faces, thus preventing violent encounters.

Some weeks later the association secretly hired two gunmen out of Wyoming, Robert "Bob" Meldrum and Willard Runnels. Meldrum was a friend of Tom Horn, the gunman who had worked for the Pinkertons and was later convicted and hanged for murdering a boy. Meldrum fashioned a reputation for killing those he was supposed to apprehend when he shot a wanted man in the back of the head. Given to bouts of drunkenness and bad temper, he liked to find an excuse for "shooting first" those who questioned his authority to arrest them. Runnels, equally cold-blooded, had worked with him. They were hired on Pinkerton recommendation and the MOA paid their wages; Rutan gave them housing in the sheriff's quarters behind the courthouse. They were deputized — illegally — to patrol Telluride at night, keeping a low profile by day. No one knew their identity until December. They and the more visible Kenley were the perfect tools to keep the miners intimidated, yet the miners stayed clear of them.

Using other subtle means to provoke the union, such as infractions of the July 1901 agreement or November wage contract, also failed. Whenever Collins strayed from the contracts made with the union, St. John and Carpenter were on top of the violations immediately, sparring, negotiating, and demanding Collins rectify the situation. For their part, the union officers adhered to the contract, even when it meant opposing friends, as in the case of illicit liquor sales in the basin.

Charles A. "Broken Nose" Smith, proprietor of the Miners' Rest near the Bullion Tunnel, was caught selling liquor to the miners without a license. The union disliked miners drinking during working hours as much as management did. The WFM officers disliked members being "seen under the influence of liquor" generally, as drunken union men were a poor image to the public and under such condition were prone to "general foolhardiness."[9] President Ed Boyce instantly suspended drunken delegates at WFM conventions. St. John was more moderate with the members of Local 63, but illegal liquor sales were not tolerated. Although Broken Nose Smith was considered a friend of the union, he was turned over to the authorities and indicted. In retaliation he and his friend Hugh H. Mullen filed complaints against every saloon owner in town, including Pheasey and McLean. All were licensed, but since the state laws required saloons to close at midnight and on Sundays, the charge was for selling liquor on Sunday. Instead of fighting the charges, the saloon men merely pled "nolo contendere" and paid their fines.[10] That angered Broken Nose Smith, who failed to escape the law himself — because he lied under oath, he was charged with perjury. He blamed the Miners' Union and, in particular, Vincent St. John, who actually had nothing to do with the perjury prosecution.

One of Smith's allies was an immigrant from Tübingen, Germany, named John Miller. Although Miller was a butcher by trade, he worked as a miner in Telluride and was a member of the union. He was frequently drunk, and in that condition was so wildly fractious that he earned the name "Crazy" Miller. Smith and Mullen were known to sell him liquor at the Miners' Rest. Neither St. John nor Carpenter cared for Miller's contemptible behavior, warning him several times to cease or lose his union membership. One day in the middle of May, Crazy Miller was teetering around town from a three-day binge, making a nuisance of himself, using his fists on a few men but getting the worst of it. St. John told him, "Get back to the Smuggler bunkhouse and sleep it off or I'll have you locked up in jail."[11] Miller bellowed against St. John, assuring him he would kill him for saying that. St. John forced him to swallow his threats. The extra night policeman, Chris Geyer, dragged Miller off to jail for disturbing the peace.

The following evening, Wednesday, May 21, Crazy Miller was staggering drunk again, having fully lubricated himself since he stepped out of jail that morning. He went looking to settle his score with Geyer and St. John, telling everyone he was going to shoot Chris and Vint and a man named Joe Gasser. Unable to find them, he stumbled into the New Resort Saloon, where he found Charley Sumner, Oscar Carpenter, everybody's favorite French chef, Frank Maumey, and a few others together among the crowd of patrons. They started joshing him about his daft brawling until Geyer walked in and Miller took to quarreling with the night marshal. Geyer threatened to lock Miller away again. Miller shut up and Geyer left. No one took Miller's noise seriously, although he had a record of bloody fights in the region. The solidly built, five-foot-nine, abundantly scarred and scrappy German[12] turned on five-foot-six Carpenter, who advised him to "get back up the hill" or he'd lose his job.[13] Utterly refusing the advice, Miller bellowed so much that one of the proprietors, Thomas Rundle, "called him down," but Miller turned his attention to Maumey, challenging the French chef to wrestle with him. Pulling off his coat, the unruly German tossed it to Carpenter, and Carpenter promptly removed the Miners' Union button from it. Maumey simply told Miller to go away. Miller snatched his coat back, noticing the union button was gone and started bellowing about that, flashing a small knife. Carpenter told him, "From the way you're acting, you are not entitled to wear the button."[14]

At that point, another union man, John Touhy and three companions came into the saloon. Miller turned his bellowing and abuse onto Touhy for no apparent reason, but Touhy, a smaller man with no tolerance for Crazy's kind of bluster and bluff, told him, "Here, I don't want you jumping on me for anything, or I'll hit you so hard you won't know where you are."[15]

Ambling to the end of the bar, Miller sulked a bit. A short time later he returned to abusing Touhy, who tried to ignore the blootered hornet. Finally he snapped, "Miller, leave me alone. I think you're a damned son of a bitch anyway."[16]

Crazy Miller stifled himself and sidled out the door. Touhy left as well, only to return about a half hour later. He was having a drink with Rundle and a few union men when Crazy stalked into the saloon, snarling, "You will call me a damned son of a bitch, will you — " and pulled a gun, firing at Touhy. Rundle yelled at Miller and another man tried to grab the gun, but Miller turned the gun on him. Touhy staggered against the wall, and as he sank to the floor, Miller leveled the pistol at him again.

"Don't!" Touhy cried. "Don't shoot me again — I'm dying, Miller!"[17]

Just as Crazy pulled the trigger, Touhy flung up his hands for protection, taking the bullet in the arm. Joe Bellino, Rundle's business partner, flew into the room, wrenched the gun from Miller, and bashed him across the skull. Miller fled the place just as Chris Geyer, with his face half covered with shaving cream, raced out of the barbershop at the sound of gunfire, and assisted Bellino in taking the German down. Geyer hauled Miller off to jail and the others took John Touhy to the hospital.

"He shot me like a dog," the thirty-six-year-old Touhy muttered several times in disbelief.[18] The first bullet had torn through Touhy's chest, stomach, and pancreas, causing massive bleeding. Touhy bucked the shadows, cheerfully telling his friends, "They think I'm going to die, but I'll fool them."[19] He died just before dawn the next day and he was given a solemn union funeral.

Forty-four-year-old Crazy Miller was charged with murder. Miller had bought the gun just a half hour before the shooting, since his own gun had been taken from him the previous day. The clerk at Tomkins-Hunt Hardware first refused to sell him a gun, but Miller telephoned the Smuggler's accountant, O. B. Kemp, asking him to tell the clerk to allow him to buy the pistol. Miller told Kemp he was going back to the bunkhouse. Kemp gave his consent, which was incomprehensible since he knew Miller had openly threatened to shoot Geyer, St. John, and others. At the trial he was not asked to explain himself.

Showing no remorse, John Miller attempted to escape the gallows by pleading insanity, claiming everything from syphilis to lead poisoning had caused him to lose his mind. District Attorney Sam McMullin told the jury, "Hang him! Hang him!" and was quickly reprimanded by Judge Theron Stevens.[20] The jury spared the accused: Miller was sentenced to "life at hard labor in the [state] penitentiary."[21] Among the witnesses in his trial was Vincent St. John and, despite the union president's minor role, John Miller muttered his revenge against him and Oscar Carpenter.

Holding St. John accountable for merited prison sentences seemed the latest Telluride pastime. In that same session of the district court, the trial of former county clerk Harry Servis came up. Although Harry pled guilty to embezzlement, begged the court for mercy, and poured out his "deepest regret" for "allowing poor judgment to master" him, he "got sent up the river" for one to three years.[22] He seemed to be taking the fall for two other official embezzlers who were left unscathed when he pled guilty. Sheriff Cal Rutan hauled him off to the state penitentiary on June 20, where he joined Crazy Miller and the two mumbled their contempt for St. John.[23] Curry and Painter, both friends of Harry, harbored condemnation of St. John for Harry's downfall, even though Vint's only action in the case was as part of the auditing committee that reported the misappropriated county funds. St. John's December appointment would rankle Curry, Painter, and Wrench for years to come.

Harry's wife, Nellie, was forced to sell everything, even her fine piano, which the Ophir Miners' Union bought.[24] In disgrace she moved in with her relatives at Creede. Harry's cousin, Sam, who had been having trouble with his wife, Elizabeth,[25] went to Denver to confront her, but he first muttered sinister tales of seeing St. John kill Will Barney in the Haines Livery. This was by way of getting even with St. John for Harry's ruin, and started as nothing more than a malicious anecdote. Sam left the district in February 1902, around the time of the auditing committee's report to the court, and nearly a year after Barney was supposedly murdered in the Pheasey (not the Haines) Livery. The 1901 tale put the murder in the Pheasey Livery, which was in the next block east of the Betts/Haines/Roth Livery. It is important to note that the murder was moved to the Haines Livery only after Sam Servis told his tale, because Sam had worked in that livery and not Pheasey's. The Haines Livery *never* figured in the initial stories and became the site only after a year had lapsed. Having made no legal statement or signed any affidavit because his "eyewitness" account was fabricated merely for revenge, Sam left his tale in Curry's kennel for its care and feeding.

While the story still had some warmth in it, Mayor Hilgenhaus, banker Wrench, Arthur Collins, and others petitioned the city council to offer a joint reward for the bodies of Will Barney and Wes Smith. No rewards for known criminals were proposed, which many businessmen thought would be more useful. Charley Sumner noted the reward for Barney and Smith was unwarranted, since no evidence indicated the men were even dead and such a fund was out of the council's jurisdiction. He suggested including embezzlers in the reward to cover the two officials who escaped the law (although these officials were left unnamed, Sumner and others were well aware of who they were).[26] Because the Smuggler and the Liberty Bell managers were pushing for the recovery of corpses, Sumner said they should be responsible for the rewards, and "if they could be induced to put up an offer of about $5,000 in each case, some energetic person might be inclined to buckle on a gun and...return with the body of Barney...as it is generally believed he is a pretty lively corpse at this writing."[27]

Curry jumped him, insinuating Sumner opposed the city council's reward and was disinterested in law and order. Sumner retorted, "*Examiner* opposed nothing of the kind.... But as violent things have heretofore been committed in this city without even so much as an arrest... we raised the point as to why the city council should be called upon in this specific instance to offer a reward." With uncanny clarity, Sumner put the Wrench/Curry/Collins plot before the public: "If the sentiment was general that the offering of a reward would be productive of good results, and was actuated by a clean wholesome spirit for the benefit of the community in the suppression and detection of crime, then all would be well and good. But that is not the case in this particular instance. There is nothing but maliciousness behind the whole affair, and a desire to throw the brunt of the responsibility on other shoulders." Since the disappearance of Will Barney the enemies of the Miners' Union "seized the opportunity" to charge the union with a crime. "We insist," Sumner said with unflinching conviction, "that the charge was dishonorable, disreputable and malicious to the very bottom, as there were no grounds for it save in an open enmity that existed. And now we maintain that an attempt is being made by two or three mining companies, their organ and other tools, to use the city council as a cat's-paw in the attempt to create public sentiment that has no basis or foundation. While these particular affairs seem to have been forgotten by the people, it is rather peculiar that a handful of agitators should omit all other crimes and center their efforts in stirring up feeling in order to throw odium on an enemy they cannot touch. In the face of this known condition we hold that the parties most at interest should add sincerity to their methods by assuming the burden of rewards. If the Smuggler-Union, Liberty Bell or other corporations so ardently desire the unearthing of a possible crime, let them pay the freight while the agitators crawl off and lay down."[28]

Demonstrating how little he comprehended what he read, Curry replied, "Our neighbor is fast putting itself irrevocably on record as opposed to the ferreting out of crime and the punishment of criminals. If this policy obtains, this will not be a desirable place of residence."[29]

Once again, amiable Charley Sumner lost his tolerance for his fellow editor, asking Curry how long he was to "abuse our patience?" He identified the *Journal* as "The sunset hallucination" that "threw a hop-dream" in uttering a libel against the *Examiner*. "It is a libel for the reason that it must embody falsification through malicious intent in charging us with criminal affiliation...our sweet-willed neighbor harbors a desire to cast a stigma on *Examiner*." He challenged Curry to prove the *Examiner* "opposed the ferreting out of crime and criminals." If Curry failed to do so, Sumner was ready for another battle. Sumner pointed out the *Examiner*'s policy was "Let no guilty man escape," and had always followed it. "Because we have defended the council against a miscellaneous collection of grafters who have been endeavoring to run this town for many years, we have been charged by our peevish neighbor with criminality, which has made a malevolent and malignant accusation — that has nothing more nor less beneath it than the sordidness of a vindictive and vicious spirit.... Do you wonder, after all, that a paper is boycotted when it everlastingly maintains an attitude of casting an odium on somebody, some organization or institution?" Sumner hinted at some dark dealings in Curry's past when he fled the district under shadowy circumstances. He asked if Curry would like that brought to the public's attention, for the *Examiner* certainly favored the punishment of "embezzlers, absconders and accessories to other unlawful performances." Curry's past was kept well buried. Sumner suggested Curry had better exercise more discretion. He agreed that Telluride had been too lenient on certain criminals and thought it was time to reveal such leniency to the public, but pointed out that no evidence justified the Barney and Smith rewards. Several members of the city council opposed rewards for Barney and Smith on precisely the same grounds as the *Examiner*. "We have shielded no one," Sumner concluded. "But in two instances have we defended, specifically, the Miners' Union and the city council from abuse and slander.... Our neighbor is enjoying another attack through its usual cowardly practice of innuendo."[30]

Sumner had identified where the real problem in town concealed itself: in the *Journal*'s offices. If Curry and Painter had backed off and no longer acted as the public horn for Wrench and Collins, the rupture in Telluride's gut would have healed. Signs of progress were everywhere. Although aloof with those in the community, Collins was a likeable man in general and he and St. John had come to appreciate one another in their negotiations. His bitterness over losing to the union in 1901 had been shoved out of the public's view for the most part, and the Smuggler and Contention Mines and Mills were booming, with nearly 500 men employed. The Liberty Bell had been rebuilt and was in full operation. The body of Harry Trobridge, "believed to be the last" of the missing avalanche victims, was dug out in June, 112 days after the slide wiped out the mine.[31] If the heavy snows had claimed Wes Smith, no one would know where to search for him. The Tomboy was expanding on an immense scale and John Herron had gone to England to consult with Director Lord Arthur Butler on further investment. The district was prosperous and agreeable, except when it got "Curried" or "Wrenched" by the "twilight hallucination."[32]

As summer exploded in wildflower days and owl-moon nights, the usual fights and stabbings in the tenderloin heated up: a prostitute carved up the owner of a dance hall called the Coeur d'Alene, and several horse thieves and footpads kept the sheriff's and marshal's officers from falling asleep. While Curry and Sumner hurled words at one another, the basin saloon owners boiled up another fight. Gio Oberto, who owned a small store and saloon near the Tomboy, was charged with selling liquor without a license. The property owner told him to move the building, so Oberto negotiated with Ed Herbert and Ike Kling to move it to their ground nearby. Herbert agreed, but Kling refused, brandishing a rifle. The five-foot-four-inch rough and robust Kling, a fifty-three-year-old man with alley-cat scars, hated the Italians who frequently got inebriated at the "joint." Having threatened some in the past, his glossy shaved head and perpetual sneer was a warning to them wherever he went. When the building was moved under the official eye of a deputy, someone put a bundle of dynamite under it in the dark hours, demolishing the place. The contents of the building, including $500 in cash and coin, belonged to the merchant running the store, Sam Ginotti, a close ally of the Miners' Union.[33]

Kling and his associates, Hugh Mullen and John McDonald, were indicted for the dynamiting, even though Kling was in town when the building was blown up. Kling belonged to the union and some said his enemies took advantage of his threats to drunken Italians to make it look as though he took out the Ginotti joint. Meanwhile, Mullen and Broken Nose Smith, on a decidedly vindictive spin, charged numerous saloon men in town with keeping a room for gambling; and once again, Pheasey and McLean came under the gavel. The Scottish-born, pint-sized Smith, his knobby nose stuck to his face like a pocked and scarred tuber, should have quit while he was a free man. He wrangled with the court on his perjury charge, eventually losing that autumn and landing in the state lockup. Broken Nose Smith once again decided the root of his misery was Vincent St. John; he had expected St. John to defend him at his trial and was disappointed. He was hauled away with Ike Kling, who alone had been convicted of dynamiting the Oberto/Ginotti saloon. At the state penitentiary Smith renewed his bond with his old friend, Crazy Miller, and their shared hatreds began to produce fanged hatchlings. St. John was oblivious to the incubation.

The Miners' Union president had attended the Western Federation's tenth annual convention in Denver from May 26 to June 7, receiving a generous donation of $3,000 from the WFM for the Telluride union hospital. In his boyish excitement, St. John was unable to wait to bring the news home, wiring Carpenter that very day. The brick hospital was expected to be open for business in September. During the convention President Boyce announced his retirement, for health reasons, after serving the WFM for six years. Boyce's personal choice for his successor was the conservative Charles H. Moyer, a member of the WFM Executive Board, yet Moyer wasn't the first to be nominated. After a round of delegates tried to convince Boyce to stay on, Delegate H. W. Fox nominated the chairman of the WFM Ways and Means committee, Vincent St. John, surprising Vint by the tribute. Boyce had praised St. John and Carpenter, telling the delegates, "It is my earnest desire that every union in the Federation should be fortunate in electing officers like the President and Secretary of the Telluride Miners' Union. Were this the case the Federation would be respected by friends and enemies...."[34] Vint didn't decline the nomination, yet he and the majority knew Moyer was the man for the job. In the end Moyer received 105 of the 201 votes cast, with St. John receiving sixty-eight, and a third nominee, twenty-eight. Vint wasn't ready for such a position and told Charley Sumner he was relieved by the result. Sumner was certain Wrench and Curry and a few other locals were more relieved than St. John, chuckling that Moyer's election spared them distressing tears.[35] Charles Moyer of Lead City, South Dakota, was the perfect balance to the flamboyant, bold, boisterous WFM Secretary William D. "Big Bill" Haywood, whom Boyce had groomed in the ways of social revolution. Big Bill was ready to take the Federation to the meadows of socialism; Moyer was not. Most of the members agreed with Moyer.

Charles Moyer, President of the WFM.
Author's collection.

For mine operators, socialism was viewed as a greater evil than organized labor. Even Sumner was glad to report that various unions had not endorsed it at the recent conventions. "Socialism is no good to Unionism," Sumner wrote, "and gives enemies of labor an opportunity to perpetuate their graft."[36] Yet one of Telluride's best-loved merchants, Andrew H. Floaten, and his brother Henry Anton Floaten, had been advocating socialism for several years. During the WFM convention, Socialist Eugene V. Debs, founder of the American Railway Union (ARU), once the country's largest union, gave a talk at the Denver Coliseum Hall. Debs had been active in the

Democratic Party, serving a term in the Indiana state legislature before giving his all to the ARU. His railway union was destroyed in the 1894 Pullman strike and he was convicted for violating a court injunction, serving six months in jail in 1895. While incarcerated, he embraced the ideals of socialism, which he had been courting for some time. On his release he was greeted by thousands of cheering workers, and soon became the Socialist Party candidate for United States President in the 1900 election.[37] Vincent St. John was so inspired by this optimistic and evangelical "servant of the people" and his conviction that socialism was more about the dignity and humanity of the individual worker rather than about abstract questions of organization of production or the distribution of wealth,[38] that St. John planned to arrange for Debs to speak in Telluride.

St. John had returned home from the WFM convention with an enthusiasm and renewed vigor that seemed to spark the union men in the entire district. Nearly two thousand of them gathered for the dedication of a memorial to John Barthell on July 3. The dedication commenced the July Fourth holiday celebrations, and the huge procession of union miners through town to the Lone Tree Cemetery impressed on anti-labor advocates the spirited solidarity of the district unions. A beautiful marble monument to the slain Finn acclaimed him with the words of Henry Wadsworth Longfellow:

In the world's broad field of battle,
In the bivouac of life,
Be not like dumb driven cattle,
Be a hero in the strife.[39]

A few men of wizened soul later screamed in their journals, "This — this is the prostitution of poetry!" and condemned Barthell as a "murderer."[40] But that day in Telluride they stood silenced while keynote speaker, John M. O'Neill of the WFM Executive Board, praised heroic workers, such as Barthell, who fell in the battle for liberty and dignity. In a speech that could have raised carbuncles on Curry, Collins, and Wrench's backsides, O'Neill told the throng gathered at the cemetery, "There will be no equality...while the Smuggler-Union corporation, under the protection of the law, can place the instruments of death in the hands of hirelings to murder manhood that struggles for a reasonable remuneration. But there will be equality...when Socialism removes the causes that breed the scab and the oppressor. When the people shall own, collectively, the land, the machinery of production and distribution, the scab-hatchery and the private corporation will be relics of the barbarous past, and no more Barthells will be sacrificed under the murderous system of profit." He noted the enemies of equality coupled socialism with anarchy, but "Intelligent men and women...have discovered that anarchy and Socialism are as far apart as the poles, and that Socialism means the pure, unadulterated doctrines of Christianity; that Socialism means the emancipation of the callous hand from the bondage of corporate oppression.... Hopeless men become anarchists; hopeful men become Socialists."[41]

Despite the call to embrace socialism, the hint that Bill Jordan was the hireling who snuffed out John Barthell's life, and the direct mention of the company considered responsible, the huge throng of strapping miners and their supporters remained composed, carrying their camaraderie into the July Fourth contests, parades, ballgames, speeches, concerts, and dances throughout the week-

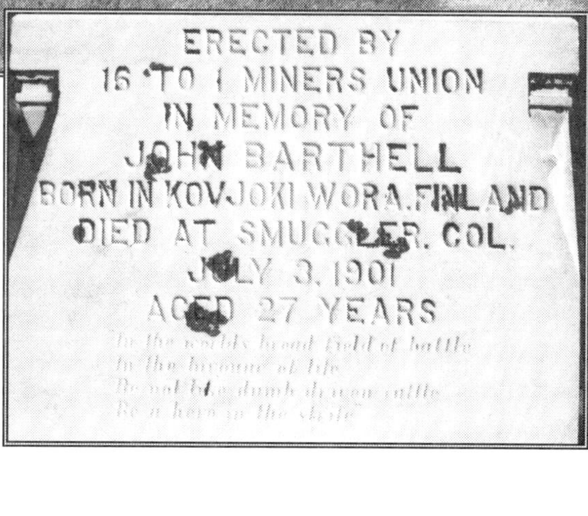

Miners' Union memorial to John Barthell. Inscription on the Barthell monument. Photos by the author.

end. The holiday was a smashing success on every level. Life was grand and Vincent St. John was walking in the clouds.

In the last two union meetings of July, Local 63 gained an unprecedented sixty new members, "thirty at a whack," according to Harry Jardine's enthusiastic report. "How's that, boys?" Shortly the Telluride union would have 1,000 members on completion of organizing the millmen, and St. John flew higher in the clouds. As the Miners' Hospital became more of a reality, numerous Telluride businessmen furnished the wards and private rooms. "It will be the best little hospital this side of the Big Muddy," Jardine beamed.[42] And St. John smiled from the altocumulus.

The successes of Local 63 and the inspiration of Eugene Debs were not the only reason St. John was hanging in the stratosphere. In June he had been introduced to Fred Pheasey's cousin, Clara Alice Meadows, the stunning and shy daughter of Lancashire, England coalminer, Peter Meadows. Clara was the same age as Vint. She was small and delicate, intelligent, and keenly interested in ideas that could change the world. She had been living in Canada's Ontario Province for the last few years and decided to visit Fred simply for a change of scenery. More than her scenery changed. When the fire of Vincent's soul met the flame of her joy, there was no turning back. She erased the pain of his damaged health, eased the sting of persecution, restored his humor, and went walking in the clouds beside him. Together they attended the packed presentations of Eugene Debs, one in town and the other in Savage Basin at the end of July.

Debs rallied the miners to abandon the Democratic and Republican parties. "If the Democratic Party is the 'friend of labor' any more than the Republican Party, why is its platform dumb in the presence of Coeur d'Alene? It knows the truth about these shocking outrages — crimes upon workingmen, their wives and children, which would blacken the pages of Siberia — why does it not speak out?" In the Coeur d'Alene district the Mine Operators' Association had implemented a nightmarish system where no known member of the WFM could secure employment. All miners and millmen were hired via an MOA employment agency where photos, descriptions, and other information on WFM members were kept on file. Debs said hope came only from the Socialist Party. "The enslaving and degrading wage-system in which [workers] toil for a pittance at the pleasure of their masters and are bludgeoned, jailed or shot when they protest — this is the central, controlling, vital issue of the hour, and neither of the old party platforms has a word or even a hint about it."[43]

The *Journal* had been issuing condemnations of Debs as an anarchist since he spoke in Denver, and now that he tainted San Miguel County, the attack included the Miners' Union. Under the heading "Some Superficial Soliloquies," the *Journal* ridiculed both, asserting that the union "paid too dearly" for the two lectures and that Debs was merely spouting his "unoriginal" agitation for the $150 demanded in advance. "He is out for the 'stuff.'"[44]

St. John set the record straight. He had written to Curry on August 1, but the editor had no interest in publishing his response. Sumner readily allowed the union president space in the *Examiner*, August 2, 1902.

St. John believed the editorial was not "the product of the buzzard brain of either" Curry or Painter, but from the pen of Deputy District Attorney Edward Howe, who was also the attorney for the TMA. Howe had stated in the article, "It is well to understand and appreciate the facts, that we not esteem too seriously the superficial ranting of the professional agitator."[45]

"I quite agree that the facts should be known," St. John wrote in reply. "When they are known abuse and slander coming from this source will be praise."

"Now as to the statement that Mr. Debs received $150 for his two lectures I will state that this is all the truth the article contains, and the strangest part of this is how so much of the truth managed to get to the editorial columns of the *Journal.* The sum of money was voted Mr. Debs by the Miners' Union...and he has returned value for value received a hundred fold." St. John pointed out the money was not "paid in advance, nor did Debs know what would be paid him until the morning he left." He defended Debs as a reflective, selfless man who "will always stand faithful to the cause of humanity." To the *Journal*, which acted on behalf of "moneyed men" to deceive the public, he said, "All *you* think about is the 'stuff'. Pity you never got higher in capitalistic circles. The world has always been full of slaves ready and willing to do the bidding of their masters."

St. John hinted at the shadow skulking in the editor's past, suggesting that investigation turned on such as Curry or any like him would reveal "they are far worse than those they try to abuse and place before the public as cranks and charlatans.... But Debs and the great work will go on despite the snarling of all such petty curs as the *Journal* editor or others of his ilk. But I owe the canine an apology, for classifying these people as such were to malign and slander the dog — one of man's best friends. Yours for the truth — V. St. John."[46]

Poking fun at those hidden crimes in Curry's closet, Charley Sumner added that St. John was preparing a book entitled, "Criminals I Have Met and Some of Whom are Traveling Incog," the proceeds from which would be "devoted to the maintenance of dope incurables" such as various regional editors.[47]

Despite the *Journal*'s slurs, Debs inspired another 200 miners to join the local Socialist Party, the leading members including businessmen August Anderson, Joseph Balliger, Andrew and Harry Floaten, and Miners' Union members Harry Jardine and the scholarly Guy E. Miller.[48] It was a mushrooming political party in San Miguel County, and Guy Miller was nominated for state representative on the Socialist ticket that fall.

With his life now more complicated by the personal yearnings of his heart, St. John was hesitant to abandon the Democratic Party for the Socialists. He wanted no stigma, no distress, no persecution to sink its fangs into Clara because of his political activism, since he had asked her for her hand in marriage. On September 1, 1902, he and Carpenter were again elected as president and secretary of Local 63. After a grand Labor Day celebration and another amazingly successful Miners' Union ball in support of the hospital, St. John went to the Democratic convention in Denver with Kinney McLean and other delegates.[49] Because the Democrats nominated E. C. Stimson, an attorney for the Mine Operators' Association, as their candidate for governor, St. John knew he could no longer remain in the party. As soon as he returned home, he signed on with the Socialist Party, immediately taking up duties as a committeeman with Guy Miller and Andrew Floaten. A few of his Democratic friends gradually distanced themselves from him.

While Guy Miller campaigned around the county, gaining substantial support due to his mildness and clarity of presentation, Floaten took to writing a socialist history column for the *Examiner* to enlighten the public that socialism was neither anarchistic nor anti-American. Sumner acknowledged the Socialists as honest, reliable, "clean and conservative men," but fully supported the Republican ticket.[50]

The Republicans had nominated conservative James H. Peabody for gubernatorial candidate that fall, an esteemed Mason and astute banker who had been elected mayor of Cañon City in 1901. He demonstrated no visible opposition to the WFM, yet he had proved himself disinterested in acting on behalf of organized labor while mayor.[51] The wealthy Republican power base of the state believed Peabody would restore the influence of business and industrial leaders. Labor leaders were very wary of this business candidate who danced around labor questions, evasively skirting inquiries on the proposed eight-hour constitutional amendment.

The possible election of either Republican Peabody or Democrat Stimson as governor hung an ominous mist over labor interests in the state that fall. Yet little dampened Telluride as autumn chased summer down from the frosty peaks in a blaze of gold. The mines were booming, the new wagon road from town to the Bullion Tunnel and Tomboy was advancing, as was the Bridal Veil road. More miners were employed than ever before. Blazer had sold the Saddle Rock Restaurant, consequently ending the boycott of that business.

Governor James H. Peabody.
Restored photo © M. J. Martin 2003.

On October 6, Vincent St. John and Clara Meadows were united in marriage at St. Patrick's Catholic Church, Reverend Thomas J. Carroll presiding. Fred Pheasey and his wife were official witnesses to the quiet ceremony, and they gave a private reception for the couple at their home near the church. Kinney and Alice McLean, Oscar and Ada Carpenter, Charley and Naomi Sumner, and a few others made up the wedding party. Pledging to always protect his wife from his enemies, St. John kept the news out of the papers. Too poor for a honeymoon, he promised to take her one day to San Jose, where his Irish mother resided. He and Clara simply celebrated the opening of the Miners' Union Hospital that week, one of the many achievements that filled Clara with unfathomable admiration for the twenty-six-year-old man she married.

The four-story brick and stone hospital was the pride of the town. Even the *Journal* deigned to praise the achievement of the Miners' Union, yet pointed out that the triumph all belonged to Dr. E. D. Allen, without mentioning St. John or Carpenter. With cold-blooded calculation Curry also inflicted discord, claiming when the hospital committee recently elected Dr. J. G. Sheldon as surgeon-in-charge, instead of Dr. Allen, that Allen was disgruntled.[52] This was false. Dr. Allen was the hospital's general physician. The *Examiner* dedicated a front-page spread to the grand opening, complete with photos of the building, the Miners' Union officers, the builder, and the two doctors in charge. He praised the work of the union, its officers,

The Miners' Union Hospital and union hall as it appeared in the newspapers on opening day.
Joseph E. Byers photo. Denver Public Library, Western History & Genealogy (#X90).

and their gratitude to Dr. Allen, and listed the numerous leading businessmen who furnished wards and private rooms, their names on brass plates on the doors. The surgery was equipped with the most modern sterilizing room, and the entire hospital was heated with hot water radiators, furnished with electric light, and served by hot and cold tap water throughout. The hospital had a full nursing staff under Miss Buchanan, with living quarters for the staff on the top floor. Harry Jardine was "steward," a position akin to warden. The ground floor provided permanent headquarters for Local 63, as well as the Federal and Carpenters' Unions, and included a large meeting hall. The miners' library and lounging room were also on the ground floor. Designed to accommodate forty patients or up to seventy-five in an emergency (twenty-one were already occupying beds), the hospital cost the union $32,000. It was indeed a "magnificent institution" and, as Sumner put it, "a most creditable thing for a single organization to so successfully carry out."[53] Inexplicably, the anti-labor gang denigrated the achievement, but no amount of manure could smother the town's enthusiasm.[54]

On the weekend, and still riding the wings of the hospital's grand opening, St. John attended the local Democratic rally, getting permission from chairman Kinney McLean for the Socialists to ask the candidates questions. Senator Buckley balked at the presence of the Socialists, believing they were the anarchists so often portrayed by the press, but here was St. John, the young union president he trusted and admired, and after some discussion and ground rules, Buckley consented.[55] With straightforward inquiries, the Socialists' sought to determine where the Democrats stood on issues relevant to workingmen and women. The *Journal* painted them as intrusive gatecrashers "seeking to convey an idea of their importance" and their questions made them "appear quite ridiculous."[56] Then the newspaper admitted it was not present at the rally, which made its entire report hearsay. The *Examiner* had a reporter present, and published the letter from the Socialists asking for permission to question the candidates, proving the *Journal* in error once again. The *Examiner* also published the actual questions and responses. A few inquiries caused distress in the candidates, particularly in E. C. Stimson. When Stimson was asked if he was the attorney for the Colorado MOA, "the effect was electrical," according to the *Examiner* (October 18, 1902). "For a few minutes it was believed he would tear up a suit of clothes and become real irate.... He of course, made a flat denial in the face of the facts."

The Socialists were barred from the Republican rally chaired by A. M. Wrench, which took place at the Stubbs and Jakway Opera House. Gubernatorial candidate James Peabody was the featured speaker. In a frenzy of enthusiasm for Peabody, Wrench stirred up the crowd in "prolonged and vociferous" cheering.[57] Peabody's campaign had enormous financial backing from wealthy businessmen and corporations throughout the state, a foreboding signal for labor.

That November the election went to James H. Peabody, who planned an inauguration so splendid it would "eclipse all others," striking some as a "royal coronation."[58] Locally in San Miguel County, Peabody supporter Herschel Marcus Hogg, a rotund attorney, was elected as congressman, but William Buckley, a Democrat, was reelected as state senator. Socialist candidate Guy Miller had a surprising showing of nearly 700 votes.

Statewide all key offices went to Republicans, yet the voters overwhelmingly approved the eight-hour amendment. Peabody and his Republicans had no real intention of seeing such legislation become a reality, for corporate interests adamantly opposed it. His excuse was that he disliked "tinkering" with the constitution. Nor did he care to continue funding for the six-year-old State Board of Arbitration.[59] His election was buckshot in the back for the WFM and organized labor.

The ascendancy of Peabody and his Republicans pumped enough bravado into Curry to turn his innuendo into blatant lies. He claimed that the night before the election, St. John gave a political speech at a saloon in Ophir, the village where William Buckley's mining company was located. Curry claimed St. John dumped scorn on the senator, showing total ingratitude and disrespect for Buckley, who had "almost stultified himself in behalf of St. John and his followers."[60] The lie was disproved, so Curry again condemned St. John and the union officers as the disturbing element, even when their actions were visibly generous, compassionate, and beneficial to the community. He wanted to prove they were criminals. As if in sudden realization of his most repulsive desire, he was given an appalling crime with which to destroy them, a crime that would rip open the very soul of Telluride.

In the icy obsidian hours of November 19, while the Smuggler and Pandora Mills pounded ore with a steady thudding roar, a sharp shotgun blast tore the night, shattered a window, splintered a chair, and ripped into the back of Arthur Lancelot Collins.

CHAPTER

15

A NOOSE FOR THE SAINT

"A deep distress hath humanized my soul."

— William Wordsworth

"I'm shot!" Arthur Collins gasped in disbelief as he sank forward against the card table, blood spilling from his mouth.

The shotgun blast had scattered window glass and splinters of Arthur's oak chair across the small table, scratching O. B. Kemp's cheek and Maude Blainey's hand. Momentarily stunned, the party of four, which included Maude's husband, A. B. Blainey, froze in horror. Seconds later, Blainey and Kemp had rushed outside and Maude helped Arthur to the couch, blood now gushing from the manager's back.

Arthur Collins and his friends had been playing bridge whist in the sitting room of the Smuggler-Union's main office and manager's residence Wednesday evening, November 19. Seated at a very small square card table about eight feet from the window, Collins had occupied a heavy oak chair, his back to the window. Directly across from him on the couch against the wall sat Maude. Her husband, the Canadian-born chief electrical engineer of the Smuggler company, sat in a chair on her left. On Maude's right was Kemp, the company's chief clerk and bookkeeper. The shade on the window was pulled down only halfway, Collins having forgotten to pull it all the way, as was his custom. He had returned to Telluride just five days earlier, and was not yet back into his Telluride habits.

Outside, the porch light was on and a floodlight shone brightly from a tower about twenty-five feet from the porch. The Smuggler Mill was going full blast 150 feet to the east of the house. Up a hill and 400 feet to the west, the Pandora Mill was hammering away. The cyanide plant, 160 feet to the south, was also running.[1] Numerous employees were in the mills, plant, and mine, and a few were on the trails when the assassin struck between 9:30 and 10:00 that night. Kemp and Blainey saw no one when they rushed out on the porch. The night was black and the tower light deepened the shadows near the house. They said they actually hadn't heard the blast of the shotgun due to the constant noise of the mills, but heard only the shattering of the glass.

Blainey ran up the hill a few yards while Kemp returned inside. Using the office telephone, he called Deputy Sheriff Bill Beattie, who was up at the Bullion Tunnel. He also spoke with Charles Becker, the mine superintendent, who was at the tunnel. He then telephoned town for a physician and Undersheriff Gus Roth (Sheriff Rutan was in Grand Junction).[2] Kemp also notified the company president, B. B. Lawrence, in Denver. Shortly thereafter Kemp or Blainey telephoned Arthur's friends at the Telluride Club. The terrible news hit town at 10:00 and the undersheriff and his men, three doctors, and numerous citizens, including Nate Mansfield and Senator Buckley, rushed the two miles east to the manager's residence.

At the moment of the blast, Arthur's other brother-in-law, Theodore Becker, company mill assayer, and Harold P. Smith, assistant electrician, were walking back to the Smuggler residence from town. They were a little over a half-mile from the residence, near the powder houses on the west side of the village of Pandora, when they heard the shot. They speculated to one another that it was an odd hour for someone to be shooting, noting it "was too sharp for a [mine] blast, and not sharp enough for a rifle shot."[3] Within the short time it takes to travel in the dark along the railroad bed another 200 feet, they met a man carrying something long and narrow. As the darkness was nearly engulfing, they didn't see this man until they were nearly on top of one another. The man stepped off the track and let them pass, then went on his way. They thought nothing of him until they reached the Smuggler residence and found Arthur prostrate on the couch and bleeding profusely from a shotgun blast.[4] Deciding the man who passed them was carrying "a shotgun in a scabbard,"[5] they telephoned the sheriff's office to apprehend the man, but no one coming from town saw any such person.

Two miners coming down the Bullion Tunnel trail around the time of the shooting said they glimpsed someone running out of the floodlight's glare near the house. This man appeared to have what looked like a short-barreled or sawed-off shotgun.[6]

Earliest on the scene were Drs. Ochsner, Sheldon, and Smith, the latter two surgeons from the Miners' Hospital with their surgical nurse, Lena Strauss. Collins told them he "did not know who" his assailant was or "why anyone would wish to shoot" him.[7] He wanted his wife at his side. They examined Arthur, finding one lead shot had gone through his arm; four into his back, piercing the lower section of his lungs; one shredding his stomach and liver; and one lacerating a kidney. They administered opiates and moved him upstairs to his own bed. If they had any chance to save his life, surgery would be necessary. They began the risky operation immediately, completing their work at 3:00 in the morning. They removed only three of the seven pellets. They held little hope their patient would live, but wished to keep him alive at least long enough to see his wife one last time.

Arthur's wife, Marguerite, and their two boys, ages three and five, were at their Denver house when informed Arthur had been "injured in a mine accident."[8] Marguerite was told to go to Telluride as quickly as possible. A special train — engine and Pullman coach — was immediately arranged, leaving at 1:15 that morning with Marguerite Collins, B. B. Lawrence, company president; Jacob Fillius, corporate attorney; one or two Pinkertons, and three of Denver's best surgeons aboard. It would cover the 422 miles of track, standard gauge to narrow, in a record-breaking thirteen hours and fifteen minutes.

Upon reaching the Smuggler residence, Marguerite asked why her husband had been up at the mine at such an hour. Maude Blainey gently informed her that an accident hadn't occurred, but that an unknown assailant had shot Arthur in the back. "[He] cannot live twenty-four hours," Maude told her, repeating what the doctors had said.[9] The truth prostrated the young mother, but she rallied for her husband's sake and remained hopeful and encouraging in his presence.

Under the sedation of opiates since the arrival of the local doctors, Arthur seemed to improve slightly after surgery.[10] He was resting easy and appeared to be relieved to have his wife, brothers, and friends at his bedside. Edgar Collins resided in the same house with Arthur. George had come over the range from Silverton on horseback and foot in a desperate effort to reach his brother's side before it was too late. Arthur remained "conscious, but talks very little," was the report, "not even saying who his assassin might be, and expressing no anger or malice."[11]

Internal bleeding could not be stopped, but the patient was rational until about 9:00 Thursday evening, "and confident of recovery, as he was not suffering in any way," according to Dr. Ochsner's report. Around midnight "he became very restless and from that time on failed rapidly."[12] Arthur Lancelot Collins poured out his soul at 1:43 in the frosty hours Friday morning, November 21.[13] Marguerite collapsed in uncontrollable sobbing; Arthur's brothers were crushed with heart-wrenching sorrow. Their grief so overwhelming, none at Arthur's side had strength to inform anyone in town of his death until after sunrise.

From the moment news of the shooting had reached town Wednesday night, law officers from both the marshal's and sheriff's offices acted swiftly, setting armed guards on all roads and trails leading out of town and over the range, and searching every street for signs of a man with a sawed-off shotgun. They searched the Smuggler property for clues, but the ground, rocky and snowless, showed no boot prints. Upon examination of the window, they found thirteen distinct holes, twelve in the glass and one in the wooden frame, all within a tight radius of eight to ten inches. In the room where Arthur and his friends had been playing cards, a rocking chair with a wood frame and reed back sat near the window. Two pellets had gone through the rocker's back and two glanced off the frame and struck a piano, falling to the floor. Seven pellets had entered Collins's body and two more appeared to be embedded in his chair. The twelfth pellet was missing (some reports said Collins was hit by eight pellets). Officers concluded the thirteenth hole in the window glass was from the wad, which meant the killer was no further than twenty feet from the window, for had he been further, the wad would not have had enough force to go through the glass.

About fifteen feet from the porch, in the shadow of a tall spruce, was a stump. Officers surmised this was where the shooter stood. It was directly in line with the window and afforded a view of the card party. The tower light was behind the spruce, creating a shadow to cloak the assassin while at the stump. Beyond the spruce a trail abruptly went up the hill, leading to the Pandora Mill. The morning after the shooting, J. R. Hogan, foreman of the freight department, and an engineer named Smith found an empty shotgun cartridge on this trail, sixty-two feet from the house. It was a twelve-gauge, smokeless powder shell, which had carried a twelve-pellet load of "No. 3 buckshot" (equivalent to the modern double-0 buckshot). According to Kemp, this empty shell "the assassin had evidently thrown away as he ran up the hill."[14] Believing the shooter went up hill indicated Kemp and the officers were already concluding the assassin was the same man who passed Theodore Becker in Pandora, merely because Becker had seen a man with what was now decided to be a shotgun. No one seemed to realize this assassin had to walk in the night's blackness along a narrow-gauge track for a half-mile in the time it took Becker to walk 200 feet. According to Becker, the man was not running. In general, criminals take the path of least resistance — running downhill in preference to up.

The only other scrap of evidence was the discovery that twelve-gauge smokeless powder shotgun shells had been sent to Ophir from a Telluride merchant on Sunday, November 16. This was meaningless. The Winchester Model 1897 twelve-gauge, pump-action shotgun was the most popular in the West.[15] Numerous hunters used it. Lawmen and gunmen were especially fond of it with a shortened barrel and no choke, affording quicker draw and certainty of "doing up" the target. Both Bob Meldrum and Willard Runnels, the illegally deputized gunmen, carried

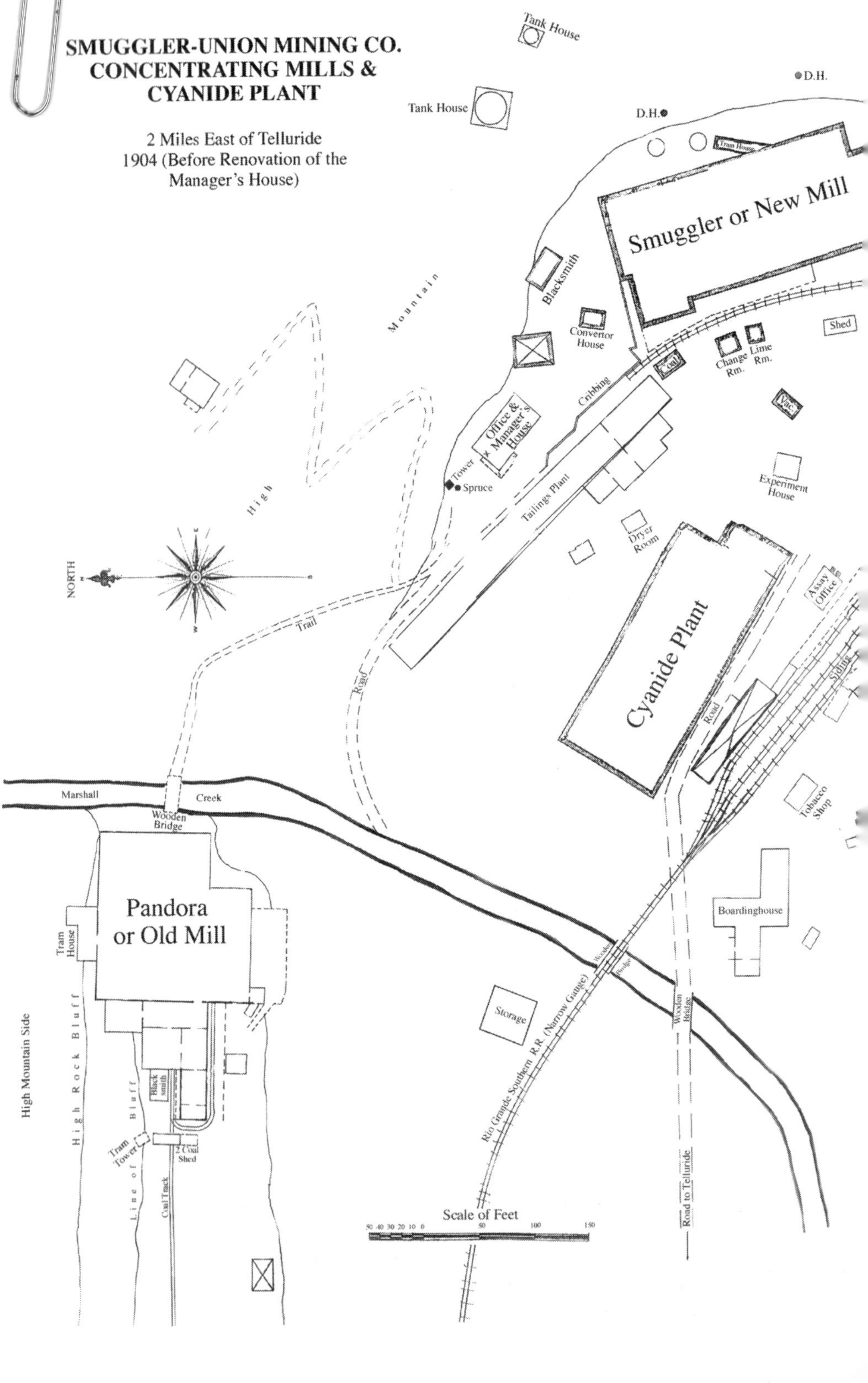
SMUGGLER-UNION MINING CO.
CONCENTRATING MILLS &
CYANIDE PLANT
2 Miles East of Telluride
1904 (Before Renovation of the
Manager's House)
Tank House
Tank House
D.H.
D.H.
Tram House
Smuggler or New Mill
Blacksmith
Convertor House
Shed
Coal
Change Rm.
Lime Rm.
Cribbing
Office & Manager's House
Vac.
Tower
Spruce
Tailings Plant
Experiment House
Dryer Room
High Mountain
NORTH
Trail
Road
Cyanide Plant
Assay Office
Siding
Road
Tobacco Shop
Marshall
Creek
Wooden Bridge
Pandora
or Old Mill
Tram House
Boardinghouse
Wooden Bridge
Storage
Wooden Bridge
Rio Grande Southern R.R. (Narrow Gauge)
High Mountain Side
High Rock Bluff
Black smith
Line of Bluff
Tram Tower
2 Coal Shed
Coal Track
Scale of Feet
50 40 30 20 10 0
50
100
150
Road to Telluride

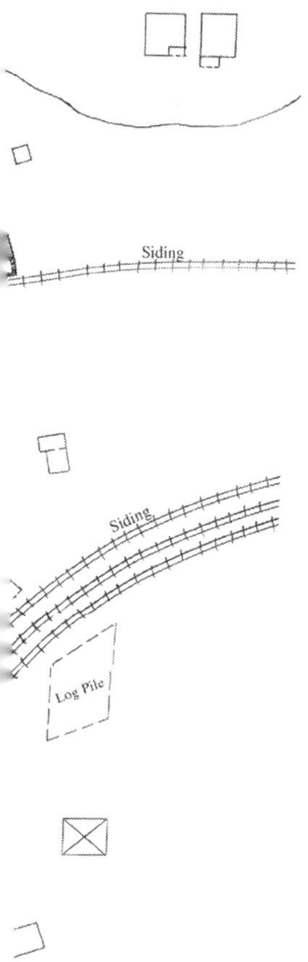

sawed-off shotguns while on horseback, as did numerous cowboys in the county. Runnels proudly posed with his for the *Denver Post*. Anyone with such a gun in the district could be suspect. The difficulty in the Collins assassination was the lack of clear motive.

Although the enemies of labor immediately and conveniently jumped to the conclusion that the strike of 1901 led up to the assassination, which meant the Miners' Union was responsible, those thinking logically quickly pointed out "the strike had been settled for more than sixteen months."[16] Most newsmen echoed the first sentiment of the *Denver Republican*, November 20, 1902: "There has been no trouble with the men at the mine of late and the friends of Mr. Collins state that feeling between employees and the management was never so cordial as now. Motive for the crime is unexplained."

Any union man knew such a crime or even the attempt of it would destroy the Miners' Union faster than any state militia could. The union had nothing to gain and everything to lose by the murder of a mine manager. A motive for the killing was more likely personal enmity, perhaps in revenge for an imaginary wrong or a recent discharge. Some actually surmised the motive might have sprung from the Bullion Tunnel fire deaths. Several fire lawsuits were still pending; some already had been lost. Many in town noted this was the anniversary of the fire, although the actual date of the fire was November 20, 1901. The shooting occurred on November 19, but both the fire and the murder fell on the third Wednesday in November. Counting anniversaries this way was fairly common to those in that era who were either illiterate or of a different culture. Because Collins had left the shade up only one night by chance, it seemed the murder was almost random in choice of victim, or at least a spur-of-the-moment decision.

When Arthur's body was taken to the train on Friday, more than one hundred Smuggler employees, union and nonunion alike, escorted the coffin in silent reverence. All the mines closed that day out of respect. His body was shipped back to Denver and buried in Fairmount Cemetery. Stunned and horrified, Telluride was bent over in confusion and disbelief. But this death did not create holy ground, for where sorrow should have been, descended black-winged, ravenous hatred.

The Denver Smuggler-Union Company office hired Pinkerton agents to ferret out the assassin, and they swarmed into town disguised as miners in slouch hats and rough clothes, instantly noticed by the locals. The Pinkerton assertion, that an evil inner circle in the Miners' Union committed such crimes, was spread to every corner of the state. Newsmen hoisted the concept and heralded it in sheets covered with raving ink. The murder allowed Curry and editors like him to declare open season on St. John and labor, socialism, and anyone connected to either, even though St. John was horrified by the assassination. He and the WFM condemned the murder as fervently as everyone else did.

Curry paraded Arthur Collins's murder as the most monstrous crime in the history of Telluride, merely because Collins was "one of the most prominent and highly respected citizens of the community."[17] Anyone who had been gunned down in cold blood previously, including Town Marshal Jim Clark and the unarmed John Barthell, were insignificant because they were not prominent citizens. Curry made Collins more important than he was. Most people knew the manager "had never associated to any great extent with the people of Telluride, and he was personally known to but a few" local citizens.[18] His residence was in Denver at 1337 Gilpin Street. He was away from Telluride frequently on business to other cities and states. Even Kemp admitted Collins's first interest was to the shareholders of the company and that he "might have been a little more politic" in his dealings with employees. "He was a strict disciplinarian and always demanded and insisted upon what he thought was right from the men under him."[19] The substance of the man was in his ability to make a mine turn a profit. The best thing Benjamin Lawrence could say in memory of him, in addition to his being a good husband and father, was that he died "a martyr to his duty."[20] The president of the state MOA eulogized Collins in much the same words: he died a "martyr to the...courageous discharge of his duty." A British mining journal went a step further and said he "died the death of a hero, fighting the battle of freedom of contract."[22] In an effort to give him more soul than just his finesse as a mining engineer and manager, Sumner wrote, "Although a man of pleasing personality and address, those who knew and understood him readily appreciate where he could make the most bitter of enemies.... Socially and personally he was inclined to reserve, and seldom moved with the people of this community...few were acquainted with him. He was a pleasant conversationalist and affable to a degree."[23]

Willard Runnels poses with his sawed-off shotgun.

Restored portrait © M. J. Martin 2003.

Yet this near-stranger's death became central to life in Telluride and if it was not avenged, and the "assassins run to ground," then Telluride could "expect a Molly Maguire campaign such as terrorized the Coeur d'Alene region."[24] The "Molly Maguire" phrase stank of Pinkerton and James McParland. The real push behind the "hunting down of the inner circle responsible for such assassinations"[25] came from that quarter. The concept of an inner circle flowed from McParland's DNA. Born in Ireland March 22, 1844, McParland grew up with the terrifying sagas of malevolent secret societies that kept English landlords or political rivals in check via murder and fear. These Irish societies had dual layers: the trusted "inner circle" members who planned outrages, their foot soldiers who carried them out, and the ordinary members who provided funds.[26] McParland's goblins and politics came with him to America, where he became involved with the Pinkertons when he was in his twenties. His undercover assignment in the coalfields of Pennsylvania convinced him the AOH members were the Molly Maguires of Ireland, even though their charter said they were the AOH and he joined them as a member of that organization. Now he was convinced the WFM was like the Molly Maguires, and the state's anti-labor editors took up the cry: Collins was killed "by emissaries of the Miners' Union of Telluride."[27] No one intended to look for a solitary assassin with a personal motive because the Pinkertons had already decided the Inner Circle (now bloated to upper case importance) was guilty of the crime. That meant St. John and Carpenter, because these two were said to be guilty of making numerous men disappear from the region. That also meant Arthur Collins died for an insignificant belief in a contract system that had nearly vanished from use at the mine. A monster was salivating at the door.

Although Curry civilly claimed "local organizations are not responsible,"[28] he filled the *Journal* with the howlings of the most rabid editors in the state, including the rancorous Dave Day who said the crime "must be charged to the Miners' Union."[29] The *Ouray Herald* bayed it was done by the "murderous gang that murdered Barney and Smith...a ring within a ring, thoroughly organized, and reminds us of the Ku Klux of rebellion days."[30] Some editors dared to name St. John as the assassin. Reflecting the more conscientious citizens, Sumner pointed out, "Looking at everything in a careful light, it would appear more like the work of some disreputable scoundrel who had a rankling enmity for the manager, and who saw an opportunity to escape by having his crime charged to others who, unfortunately, at one time disagreed with the company of which the deceased was manager. In the face of present peaceful conditions this seems to be the only logical conclusion."[31]

Logic was scarcely a function of those who saw this murder as their excuse for annihilating St. John and Local 63. The prominent citizens of Telluride gathered behind locked doors at the Telluride Club on Friday afternoon while Arthur's embalmed corpse was en route to Denver. Edward Howe, currently the county attorney, had already orchestrated a reward for the assassin's apprehension, but it was not enough. Addison Wrench wanted blood.

The meeting at the Telluride Club was informal, convulsed with hatred and punctuated by demands for vengeance on those who stood accused without foundation. Some called for a vigilance committee to rid the town of criminals, and a few screamed for St. John's body hanging from the nearest telephone pole. It was no wonder reporters were unable to discover where the union president went. Other businessmen demanded that the Miners' Union be driven from town, for it had obviously conspired to employ a man-killer to assassinate Collins. Proof of this was the union's murder of Barney and Smith, they railed. The rewards for the two missing men were suddenly advertised in both local papers, the total now $6,000 — $1,500 for each corpse and for each conviction. The longer they were "dead," the more they were worth. Wrench raged that as long as the criminals were allowed to operate, neither life nor property would be safe. It didn't matter that the TBA and MOA's pet cowboy, Deputy Walt Kenley, had recently shot off a man's finger in an argument at Placerville and now was charged with attempted murder. He was out on a very low bond and still working as a deputy.

Fortunately, rational men like Senator Buckley, John Herron, and the recently returned Nate Mansfield, were present in the obsessed crowd of leading businessmen, and were able to bring reason into the picture. They proposed a larger county reward than the existing $2,500, believing huge sums would induce the best detectives in the nation to hunt down the assassin. They also proposed that Howe file a petition with District Court Judge Stevens to empanel a grand jury to investigate the Collins murder as well as all the other murders and lawlessness in the region. This was agreed upon, with thirty-three men signing the document, including Wrench and Mansfield. Many, however, still wanted a hanging.[32]

The Smuggler-Union Company, represented by Charles Becker, was to be shut down and possibly never be reopened as a punishment to the community. Word from the Denver office claimed too many Telluride citizens "are too indifferent to the murder of its mine manager," and the company resolved to use the "most stringent measures in punishing the miners as a class." [33] Shutting down began that afternoon, with the stock animals removed, and men told to come collect their pay. The Smuggler was "one of the least important holdings" of the parent company and a stockholder informed the *Journal*, "If we are not allowed to go ahead unmolested, we shall shut down the property and abandon it...possibly permanently." [34] The company pushed for other large mines in the district to shut down as well, suggesting they throw 1,000 miners out of work in retaliation. Sumner reminded the naysayers that the mine was shut down for a mere three days when twenty-four men died in the fire because of inexperienced management. Where was justice?

The TMA held a meeting to discuss a general shutdown. Profit loomed larger than vengeance. None of the mines would close. It served no purpose since nothing remotely pointed to the miners as killers. General Frank Hall, special mining columnist for the *Denver Post*, remarked, "It would be absurd to charge the Miners' Union with this murder.... It was the despicable deed of an individual, not of an organization." [35] But Theodore Becker, while in Denver, cast the miners in an ugly light by claiming the men, "foreigners mostly" showed no sign of sorrow or regret at Arthur's death.[36] Becker had not even been among the miners during those dark hours. He simply was striking out against the outrage that had no focus.

Fueling the cry for blood, Theodore's older brother, Charles Becker, informed the press that Collins had received "anonymous notes warning him to" leave Telluride that were signed with "a crude drawing of a stiletto." [37] It was well known Collins disliked Italians, union or otherwise. Initial reports of the death notes claimed two or three had been sent. When the anti-labor editors acquired this scrap of carrion, the notes became "numerous," and instead of having been sent in the "last few weeks," they were sent constantly since the strike of 1901. Additionally, they claimed many mine managers in the district were in receipt of similar death threats. Sumner questioned the validity of this. He had had numerous conversations with Arthur Collins and had spoken specifically about such threats, because Collins had received a few in the heat of the strike in 1901. At that time, Sumner had also received death threats and likewise did St. John and several other people in town. Sumner had kept his and one of the St. John letters on file. In December 1901, Collins told Sumner he had not had any more anonymous notes for some time.[38] In an effort to determine if the report of recent death threats was legitimate, Sumner personally asked Kemp, a close friend of Collins as well as the company's chief bookkeeper, if he knew about the notes. Kemp had no knowledge whatsoever of such notes. Nor had he heard anything of the sort from Collins. It was possible Becker invented the tale.

Sumner also questioned the mine managers of the leading properties and they said they never received any anonymous letters threatening death. John Herron, manager of the Tomboy, told Denver reporters he had not "at any time had threats of any kind" toward him, and as far as the report went, "I do not believe there is anything at all in the statement." Senator Buckley, manager of the Ophir Consolidated, denounced the hearsay and said "publication of such reports are making matters, which are already serious enough, a great deal worse."[39] The only manager anyone could find who personally claimed he had a threatening note was Hallet at Silverton. Suddenly Curry came up with such a note and had to feed the flames by publishing it on December 4. This was allegedly to Manager Charles Hooper of Rico, twenty-nine miles distant, which was nicely dated November 22, telling him to leave town or "take your medicine, as Collins got" and was even amiably signed, "Yours truly."[40] The majority of local businessmen denounced it as a hoax. Unconcerned, Hooper said he had not seen it.

The rumor that St. John had fled in terror proved a disappointment to Wrench and his gang when Vint returned from Denver on Saturday, November 22, after consulting with the WFM's executive board. He was determined to protect Local 63 from the libel heaped on it and had the board's full support. Most of the town's citizens had cooled off and the cry for indiscriminate hangings subsided as a heavy snowstorm cloaked the San Juans. The latest demand from numerous editors and businessmen was for the Miners' Union to put up money for investigating the assassination, saying they owed it to the citizens to assist in bringing the criminal to justice. Strapped by the recent expenses of building the hospital, the union had no funds to contribute. In general, the members were so thoroughly disgusted with the vile accusations and extreme threats of unjustifiable vengeance that they were inclined toward indifference. Resolutions drawn up by Guy Miller and signed by the committee, St. John, and Carpenter were presented to the public, denouncing "the most damning insinuations against a part of the Telluride Miners' Union and the Miners' Union in general, implicating them in the assassination of Arthur L. Collins." They also condemned "the cowardly assassination of Arthur L. Collins as being wholly opposed to the policy and spirit of organized labor, as well as the laws of civilized society, and we further condemn the dastardly attempts to cover a community with opprobrium and our brothers with disgrace."[41] But they offered no financial support.

The county announced a guaranteed reward of $10,000 for the apprehension of the assassin. Governor Orman promised a $2,000 reward from the state and the Smuggler-Union Company offered a secret reward of $10,000 to $20,000. The company said it did not "believe in offering public rewards...it believes in hunting this class down without noise and bluster." They preferred to hire the "shrewdest criminal hunters" available and they were certain the enormous reward money made failure impossible. "Immunity from punishment will probably be guaranteed to the accomplice who turns state's evidence, as the company most desires to hang the man who did the actual killing."[42] Nothing indicated an accomplice existed in this crime, but the company, on Pinkerton guidance, was certain St. John and the Miners' Union were guilty. This was the secret information reserved for the man-hunters. It guaranteed the real assassin a walk to freedom.

While Telluride scrambled to regain her balance, dancing around the host of comically disguised Pinkertons and numerous reporters picking over the carrion, a "special" reporter blew into town. Known by the penname Polly Pry, the woman was the *Denver Post*'s scandal columnist, a haughty meddler whose real name was Mrs. Lionel R. Anthony. "I have not hesitated to attack the great or ridicule the mighty,"[43] she proudly proclaimed. On Sunday, November 23, she interviewed O. B. Kemp, temporarily superintendent of the Smuggler during its shutdown operations. She had him recount the details of Arthur Collins's murder. He showed her the bloodstained, splintered chair. He escorted her around the plant and told her how Will Barney went missing. He praised Collins's contract system and let it be known that the squalor in the mine bunkhouses was due to the disgusting habits of the miners and not the company.

Armed with Kemp's inside knowledge and a copy of the *Miners' Magazine* from September 1901, Polly Pry then went to see the president of the Miners' Union at the office in the new hospital. She had requested an interview in the morning, and at 3:00 in the afternoon received a reply stating St. John was available for only a few moments. She disliked the cold desolation of this mountain town and "thanked God" she had not been "condemned to such a place." She thought it "oppressive beyond words, a fit stage for the terrible drama of senseless hate."[44] Polly was all too ready to add to that senseless hate.

"I was still trembling with the memory of that cowardly murder when I shook the snow from my clothes and knocked at the door of the union headquarters." She was already predisposed to meeting a monster since she filled several paragraphs describing Telluride as if it were in the Dark Ages and depraved gangs were whisking away men on a daily basis and sending death letters to others. She wrote, as if on authority, "That the assassin of Mr. Collins is connected with the Miners' Union of Telluride in some way no one can doubt." Vincent St. John was about to be cooked.

Polly Pry served up her questions with customary sarcasm and polished ridicule, condescending to his youthfulness. In his usual candid manner he answered her, unaware of her goal. She prodded him, asking his opinion of the causes that led to the assassination of Collins. "I have no opinion to express," he replied. "So far as I know, there was no cause." She pushed for evidence of enmity between the union and Collins. He said there was none. She asked if Collins was "very objectionable to the miners." He replied, "Not any more than he made himself. If a man decreases the earning power of his men, he is liable to make himself objectionable."[45] She probed him on the mysterious disappearances in the camp. He told her Barney had not disappeared but was involved in a divorce. He knew nothing of Wes Smith because Smith was not involved with the union in any way. He then showed her through the hospital (which was of no interest to her[46]) and saw her out, returning to his business without a thought about her.

The following day St. John went to Denver again to confer with WFM Secretary Haywood on business matters. He was still in Denver on Thursday, November 27, 1902, when he saw the *Post*'s front-page banner headline: COLLINS MURDER NOT THE END OF THE SAN JUAN OUTLAWRY. Just below it startling words proclaimed, "'Firebrand' St. John's Pronunciamento" and a quote contributed to him about "removing" men from Telluride. The young man was thunderstruck. As he read the so-called interview between him and Polly Pry, he became aghast, then angered by the deliberate lies.

Polly painted him as too rude to even lift his hat or put aside his small pipe for a lady. Describing him first as a "boyish-looking individual" who interrupted her, she then expanded it to, "A slender little fellow...roughly clad...a smooth-shaven face, a thin, sarcastic-looking mouth, a straight, cruel nose, and a pair of steady dark gray eyes of a peculiarly penetrating character.... In his gait as in his appearance he looks the typical anarchist. Alert, intelligent, magnetic, and dangerous. A youth to lead men, a man to fear." Oddly, the *Post* published a photograph (November 27) of St. John that had no resemblance to Polly's description. The young man was in his "Sunday best," looking dignified and boyishly handsome, his thick dark brown hair combed back, his eyes large, gentle, and shadowed by sorrow, his nose neither straight nor "cruel," and his mouth full-lipped. The caption claimed he "asserted that more 'removals' were needed in Telluride."

Vincent St. John. Portrait created from the Denver Post, *November 1902 photo, believed to be his wedding photo, but used to illustrate Polly Pry's sinister tale of him.*
Restored portrait © M. J. Martin 2003.

Filling column after column with quotes from her alleged anarchist, Polly wore herself out, wrapping up, "In fact, he made the regulation anarchistic harangue until I asked: 'How old are you, Mr. St. John?'"

"I am nearly 26," Polly had him say with a "flush" that November in 1902. Her anarchist didn't even know his age, for Vint was actually well on his way to twenty-seven. He would never forget his twenty-fifth birthday July 6, 1901. She also had him say he went to school last in Pueblo and before that in New Jersey, when he actually last attended school in San Jose, California, and prior to that in Seattle. Polly was so obviously repeating information from a source other than the man she interviewed, that it appeared imaginary. She even insinuated he had become president of the union because it was "rather more profitable" than being a miner, which was utterly false.

Her lies twisted into malicious libel when she claimed he told her, "There are two or three more people in this place that ought to be removed, and then we'd get along all right." Supposedly she asked the names of these and, like a true guarded assassin, he took her into his confidence, naming his next victims: Painter, Wrench, and Van Atta. Such statements certified him and the union he represented as murderers in progress.

She followed with paragraphs about Will Barney as a "miner" who was last "seen to enter a stable" and was never heard from again; and Wes Smith, supposedly "one of the famous guards" in Coeur d'Alene, who had come to Telluride and was "persecuted in every way possible" until he "joined the union."[47] Then he disappeared.

Polly's Barney and Smith tales were used only to establish the union's earlier "removals." Telluride businessmen who knew St. John said her "inflammable gush" was invented and suggested she had been speaking with Wrench, Painter, and others before she wrote it. Pheasey, Buckley, Alderman Bob Rogers, Marshal McIntosh, and several others pointed out, "St. John would think a long time before he formed the ideas credited to him by Polly, and he would think a blamed sight longer before he thought he would say anything about them to anybody."[48]

Chortling in delight, Sumner said, "St. John made a mistake by not opening the interview with a small bottle and a cold bird." He could have charmed her since "he looked good to Polly, for she remarked that he was magnetic" and she was fascinated by his "penetrating" eyes. "But this is where St. John went to the bad — he lit his pipe, which was filled with nitro glycerin and lyddite, settled back on a box of dynamite...and unfolded to her trusting ear a deep and dark and terrible plot of how he was going to change the scenery about Telluride, tie the townsite up in a sack and dump it into the creek."[49]

Harry Jardine guffawed over the "anarchistic gait" and wondered if it was anything like a duck's waddle. He said Polly expected St. John to be hypnotized by the mere announcement of her name, and when he wasn't, her "vanity was slighted." He was sure St. John went wrong by not wearing a "claw-hammer coat and a vest that would have exposed an acre of starched shirt front."[50]

Far less amused, St. John, Carpenter, Guy Miller, and others, including the WFM leadership, found the Pry article appalling because it wasn't merely a personal misrepresentation of St. John, but a libelous and malicious attack on him and the union. He was advised to seek a retraction from the *Post* or sue for libel. St. John fortunately had had four witnesses present during the entire interview, including President W. A. Jackson of the Federal Union who shared the same office. The witnesses signed sworn statements that St. John never said anything about removing anyone, and that all but two sentences in the published interview were false.[51]

Polly Pry insisted she was correct in every word. The *Post* stood by her, offering St. John and WFM Secretary Haywood an opportunity to answer the story. The *Post* continued filling issue after issue with Polly's condemnation of the Telluride Miners' Union. She claimed Telluride was in a state of terror, paralyzed to do anything against these Molly Maguires, "its merchants in receipt of death notices, its leading banker and its only newspaper boycotted."[52] Nate Mansfield was none too pleased, since his bank was on an equal footing with Wrench's. Neither was boycotted nor could any merchant be discovered who had death notices. Curry was delighted the *Journal* was the "only" newspaper — the *Examiner* was a turnip. The *Post* ran photos of Painter and Curry under the heading "Two of Telluride's alleged 'marked' men." While in Denver, Painter made a statement to the *Post* that he had no fear of St. John. "I have never been threatened before by St. John or any other member of the Miners' Union," he said. These newly revealed threats were not going to drive him from the town he loved, where he and his wife had raised several children.[53] Apparently the reign wasn't very terrifying and the Firebrand not too hot. Painter was disgruntled the boycott hurt his paper, yet he knew St. John and the Miners' Union had followed the letter of the law in the boycott. He found no legal means to stop the union, now the biggest local in the WFM, second to Butte.

The *Post*'s sensationalism created a free-for-all in the anti-labor press, the screams against St. John and the union becoming vicious again, one headline demanding, "Kill this viper."[54] The *Rico News* sneered, "St. John's talk cannot be considered as 'nervy.' It is the utterance of a bigoted ignoramus."[55] The "bigoted ignoramus" term was more fitting for Polly Pry. Had St. John been as stupid as she portrayed, the union would have run him out long ago. The press published rumors the union would do just that or ask him to resign. Once again, the truth had escaped the press. Local 63 not only stood by him, but also said they would elect him again when it was time. Some members did regret that he had allowed the interview; St. John regretted it, too. He'd have to learn to trust less.

The *Rocky Mountain News* told him they would be pleased to set the record straight. Sick of reporters, he said he would answer questions only if guaranteed a preview of the article before it was printed. That was a courtesy given to most interviewees, and it was readily granted. St. John bluntly told the *News*, "The miners' unions all deplore the taking off of Mr. Collins, and would never countenance violence upon anyone. I have never advocated the removal of anyone from Telluride, and such statements as attributed to me I never made."[56] He explained he had mentioned Painter, Curry, Wrench, and Van Atta's names to Polly when he told her these men and their clique had circulated reports about so-called crimes and disappearances to discredit the Miners' Union. He had told Polly actual brutal crimes had occurred in Telluride, but no one did anything about them. By way of example he had mentioned how Marshal Jim Clark was gunned down in the street at night while on his way home with August "Mexican Sam" Baral in 1895. Several prominent businessmen of the day didn't want Clark's murder investigated and they told "Doc" Shores, who was asking too many questions, to leave town. Some of these men were still running things, such as Wrench, Painter, Hilgenhaus, and Curry.

Polly hadn't bothered printing the Clark story, but she did admit St. John told it to her. Flinging himself on the tale as if to throttle rising ghosts, Curry denied he was in Telluride at the time of Clark's death, and that there were no parallels between the murder of Clark and that of Collins. He painted Marshal Clark as an evil desperado who had been "implicated in stage hold-ups, bank robberies, etc.... He had 'it in' for several prominent citizens of the town because he had failed of re-election as city marshal.... Threatened citizens felt a relief when he was gone."[57]

Curry lied. Clark was no desperado. Oscar Downtain knew the truth. He had met Jim Clark in 1870 at Ft. Lyon when Clark was a cowboy. He was friends with him and was a town trustee when Clark was gunned down. The tales that Clark was an outlaw came from a *Rocky Mountain News* article at the time of his death, claiming he was actually the James Gang associate Jim Cummins.[58] Cummins refuted this simply by writing his memoirs years later.[59] Clark's friends in Telluride knew he wasn't using an alias, knew his aged mother, Margaret Clark, was still alive in Jackson County, Missouri, and had wired her about his death. Clark's enemies used every lie imaginable to blacken his name and make discovery of his killer impossible. They claimed he had used his position as marshal to allow Butch Cassidy's gang to escape after robbing the San Miguel Valley Bank on June 24, 1889, and was paid $2,000. Charles Painter was cashier of the bank. The lie stood pat despite the records showing Clark was appointed town marshal *after* the robbery, on July 11, 1889, *because* of such gangs. He was the only one tough enough to stand up to lawless cowboys in the new camp. Cyrus W. "Doc" Shores, ex-sheriff of Gunnison, and at the time of Clark's death, supervisor of the Denver Rio Grande Railroad's detectives, said, "I suppose many mean things will be said about him, now that he's dead." Shores said Clark was tough, brave, and "the most fearless man that I ever knew." He couldn't have been killed in an open fight, Shores said; someone would have to ambush him. He had a soft side for stray dogs and little children, who loved him. Because he supported his mother in her old age, he developed a tender concern for old widows.[60] The deceased

The photo of Marshal Jim Clark (right, courtesy © E. N. Wright; used with permission) *was the obvious source of the drawing (left) used in the* Rocky Mountain News *at the time of his death. Although the* News *printed a completely fabricated tale of Clark as the outlaw Jim Cummins, the accurate drawing accompanying the story gave credence to the lies.*

Clark could not defend his name; the repulsive tales making him an outlaw stood. Downtain remained strangely silent, as did his other friends in Telluride, including Mexican Sam Baral, who worked at the Liberty Bell Mill. Was it fear? At the time of Clark's death, Cal Rutan was sheriff and rumors flew that city officials were deeply involved in the murder.[61] Everything was put out of the way when Clark's corpse was put into the ground and his memory trampled. His murder being brought up by St. John rattled a few cages. Something dark and ugly was hiding in the shadows of Telluride, and it wasn't the Miners' Union.

Wrench's grand jury had no intention of looking into the Clark murder. Ever. District Judge Theron Stevens had granted the petition for a special grand jury and Sheriff Rutan "was given his own discretion in selecting men to serve on the jury."[62] Addison M. Wrench just happened to be one of them. The grand jury was sworn in on Monday, December 1, and given instructions by the court "to investigate any matters" brought before it, particularly those cases "called to the attention of the court" by the petitioners: the disappearances of W. J. Barney and Wesley E. Smith (note John Mahoney's name was no longer on this list), the riot of July 1901 at the Smuggler-Union properties, the assault on James Phillips, and the assassination of Arthur L. Collins.[63] Clark's murder, John Barthell's slaying, Norbert Savignac's murder, even the death of nonunion trammer Joe Lujan, and the numerous assaults on union men by Hyde, Jordan, Robbins, and others were strangely left out of the grand jury's investigation list as were the brutal assaults by cowboys in the west end of the county and the slaughter of hundreds of valuable sheep, which sheep ranchers wanted investigated.[64] The intention was to investigate only those crimes Wrench wanted to hang on the union, despite the court's instructions.[65]

"I am satisfied," Judge Stevens said to the twelve jurors, "...that you will obey the obligation which you have taken that no one will be presented through prejudice, malice, ill will or hatred...."[66]

Curry was beside himself, proclaiming the jury an excellent one because every man on it was known to be anti-St. John, with the exception of Nate Mansfield and Frank Barnes (Barnes had just offered to loan the Miners' Union $5,000 to pay off the hospital construction costs and had been one of the businessmen who contributed hospital furnishings[67]). Although miners numbered above 1,800 in the county, not one miner was chosen as a juror. Not one friend of St. John, such as Alderman Rogers or Marshal McIntosh, was chosen. Ten of the jurors were members of the TBA.[68] Sheriff Rutan had done his work well. No prejudice? No malice? At least Mansfield was elected foreman, the only scrap of hope the courthouse wouldn't become a noose factory.

For the entire two weeks the grand jury was in session, many of the state's editors "indulged in a great deal of senseless rot, sharp criticism and prejudicial comment on the Collins murder, and have been very malicious in charging the Miners' Union with responsibility for the deed."[69] Local 63 was fixed in print as the Molly Maguires of the San Juan. The *Denver Post* kept the wounds bleeding by printing Polly Pry's inventions and the Pinkerton yammerings of the Inner Circle, although they did send a reliable reporter, Paul Thieman, to Telluride for official news reports.

Thieman managed to convince St. John to grant another *Post* (December 6, 1902) interview, although once again Vint requested a preview before the story was printed and had several witnesses on hand. Thieman told him he should be the one to put an end to the reign of terror in Telluride and Vint retorted, "It is a manufactured reign of terror." The man was growing weary of this persecution and battering of the Miners' Union. "The idea that there are some persons among us who are assassins is absolutely false."

St. John said he believed both Barney and Smith were alive, which meant there was nothing to validate any reign. How could two missing men be the sum of a reign? "No one has yet mentioned the peculiar possibilities of offering a large reward for the production of the men's corpses," he said. Why weren't rewards offered for their whereabouts? Someone might be inspired to kill them for the cash. Similarly, posting enormous rewards for Collins's murderer might produce false testimony that could convict an innocent man.

This situation, St. John told Thieman, "is a fight against organized labor, and everything that occurs which is wrong is ascribed to the Miners' Union for the purpose of harming the cause of labor.... We have no reason to apologize because we have not done wrong." They could only wait calmly for the results of the grand jury.

"They feel that the evidence being taken there is one-sided," Thieman reported.[70]

They were right.

The man conducting the investigation was Deputy District Attorney Howe, a lawyer retained by the TMA. Among the scores of witnesses examined, the only union miner called was Mose Farragher. The only merchants examined who were blue card union supporters were Abe Nathan, Billy Selby of the National Club, and Charles Peterson of the Brunswick Saloon. All the Smuggler-Union Company's management personnel were called, including Edgar Collins, Blainey and his wife, the Becker brothers, and Bob Wright. Most questions pertaining to the riot centered on the death of Ben Burnham and the firearms of the miners. No one mentioned how the company had supplied arms to nonunion employees. Jack Hyde, who loudly proclaimed his desire to "do up" miners and had committed assault after assault, was called as a witness against them. Likewise was Mel Robbins. Bill Jordan was left out of the picture completely. He was currently in Cortez, arrested for a brutal assault on R. R. Smith of that city and staring at time behind bars. His record of violence had to be avoided at all costs.

The Miners' Union saw the bloody writing on the wall. The men believed this very lopsided grand jury would produce wholesale indictments, and those indicted would all be members of Local 63.

They were right.

Vincent St. John was unruffled by these probable indictments. He knew none had any foundation or evidence to stand up in court. He told Thieman, "I am ready to face the charges. I know there has been talk about driving me and some other union men out of town. There is an organization here which would do it if they dared, but the fight on us will take the form of arrest for conspiracy."

He was right.

"I am going to stay right here to the end," said Vint, "and when everything is cleared up and I am ready to go, I will stay another six months for good measure."

"Everybody believes that he means what he says," Thieman wrote, "and I believe that he has great courage."[71]

CHAPTER

16

A PINKERTON ON EVERY CORNER

"Never interrupt your enemy when he is making a mistake."

— Napoleon Bonaparte

Courage and absolute patience were required of every union member in Telluride during the grand jury's inquisition. Although the court had sentenced Deputy Kenley to sixty days of hard labor in the county jail for closing an argument with his gun,[1] his counterparts — the deputized and shadowy gunmen, Meldrum, Runnels, and a man named Austin — were itching for the miners to furnish reasons for killing them. The TBA had swayed more merchants to join them and some were pushing to carry out Wrench's radical plans to rid the town of all union men. Some cowboys from the west end of the county swaggered into town, offering their services to Sheriff Rutan to "wipe out the Miners' Union" in much the same manner they had taken care of the sheep "problem." Slaughtering sheep and slaughtering miners came in the realm of entertainment, since the cowboys said no payment was necessary.[2]

Each time new witnesses emerged from the secret proceedings of the grand jury, rumors were unleashed on the streets like the devil's hounds seeking flesh, causing more frenzied denouncements. Some astute observers credited the Pinkerton spies as the source of the rumors. At this point the Pinkertons were the butt of jokes in town, but reporters quickly pointed out that the visible spies were to distract everyone from the real detectives on the job who were working secretly.

One dark rumor unleashed during the Wes Smith inquiries was designed to implicate St. John. Although Smith was last seen at Nathan's general merchandise store, newspapers suddenly claimed he was last seen in the dining room of the National Club, a popular union restaurant. Soon word leaked out that he was indeed last seen at Nathan's, but that "blood was washed from the back steps of the store the next morning."[3] Nathan's alley door was a few yards from where St. John lived. John Haines, a liveryman, supposedly testified that St. John and Carpenter had rented a rig the night Wes Smith disappeared. Gossips forgot both those men were in Denver at the time.

Hot on the Smith rumor came a resurrected version of Sam Servis's tale, that he saw St. John kill Will Barney in June 1901. This was dragged into the grand jury for dissection, and Undersheriff Gus Roth flew to Denver to Sam's last known address to bring him to testify. Sam had moved on. A few editors claimed Sam's information had been sworn to in an affidavit before Sam left Telluride in February 1902. Charley Sumner scoffed, "If such an affidavit exists it is rather peculiar it was not acted on at the time, or that the grand jury did not do something with it."[4] Such an affidavit did not exist. The mutation of Sam's tale expanded to include Carpenter and five others as St. John's accomplices. It said they beat Barney to death in the Haines Livery, although that livery had never figured in the original report. As of November, the Haines Livery was sold to Undersheriff Gus Roth and was now known as the Roth Livery.

Everyone wanted a piece of the corpse. A wealthy citizen with Telluride mining investments sent a wire from Cañon City claiming he had information that would reveal the "exact spot where Barney and Smith" were buried.[5] He had gleaned this from a penitentiary inmate. He refused to say where this spot was for fear those who did the burying would dig up the bodies and put them somewhere else. Had this been true, the published scrap in itself would guarantee such a resurrection, whether or not the man revealed the location. The silly piece of reporting was a Pinkerton smokescreen — designed to motivate St. John to go digging. The Smuggler-Union Company's attorney, Jacob Fillius, and director Bulkeley Wells traveled to the penitentiary to interview the inmate. It turned out to be convicted murderer, Crazy Miller.

Charles "Broken Nose" Smith, state penitentiary photo.
Author's collection.

Miller denied he knew where the bodies were, but his revelations were astounding. Twisted by so deep a hatred for St. John, Crazy brazenly told the visitors the union president was responsible for landing him a life sentence. He spun a tale that the union killed Will Barney and Wes Smith and he was told by a friend named "Bob" that Broken Nose told him where to find the remains. It was supposedly the Nose's idea to dig up the bodies and turn them in for the reward money. If Broken Nose would talk, the union's "carnival of crime" would be exposed.[6]

Broken Nose Smith was hauled before the grand jury. He failed to deliver on Barney or the carnival, but spewed out all the names of the union men he saw during the riot of July 1901. Louis Lomax and Ed Lavender and others had already provided a similar list — the names of any union man on the mountain on July 3, 1901. The union had used Smith's Miners' Rest as their headquarters and supposedly the Nose knew all their plans, although all he delivered were names. He hated St. John for not defending him on his perjury charge and was all too willing to sing if the court would lighten his sentence. The court did nothing to relieve him, and his hatred ulcerated.

His venom mixed with Crazy Miller's at the penitentiary and the two cooked up a tale together. On December 10, Crazy wrote a letter to Sheriff Rutan, hoping he might get his sentence reduced or get an opportunity to open his wings. He claimed, "I could find the murderer of Mr. Collins if I were there. I have made a statement to Jake Spellman with regard to Barney's body and about where it is buried. If I had not been sentenced here for life the union would have killed me, too."[7] Rutan decided to have a chat with Crazy himself and came away "satisfied that the fellow was simply talking to hear himself."[8]

Counterfeit silver dollars suddenly bubbled from the muck of Crazy's and the Nose's sinister tale. Crazy told Deputy Spellman that Barney's remains were in the same stable where some children found a cache of counterfeit coins in January 1902. When the *Denver Republican* and other papers snatched up the tale, Barney became a counterfeiter ripping off the gambling houses in Telluride and was murdered for it. Around the same time a man named Foster wrote the Miners' Union a few letters claiming Barney was involved in ore stealing and was likely killed for the $200 he received from a disreputable fence. Babblers forgot Barney the stonemason never had anything to do with mining. The tales were getting so fantastic, Sumner quipped, "Truly some of these press stories put the Arabian Nights out of business."[9]

No matter how obviously ridiculous the tales became, each served to fuel the malice of the TBA. As the grand jury wound down its inquisition, the TBA screamed louder for blood. Between their yowling, the hysterical press, and the threatening presence of the cowboys and deputized gunmen, the tension in Telluride was a tangible thing. Yet on the surface all was peaceful, the wide open gambling keeping men occupied, and no more than the usual knuckle fights rattling the saloons. The most radical TBA men pushed a proposal "to forcibly lead certain labor union leaders from town and threaten them with death if they came back."[10] Others demanded sweeping indictments and the arrest of at least 200 union men.

Two New England Exploration Company officers, Bulkeley Wells and Halstead Lindsley, quietly arrived in town on December 8 to look over the Smuggler-Union properties. They conferred with O. B. Kemp and the MOA, requesting the MOA issue an "ultimatum to the Miners' Union that boycotts and other fights of the union here must be called off and stopped or all the mines here will close."[11] Most managers employed union miners and had no problem with them; the Smuggler was the only company with any lasting difficulties. Tomboy foreman Billy Hutchinson was an example of the excellence of union employees. The Ophir Consolidated's entire crew was union; nearly all miners at the Tomboy, Liberty Bell, Nellie, San Bernardo, and Butterfly were union. Only T. Walter Beam, president of the Japan-Flora Mining Company, fully supported the idea of punitive closure. The majority of Beam's employees were union and Italian, and he made it clear he disliked both. The MOA rejected the ultimatum.

At the state MOA meeting on December 20, Walter Beam and Bulkeley Wells pressed for a statewide ultimatum in support of the Smuggler-Union. Wells had told Denver's *News*, "There is an element at Telluride that must be driven out before owners of mines will feel safe to continue operations."[12] Arthur Winslow of the Liberty Bell, who had not been at the TMA meeting, wanted to inflict punishment as well, saying, "The burden of proof rests with the Western Federation of Miners to show that it is not directly or indirectly responsible for these deplorable conditions."[13] Nevertheless, the state MOA rejected closures. Profit had precedence.

Despite his own opinion that "there is no inducement to resume operations,"[14] Wells informed Sumner the Smuggler would resume on December 27 on a small scale and likely with a full crew by the first of the year. Wells himself would replace Arthur Collins as general manager. Kemp was to take the place of Charles Becker as superintendent. All the threats were merely an attempt to bully the Miners' Union. They failed.

Not only locals of the WFM, but also other labor unions throughout the state rallied behind St. John and Local 63, issuing resolutions against newspapers, including the *Denver Post* and Polly Pry, for misrepresentation and vile attacks. President Sullivan of the State Federation of Labor in Denver and President Frank Schmelzer of the Silverton Miners' Union came to Telluride to confer with St. John and Carpenter. The WFM attorney, John Murphy, was already in town, representing the union in the Blazer restaurant boycott case. He, too, conferred with St. John, advising him to keep the union men as calm and cooperative as possible, no matter how many indictments were handed down. He told St. John he believed the court, the sheriff, and the grand jury had performed some serious legal blunders, which would eventually free any union man from whatever the grand jury might cook up. St. John was to keep a lid on this information. Murphy advised him the grand jury could not legally indict anyone for the murder of Will Barney, Wes Smith, or even Arthur Collins without proper evidence. He believed any indictments, which would be flimsy indeed, would be for conspiracy in the riot incident of July 3, 1901. The dedicated, ailing John Murphy was an angel of hope.

Mine Manager Bulkeley Wells.
Restored photo © M. J. Martin 2003.

With all the witnesses examined, the grand jury began balloting on December 13. Paul Thieman looked in on them, finding the majority still clamored for wholesale indictments. Nate Mansfield and Frank Barnes insisted, "Opinion and hearsay do not establish cause" and "indictments must be drawn only on facts and evidence." As Thieman phrased their argument, "[Judge Stevens] did not convene the grand jury as a vigilance committee, he did not intend that the machinery of the law should be used by one faction of the people to crush another element."[15] Cerulean-eyed Wrench argued his prejudice, demanding the indictment of at least 200 union men. Although Mansfield was the foreman, Wrench attempted to dominate the jury, but the sharp-minded and fearless Mansfield refused to allow him to bring any action on mere theory. Mansfield was almost as small as Wrench, but he was leaner, bronzed from the sun, quick on his feet, and powerful in his words. The clash between the two bankers left the bantam Wrench in the dust, as Mansfield forced the jury away from wholesale indictments.

On Monday, December 15, the grand jury delivered its report to District Judge Stevens. They found and returned "fifty True Bills for various offences, charged against twenty-two different persons, arising out of the riot that occurred at the Smuggler-Union mine on July 3, 1901."[16] They also returned one "True Bill against John Doe for the murder of Arthur Collins" and one "True Bill against John Doe for assault with intent to murder James Phillips." Regarding W. J. Barney and W. E. Smith, they failed to determine what crime, if any, was committed against the two men, and "laying aside their personal views" the jurors were "unable to charge any specific person with any specific offense" in these cases, proving the chatter of everyone from Crazy Miller to F. E. Curry absolutely false. They also found one bill of assault with intent to commit robbery and four for illegal liquor sales. The names of the twenty-two indicted were not released until warrants were issued for their arrests.

Having a general idea who the twenty-two were, St. John held a meeting at the Tomboy that night to encourage the union men to hold fast and cooperate with the law. When arrests were made, no one was to resist. He also stifled talk of rescuing anyone from jail — a few men vowed to take the jail down if St. John was arrested. He assured them all would be vindicated. Bob Meldrum and his sidekick Runnels were being given the arrest warrants and it was said they "shoot first and arrest afterward." St. John wanted no one to give the gunmen cause. The union president held a similar meeting at the Liberty Bell the following day, while Rutan arrested Harry Jardine at the hospital and rumors flew that St. John was about to skip town.

At 9:30 a.m. Wednesday, December 17, St. John walked into the courthouse as Sheriff Rutan was coming out. "They greeted each other cordially," reported Paul Thieman, who witnessed the scene. "St. John said he had come to surrender, having understood the sheriff had a warrant for his arrest."[17] Rutan, who apparently didn't know what day it was, wrote down "Dec. 16" when logging him into jail. The sheriff's cordiality had stopped at the door, for once out of Thieman's view, he treated St. John with disdain. He measured Vint and wrote down his "particulars," asking him his occupation as he did anyone who was arrested. St. John answered, "I am president of the Miners' Union," but Rutan wrote, "Robber and Agitator."[18] All other union arrestees' proper occupations were listed.

Charged with one count of murder (the victim was Ben Burnham), two counts of assault to commit murder (the victims were Charles Becker and John Nicholson), and one attempt to commit highway robbery, his bail was set at $7,000. In a matter of minutes St. John's friends came to post his bond. They were Alderman Robert "Bob" N. Rogers, owner of the largest livery in town; J. S. Obert, owner of the Vienna Bakery; and Andrew Floaten, president of the People's Supply Company.

Ignoring the significance of three leading merchants readily posting bond for the Saint, Curry gleefully detailed the charge of highway robbery against him. Alderman Harry S. Taylor, owner of a barbershop, alleged St. John and Richard Wright had stopped him at gunpoint on September 18, 1902, near the cemetery late at night. They rifled through his saddlebag, taking nothing. Curry pointed out how Taylor always defended the union and that St. John "was never suspected of being capable of such work."[19] The tale was peculiar and Sumner discovered Taylor's testimony was "not voluntary."[20] Something was amiss. The *Post* reported "the three rode to town together" afterward and had a drink.[21] How was this a robbery? Wright was neither a miner nor a union member, but the city frequently engaged him as an extra policeman. Sumner questioned if Taylor had correctly identified the parties in the dark, since St. John did not own a gun — the only time anyone ever saw him with one was when he was advised to wear one during the battle on the mountain. Was some ordinary event twisted into a case of highway robbery? Why was a barber riding down from the mountain so late at night? The mystery dissolved later when Taylor refused to carry on the charade.

When St. John walked out of jail with Rogers, Obert, and Floaten at his side only minutes after his arrest, numerous friends greeted him. A few bystanders jeered. Later that evening, Sumner's printer dropped into the National Club on his way home. While he was in the basement lavatory he overheard two men in the hall talking. He clearly heard one say, "Yes, and when I catch that son of a bitch St. John just right, I am going to kill him."[22] The printer hurriedly finished his business and went out to try to see who they were, but they had gone. He reported the incident to Sumner who boldly put it in the *Examiner*. Sumner also printed the warnings Paul Thieman, the *Post*'s reporter, received for his resolve at fair reporting during his stay in town. Enemies of the union had "made it a business to point out 'criminals' to him" and told him not to go down alleys after dark.[23] Thieman deliberately sought trouble after dark, failing completely. Even the indictments and warrants in Meldrum's hands failed to stir the miners to violence.

Like St. John, Richard Wright, Sam Reed, John E. Conn, William Brennan, Louis Macari, James Roner, and several others voluntarily turned themselves in once they heard warrants were issued for them. Only one man, George Washburn, was fearful and tried to avoid arrest. Meldrum grabbed him at the Four Metals Mining Company's office when Washburn came to collect his pay.

Eleven of the twenty-two indicted union men had left town as much as a year before, their whereabouts unknown. By the weekend the remaining eleven had been arrested, some bonded out within a few hours. The men posting bond for them were all reputable citizens — Gus Brickson, owner of the Sheridan Hotel; J. P. Olson, Richard Treals, Billy Hutchinson (now owner of a livery and wagon hire), and Joseph C. and F. G. "Jerry" Barnes among them — which prompted Sumner to write, "The business community has entire confidence in the integrity of Union men."[24] Sheriff Rutan, on the orders of Deputy District Attorney Howe, refused a few as bondsmen, including Selby and Hoskins of the National Club, even when Selby offered $1,000 in cash for Jardine's bail. Rutan was merely following the letter of the law, he said, which required bondsmen to hold clear title to real estate. Harry Floaten, Andrew's brother, was refused as well, although their business had a stock of $25,000 and the company owned considerable real estate. Other businessmen came forward, including Fred Pheasey, offering their houses as surety, but were refused because homesteads were exempt. In past cases Rutan accepted cash, but he was making it difficult for the union men, even though Judge Stevens had demonstrated leniency in the grand jury cases by admitting the men to bail in the first place.

"The crimes with which they are charged are offences which are not bailable," Stevens told Thieman. "I did this in consideration of the peculiarities of the case, and also because of my belief in the general integrity of union men.... There exists, I believe, a prejudice against the Miners' Union."[25]

Judge Steven's example had no influence on Rutan. The sheriff especially found thirty-two-year-old Harry Jardine disagreeable. The blue-eyed, black-haired grinning Scot had a sarcastic tongue that assured Rutan's rejection of the best of bondsmen until someone inaugurated mandamus proceedings. Jardine remained in the cold cell until January 4, 1903, time he called "twenty days in a first-class banker's resort."[26]

The eighteen-by-twenty foot brick "resort" consisted of two cells with six "stacked" hammocks in each and no toilet or water. At night a heavy safe-like door was bolted shut. If serious illness or fire occurred, no one would know until they found dead bodies in the morning.[27] The men were charged for the meals they ate while incarcerated, guilty or innocent.[28]

Curry burbled with genuine joy at the indictments and incarcerations, suggesting charges alone proved the men were guilty. He published commentary from only slanderous editors, none in defense of labor or justice, which were many. He especially liked the dribble from the *Durango Democrat*, fast becoming Dave Day's strange bedfellow. Day, who never met St. John or any man in Local 63, condemned them as "debris from the pauper centers of the land of Dago, Slav, Hun, Crotian [*sic*], Hobo and illiterate chimpanzee [who] pay no taxes, their personal holdings rarely go beyond a sawed-off shotgun, bulldog and syringe."[29] Curry filled the *Journal* with so much of this repugnant material from other newspapers that Sumner dubbed him "the evening plagiarist." Curry spit back, calling the *Examiner* "the local organ of revolution" and accused Sumner of "defending anarchy."[30]

"The lost memory specimen who edits that sheet," retorted Sumner, "...seems to be endeavoring to stir up more anarchistic effort than comes from any other source."[31]

Paul Thieman, who did his best not to take sides, chuckled to Charley Sumner, "One of these days...that competitor of yours will accidentally blunder into a news item and print it."[32]

In an effort to counter the *Journal*'s malice, Sumner reprinted commentary in defense of the union. Many editors saw the grand jury indictments as a means to condemn the Miners' Union on general principles and nothing more. "It is about time," Sumner pleaded (December 20, 1902), "people of this community sat down on this rag-chewing amalgamation, and once and forever put an end to this strife-stirring business that is in the hands of a very small coterie."

David F. Day, Editor, was fond of classifying those he considered stupid, "Boobus Americanus." He disliked nearly everyone, advocated lynching as the best form of justice, and preferred Utes dead rather than alive. They probably preferred him the same way. Cartoon from the *San Juan Horseshoe* cover, 2001.

The coterie, headed by Addison Wrench, refused to cease persecution of the union and especially of St. John. They so constantly identified Vint as the murderer of Will Barney that they went livid when a Denver carpenter named Frank Mount told reporters in that city Barney was alive. The carpenter had met him around the middle of November on the corner of Fifteenth and Larimer Streets and engaged in a long conversation because of all the uproar in the papers about Barney being murdered. Mount said Barney told him he had been in "Boston settling up his grandmother's estate" and would return to Telluride to set things straight.[33] Because it was specifically the "grandmother," this information was convincing, with only the name of the city incorrect, due to Mount's not paying attention. Barney had been in Boulder, not Boston. He was indeed trying to settle up the estate of his grandmother, Sarah Rannells.[34] He had a deep interest in gaining a portion of the land he loved so well.

Another piece of information that shook the coterie actually appeared in the *Journal.* A miner in another town wrote to San Miguel County officers that he had been coming down the hill a few days before Collins was shot when he met two Italians on the trail. "They had been up to the mine in search of work and were refused, and appeared to be very bitter." They denounced Manager Collins, "declaring that he was not long for this place, one of them remarking that he would kill him himself, if no one else did." The miner said the Italian left town the day after the murder.[35]

A few weeks later a thin, droopy-eyed Eastern European was arrested at Idaho Springs for pulling a knife when asked to pay his bill at a restaurant. Once in custody he began muttering about knowing who killed Collins, and claiming he had been working in Telluride for two weeks or six months as a trammer or mucker or cook. He gave his name as Ivoin Ficko, Evania Z. Glaso, Ivana Glavo, Tovino Glavo, and other variations on the theme. Admitting to being an anarchist for nine years, he had come to America in 1900 and was known as "Number 3," his current assignment in Idaho Springs to assassinate a mine foreman. He babbled in different languages, sometimes pretending he didn't understand English or German, speaking a Slavic dialect rapidly and incoherently. Officers said he was either demented or very cunning by pretending to be demented. Yet he persisted in telling them of the Collins assassination, claiming his cohort, Kenyen Sidi or Canun Siti, known as "Number 2," did the actual shooting. He said Sidi had worked at the Smuggler-Union as a digger. Photos of the Slav were sent to Sheriff Rutan. After checking various mine employee lists for the two men, and showing the photos to Kemp and others, Rutan decided the story had no foundation. No mine had employed either a Ficko or Sidi. The matter was dropped, merely because the names weren't on any payroll list — they hadn't bothered looking over the range in Ouray where a C. Siti worked in the Revenue Mine. The man was written off as mad.[36] Were the authorities dismissing the real killer of Arthur Collins just because he babbled, just because his name wasn't on a payroll, just because he had no connection with the Miners' Union? He claimed affinity with the anarchist assassin of President McKinley. No one listened.

Meanwhile, the WFM's attorney Murphy and his team carefully combed through the grand jury's proceedings and indictments in preparation for the trials set for the May term of district court. After being released on bond, the union men remained in town, quietly absorbed in their work while the heavy winter snows silenced the San Juans and the Smuggler-Union roared into life. Bulkeley Wells promoted Ike Carruthers to Kemp's old position of chief bookkeeper, Owen Johnson moved into Ike's position as timekeeper, Mel Robbins was promoted to foreman, and the serene Scot, Will McKenzie, to night foreman[37] (McKenzie's promotion was fleeting since he died June 11 of stomach cancer). No union men were promoted. The city council decided to no longer advertise the rewards for Barney and Smith's corpses, but to keep them on the books. Curry and Painter had hoped the district court's permanent injunction against the Miners' Union's past picketing of Blazer's defunct Saddle Rock might have some bearing on the boycott, but it was strictly related to pickets at the restaurant and the *Journal* had no legal redress to escape the boycott. Curry simply snuggled up with Dave Day, and Sumner slung reproach in both their faces. Oscar Downtain was appointed to replace a resigned county commissioner and planned to run for clerk in the upcoming city elections. Dr. Sheldon had repaired the old injury on Downtain's foot, and the man was ready to face the Wrench gang again. Bob Rogers decided to run for mayor, although he would do no actual "running," since a rather large Finn had wrangled with him at the livery and broke his leg. Fred Pheasey jumped in the race for city alderman, third ward. He had sold out his interest in the Clipper Saloon and became an ardent Socialist. He pushed St. John into running for alderman in the second ward. Vint had enough of politics and was "completely sick of being in print,"[38] but Pheasey and fellow Socialist Harry Floaten (running for city treasurer) were persuasive.

As if the indictments were fictional and the threats of hanging him were bedtime fables, St. John continued his work with the local and district unions fearlessly, traveling unarmed. In January 1903, active union members numbered 1,008, with over 300 inactive members. No fraternal organization in the region came close to the enormous membership of Local 63, which, Sumner pointed out, required a "large amount of executive ability" to hold it together.[39] Between St. John's compassionate idealism and Carpenter's no-nonsense financial ability, the union was benevolently, scrupulously, and equitably directed. In 1902 thirty-one members had died from mine accidents or illness, and the union paid $2,124 in funeral benefits and $3,324 in sick benefits. The hospital cost between $1,300 and $1,700 per month to operate, and the only way to meet those bills was to collect special hospital dues of $1.50 per month from each member. Given the enormous membership, it was impossible for the officers to personally collect dues from each man; thus St. John and Carpenter asked the mine managers to deduct the miners' hospital dues from their checks. This was a customary arrangement the managers had previously had with the privately owned hospital, but declined to do with the Miners' Union Hospital, unless the union accepted a mine manager on the hospital's Board of Control. In the spirit of genuine reconciliation, St. John agreed. When he thought negotiations were complete, the TMA reneged, demanding they hold all board positions. They wanted control, which they had absolutely "no right to ask."[40] Some TMA members would prefer the hospital leased to a different operator, simply because they hated the union. Others knew the hospital was a benefit to the employers as well as the employees, and, despite the noise from Walter Beam and a few others, negotiations soon resumed, concluding in an amicable agreement. The TMA would select one manager to serve on the hospital board — the other four would remain Miners' Union officers — and the managers would deduct the hospital dues from the miners' time checks. The hospital would continue to furnish surgical and medical treatment to miners and their families.[41] The agreement illustrated St. John's astounding ability to move people beyond their personal animosity and narrow vision into embracing the greater good, for he was not only dealing with men who despised him, but he also was under the cloud of a murder indictment and out on bond. Had he used threats of any sort, the TMA would have tossed him in the creek. It was his honesty and compassion that gained their cooperation.

This success came in concert with a letter from the owner and publisher of the *Denver Post*, Frederick G. Bonfils. The WFM's Secretary Haywood had been diplomatically stroking the *Post* on behalf of St. John since Polly Pry painted the man as a fire-eating assassin, telling them, "a great injustice was done" to both St. John and the union. From editors to managers the affair drifted, finally thudding into the publisher. After getting Polly Pry to admit she used information from Wrench and others, and discussing the matter with Paul Thieman, who spent two weeks in Telluride, Bonfils wrote a personal apology to Haywood and St. John, saying the paper "was willing to make any reparation." He said if he had had knowledge of the content of Polly Pry's articles they "would never have appeared in the paper."[42] All Haywood asked was a public retraction.

This information was published in several small papers and Curry labeled it utterly false. The *Post*, one of the largest newspapers in the state, would never dare repudiate Polly Pry, he fumed; it was "contemptible and cowardly."[43] He was certain no public apology would appear.

Graciously concurring with Haywood, Bonfils published a lengthy and signed public apology on February 21, 1903, making it clear the *Post* "believed in the absolute righteousness of the cause of labor." He excused Polly Pry as being influenced by the Telluride atmosphere at the time, which was "bordering upon frenzy," but he admitted the *Post* "published statements concerning St. John, which neither his conduct before nor since has warranted.... The enemies of St. John carried to...extremes their charges against him.... We could have no malice against him, and certainly he is entitled to and will receive at the hands of this paper every consideration that justice and fair play demand."

As the miners put it, F. E. Curry "busted an intestine." From that point forward, he snarled that Polly Pry's article on Firebrand St. John was the only truth the *Post* ever published. His bitter attitude drove a few employees from his shop and one hired on at the *Examiner*. Could it get worse for the editor of the "little daily pancake of hate"[44] and his chief, the bantam banker? It could.

The Miners' Union reelected St. John and Carpenter and put them, as well as Jardine and Ed Deutsch, all Socialists, on the hospital board. The district union reelected the officers as well. When St. John went to Durango and Silverton on district union business in March, editors made a point to meet the infamous Firebrand. The *Durango Herald* said, "He certainly impresses one as being a cool, clear-headed businessman, perfectly fearless...."[45] The *Telegraph* chortled, "He's a mild-mannered little fellow to have caused such consternation among the mountains of San Miguel. Don't look as if he eats more than a bushel of fire a day."[46] The *Durango Journal* was impressed by his "modest but candid, generous but consistent" nature, calling him "a good listener" who "argues little."[47] Curry was unable to stomach so much honest praise and muttered his favorite lines about the disturbing element, prompting Sumner to offer him a list of adjectives to use in place of "disturbing." "The word has become quite threadbare," Sumner moaned.[48]

Worse still, Curry's masters, the TBA "Leading Six," were once again losing members. Many businessmen declared their distaste for the *Journal*'s attitude; hotel-operator J. C. Barnes publicly pointing out why: "...the paper has disgusted all those who stood by it. It has never hesitated to sacrifice the welfare of the community in the endeavor to strike a blow at organized labor, and it seems that now the only tonic that keeps it alive is its intense hatred for the man who has outgeneraled it — Vincent St. John."[49] The TBA Six demanded associates remove advertising from the *Examiner* — Sumner had to be punished for his impertinence. But instead of "Wrenching" the *Examiner*, the associates informed amiable Charley. He promptly renamed the TBA the "Businessmen's Boycott League" and blasted it for starting its underhanded "boycott before the Union ever dreamed of" legally boycotting the *Journal*. He called the leaders vicious and invited them to "boycott until doomsday." He didn't "care a tinker's damn."[50]

Curry sulked with his tail tucked up. The Six (or more accurately, the Five, since Wrench went East at the end of December) glowered. And still it got worse. St. John's friend, Bob Rogers, was voted mayor in the spring election. The old guard was out. Although St. John failed to gain the alderman seat, Pheasey won his. A "flaming Socialist" now sat on the city council. The new, improved Downtain was city clerk and county Judge Wardlaw was getting cozy with Democrats. Telluride courted revolution. Something had to be done about the ghastly harmony, the prosperity, and the looming good-fellowship.

A brief respite for the "better element," as Curry called himself, Wrench, and those who despised unions, came when a seventeen-year-old day laborer and his pal wandered up Royer Gulch on April 15. The young men, Grover Skelton and William Webb, discovered a shotgun partially hidden under a log about 150 feet above the railroad track. This was near the ruins of an old highgraders' shack a short distance west of the Pandora village. Skelton, who lived with his parents, Tom and Mary, at Pandora, and Webb, who worked a lease on the Smuggler, liked to pan for gold at the shack. They immediately brought their find to Bulkeley Wells at the Smuggler office, thinking it would be of more interest to him than to the sheriff. They knew he was secretly immersed in hunting Collins's assassin. Wells immediately declared this the murder weapon, for it was a Winchester, twelve-gauge, pump-action shotgun with a sawed-off barrel. Wells surmised the badly rusted gun "was not in working order" when abandoned, as one cartridge was jammed in the chamber. Two others were still in the magazine. With difficulty Wells removed one, finding the cartridge the same as the empty shell found near the Collins house. It contained twelve pellets of buckshot.[51]

Wells took the gun to Sheriff Rutan and Deputy District Attorney Howe. All three examined the area where it was found. They were absolutely certain it was the murder weapon and had been tossed in the bushes November 19. For five months it had remained, rusting so badly as to freeze up the breech, hammer, and slide. They wrapped it in paper and had William Wheeler store it in the bank vault.

The shotgun raised a dark hope in the law and order crowd who believed at least one good hanging would come out of it. All they needed was a scrap of evidence to link the gun to a union man, and the whole lot could be swinging. They eagerly anticipated the upcoming murder trials in district court. Now tail up and tongue flapping, Curry harangued the "agitators." In a surprising turn, Sumner answered him with, "Jesus Christ was crucified for being an agitator" and listed all the great agitators and revolutionaries in history, including George Washington.[52] Wherever Curry unloaded, Sumner planted roses.

Sumner donated dozens of books to the Miners' Union's new library, causing constipation in his neighbor. Word on the street was that dyed-in-the-wool Republican Charley Sumner was turning into a socialistic revolutionary agitator himself because he was getting too cozy with that sort. Sumner just happened to be inspired by St. John's belief in uplifting the worker through education and higher mental pursuits. The Saint was inspiring far too many, much to Curry's chagrin. As Harry Jardine put it, the library gave the "brothers a good clean loafing place and good reading, where there is no temptation for them to blow in their money in fighting booze or bucking the tiger."[53] Revolution loomed.

One week before the May term of district court opened, Sumner started receiving anonymous letters again. He laughed at the absurdities, printing the best for the public. The latest threatened, "If you do not attend to your own business and stop encouraging St. John in his meanness you will be taken care of after court adjourns."[54] He was fond of this one because it was not only typewritten, it also used clear, concise English for a change. No garbled verbs, no crude drawings. All it lacked was a banker's stationery.

At 10:00 Tuesday morning, May 12, district court convened with Judge Theron Stevens on the bench. The first day generated little interest, except for the discharge of Walt Kenley. He had completed his jail sentence for assault with intent to kill and Kenley was returned to the saddle, complete with his deputy star and gun.

Because the WFM attorney John H. Murphy, representing the union men under grand jury indictment, had filed pleas in abatement, none of the prosecuting attorneys was prepared to proceed until the weekend. With Murphy was esteemed Denver lawyer Milton Smith, an expert in grand jury procedure and related issues. On Saturday the courtroom was crammed beyond capacity, the intense interest of the crowd as physical as the strong scent from overheated bodies. A number of "noticeably shadowy strangers," identified by observers as Pinkertons, kept their eyes on St. John, Jardine, and others. Murphy alleged the grand jury had been "selected in a manner contrary to the statutes." Deputy District Attorney Howe demurred and the arguments proceeded. The grand jury was "illegally drawn on account of an open venire," Murphy maintained, contending "that the sheriff was prejudiced and under the influence of the Businessmen's Association and Mine Operators' Association; that this prejudice actuated him in summoning men who were biased or prejudiced against the Miners' Union; that one member of the jury had said, 'several members of the Union should be hung...'" and that five of the men selected as jurors had signed the petition requesting a grand jury in the first place, "which made them complaining witnesses."[55] A. M. Wrench was one of these. Murphy illustrated each allegation with details and it seemed for a moment the spectators were holding their collective breath. "Those arrayed against the Union," he said, "sought to charge them with every offense that was committed in this community, which could not be directly laid to someone else."[56]

Edward Howe countered with every legal argument possible until he ran out of steam. Newly elected Congressman Herschel Hogg, retained by the TBA and TMA, took over, puffing out words for more than an hour. Judge Stevens had to call a night session to conclude the arguments. Finally, at 9:00 p.m. Stevens delivered his decision.

Although he refused to consider the arguments of prejudice of either the sheriff or the jurors and declared all as "qualified," the district judge quoted the statute pertaining to the selection of grand juries, Section 2611, and ruled, "...the Court is unable to reach any other conclusion than that this grand jury is not properly drawn, and if not properly drawn, the indictments are illegal." He sustained the pleas in abatement, nullifying every indictment issued by the grand jury. Stevens discharged the eleven defendants and their bondsmen.[57]

Thundering against the unexpected and shattering defeat, some TBA and TMA members, including Bulkeley Wells, guardedly consulted Deputy District Attorney Howe and District Attorney McMullin. The two associations had offered the assistance of their best attorneys, including the Smuggler-Union's Jacob Fillius (presently in town to defend the company in the continuing fire suits). They insisted on only one course: file new informations against the men as soon as court convened on Monday. They even suggested charging St. John with the murder of Arthur Collins — they had the shotgun as evidence; they would build a case. Prosecution in the hands of Wells and the corporate attorneys twisted into patent persecution. Wells's determination to see union men hang was far more profound than Wrench's and deadlier, because it was closeted. Although charming on the outside, Wells covertly did everything possible to take down the union. The corporation money behind him, along with his retinue of Pinkertons and attorneys, gave this thirty-one-year-old Harvard graduate complete confidence in success. His arrogance outshone Wrench's. As if claiming Cal Rutan as his own property, Wells presented him with a special sheriff's badge that very week. The solid gold star was engraved on the back with Rutan's name and on the front with "Sheriff of San Miguel County, Colorado." Rutan, "one of the best pleased men in town," wore it proudly.[58]

Believing the quashed indictments and the court's removal of every legal encumbrance ended the ugly affair, several of the eleven union men left Telluride Sunday evening on the Durango-bound train and under the watchful eye of Pinkerton shadows. Practically prostrate from the stress of having a noose dangling before him, George Washburn took ill en route and his companions feared he might die.[59] Vincent St. John, Harry Jardine, and Oscar Carpenter (who had not been indicted only because he wasn't on the mountain July 3) intended to go to Denver on business and to attend the WFM convention. Hearing the district attorney was preparing to file new informations, Vint and Harry made Howe a courtesy visit Sunday afternoon. St. John told Howe he was "ready to be served with any kind of warrant and asked him if he cared to get service before he left" for the convention. Howe replied, "There's no hurry," and thanked him for his cooperation.[60] St. John and Jardine went to Ames and Sawpit to confer with union members in those villages. Monday morning they boarded the train at Sawpit and went on to Denver, conscious of their Pinkerton shadows. Carpenter joined them later.

Monday afternoon Howe hastily filed new informations in district court at the behest of Wells and the TMA, the corporate attorneys assisting him, giving it the aroma of a company witch-bake rather than a public proceeding. The first two were against Vincent St. John for the murder of Ben Burnham and of Arthur Collins, despite a complete lack of grounds.[61] The original spurious charge of highway robbery had evaporated weeks before with Taylor's refusal to back it up, but the "assault with intent to murder" Becker and Nicholson was revived. To Wells, St. John was the principal assassin and had to be dealt with quickly. The court issued a criminal capias and Sheriff Rutan put the warrant in Bob Meldrum's hands. In addition to the usual fee for serving a warrant, $400 was offered the gunman for the "swift apprehension of the assassin," which translated to "dead or alive." The public authorities did not put up this money.[62] Horrified, Charley Sumner demanded to know who did.

Sumner had believed the decision of Judge Stevens had put an end to the persecution of the Miners' Union. He believed Telluride could now repair the rift. Guy

Miller told him the "brothers who had been under indictment were willing to show a magnanimous and Christian spirit by forgiving their enemies, and pave the way to paths of peace."[63] Sumner had asked in print, "Do you suppose their enemies will be as magnanimous?"[64] When those enemies demonstrated "a spiteful and vindictive spirit" by hastily filing the new informations against St. John, Brennan, Roner, Washburn, Conn, and Jardine, and tossed in C. H. Carlson just because he was an easy mark, Charley's wrath plunged so deep it ripped tears from his soul. He loved Telluride. He did not want a handful of kickers to destroy it. "The community is not sending up a cry that the fight must be continued," he wrote. "It is being kept alive by a few men who have a grudge to settle and a fee to make."[65] At this point he was unaware the glamorous Wells was involved: Wells was a master of discretion.

The May 23 edition of the *Examiner* spelled out the situation in an article filling most of the paper. Sumner titled it, "Brief of Appellants; The People of the City of Telluride versus Ribbers, Jobbers and Hate-breeders; In the Court of Common Sense." Detailing the persecution of the Miners' Union since the strike, he gave the facts in the grand jury case from start to finish, and asked why the TBA had pressured for convictions of union men; why the TMA had employed four lawyers to assist Howe in filing unsupported new informations against union men; and why these associations felt it necessary to hire gunmen with reputations as killers to serve the warrants on union men. The only reason was a desire to "break the union."

"There is no labor trouble in the camp," Sumner wrote. "Why, then, can we not join issues and smooth over our other differences?" He wasn't writing in defense of the Miners' Union, but for the survival of the community. If the persecution and hatred were not ended now, they never would be.

"We would rest content to see conciliation promoted by any means whatever," Sumner pleaded, "to see this damnable prejudice abolished, to see prosperity and peace again settle over our troubled community."[66]

The hate-breeders refused to hear him. When Addison Wrench returned from a trip east and Curry from his holiday, their venom surged like that spring's snowslides and the flooding waters of the San Miguel. Wrench stepped up TBA efforts to punish the *Examiner*. Sumner laughed off the miniature Caesar's attacks, suggesting the "tender blond beauty of Brother Andronicus Mauritius" Wrench was more suitable to the White House.[67] In the other corner, Curry savored the absence of St. John, taking the opportunity to mock the young man's wife, Clara. Sumner called Curry a malignant spirit and growled, "A man's resources are limited indeed who would attack a defenseless woman as the only means of casting discredit and slander upon her protector."[68] Curry was hoping Clara's protector was gone for good.

With Meldrum unleashed and keen on planting a bullet in the back of St. John's head for $400, the union president was in danger of being "gone for good." He and Jardine, Carpenter, Ed Thomas, and Charles Trimble were Telluride's delegates at the WFM convention in Denver. Not hiding from anyone, they were simply absorbed in union business with several hundred other delegates, until the invasion. The *Denver Times* account of June 3, 1903, caught the theatrical atmosphere:

> *The headquarters of the Western Federation of Miners was yesterday infested by an army of Pinkerton detectives who had discarded their stylish gray raiment and palmed themselves off as reporters. The miners could not guess why they were so persistently besieged by reporters, as they knew of no news of importance which could be given out. Later in the afternoon some*

> *of the officers, who almost know the scent of a Pinkerton, became wise, and then it was discovered that one Robert Meldrum...was hunting for Vincent St. John....*

Once exposed, the stocky, neatly dressed Meldrum failed to locate his quarry. Usually very profane when angry, Meldrum grinned at reporters as they gobbled up his hot air.[69] He merrily boasted about his intimacy with killer Tom Horn and the large price on St. John's head, informing reporters he had a warrant on St. John and Jardine for the murder of Arthur Collins.

In alarm, the WFM delegates scrambled to protect the two men. They feared Meldrum would use any excuse to kill his quarry. Attorney John Murphy strenuously advised Vint to keep out of harm's way, and the officers made arrangements to lodge him and Harry in a secure place. They weren't trying to hide them from the law, but protect them from being shot down like dogs. Murphy was convinced a convenient fatality was the only reason Meldrum was chosen to execute the warrants, since they were illegally issued. He firmly believed the new informations "were not filed in accordance with the law, and I feel certain that we will not only be able to have the one against Mr. Carlson set aside or stricken from the files, but we will also be able to do the same with respect to the others."[70] Carlson, who had been a victim of Jack Hyde and Jordan, was recently arrested at Ophir, where he had gone into the saloon business with John Conn. His case was already in the hands of WFM attorneys.

Murphy refused to allow St. John or Jardine to return to Telluride until the situation was settled in court. The risk was too great. Murphy didn't want his boys dead. St. John protested, to no avail: the WFM board members would sit on him in Denver to assure his safety. Carpenter left the convention immediately to safeguard Clara in Telluride.

Returning to Telluride empty-handed, Meldrum reported to Rutan and the TMA that he found no trace of St. John or the cheeky Scot. While lawyers took the battle to court, Guy Miller blasted the TBA, the TMA, their hired killers, and the persecution that drove innocent men into exile. His fiery indignation was certain to mark him for the next round of persecutions, for in criticizing Deputy District Attorney Howe's role as the TMBA and TMA's "tool," Miller said Howe would be "a nice thing in a glass case." Pinkertons stored the line for future use. Miller didn't care. Despite the manipulation of the law by Howe and his "masters," Miller was confident, "Truth and Justice are mighty and will prevail against all machinations of any clique of men and shyster lawyers."[71]

For the moment he was right. On July 7, 1903, Murphy and his tenacious colleagues forced District Attorney McMullin to the wall. The case of the People versus Vincent St. John quietly suffocated, its end briefly stated in the record: "On motion of district attorney ordered that this cause be stricken from docket." McMullin wanted it, "with leave to reinstate," but Murphy refused to allow it, and the proviso was quashed.[72] This same order was entered in all the cases, including Harry Jardine's. The TBA and TMA couldn't continue bashing union members over the head with the same old hammer.

Bulkeley Wells resolved to find a new one.

CHAPTER

17

PEABODY SELLS THE MILITIA

"If voting could really change things, it would be illegal."

— Revolution Books, New York

James H. Peabody had ascended to the statehouse January 13, 1903, on a cloud of pomp and splendor befitting a monarch. "The attendant exercises were elaborate in the extreme," newsmen reported.[1] Peabody himself led the grand march into the inaugural ball and the throng of bankers, businessmen, railroad magnates, industrialists, and mine owners attending him hinted of things to come. He had promised to make Colorado safe for investment, to use all the power and authority of his office to protect life and property, and put an end to the activities of the criminal classes. This spectral statement implied the "criminal classes" weren't only those who committed actual crimes, but also those who engaged in disruption of industry through demands, strikes, and agitation for social change. And the National Guard, he asserted, would be his instrument to control these classes and maintain law and order. He planned to reorganize the guard into a force with the efficiency of a regular army. He appointed Sherman M. Bell, a mine manager from the Cripple Creek district, as his adjutant general, and John Q. MacDonald, a mill manager with the huge United States Reduction and Refining Company (USRRC), as secretary of the State Military Board.[2]

Immediately upon entering office, Peabody demonstrated his readiness to use the National Guard to establish order when senators and representatives attempted to unseat their rivals in the state legislature. Before the near revolt was ironed out, Peabody called on the state militia to be prepared for action, prompting an editorial cartoonist to depict him as "The No Revolution Governor," with a firm stranglehold on the neck of a dynamite-wielding salamander.[3] Since dynamite wasn't standard gear for legislators, many WFM members interpreted this as a caution against miners.

Perhaps it was. On February 14, when millmen at the Standard Mill of the USRRC in Colorado City went on strike for living wages and the right to organize, Manager Charles M. MacNeill requested troops. Without conferring with the WFM or investigating the need for troops, Peabody consented. Later he appointed MacNeill as well as the USRRC's treasurer and metallurgist to his military staff.

Although Peabody's mayoral term at Cañon City had been marked by excellence and accomplishments beneficial to the common good, his role as governor was distinctly sympathetic toward those opposed to organized labor. Historian George Suggs, Jr., said, "his administration was joined so firmly to corporate interests that the two acted as one against the WFM."[4]

In the November 1902 election, an overwhelming majority of 46,714 Colorado voters mandated the legislature to enact a constitutional amendment creating an

This cartoon was in the Denver Post *in January 1903, just after James Peabody's inauguration.* Author's collection.

eight-hour law.[5] Peabody did nothing to assure the will of the people was carried out. As soon as bills were introduced in the legislature, corporations immediately crushed the legislation in a massive campaign of confusion, multiple eight-hour bills, endless anonymous circulars, charges, and counter charges. By the end of the session that spring no bill had been passed. Peabody's stipulations for calling a special session to pass such a law were designed to defeat the purpose. As Suggs put it, "His refusal to be bound by the mandate of the people and his party and the Republican platform of 1902 was a major cause of the strikes which beset his administration.... The enactment of an eight-hour law would have removed a major grievance of the millworkers and the miners, possibly precluding the strikes" that erupted that year.[6]

The governor's action at Colorado City, where strikers had violated no laws and not even a hint of riot was observed, stirred up the ire of many Republicans who anticipated an impartial and just administration when voting for this successful businessman. What they got was horseplay. As Charley Sumner observed, "There is about as much reason for the presence of troops at Colorado City as there would be in sending ice up the Yukon.... Governor Peabody is spending the people's money helping operatives in an attempt to break the Union, and irrespective of politics, is deserving only of condemnation for it." Because the governor was so quick to send troops at MacNeill's request, it seemed to many that Peabody was, as Sumner put it, "carrying out an obligation made before election."[7] He appeared to be a corporate tool even to those who had voted for him and before his first three months were complete.

Peabody was an intelligent, shrewd, aggressive businessman, but, according to Suggs, "he was also easily swayed, compromised, and flattered by those who wished to use the powers of his office for their own purposes."[8] Yet he was more than just a corporate puppet: he was a willing collaborator. Although the military power of the State was meant for the protection of all its citizens, Peabody firmly believed in using "that power to war against a labor union" because "it dared to challenge the supremacy of the class and the values he represented."[9] He had a clear set of principles to defend, the same principles he shared with industrialists. Perhaps his readiness to use force against those who simply held different ideas than his sprang from his times, yet other Colorado governors of the same era, such as James Orman and Alva Adams, stepped outside the narrow confines of popular belief, corporate pressure, and their own personal circumstances to create a more humane and just government. Adams, who served two terms, was a man of distinctly broad vision, instrumental in prohibiting child labor and physical punishment at state institutions, ending public executions (he fully opposed capital punishment), and establishing the State Board of Arbitration in 1897.

Peabody's values were not so altruistic. Like the industrialists, his inspiration was prosperity and profit, and whoever threatened those threatened life, property, and peace. Neither puppet nor victim of his times, Peabody identified his principles as "right," and what was right he had to defend. He stood true to his promise to uphold law and order, but only for those who shared his values. The rest were the "criminal classes" who were attempting to overthrow the existing order, and he felt moral righteousness mandated he use force to subdue them. Any means were justified in eliminating the danger. The WFM had reason for alarm.

Governor James H. Peabody.
Restored photo © M. J. Martin 2003.

Peabody was not so much opposed to the WFM's commitment to the rights of labor as he was to the organization's dance with socialism. Socialism was the intrinsic "evil" that threatened his values. As more WFM leaders became Socialists, the organization became more evil. It was a menace to everything American, to private property, and social order. Even the Catholic Bishop of Colorado, Nicholas C. Matz, condemned it as evil, thus becoming an ally with Peabody and his conservatives in the drive to crush the Federation.

Bishop Matz joined the war in May of that year while soliciting wealthy capitalists for funds to build a cathedral in Denver. In some letters to potential donors, he made a point to remind them of his stand against the "lawless laboring element" that had "done incalculable harm to the business interests of Colorado."[10] His alliance with Peabody and conservative industrialists was significant, since he was the leader of the largest religious denomination in the state.[11] Many WFM members were Catholic, and in some instances, as in Local 63, the membership was overwhelmingly Catholic.

While his first attacks on the WFM went unnoticed, as the bishop became more vociferous in his condemnation, WFM leaders felt it necessary to defend their position. They saw socialism as fundamentally Christian and directed Matz to read the "Acts of the Apostles" wherein members of the early Church had sold their property and held all things in common.[12] On June 14, 1903, the *Rocky Mountain News* published the bishop's harangue in which he said he was certain socialism "will burn the souls of such reckless leaders [of the WFM] for all eternity in hell." The WFM replied, "Judge not."[13]

Matz pretended to be labor's friend and was apparently determined to save the miners from sulfur by instructing them to accept earthly starvation under the oppression of mine owners, for they would have their porterhouse steaks in heaven. Secretary Haywood lashed back, but in the end, Matz had the power of hell to flog the WFM's Catholic membership into line and keep the capitalists pouring in the donations. He threatened, without due process: "Catholics, beware! And bear in mind that you cannot become Socialists without placing yourselves under the ban of the church. Socialism is condemned and you are excommunicated the moment that you enter the ranks of socialism."[14] Since the WFM was seen as a vehicle of socialism, Matz demanded miners choose between the WFM and their church. Armed with Pope Leo XIII's encyclical, *Rerum Novarum,* Ed Boyce quietly assured members no such choice was necessary — Catholic Socialists should remain in their church, working for freedom and human dignity.[15] Matz was one small bishop gabbling, not the Catholic Church, and thus, devout Catholics such as Vincent St. John, Guy Miller, and others remained equally dedicated in their faith, their union membership, and their socialism. For men such as these, the three things were inseparable — they believed faith meant action, and action meant lifting the oppressed from the gutter and sharing with the downtrodden whatever one had.

No amount of hellfire from a bishop who believed force and Gatling guns were justified in subduing those who asked for living wages would deter Vincent St. John from his church. His God did not own Gatling guns. His was not a god of violence, oppression, hatred, lies, or condemnation. Yet those with small thorny gods used such things in an attempt to destroy him. Although they sadistically bludgeoned his spirit until frustration fired his temper and he fleetingly stumbled in his commitment to nonviolence, he refused to meet them on their terms because he believed, "Brutality done in the name of justice is still brutality."[16] He always told his brothers, "Let agitation, education, and organization be the battle cry of the present and the future; let us be known by our deeds."[17] Vint was no saint, but his genuine self-sacrificing commitment made him anxious to return to Telluride to continue his work, despite Meldrum's itch to shatter his cranium.

Those who despised him despised anyone who stood by him. When Charley Sumner's horror burst out to ask, "Who is paying the price?" for Meldrum to take St. John's life, Vint's enemies attacked Charley. A letter signed by a "John Smith" bluntly informed Sumner, "Do not worry about 'who is paying the price.' When the murderer of Arthur Collins is swinging from the scaffold in your town, and his accomplices and associates are in the penitentiary, you will know 'who is paying the price.' Look out for yourself."[18]

This was a mild sample of many sadistic letters flooding Sumner for his stand on behalf of St. John and the Miners' Union. In disgust he branded "John Smith" a "chicken-livered milksop." Although Sumner knew St. John was "absolutely innocent" of the heinous crimes charged against him, he wrote, "But were he today to stand convicted of every crime, I would have more respect for his courage and character than I would for the individual who wrote the letter." Sumner cried for an end to the "vindictiveness, viciousness, cowardliness, malice, hatred."[19]

In the last few months Sumner's life had tumbled. His business partner and friend, Charles Fluke, had sold out and left due to Fluke's son taking ill. His marriage to his high school sweetheart collapsed and Naomi divorced him. His health was shattered by what was diagnosed as "enteric typhoid," landing him in the Miners' Hospital. Once back on his feet, he concentrated his energy on saving what he had left, the community he loved. He was fighting for Telluride's soul, but his desperate pleas for an end to hostility fell on stone ears.

The Miners' Union wasn't continuing the fight, he said. "The persecution, the threats, the fight, is continued from the same sources which incubate anonymous letters and promote vicious ideas, and if such sentiment is kept alive, the community will eventually be 'paying the price.'"

Vindictive men wanted St. John's head on the city gate because he had become something too big for them to comprehend. "St. John the individual is not being fought by enemies, but St. John the president is being knifed and persecuted in the hope that it will throttle the Union. It won't." Ideas were bigger than one man and persecution would only make them blossom.

"Since St. John has been absent, his enemies have been active," Sumner pointed out, "and reports have been circulated that he is wanted for everything from the murder of Arthur Collins to the mudslide over at Ames last night.

"St. John is a much wanted man. He wasn't wanted at all while he was here. In fact some preferred that he should be placed in a position where he never would be wanted." Charley scratched at the consciences, pounded at the hearts of those who dared to put a price on a man's head, asking, "Why wasn't he arrested for Collins's murder and a number of other things before he left?"[20]

The WFM shared Charley Sumner's horror. Despite St. John's assertion that Meldrum would be unlikely to shoot him with the plot exposed in the newspapers, Murphy, Haywood, and the rest were unconvinced. Meldrum carried "a death warrant." What other reason existed for an illegally deputized hired killer to be given an illegally executed warrant and offered the additional price of four month's wages to illegally serve that warrant on a man who had volunteered to be served? No one had asked for the assistance of Denver officials in carrying out the warrant. Instead, Pinkertons, who were not law officers, assisted in the farce as if trying to flush out game. Someone had carefully calculated St. John would flee in fear, at which time he could be shot in the back while "resisting" arrest. The plan's failure didn't end Meldrum's assignment.

Even though all charges against St. John had been stricken from the docket and he was under no legal warrant, the WFM felt certain if he returned to Telluride he would be slain. They put him to work at WFM headquarters, but not until August 14, after every legal shard had been cleared, did they dare take Clara out of Telluride to join her husband. Although a Pinkerton had been shadowing Oscar Carpenter since May, the operative didn't know what St. John's wife looked like. Carp bought a ticket for her as "Mrs. Baugh" and escorted her to Denver under Pinkerton nostrils. Local 63 paid the train fare.[21]

In St. John's absence, union vice president Charles Trimble assumed his duties. Vint remained president of Local 63 and of the district union, communicating with them via letter.[22] As long as the TMA kept Bob Meldrum on the payroll, the WFM refused to allow St. John to return to the San Juans. Friendships had to be gathered into memory, achievements and dreams had to be released to those who would come after him; he had to say goodbye to the mountains he loved, to Fred Pheascy, Oscar Carpenter, Charley Sumner, Bob Rogers, Billy Hutchinson, and all the people who had steadfastly stayed at his side, believed in him, shared his triumphs, his darkest hurts, his commitment to making a more humane world. So many miners didn't want him to be gone that they kept him symbolically as president even after they knew he wasn't coming back.[23] This quiet "firebrand," this "boy" with a soul so ancient as to absorb relentless, pummeling hatred and still remain unbent, had brought them hope and dignity. They didn't want to let go.

St. John worked with Big Bill Haywood, out of the limelight, but free to walk the streets of Denver unmolested. The MOA did not own Denver's police; no hired killers hiding behind badges were allowed to strut about boasting how much their quarry was worth. Nevertheless, if St. John's name appeared in print even on some obscure back page of a Denver paper, Curry axed the man, now labeling him a fugitive. Curry couldn't let go. Even after St. John and Jardine were sent to the booming mining camps in Utah and Nevada, Curry hacked at the shadow on the wall. The WFM knew St. John's unique talents were wasted at headquarters, so they sent him out as an organizer. "Oblivion was too dense, too dark and too appropriate for this creature to rest in," Curry spewed, "so he broke out, and will hereafter strive to stir up strife and discord, which but for him might have been settled years ago."[24]

Despite the Miners' Union and the Federal Union continuing their boycott of the *Journal* long after St. John's departure, Curry still blamed "the creature" for it. In an address to the Editorial Association, Curry had turned the general topic of boycotts into a personal diatribe against St. John, saying he was directly responsible for the troubles in Telluride, claiming because of the "absolute power with which labor unions endow their leaders," St. John was able to exact his own personal revenge with a boycott.[25] Documentation proved Curry wrong, since the Federal Union was first to boycott the *Journal*, and the Miners' Union didn't follow suit until the members voted to do so. Most editors in the association knew union officers didn't have absolute power under any circumstances — the members were the power. They shrugged off Curry's blathering as "unqualified rot." State Senator Frank Moore, publisher of the Florence *Ex-Parte*, pointed out "a boycott, to be successful, must have the endorsement of the community, and if that endorsement exists, it is ample indication that something must have been done in direct opposition to the community and its best interests."[26] Curry refused to acknowledge that more than ninety percent of the community had, and still did support the boycott. Boycotts were an ugly thing, and perhaps a "public evil," but citizens resorted to them because of some extreme condition that could not be righted any other way. "Let the evil alone," Charley Sumner said. "Go after the cause."[27]

Telluride was not so much a town divided as a town being whacked by a handful of unbending plutocrats who clearly wanted organized labor to crawl away. No lawlessness marred the quiet, "no desperate characters [were] running loose. There is nobody anxious to make a gun play except those hired for that purpose," Sumner noted.[28] He had fought his fight for justice, believing "the course of the Union to be right, and in its official acts has never been wrong."[29] Charley was spent. He needed a rest. He wanted to breathe without sucking in the bile of his "angelic neighbor." He knew it had no end. In July he sold the *Examiner* to John Wardlaw. He was once again a friend of the Republican county judge whose pendulum was swinging away from Wrench. Wardlaw's first love was the newspaper business. Sumner believed Wardlaw would uphold the just demands of labor and would not bend to the threats and corruption of any gang. Sumner's younger brother, Ernest, remained with the newspaper as printer and part-time reporter.

On July 11, 1903, Sumner wrote his last column, bidding the community his best. "I would like to see some things smoothed over in peace," he said. "And I hope to see my boys always successful in any contests where they are striving for decency, honesty and fairness." Sumner had the greatest regard for the Miners' Union, knowing Local 63 had never adopted an offensive attitude like the Cripple Creek local had by driving five nonunion miners from that district just because they thought the men were Italians.[30] St. John and Carpenter had welcomed men from every country and Local 63 stood the test of integrity and fairness. "They have been honest in every way with me," Sumner said of the entire union, "have never offered suggestion or dictation as to the attitude of the paper, and have never offered counsel. My position was entirely voluntary, and the unions have been open and above-board in all dealings. Friendship with its officers has failed to reveal to me the 'irresponsibility' so much prated about, and the 'agitator' is unknown.... Stay with your principles — or lose always.... To our enemies — forgiveness and forgetfulness, that's all."

His heart always on the side of hope, Sumner concluded, "The sun shines just as bright for Telluride as it ever did." He believed that one day "hate and envy and malice" would die unmourned and "from the burial place a rare flower of peace and contentment" would spring. He closed with, "Adios."[31]

His adios fell hard on the Miners' Union. At the district union convention in August, Oscar Carpenter drew up resolutions commending Charley Sumner for the years he "steadfastly upheld our principles, fought our battles by means of his paper...[and] proved a thorough and trusted friend.... He did not hesitate to stand by us, even though he sacrificed business and made life-long enemies.... In retiring from business in Telluride we realize and appreciate we have lost one of our best friends, and wish him success wherever he may locate."[32]

For a few weeks Charley lolled about Telluride in his flashy neckwear, trying to get the dandy back in his step. He lolled too long. When the Durango *Trades Journal* went belly up, the district union decided to launch a newspaper of its own, the *San Juan District Miner*, and made arrangements to purchase the *Trades Journal*'s presses. The paper needed a competent editor and manager. Charley let them rope him in. Telluride's miners had become his brothers; their struggle bound the town's future.

At the end of summer, the mines were running smoothly. Despite the TMA's posturing, they did see the value in the union and adopted the standard system of bell codes drawn up by the WFM in June, another step toward ensuring the safety of miners. After Local 63 presented a hospital report to Manager Charles Chase, showing him that Liberty Bell employees had 200 percent more illness than any other mine's, he paid to clean up the squalid conditions and put better food on the tables.[33] Due to its precarious location, the Liberty Bell dumped kitchen slop, offal, rubbish, and dead stock off the side of the mine complex. The creek supplying the town's drinking water washed through it, dragging disease down the mountain. Chase contracted men to clean up this mess as well. Other than the usual appalling accidents and deaths inherent to the industry, the Tomboy, Smuggler-Union, and all district mines were pounding away without incident. And it might have stayed that way if Governor Peabody had taken it as his duty to see eight-hour legislation enacted; if the huge USRRC had agreed to raise millworkers' wages to the standard in Colorado City, thereby settling the spring dispute with the WFM once and for all; and if Peabody hadn't stuffed his military staff with USRRC officials. It might have stayed that way if the WFM convention of 1903 hadn't empowered the executive board "to call out every member of the Federation when they see fit in the interest of the Federation," circumventing the vote of the locals as had previously been required.[34] The WFM was swiftly preparing to meet the rising storm of Governor Peabody and his corporation allies, and gathering like an army behind the latter was the national employers' movement, which spawned Citizen Alliance groups patently opposed to organized labor.

Unsuccessful strikes in the early part of 1903 had resulted in the Citizens' Alliance gaining strength in every mining town in the state. At Idaho Springs the Sun and Moon Mine was blown up. Union men were blamed and thrown in jail. The Citizens' Alliance rounded up other WFM miners, escorting them at gunpoint out of town.[35] The dispute with Manager MacNeill of the USRRC boiled into a multi-faceted, complicated strike that swept the mills, smelters, and mines from Denver to Colorado City to Cripple Creek. On August 11 more than 3,500 men walked off the job in Cripple Creek, the largest gold-mining district in the state. The MOA and Citizens' Alliance demanded Peabody send troops. This new consolidated power began a massive propaganda campaign to "demonize the enemy," to make the public across the nation believe the WFM was worse than criminal.

Anticipating a difficult contest and maneuvering to bring the full power of the WFM into the struggle, President Charles Moyer sent a telegram to the San Juan District Union, urging them to "take immediate and decided action on [the] eight-hour proposition for Mills and Smelters.... Your action will greatly strengthen the situation in Cripple Creek District and Denver."[36]

The directive was presented at the district union meeting and the members voted to submit resolutions demanding an eight-hour day for millmen beginning September 1. Charles Trimble, president of Local 63, Oscar Carpenter, and other union committee members held a conference with the mine managers, but the TMA flatly rejected the union's proposition on August 27. The TMA claimed millmen were under the November 1901 wage contract, good for another year. Nor had the union proposed a wage adjustment to compensate for the hour reduction. As it stood the proposition would be too costly, the TMA said.[37]

The union countered that, since the millmen had not yet organized at the time, they were not included in the 1901 contract. The contract bore this out. Furthermore, a number of mine managers broke that contract themselves within the first thirty days — the union had documentation. Nevertheless, hoping to prevent a strike, the millmen said they would accept a wage reduction of fifty cents per day. This offer was submitted to the managers' committee and their spokesman, Bulkeley Wells, said he'd get back with them.[38]

Instead, the response to the modified proposal was the hasty organization of the San Juan District Mining Association (SJDMA), consolidating the power of the managers of thirty-six properties in San Miguel, Ouray, and San Juan Counties. Bulkeley Wells, who had no intention of bargaining with unions under any circumstances, was on the executive board. Mine managers "pledged themselves to obey the orders of this board." The new association resolved to "stand together as a unit in all matters pertaining to the operations of mines" and "that no demands would be received and considered that tended to a reduction of hours and an increase of wages, or in any way curtailing the production and adding to the tonnage cost of mining and milling ore." This meant that any individual manager belonging to the SJDMA could not consider any propositions apart from the association. The millmen were immediately notified their requests were denied and not open to negotiation.[39]

That night at a meeting of the Miners' Union, the vote to strike was left in the millmen's hands, miners retiring from discussion. The millmen unanimously voted for a walkout. Knowing the obdurate stance of the SJDMA, Local 63 prepared for a grueling contest, fully aware the mines would shut down within hours after the mill ore bins were filled. The union arranged to lease the Belmont and Standard lodging houses to accommodate strikers that had no families. Andrew and Harry Floaten offered their Peoples' Supply for a commissary and the Cosmopolitan Restaurant agreed to serve meals at cost. All proprietors would be compensated with union strike relief funds.

On Tuesday morning, September 1, 1903, about a hundred millmen and cyanide men in the Telluride area walked off the job. All mills closed, except the Smuggler-Union's new mill and the Alta, which the foreman and his wife continued operating. A strike committee visited Bulkeley Wells, requesting he shut down the Smuggler Mill, now operated by his foreman, his office staff, a few nonunion employees, and himself — donning overalls and getting his hands dirty.[40] He refused the committee, but they convinced two of his nonunion men to cease work. Later in the afternoon the committee returned and Wells became incensed. Informing them, "No one will dictate how I operate this property," Wells bluntly ordered, "Now, clear off. Don't bother me or my employees."[41] He rang up Sheriff Rutan who sent Meldrum to keep the offenders away. Wells roped off the road to the mills and office, posting signs warning those with no legitimate company business to keep off the ground or be arrested for trespass. His cyanide workers asked permission from the Miners' Union to clean up the plant before going out, and as it was readily granted, they did not leave until that night. Miners at the Bullion Tunnel continued sending ore to the mill.

With the Liberty Bell Mill's stamps hung up, by Tuesday evening Charles Chase was forced to close the mine, throwing nearly a hundred miners out of work. The Nellie foreman asked his night boss to help him clean up the mill, but the boss told him he would do so only if he was granted an eight-hour day. The Nellie Mine shut down that night, Manager Cooper Anderson discharging over a hundred miners. The following morning the Tomboy Mine was forced to close down, except for development work that employed about twenty-five men. Another hundred miners went down the hill.

The Miners' Union solicited Mayor Bob Rogers to help maintain peace by closing the saloons and dance halls at midnight each night and on Sundays and by shutting down the gambling houses completely, according to law. After consulting with the city council on Wednesday, Rogers willingly issued the order as requested, all but two saloons complying. Rutan saw to it the two complied as well, by order of the mayor.

That evening, a few hours before the saloons shut down, one of the striking Liberty Bell millmen was chatting with policeman Kinney McLean outside the Silver Bell dance hall, a union hangout. The man was Mexican Sam Baral, also nicknamed Sam Houston on account of his features. As McLean stepped into the dance hall, a gunshot suddenly shattered the calm. He rushed out to find Baral dead on the sidewalk, a bullet through his skull. A witness had seen a man coolly shoot Baral then walk away. No other clues were discovered. Baral had been with ex-Marshal Jim Clark eight years before when Clark was gunned down in much the same. It was a baffling mystery, and no one pursued it beyond the coroner's inquest. Wrench's group was unconcerned, since Baral was a union man. Curiously the union raised no cry, the strikers remaining under control.

As if the angel of death were snatching one from each side, that same night Ike Carruthers, the Smuggler's bookkeeper, died from complications of an appendectomy. Manager Wells and Wells's wife were at his bedside. It seemed a shadowy omen.

By Friday the only mines still in operation were the Smuggler and Alta. In support of the Miners' Union, the Federal Union called out all boarding house staff in a sympathetic strike, which protected the Miners' Union from violating its 1901 contract. The Carpenters' Union was poised for similar action. Since the "legislature willfully refused to pass" the eight-hour law, and enforce the "will of the people," the unions signed resolutions with the Miners' Union to support the millmen in their fight to reduce their workday from twelve hours to eight. The millmen's fight was every worker's fight, "because we...can never hope for just and fair conditions unless our fellow workers in every industry can have the same."[42] What St. John had begun as a consolidation of unions in Telluride became a reality, the full power of organized labor coming to bear on the mine managers. The sympathetic strike effectively shut down the Smuggler and Alta Mines, putting another 300 miners out of work. By September 6 around 700 men left Telluride, seeking employment elsewhere, for they feared the strike would be lengthy.[43]

Many citizens thought this peaceful and complete shutdown would push the managers for an early settlement. They dreaded the recent turn of events at Cripple Creek, where the MOA and Citizens' Alliance's demand for troops was heard: Peabody had sent a force of about 1,000 soldiers under Adjutant General Sherman Bell's command on September 4. In that district the strike was complicated and marred by violence, but all was peaceful in Telluride, and the demand for an eight-hour day for mill and smelter workers, when all other mine employees enjoyed such, seemed just. The Miners' Union submitted a full wage scale with the fifty-cent reduction on all twelve-hour positions currently receiving $3.50 to $4.50, no reduction on ten-hour positions unless another shift was required, no reduction on any position receiving $3.00 per day, and time and a half for the first four hours of overtime.[44] No worker would be reduced below the minimum. The SJDMA refused to even consider it.

On September 6, the Ophir Union submitted the same demands to the mill managers in that village. Much to the displeasure of the SJDMA, two of the largest mills granted the eight-hour day and wage scale readily. The first was the Silver Bell Mill, part of the Ophir Consolidated Mining Company, whose general manager was Senator William Buckley. Not one to join dictatorial associations, he almost gaily pointed out that the change required only one additional employee at his mill, and that he was able to "operate the plant at a reduction instead of an increase in expense."[45] His neighbor, Captain Davidson, general manager of the Carribeau-Montezuma, followed suit, both properties continuing operation without interruption. They demonstrated good business sense as well as integrity, causing citizens to question the validity of the SJDMA's stubborn stand. Manager J. F. Keating of the Butterfly Mine and Mill at Ophir refused the millmen's request. He was a member of the SJDMA. The men called a strike, and the Butterfly shut down.

With the strike situation promising only to get worse, the Miners' Union sought a stronger leader than Charles Trimble. At the election on September 8 they reelected Carpenter as secretary and put Guy E. Miller at the helm, a man so much like St. John that the mine managers and Wrench's TBA cringed. In an effort to curtail Miller, the managers put their most self-assured member on the Miners' Hospital board: Bulkeley Wells.

Like St. John, Guy Miller was a miner, working for the last few years in the belly of the Tomboy. He was a Catholic, a Socialist, well read, and discerning. But this brown-eyed, dark-haired, big-eared man who stood an inch taller than St. John, couldn't be degraded for being an "inexperienced boy" — he was forty years old. He had a gentle, steady gaze and a humble manner that threw the arrogant off guard. Like Vint, he had a commitment to his brothers in their struggle for dignity, and the same deep integrity unmoved by flattery, unpurchasable. Unlike Vint, he enjoyed addressing large crowds, was an "eloquent orator" and held a university degree in economics.[46] He didn't have Vint's extraordinary ability to inspire individuals or to cultivate uncanny personal bonds with each, but he had the full confidence of the membership. Once again, the TBA and TMA had a formidable obstacle in their path, but they believed the glamorous and forceful Bulkeley Wells could handle him.

Guy Miller, President Miners' Union Local 63.
Restored photo © M. J. Martin 2003.

Behind Wells stood banker Wrench and the TBA "Six." They and an assortment of allies cooked up a proposition to fuse the TBA with the Citizens' Alliance movement, allowing for a larger membership, which could include anyone who detested the union. On September 21 the TBA transformed into the San Miguel County Citizens' Alliance (SMCCA)[47] and by October 27 would become a charter member of the State Citizens' Alliance. The organization claimed its purpose was for protection and to resist the unlawful demands of the unions, the word "unlawful" meaning anything the unions requested. The thinly veiled objective was the eradication of the Miners' Union and, on the state level, the obliteration of the entire WFM. With an initial membership of fifty, including Painter, Curry, Van Atta, and several city councilmen, the SMCCA appointed as chairman William E. Wheeler, cashier of Mansfield's Bank of Telluride. Wheeler, whose "frown was terrible to see,"[48] could out-wrench Wrench. The executive committee consisted of Wheeler and four others, one of whom was Addison Wrench.

The Citizens' Alliance was adamantly opposed to the TMA settling with the Miners' Union in any way, urging the managers to refuse to negotiate. When a brief window of resolution opened the first week of October, Wrench, Wheeler, and their SMCCA stormed into action. Guy Miller, Herron, Wells, and Chase had gone to Denver to confer with WFM Secretary Haywood, attorney Murphy, and the governor's personal representative, Assistant Attorney General Irving Melville. They reached an informal agreement, subject to acceptance by the SJDMA. Good feeling prevailed, according to Miller, who was somewhat dazzled by Wells's style. Wells seemed to indicate his only resistance to an eight-hour day for millmen was the additional cost it added to the product. Most Telluride ores were too low-grade to support higher costs in milling, he contended. Wells proposed lower wages — $3.00 per day for nearly all positions. This tentative agreement was presented to the union, the SJDMA, and SMCCA. The Citizens' Alliance committee directed the managers to reject this and all proposals. Their idea was to take the opportunity to rid the town of the anarchists once and for all, with help from the state. It was working well on a grand scale in Cripple Creek. It would be much quicker and simpler a task in Telluride, where only a few hundred union men were left.

The Citizens' Alliance executive committee, Sheriff Rutan, and the directors of the TMA approached Governor Peabody on October 14 to discuss the mine managers' plans to resume operations on November 1 with nonunion labor. They said without protection from the state their plan was impossible. With troops already in Cripple Creek, and a potential need for them in the state's coalfields, the governor was hesitant, but he asked the committee to submit a formal petition. Since disapproval of his Cripple Creek actions was mounting, Peabody now wanted legal documentation to defend the use of military intervention.

In the meantime, several of the larger mines rehired a number of miners for "development work," but if any man spoke in favor of the union, he was discharged. The Tomboy hired and discharged its entire development force periodically as if to shake off "union lice," refusing to rehire specific union miners when the next round of development work began. This ingenious scheme made it appear Herron was not in violation of the November 1901 contract. Guy Miller became wise to it.[49]

As the development work continued, it became obvious it was no more than normal mining operations, since the mill bins were being filled. The mines were preparing to start the mills with nonunion labor. The Tomboy union miners called a strike on October 21, charging Herron with discrimination. Although Herron and the TMA accused the union of violating the 1901 contract, it was clearly the managers who were guilty. The Tomboy cranked up its mill with nonunion employees. Thus, valued union Foreman Hutchinson and all union shift bosses went one better than a strike: they turned in written resignations, a resounding shock to Herron and a victory for the union.

Union men discovered the Liberty Bell foreman was using a blacklist when he informed an applicant that "his name was on the book" so he couldn't hire him.[50] This brash discrimination violated the 1901 contract and the Liberty Bell miners voted to strike that company as well, walking out on November 7. The contest for an eight-hour day for the millmen now splayed into a fight against discrimination and for the right to organize, a fight for the very survival of the union.

Pickets were placed near the mines, a number of them at John Conn's Japan store, which was surrounded by Tomboy property. In turn, Sheriff Rutan stationed deputies near the pickets and deputized more cowboys hired by the TMA and SMCCA. Attempting to provoke a fight, the deputies hurled stones at the pickets at Conn's store, used abusive language, and took potshots at them.[51] Knowing they could be charged with trespass, the deputies made certain not to cross onto Conn's property. Meldrum and his sidekick, Runnels, enjoyed a few fistfights in the name of law and order, claiming the miners or millmen instigated them. Curry reveled in it, drawing blood whenever he set words in type. He praised Meldrum, castigated Carpenter as a brawling drunk, made Guy Miller appear to be a skulking imbecile, and produced the most outrageous defamation against Local 63 imaginable.[52] Charley Sumner no longer held him in check and Wardlaw seemed timid as editor of the *Examiner*. He never raged against Curry's venom, nor defended truth. Curry even printed an outlandish libel claiming the Telluride Miners' Union was missing $40,000 and was refusing to let anyone examine the books. Wardlaw let out no sound from his corner. The truth was on record, documented, and available for anyone to examine: Local 63's books were audited and found to be in order to the penny.[53] The ominous, bloody specter of 1901 seemed to be rising again, this time with no one to drag it back to earth.

The Citizens' Alliance wanted to make it rise faster. Wrench, Wheeler, and an executive committee produced an official petition for the governor in record time, detailing how the Miners' Union men, led by St. John, had attacked the Smuggler company's employees in 1901, shot them in their bunks while they were sleeping, and blew up the office building.[54] Brazen lies seemed to be the order of the day. The petition claimed Arthur Collins was forced into settling with the union because the governor failed to heed the call for troops. Being sure to include the reign of terror issue, Wrench and associates said W. J. Barney was "without question, murdered by the Union leaders," St. John and Carpenter. The petition boldly flew in the face of the grand jury's statement that it was unable to establish if any crime had been committed against Barney. The petitioners added Wesley Smith and James Phillips to the list of St. John's crimes. They said, because the "same clique" still ruled the Miners' Union (Carpenter, the only remaining officer from 1901, had now become a clique), they could expect the same terror and violence as in the past, "even though no threats had been made" at this point. But they were certain if the mines and mills resumed operations with nonunion labor, as was their intention, then the resulting violence and tumult would be beyond the control of Sheriff Rutan and the civil authorities. They clinched their plea by citing the recent murder of Mexican Sam Baral, as if the union had shot him. They neglected to tell the governor Baral was a union millman out on strike. Peabody wasn't interested in the truth and they knew he would not question their patent lies. They spelled out what they thought might be an all-out guerilla assault on mines, mills, power company and water supply, and "respectfully requested" a mere 300 soldiers or less to clean out the raging anarchists.[55] The petition carried an impressive twenty-three signatures out of a town of over 4,500,[56] with three of those from one man. Cooper Anderson signed thrice, once as the manager of the Nellie, once for the power company, and once for San Miguel Consolidated. Eight signers were mine managers, including Wells; five were Citizens' Alliance, including Wrench; the remainder were city and county officials, including Howe and Rutan. Mayor Rogers's signature was noticeably missing, as was county commissioner Downtain's.

Peabody was still hesitant. He now had documentation to cover the legal questions, but the state treasury was depleted. He informed the petitioners he had made arrangements for the Cripple Creek MOA and Citizens' Alliance to underwrite the military campaign there, and could do the same for Telluride. If certificates of indebtedness were issued and sold, then the soldiers could be paid. Vice President Wrench of the First National Bank and Cashier Wheeler of the Bank of Telluride most likely provided for the purchase of enough certificates to finance the troops. Guy Miller said both the mining association and the Citizens' Alliance had to agree to "put up $155,000 in collateral securities to guarantee their payment" before Peabody agreed to order the National Guard to Telluride.[57]

The Colorado National Guard, meant to protect all the state's citizens, was on offer to Wrench and Wells as a means to crush whatever stumbled into their path. It was an offer they couldn't refuse.

CHAPTER

18

TO HELL WITH THE CONSTITUTION

"Military necessity under present conditions... recognizes no laws, either civil or social."

— Sherman M. Bell, Adjutant General, Colorado National Guard

They knew danger loomed. The National Guard's presence in Cripple Creek had already been defined by brutality under the command of General Sherman Bell, the "little tin god on wheels."[1] Union men were being beaten and deported, they and their sympathizers arrested without due process — even the *Victor Daily Record* newspaper staff was seized simply for printing the facts. Supposedly the guard was sent to assist the civil authorities and was to be subordinate to them. Instead, the civil authorities in Cripple Creek were threatened and forced to resign because they were pro-labor. The words of Major Thomas E. McClelland, military judge advocate under Bell, epitomized the military campaign and shook all of Colorado when the press quoted him on October 4. After the *Record*'s staff was kidnapped and the offices ransacked, attorney John Glover said to McClelland, "Your people apparently have not much respect for the Constitution...." McClelland replied, "To hell with the Constitution! We are going by the governor's orders."[2] Union members knew if the militia came to Telluride nothing would protect them. It seemed the state, under the administration of James H. Peabody, had ceased to be "republican in its form of government, and [had become] a military oligarchy."[3] Even the *Army and Navy Journal* denounced Peabody for sending the troops as "hired men to the mine operators...a rank perversion of the whole theory and purpose of the National Guard."[4]

Although Governor Peabody had already promised the Telluride petitioners they'd have military assistance, he became more cautious. Instead of issuing an order immediately deploying troops, he sent Assistant Attorney General Irving B. Melville and Secretary of the Board of Pardons Clarence E. Hagar to investigate the need for troops in the region. Melville and Hagar arrived in Telluride on November 8.

The two men interviewed the Citizens' Alliance leaders, the mine managers, and those men who had signed the petition. They also spoke with union men and their supporters. They requested officials from all the local unions to advise Peabody of their position. The union officials immediately did so, notifying the governor no need for military intervention existed. The strike was orderly, the district peaceful.

Knowing the governor was willing to send troops, Wheeler of the SMCCA simply gathered affidavits from members of the alliance, including letters from Wrench, Painter, Curry, and Howe, submitting them with another copy of the petition. While pointing out the peaceful conditions that "may seem to a stranger...that there is no occasion for alarm," Painter wrote, "the situation is most alarming."[5] Curry also noted that "everything is apparently quiet and peaceful," but he knew with certainty

Adjutant General Sherman Bell and General Chase's efforts to silence editors and civil authorities led to enormous public indignation.
Author's collection.

the Miners' Union was planning "to do violence to property and workmen."[6] In his official capacity as county attorney, Howe assured the governor that the leaders of the Miners' Union had "no regard for human life, property rights or any other rights of anyone."[7] Wrench said every businessman "not in sympathy with the union, feels that his life is in constant danger."[8] The idea that the union was planning atrocities and the destruction of power lines, waterworks, mills, and trams had caught the attention of several affiants who explained the line of attack as if they held the blueprints. Sheriff Rutan was certain such was the union's strategy and noted that "so many murders have been committed in this county by and under the direction and command" of the union leaders, that they needed to be eradicated from the landscape for normal life to continue.[9] Using only the disappearances of Barney and Smith as proof of the reign of terror, not one affiant presented any real evidence to back his claims. Even the sheriff failed to elucidate his "many murders."

Rutan told the governor 250 well-drilled soldiers could clear out the lawless element in ninety days. The situation was drawn the same in every affidavit: everything was *orderly and peaceful*, but as soon as the mines and mills started up with nonunion labor, as they desired, the affiants were certain mayhem would break out. Troops were "an utter necessity," and with them, the district could be cleaned out of union men forever.[10]

Believing in arbitration, Assistant Attorney General Melville held a meeting with Bulkeley Wells, Charles Chase, Guy Miller, and WFM attorney John Murphy. This meeting was businesslike and cordial. Resultant proposals were presented to the SMCCA, but Wrench would have none of it, haranguing the Citizens' Alliance with such intensity that Melville remarked, "That man is a genuine anarchist. Take away his polish and education, put him at work among the miners and he would blow up the county roads."[11] Melville was exasperated by the "animosity manifested by some members" of both the SMCCA and the union. "If this bitterness could be eliminated," he said, "it would not be difficult to reach an amicable settlement."[12] The word "settlement" galled Curry, an SMCCA member, so much that he blasted Melville as "simple minded" and "full of socialistic ideas."[13]

Melville failed to realize the mine managers and the SMCCA had no intention of reaching any sort of settlement. Bulkeley Wells was merely playing the game.[14] He hired a few deputized cowboys as sharpshooters and installed an immense revolving searchlight on top of the Smuggler Mill as a show of power. The same was done at the Tomboy. Tired of games, Wrench spread his animosity on the table. More caustic and scowling than Wrench, Wheeler usurped the bantam banker's throne, ruling the SMCCA with demands for permanent action. Both hammered at Mayor Rogers until he caved in and began believing Italian-born miners were dangerous to the community.[15]

Some citizens supposed the Italians *were* the disturbing element and they needed to be removed from the district. Others said Socialists were the problem and a campaign against them sidled out of town hall. Several were arrested, including Andrew Floaten, for "disturbing the peace" when they spoke on behalf of labor in the streets. Five Socialists were charged; the cases ended without convictions. Pickets were also harassed, arrested, and charged with anything possible. Tom Corra, at one time a marshal, was hired by the union to protect the pickets from gunmen. Stationed at Conn's Japan store, Corra wore his revolvers strapped on the outside of his vest to ensure he could not be arrested for concealed weapons. When he rode into town he was quickly arrested for "carrying concealed weapons," although every witness at the scene said his guns were visible.[16] He was later acquitted. John Conn, the prosperous Finn who owned several miners' supply stores, was himself arrested on the old information stemming from the grand jury indictments. The charge was reworked and refiled, the original $6,000 bail intact. At Conn's basin store some pickets lost their patience and got into scuffles, yelled obscenities, and raised their fists and guns in defiant gestures, giving gunmen Meldrum and Runnels a reason to work them over and drag them into jail. The city council began passing ordinances clearly designed to kink the union's activities. One ordinance required anyone distributing handbills or posters to have a license — licenses could be denied — and those doing so without license could be fined. Still the campaign failed to rile striking union men to violence, other than individual scraps. The district remained peaceful, yet the clamor for troops was incessant, Wrench's Citizens' Alliance telegramming Peabody in the middle of the night. The last of these telegrams arrived on the anniversary of the shooting of Arthur Collins, as if his blood could command.[17]

Peabody's Attorney General Nathan C. Miller advised him that actual crimes, violence, or riot weren't necessary to warrant troops, but merely the threat thereof was sufficient. He strongly urged troops be sent. Peabody signed the order on November 20, causing a public outcry against the move. At the same time, the actions of General Bell in Cripple Creek were being denounced by the press, an outraged public, and from pulpits across the nation. Guy Miller begged Charles Chase, secretary of the TMA, to push the association for further discussion. The TMA refused. The managers planned to open the mines and mills with nonunion labor, and they told Miller there was nothing to negotiate.

With Mayor Rogers now unable or unwilling to speak out on behalf of his old union friends, and all other members of the city council, except Fred Pheasey, on the side of the TMA and SMCCA, Guy Miller had few allies in the local government. Union friend Marshal McIntosh was out of commission, flat on his back after a bad fall from a bridge. Attorney Luther Kinikin, who lost his seat in the spring city election, became the union's local counsel, but he had little influence, even with his close friend Bob Rogers. Oscar Downtain had lost his seat as county commissioner to

Manager Herron in the election on November 3, removing the hope he had offered. He was still city clerk, yet had no power in that position. Many union sympathizers were mute behind trepidation. Did Wrench and his cohorts have them on grappling hooks? With trains of soldiers en route, the town seemed paralyzed to reclaim its sanity. Even the Miners' Union Hospital closed for lack of operating funds. Not enough union men were left at local mines to generate the revenues necessary.

The evening of November 21 five pickets at Conn's Japan store were ordered by deputies to relinquish their guns. An employee at the Tomboy accused them of threatening him when they called out to him to join their cause. He reported no physical contact, only verbal threats. Nevertheless, they were arrested for "congregating for riotous purposes."[18] Six deputies in the employ of the TMA and SMCCA, including Meldrum, Kenley, and Runnels, rounded up the pickets. "They pulled one man off his horse and struck him over the head with a revolver and struck another a vicious blow in the back with a Winchester," although he had dropped his gun and thrown himself on the ground.[19] The deputies tied the men's wrists to their thighs, tied them together round their necks and marched them down the trail. Curry gleefully told how the agitators were hogtied and others would soon experience this law and order if they weren't careful.[20] The Citizens' Alliance knew soldiers were soon to arrive and they gave free rein to their deputies.[21] Lights were going out for the Telluride Miners' Union and the night was falling savage-winged and sharp-clawed, a death cry on the wind.

On November 24 at 2:15 in the afternoon, the Colorado National Guard arrived. Major Zeph T. Hill was in command of the two cavalry troops and six infantry companies, about 500 men and officers.[22] Sheriff Rutan, Bulkeley Wells, and Cooper Anderson had earlier met the train at Placerville in order to brief Major Hill on the localities of the mines, mills, power stations, reservoirs, and other strategic properties where they wished soldiers to be stationed. The TMA had mapped this out precisely for Rutan, who put his signature to the document to make it appear the military was there to assist him and not to establish military rule. All railroad bridges in the county had been under the guard of deputized cowboys and gunmen before the arrival of the train, to prevent the insurrectionists from destroying those structures. Had the union any such plan, it would have been simple to merely bomb the bridges just outside the county, as they were not under guard — thus this little exercise had the suspicion of posturing. The welcoming committee rode into the Telluride depot with the soldiers and an escort of military sharpshooters, but the train only briefly stopped here. It went on to Pandora, where Wells had a fine dinner waiting for the hungry soldiers. As soon as the men were fed they were marched to their posts. More military trains followed, with livestock, artillery, and personnel.[23]

Wrench's First National Bank offered accommodation for Major Hill's headquarters. Ed Lavender, SMCCA member, offered his stables for the cavalry's horses. Every major mine and mill and the Ilium and Ames power stations offered room and board for the soldiers, making it perfectly clear in whose service they were engaged. The goods and services were provided without obligation to the State, reducing the expenses of the campaign. When the military did make purchase orders, only SMCCA members benefited: Hill ordered 400 "ticking bags" for mattresses from Van Atta, although both Nathan's and the Peoples' Supply Company could have supplied the same. No businessman who supported the union was given consideration for any military contracts.[24]

Established in 1883, Van Atta's outfitting store later had to compete with the hugely successful People's Supply Company run by the Floaten brothers.
Courtesy William "Senior" Mahoney.

Major Hill's orders were "to properly assist the said sheriff of the said county in the enforcement of the Constitution and laws of this State."[25] That apparently was to be interpreted as applying only to those who were neither union members nor sympathizers; for those, the Constitution no longer existed.

Governor Peabody personally had suggested to Major Hill that he take advantage of the state vagrancy statutes and have Sheriff Rutan arrest strikers as vagrants. He further suggested the judge might offer the offender a choice of punishment: fine, jail time, or banishment from the county.[26] This was meant to either force the strikers to leave or encourage them to go back to work. At this point, the large mines and mills refused to hire any known Local 63 member. The plan was clearly designed to rid the district of union men.

The day after Hill arrived and established camps at Placerville, Ophir, Ames, Telluride, and the mines, he made arrangements for Sheriff Rutan to begin the roundups. As Major Hill expressed it himself, "It is my intention to clean this county thoroughly of the disturbing element."[27] The "disturbing element" was defined by the TMA and SMCCA as anyone belonging to or supporting the Miners' Union. Further, "disturbance" or "aggression" was defined by Hill as "continuous use of insulting or abusive language towards troops," since no other disturbance of any sort was reported. Even Hill himself consistently reported to the governor that "Everything is quiet" in the district.[28] Guy Miller had succeeded in persuading strikers to offer no resistance, which frustrated Hill and Rutan in exercising aggression against the law-abiding citizens. Nevertheless, Hill and Rutan went forward with the plan to "clean out the county."

On November 25 under advice of Hill, Sheriff Rutan deputized a substantial number of Citizens' Alliance members and several mine managers, Bulkeley Wells, Herron, and Chase among them.[29] They then went after the union leaders and committeemen with a blanket warrant charging that the men "Did unlawfully conspire to do unlawful acts" and "disturbed the peace and quiet" by "offensive carriage" — not "abusive language" but the way they stood and walked was now an offense — and if they stood together it was conspiracy.[30] Such charges were not only ludicrous, but had no chance of conviction in court, even with perjured witnesses. Under this warrant seven men were arrested that night, including Oscar Carpenter, John Conn, Gus Mohrhardt, and Martin Marchiano. While Rutan subjected the men to a rough search by Kenley in the office, Carpenter quickly tore up a letter from WFM Secretary Haywood, tossing it on the floor. Rutan seized the bits as evidence, reassembling the letter. Concerned for the safety of union men, Haywood had merely warned Carpenter to get anyone prominent during the 1901 strike out of town, since the Citizens' Alliance was very likely to mob them. Rutan took it as an admission of guilt.[31]

Also among the union arrestees that night was Pinkerton Operative George Riddell. Neither Rutan nor Hill knew Riddell was a Pinkerton, but Riddell's instructions were to stay with the arrested men in order to learn their plans. He went to jail with them, as a brother, and made the astounding discovery that they had no plans.

Twelve men were now incarcerated in the county jail, twelve whom Curry scorned as "creatures in the form of men" simply because they were union members.[32] They had committed no crimes, except perhaps the use of words in a similar manner as Curry. They were made to remain in jail on Thanksgiving Day, away from family and friends. While Wells entertained the troops with a Thanksgiving feast, Ada Carpenter and her two children prayed for Oscar, worried he might not come home again. Wives and brothers waited in fear. In the growing dark and troubled hours it was easy to feel alone, shattered, and powerless while the icy winds of winter howled at the back of night.

If Rutan actually possessed the evidence which he told Governor Peabody proved Oscar Carpenter was a murderer, it was remarkable Carp was charged with little more than conspiring to disturb the peace. If any of the union men were dangerous criminals, it was odd vagrancy was the most popular charge against them. "Vagged" men were those falsely charged with vagrancy. Their cases were bound over for district court and bond was fixed as steep as possible. Their only hope was habeas corpus proceedings. Since Murphy's health was failing, the WFM sent former state Attorney General Eugene Engley to defend the men. Engley had handled similar cases at Cripple Creek. He and Guy Miller were in Judge Wardlaw's office on Thanksgiving Day discussing strategy. From Wardlaw's window Miller spoke across an open lot to the arrested men who were looking out the jail window. Deputy Runnels hollered to Miller if he "desired to talk to" his comrades.

"Yes," Miller said.

"All right," Runnels said. "You come on over here and I'll let you talk to them long as you want."

Miller went to the sheriff's office and Runnels escorted him into the jail, shutting the heavy iron door with a snort and turning the tumbler. Miller was kept in the cell until Friday morning. Incarceration was nothing more than an object lesson.[33]

Such object lessons were numerous, many being "taught" by the deputies whose violations of miners' civil rights got a wink and nod from Rutan and were virtually ignored by Hill. The major merely wanted the campaign ended as soon as possible. His strategy was to make the harness so tight as to cause the horse to bolt. Strikers certainly would give up and depart if their freedom of movement was thwarted.

Major Hill implemented a pass-card system immediately. Citizens were required to apply for the color-coded passes signed by Hill or Lieutenant W. R. Eaton in order to go anywhere within the county. Special "trip passes" were for persons who left the county but wished to return. Sentries posted along all trails and roads near the mines and mills, as well as at the county's boundaries, had orders to shoot anyone who tried to sneak by them. No one was allowed to enter or leave San Miguel County without military permission.[34]

Just over the range in Ouray County, union men set up a picket line to prevent scabs from going to work at the Telluride mines, much to Hill's disgust, as he could not control anyone outside his little kingdom of San Miguel. Hill warned Peabody this would only lead to trouble, yet, once again reporting the district quiet, he said he anticipated no disorder. He had no clue what the union men were up to, other than what he could see; unlike General Bell, his brain failed to simmer with the paranoia of mine and railway bombings. Wanting to discover plans of any sort, he suggested to Governor Peabody that "a couple of Pinkerton men" could be useful to mingle with the strikers. Peabody liked the plot, but wanted the TMA to foot the bill for spies.[35] Discussing the plan with the TMA, Hill discovered the Smuggler-Union Company already had spies in the union. They requested Hill keep the matter quiet.

The mines were beginning to resume operations under military guard. The Smuggler, Tomboy, and Liberty Bell had sent employment agents (also called "scab herders") out to Missouri and other states to bring in miners, offering free rail fare to anyone willing to work. The agents did not inform the men of the military

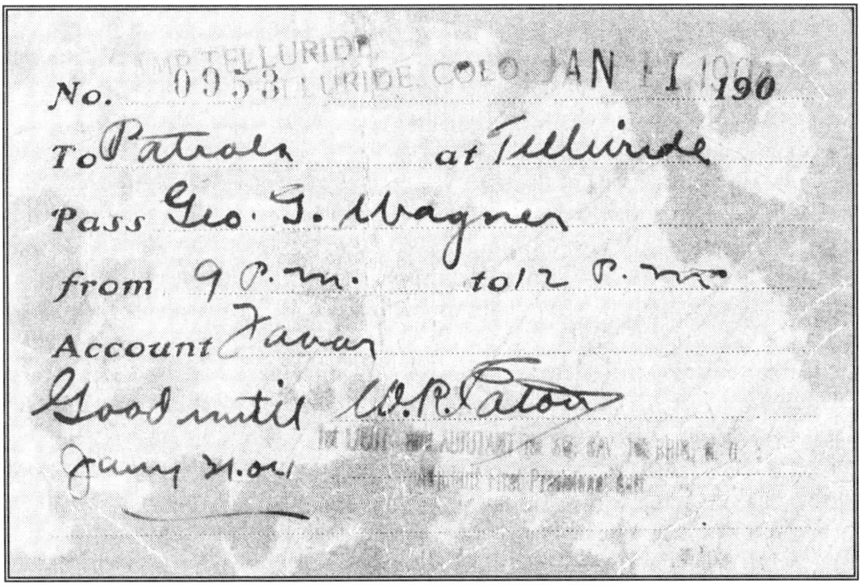
No. 0953 TELLURIDE, COLO. JAN 11 190
To Patrols at Telluride
Pass Geo. G. Wagner
from 9 P.m. to 12 P.m.
Account Favor
Good until Jany 31.04 W. R. Eaton

Color-coded military passes were issued in many cases as "favors" to certain citizens in good standing with the military command. This one was for George G. Wagner, issued by Lt. William Eaton, who was liberal in giving passes to his friends for after-curfew hours.

Courtesy William "Senior" Mahoney.

occupation or the strike. Hill knew the nonunion miners were en route and he needed to clean out the strikers as quickly as possible. Those using the Cosmopolitan "soup kitchen" or the relief sites operated by John Conn's partners in the basins, were the first to be targeted.

Although a few more arrests had been made in regard to a drunken soldier at Ophir — the dance hall proprietor and his "kitten" who sold him liquor and "favors" were thrown in jail — the actual sweep began Monday morning when Hill encircled Telluride with sentries and patrols. Squads of soldiers "in charge of a deputy sheriff" banged through town, raiding the Cosmopolitan and any other places where strikers might be found. Wells and Night Marshal Geyer led the raids, Wells pointing out which union men he wanted taken.[36] By noon thirty-eight men had been arrested and hauled to city hall before Police Magistrate Warrington Robinson, a pro-labor Democrat.

Soldiers bristling with bayoneted rifles lined the walls of Robinson's office, as if vagrants were the most dangerous creatures on earth. Vagrancy was the only charge brought against the union men. Many were Italians or Austrians who spoke insufficient English and needed an interpreter. Curry labeled them "a motley collection of filthy, ill-smelling, uncouth, ignorant foreigners."[37] Entering a plea of not guilty for all the accused, WFM attorney Engley informed them not to answer any incriminating questions put to them by Deputy District Attorney Howe, but to answer all questions put to them by the judge. One man misunderstood the instructions and, refusing to answer Robinson, was fined for contempt of court. Each man was asked if he was working. If he answered, "No," the justice asked him, "Do you want a job at the Tomboy Mine or Mill?" Andrew Floaten, who attended the hearings, reported, "If the prisoner would say No, he was fined $25 to $35, according to how firmly he said no."[38] In the end, Robinson discharged twenty-one men who had proved they had means of support, possessed substantial funds, or were too ill to work. He fined the rest, giving them two days to either pay the fines, leave town, or go to work. None were jailed.

By Wednesday none of the seventeen fined for vagrancy had appeared at city hall with payment. Rutan ordered Runnels to arrest the men. He set to the task with a squad of soldiers. The next morning the arrestees were forced to clean ditches and fix roadways under military guard. As the *Examiner* put it, "...the glistening of bayonets combined with the raw, chilly atmosphere was an incentive to cause the men to work hard."[39]

As more men arrived by train to work the mines and mills, more union men were arrested, including the vice president of the Ophir Union, Norman Young, and Local 63 committeeman Stewart Forbes, an Englishman. Sheriff Rutan arrested Tony Langeri at the Cosmopolitan for informing nonunion men who had arrived to work at the Tomboy that a strike was on. Rutan accused Langeri of "making false statements" (no strike was on) and by such, "intimidating" the men.[40] It was a bit of a stretch, but Langeri was locked up. Tom Corra was arrested again as well, supposedly for jumping bail. He hadn't jumped bail, for he was not forbidden to come and go from the county. He was returning for his court date when Deputy Kenley spotted him on the inbound train. As Kenley barreled toward him, Corra jumped off the train near Ilium to avoid a beating. After a foot race down the mountain, Kenley grabbed Corra and marched him into Telluride. Curry acclaimed trigger-happy Kenley as a brave young deputy because he didn't use his gun on Corra.[41]

A 1904 cartoon of Gen. Sherman Bell lists penalties illustrative of his megalomania:
— For staring rudely at soldiers-ten days in the bullpen.
— For making remarks against soldiers-bullpen for life, heavily ironed.
— For thinking in the presence of the General-solitary confinement for life on bread and snowballs.
— For taking the General's name in vain-instant death!

(Author's collection.)

Since nineteen bodies in the county jail taxed the twelve-occupant limit, Rutan housed the additional prisoners in the new city jail. Petitions for writs of habeas corpus were filed in county court in all the cases. The writ's function was to release the men from unlawful imprisonment, testing whether a prisoner had been accorded due process, not if he was guilty. In the cases of Young and Langeri, no complaint had been filed against them and no warrants had been issued, thus they were illegally imprisoned. Wardlaw ordered them released. Martin Marchiano and Louis Macari were also released. These men and Guy Miller filed unlawful arrest and false imprisonment suits against Rutan, raising the possibility of legal action against Major Hill.[42]

In the conspiracy and vagrancy cases, the district attorney argued the petitions for writs of habeas corpus were insufficient due to technical details. Judge Wardlaw quashed the writs and dismissed the petitions. Sixteen "vagged" men had to continue roadwork under military guard, but Police Magistrate Robinson released most of the conspiracy men on bond, including Carpenter, whose sureties were Harry Floaten and the president of the Federal Union. John Conn and Pinkerton Riddell were also released, bailed out by Andrew Floaten and Joe and Jerry Barnes. Riddell turned around and signed bonds for Gus Mohrhardt, which sparked suspicion in several union men. Riddell seemed to come up with cash out of nowhere. Guy Miller kept a close eye on him.

With more strikebreakers pouring into town, a few fistfights broke out, the strikebreakers badly beating several union men in separate incidents. None of the strikebreakers was arrested and Curry lauded their violence. He also extolled the virtues of the soldiers, failing to report numerous military infractions and misconduct on their part. Although kept out of the papers, the accounts of drunk and disorderly soldiers, their acts of intimidation, assault, and theft against citizens, their frequenting of brothels, as well as incidents of sentries sleeping at their posts and soldiers absent without leave were detailed in court martial records.[43] The only hint of military misconduct in the papers was Hill's frequently published notice warning people not to sell or give liquor to soldiers. He also ordered Rutan to close dance halls and gambling houses until the military occupation ended, since soldiers were unable to stay out of those places on their own. One small report concerning Private Thomas Clagett being asleep on sentry duty did slither into the news because General Sherman Bell bellowed to the world that "under the law death could have been meted out to" Clagett.[44] The arrogant Bell, who strutted around in enough gold braid to choke himself and who admitted freely to torturing miners in "the sweat box" to obtain confessions,[45] was merely trying to impress the "boys" to perform their duty or die. "Oh fudge, Shermie, it can't be true!" was the *Examiner*'s response (December 19, 1903). Yet for the most part both the *Examiner* and *Journal* portrayed the military occupation as just and right and the soldiers and officers as faultless. If either soldiers or strikebreakers derided, threatened, or assaulted citizens, it was no crime; yet let a union man, sympathizer, or Socialist open his or her mouth, and speaking became a felonious act. The nightmare had only just begun.

The Citizens' Alliance advertised in outside newspapers for miners and millmen, claiming there was no strike in Telluride; "union men encouraged."[46] The SMCCA and the militia were acting as a fortified employment agency for the mines. Actual agents hired by the various mines and the TMA began bringing in trainloads of strikebreakers. Ed Droddy swept into town, boasting loudly he was a "scab herder." He delivered fifty inexperienced miners to waiting wagons that whisked the men under guard to the mines and mills. By the second week of December between forty and fifty men arrived each day and were taken to the mines. The Tomboy asserted not one former employee now worked there, thus the new men would have no pressure from men who knew the mine regarding the better stopes. This translated to accidents and deaths almost as soon as the Tomboy cranked up operations.

"It looks something like old times in Telluride again," Curry chortled.[47] The new miners, proudly calling themselves Scabs from Scabville were proclaimed "genuine Americans" by the *Journal*, and Curry was certain they would flatten any skulking foreigners. "Really, it looks as if emancipation approaches."[48] He felt so confident, surrounded by soldiers and the well-armed SMCCA, that he published threats against the union's new pressman Charley Sumner and filled the *Journal* with Polly Pry's latest calumny. Since the *Post* no longer restrained Pry, she freely attacked the Telluride union leaders as the most vicious degenerates on earth. Damning Vincent St. John as the instigator of riots and murderer of Arthur Collins, she claimed he was now a fugitive from justice, although no court record existed to support that claim. She said the union men were "ignorant, inflammable" foreigners; Guy Miller was "a shrewd, crafty and altogether unscrupulous man;" and Carpenter "a cringing, cowardly schemer." Under the WFM "in both Cripple Creek and Telluride mine owners and managers walk in fear of their lives...."[49]

The only persons actually walking in fear in both places were union members and their supporters. Martial law had been declared in Cripple Creek in order to give General Bell, the Citizens' Alliance, and MOA complete control, and Governor Peabody graciously offered to do the same for Telluride. Major Hill coveted the offer, but since the district was and had been peaceful from the start, he would wait until he felt it was "necessary."[50] His "vagging" operations were working well. He did suggest that the "Military District of Telluride" should be extended to include two miles on each side of the railroad to Ridgway and the section of the Denver & Rio Grande from Ridgway to Montrose. If the governor designated this as part of Hill's district, "no possible question can then arise with the civil authorities in the counties through which the railroad passes."[51] Both Hill and Peabody were aware numerous officials in the state viewed the military occupation with disapproval. General Bell's highhanded approach at Cripple Creek had caused extensive criticism of Peabody's administration. Caution was demanded. The extension was to give Hill jurisdiction over the strike committees that rode the train with incoming strikebreakers from Montrose to Telluride. During the trips the committeemen informed the strikebreakers a strike was on at Telluride and the town under military occupation. They hoped to persuade the men to join them or at least not to go to the mines. Guy Miller had instructed all committeemen to use neither abusive language nor threats. How well they held to the instructions is unknown; nevertheless, the mere act of informing the strikebreakers of the situation was construed as "intimidation" according to Howe, Rutan, and Hill. Thus after Howe consulted with Peabody and the railroad officials, Hill assigned an officer and several soldiers to accompany deputies on the daily rides with instructions to arrest anyone attempting to "persuade" strikebreakers not to work.[52]

Some of the strikebreakers became outraged when they arrived in Telluride and discovered the conditions, a number of them refusing to take employment. A few actually joined the strikers. One affidavit, published in the *Examiner* (December 19, 1903), clearly spelled out the tale. "We were told...there was no trouble, only a scarcity of men in Colorado," said J. W. Wilkerson from Joplin, Missouri, who had come with a group of forty-six men. "After arriving in Telluride we found the conditions had been misrepresented to us." He protested being "used as a strikebreaker" at the Tomboy, where a military patrol kept guard on the property. In order to walk to town he had to show a pass to five different sentries. He wasn't about to "work under a bayonet," thus he quit and informed his fellows back in Joplin about the situation. The WFM locals in Joplin responded by placing pickets in front of employment agencies there and sending out circulars informing every union in Missouri of the deceptive employment agents from Telluride and Cripple Creek. Handbills with similar information were left all over town, much to the chagrin of Hill, Howe, Rutan, and the TMA. If they could trace the source, they declared they would throw the book at the distributor for violating the new city ordinances.

These were only miniscule setbacks for the mine managers and Citizens' Alliance. When Joplin "dried up," Bulkeley Wells simply sent a man named Donald Cameron on a hiring expedition to Michigan. The Smuggler-Union was still well under a full force and operations limped along. Superintendent O. B. Kemp had departed for California some weeks before military occupation. He had been with the company since the golden days of Nate Mansfield's management and was growing weary of the manipulations and unmerited dealings. On his return he formally resigned from the company, having made arrangements to manage a mine in Mexico. This was a blow to Wells. He quickly promoted Donald Cameron to Kemp's position.

Focused on seeing a permanent end of the union, Wells approached Major Hill with a scheme to create a local cavalry troop that could take over the protection of Telluride when Hill and his troops departed. Wells's offer would make it possible to have men on hand who could be called out instantly whenever the sheriff required military assistance. No soldier would be far from home. Further, Wells said the cavalrymen would supply their own horses and equipment, saving the State such expenses. Paying the men would be a problem for the State, Hill pointed out, but Wells was certain he could sign up at least twenty men who were willing to serve without pay — the men offering their services were cowboys anxious to bust union skulls. Hill sent the proposal to Peabody.[53]

Major Hill was anxious for a quick and tidy end to the campaign, hoping the heavy snows of winter would arrive and "stop up all the passes," ending the need for patrols at the county's high borders.[54] He had already sent forty men home. Yet to wrap up matters strikers would have to be eradicated. In order to break the strike, the TMA and SMCCA needed a way to legally carry out the systematic expulsion of union leaders from the district and to prevent their return. Attorney Howe feverishly devoured the state statutes in search of material to use against the strikers, particularly the union leaders and committeemen. "Vagging" was reserved for the soup house lads. Howe couldn't file the same information against those out on bail that was previously used, so he suggested Section 3165 of *Mills' Annotated Statutes,* which prohibited anyone from interfering with the operation of a mine or conspiring to seize mining property by force.[55] He consulted with Peabody and the state attorney general, who approved the plan. He and the TMA carefully plotted a Christmas package for the union, preparing warrants for twenty-two of the most "dangerous agitators," which included all those previously arrested for conspiracy to disturb the peace. Legal harassment had no plainer grin. A committee of TMA and Citizens' Alliance members furnished the names.

On December 22 in the forenoon, a special train was surreptitiously sidetracked and put on standby to transport the twenty-two agitators out of the county. Later that day Rutan and his deputies began quietly rounding up the men, arresting seventeen. Charley Sumner was arrested on the same warrant, charged with interfering in the operation of the Tomboy Mine on November 20. Sumner wasn't anywhere near Savage Basin on that date, but he was a dangerous man due to his defense of truth and his position as managing editor of the union's district newspaper. He was searched, measured, and tossed into jail with Carpenter and the others. When Runnels brought in Guy Miller, Rutan was amused to see him again. Referring to the wrongful imprisonment suit Miller had filed against the sheriff, Rutan remarked, "You must have had your feelings pretty badly hurt" as if Miller was supposed to laugh at being locked up for no reason.[56] Now Rutan had a "reason" although the charge had no evidence to sustain it. Miller suspected bringing the men to trial wasn't the goal.

The Citizens' Alliance and TMA were disappointed in Police Magistrate Robinson's performance — he discharged too many vagged men — thus the arrested men were taken before Police Justice Albert Holmes at 8:00 p.m. This time soldiers weren't present, since Peabody cautioned Hill that a military display made it appear the militia was dictating to the civil authorities. All the prisoners entered a plea of not guilty and Holmes set bail at $250 each, the preliminary hearings scheduled for Tuesday, December 29. Oscar Carpenter and Joe Barnes asked to give bond, but Holmes appeared too drunk or drowsy to conduct the action.[57]

The seventeen men were taken back to county jail where five vagged men already resided. This put twenty-two men in two cells meant to hold six each, nearly half being forced to sleep on the floor with no bedding. Most stood all night, trying to stay warm. No one was allowed visitors. At 2:00 in the morning Sheriff Rutan applied to Holmes to move eleven of the prisoners to the Montrose County jail, citing the crowded conditions as his reason. Holmes objected. The men could easily be placed in the city jail and Holmes also believed most would be bonded out by morning. Rutan insisted and Holmes signed the order.[58]

Around 4:30 a.m., while Telluride slept, Rutan, Meldrum, and Kenley pulled eleven men from the jail and cautiously marched them to the sidetracked special. The two deputy gunmen escorted the prisoners to Montrose and, en route, Meldrum made an offer to each man that all charges would be dropped, including the conspiracy charges, if the men promised not to return to Telluride. None promised. Kenley threatened to shoot a few. The men were given into the custody of Sheriff Taylor at Montrose. Later it was learned the plan was to deport all those arrested, but Holmes and the Montrose sheriff spoiled it simply because the Montrose jail was not large enough.

The plot was exposed Wednesday afternoon when several of those remaining in San Miguel County jail, including Guy Miller, were released on bond. Dr. J. G. Sheldon, several merchants, and Ernest Sumner, who were sureties for the men, vehemently protested the secretive proceedings and deportations. Miller tried to arrange for the release of the deported men, but discovered the law would not allow it until their preliminary hearing. They had been specifically removed to Montrose "under cover of darkness" precisely "to prevent them obtaining bonds" and to keep them jailed over Christmas.[59] Among these men were Charley Sumner, Joe Barnes, Andrew J. Tyner, and Oscar Carpenter. Since Carp had been imprisoned on Thanksgiving and was once again ripped away from Ada and their children, it seemed to union men and their supporters the move was calculated to "break Secretary Carpenter's spirit."[60]

Christmas passed in dreary silence in Telluride, union families praying they might endure. Wheeler, now the president of the Bank of Telluride (Mansfield had resigned and moved to Connecticut) scowled his happiness and the gang of saints raised a glass to law and order, while Wells indulged the troops. By Monday morning, the SMCCA deportation and incarceration plot was blatantly confirmed: the deputy district attorney had the cases dismissed against all those arrested on December 22. Those in Montrose were released immediately, but told they "will not be allowed to return to Telluride."[61] They were charged with nothing, yet banished from home and family.

Hill ordered Lieutenant Ranger and a detachment of soldiers to Placerville with instructions "to remove people not desired as citizens of Telluride from all incoming trains."[62] A few deported men attempted to return, but were taken off the train. A few others, including Gus Mohrhardt and Joe Barnes, slipped back to town via the hills. Barnes, Carpenter, and Sumner, all Telluride property owners, initiated lawsuits against San Miguel County. Any of those arrested but released in Telluride were informed that once they left town, they, too, would not be allowed to return.

Pinkerton Riddell hurried to Denver to consult with James McParland at the agency's office, and on his return, he was removed from the train at Placerville.

WFM Secretary Haywood had appealed to President Roosevelt for protection of those in Cripple Creek and Telluride whose civil rights were being violated. Roosevelt sent a friend of the Mine Operators' Association to investigate. Haywood also appealed to Senators Teller and Patterson to initiate a federal inquiry into the outrageous violations committed by Peabody's militia, the Citizens' Alliance groups, and the MOA. Unfortunately, the wheels of government turned on a rusted axle.

In the meantime, the WFM dispatched Vice President John C. Williams from Grass Valley, California, to Telluride to direct the union men and brace their lagging spirits. Several had written the WFM asking for St. John's return to Telluride. Even if they only thought he might come, it gave them hope. They were backed to the wall, now denounced as conspirators whenever they conferred with one another, as agitators whenever they opened their mouths to speak, as anarchists whenever they protested abuse from TMA deputies, and as vagrants whenever they refused to work under unjust conditions. No sign of hope, no light of settlement glimmered in the wintry night.

By January 1, 1904, 120 soldiers and officers had been relieved of duty in the Telluride campaign, since snowstorms were "stopping up the passes." Major Hill was pleased. The "military district," according to Hill, was "peaceful," always had been "peaceful," and still showed no "sign of disturbance."[63] He reported such to Peabody almost daily. But the major was losing patience with all the peacefulness and was chafing for an end to the campaign. Howe, Wells, and the leaders of the SMCCA insisted more pressure needed to be applied to the strikers before the troops departed. The SMCCA noted legal efforts failed to keep the conspirators, agitators, and anarchists out of the district. They popped up like rats out of the sewer. Hill thought perhaps a bigger rattrap, one that would terrify the union men with the prospect of execution, might accelerate matters. When one man, who was not even a union member, had verbally threatened strikebreakers and soldiers, Hill asked Peabody to look into the legalities of having such vocalists arrested for incitement to rebellion or treason, punishable by death.[64] The state's attorney general flatly barred such action, knowing it would be too obvious a violation of the Constitution.[65]

Hill simply needed something that would give him the power to drive the remaining "law breakers" from the district. He needed the iron fist of military authority to crush the rebellious lot in a permanent blow. He needed martial law.

CHAPTER

19

REIGN OF TERROR

"Habeas corpus, hell! I'll give them post mortems."

— Sherman M. Bell, Adjutant General

No shattering event or violent incident had occurred to justify the sudden executive proclamation on January 3, 1904, giving Major Zeph T. Hill jurisdiction over civilian authorities in San Miguel County. It was simply Peabody's gift to the Citizens' Alliance and the Telluride Mining Association. Both groups had pressed for it and Sheriff Cal Rutan, their ever-willing servant, consented, his compliance necessary for the record. He submitted a letter to Governor Peabody calling for martial law, claiming he was unable to satisfactorily cope with the "lawless" conditions in the county. To back up Rutan, Hill cited the failure of local courts to restrain the "disturbers and law breakers." Martial law would bestow upon Hill the power to carry out the SMCCA's and TMA's mission to "clean out the county."[1]

On paper the governor's proclamation almost looked defensible. He declared San Miguel County contained a "certain class of individuals who are acting together...offering violence to citizens and to property," and that these "vicious and lawless persons" had directed "sundry murders and other crimes" in the district. Since the civil authorities had tried but failed to control this "reign of terror," Peabody declared, "a condition exists in said county bordering on absolute insurrection and rebellion," and therefore directed Major Zeph T. Hill "to restore peace" under military authority.[2]

Reality refuted the governor's declarations. Senator Buckley put it on record that he saw no trouble in the district; his mines and mill had been running with union labor and shortened millworkers' hours since the beginning. He saw no reason for martial law. Major Hill himself reported peaceful conditions with tiresome regularity. If some murderous band existed in this town of 4,500, they weren't leaving enough corpses to be noticed. They failed to blow up mines, derail trains, kidnap officials, or even spit on the sidewalk. Peabody's order gave the appearance of terrible events while remaining utterly vague. The phrase that conditions were "bordering on absolute insurrection" was particularly bothersome to Attorney General Nathan Miller who cautioned Peabody that martial law wasn't actually declared if conditions were merely "bordering;" thus should Hill attempt to suspend habeas corpus, such action would be unconstitutional.[3] Having not considered if local courts might issue writs of habeas corpus, Peabody quickly informed Hill, "You have permission to either deport...any undesirable persons, or file informations against them."[4]

Hill assured the governor he would use deportation exclusively, for if he filed charges, he said, "it would mean my referring to the civil authorities and acknowledging their jurisdiction in such matters, which would mean that if an injunction or restraining order were issued against my taking them from the county, I would of necessity have to obey."[5]

If Peabody and Hill actually believed the strikers were dangerous insurrectionists, it was more than astounding that all they planned to do with such criminals was dump them in a neighboring county. Both Peabody and Hill and the others involved knew the game of martial law had nothing to do with any real rebellion. Captain H. M. Burge of the United States Navy specifically sought signs of insurrection in Telluride and not only failed to find any, but donated $500 to the Miners' Union.[6] To have an insurrection, Peabody needed a group of citizens revolting against the government or the established civil authorities. Nothing even bordering on this existed in Telluride; the courts were open, process being served, and peace officers were discharging their duties without any interference. No one was breaking the laws; had they been, lawbreaking itself did not constitute a rebellion. Even a riot was not a rebellion. If such events were, then "every city in the union and many rural communities are exposed to the alternative of martial law," former Governor Charles S. Thomas, a prominent attorney, wrote, denouncing Peabody's actions.[7] The Miners' Union was not revolting against any authority, not even the authority of the mine managers. Where was the insurrection?

Major Zeph Hill knew martial law was unjustifiable. He knew the most serious legal infractions were no more than a few cases of drunkenness.[8] He knew the mines and mills were operating without interference; the strikers conducted themselves "within the letter of the law." He stated it in his reports and later admitted it in no uncertain terms in the *Biennial Report of the Adjutant General of Colorado.*[9] Telluride citizens were not in fear for their lives, nor did anyone actually believe in the predictions that violence would break out once the military withdrew. Wheeler, Wrench, Painter, Curry, and their allies' "fears" were maintained as a means to achieve a goal. Martial law was simply the quickest route to that goal: the destruction of the Miners' Union.

Under Governor Peabody everything the Mine Operators' Associations and Citizens' Alliances desired was becoming reality. Peabody granted them troops. He granted them control over the troops. And now he granted them martial law, which gave them absolute control. By no stretch of the imagination was martial law declared for the protection of all citizens. It was a tool granted to a minority to use at their own discretion against the majority. Major Zeph T. Hill was in the employ of the TMA and SMCCA, and his actions, directed at only those citizens specified by the TMA and SMCCA, verified it.

As soon as Peabody issued the written declaration, he telephoned Hill to implement it. Hill in turn called Wells, Wheeler, and Howe and suggested to them to encourage a number of the SMCCA members to enlist in Company I, Second Regiment, so Hill "would have an additional force to call upon." He intended to discharge the men the moment the troops were withdrawn, their service meant for use against the Miners' Union specifically, not for the State. He also told Wells to have his cavalry aspirants enlist with Cavalry Troop B, which would give Hill more men as well as give the recruits training before they were transferred to Wells's new cavalry troop.[10] Hill reported to the governor that Wells had signed up about sixty men for the proposed local cavalry troop, pushing Peabody to expedite approval of the troop "at once."[11]

That Sunday evening, just after 8:00 the bugle sounded, bringing the soldiers from the barracks to the street in front of the First National Bank, Hill's military

The First National Bank housed military headquarters.
Courtesy William "Senior" Mahoney.

headquarters. Curious onlookers flocked to the scene. Lieutenant Eaton read the proclamation of martial law, causing some in the crowd "to rejoice and others to fear. The declaration was read from all principal street corners of the city, and shortly afterwards, copies" were posted in conspicuous places throughout Telluride. As soon as word was put out that wholesale arrests were to be made, members of the SMCCA gathered in the street, anxious to witness the crowning spectacle.[12]

At the Miners' Union Hall in the basement of the hospital, Local 63 was conducting a literary program for miners and their families. Captain Harry Gamble swooped into the hall with a squad of twenty-five soldiers, accompanied by Sheriff Rutan, who was needed to identify the specific troublemakers. With their bayonets pointed toward the people, the soldiers surrounded the crowd. Women and children scrambled to their feet, clinging to one another in fear. Union officers asked everyone to remain calm. The soldiers put their guns down "with a bang" and Gamble read the declaration of martial law, "in a voice trembling with emotion." In recording the event, Guy Miller wrote, "Why he should have trembled I could not imagine. Was it fear? A large portion of the audience was women and children. Was it a sense of his own importance in the game being played? Or did the ghost of his dead manhood rise up and protest against the brutality of the act?"[13]

No harm was intended, Captain Gamble assured the crowd. He only wanted those "whom the civil authorities considered a menace." Eighteen men were arrested, including Guy Miller and WFM Vice President John Williams. Leaving a few guards in the hall as if the women and children were also a menace, Gamble told the audience to carry on with the entertainment and marched his prisoners out. In protest, the remaining women and men sang songs against tyranny. A few soldiers joined in.[14]

Under military bayonets, the union men were paraded through the streets to headquarters, Citizens' Alliance members heckling them. From then until 2:00 a.m. other squads of soldiers rounded up numerous union men from various locations, bringing all before Major Hill for questioning. Hill took down their particulars — height, weight, age, and description — compiling a list later used by the MOA to prevent the men from gaining employment in the state. Hill also demanded to know each man's country of origin, how long he had been in the district, his marital status, and other details.[15]

The last man to be seized was former Attorney General Eugene Engley, who had been doing nothing more than conducting the defense of union men in the courts. Hill sent Sergeant Gale and a squad of soldiers to Engley's room at the hotel at 2:00 a.m., waking the man from a sound sleep. Gale demanded Engley come out at once, but the lawyer refused to open the door. The soldiers pried the door open, taking Engley to Major Hill's private office.[16]

Playing the benevolent dictator, Hill told him, "I consider it to the best interest of the district that you should leave tomorrow morning." Hill promised to allow Engley to remain in his own room until departure if Engley agreed to go peacefully.[17]

Engley was outraged. He knew his constitutional rights. He demanded to know by what authority Hill was ordering him out of town. Hill replied the governor's declaration of martial law gave him the authority to do what he considered in the community's best interests. Engley said the governor "had no right to issue such an order without an act of the legislature."

"I will not go unless I am compelled by force," Engley declared.[18]

"As you wish," Major Hill replied.

Soldiers seized the distinguished attorney and hauled him to the county jail. Thirty-one men had been arrested that night and seven previously vagged men were already in the jail. Serving up another object lesson, Hill refused to put any men in the city jail — all thirty-eight prisoners were sardined into the two small cells.[19]

At dawn the men were given "a sandwich and a cup of coffee" and taken into the courthouse. A squad of bluecoats under Captain Sholtz assembled in front of the building, while a crowd of friends and families pressed forward in the bitter cold to see their loved ones. SMCCA members came out to cheer. Without civil or military process, the prisoners were brought out one at a time between two soldiers and forced into a line. Once all thirty-one men were on the street, each in the custody of two guards, they were marched to the depot and shoved into a waiting train. Family members cried out their goodbyes, some trying to touch the outstretched hands of their fathers, brothers, and sons. Soldiers pushed them away.[20]

Under military escort, the men were deported to Ridgway, where they were removed from the train and warned not to come back or attempt to enter the county as long as martial law was in force. They had nothing with them but the clothes on their backs. The men took the next train to Montrose.

Guy Miller and John Williams contacted WFM headquarters and action was immediately taken to set up housing and a soup kitchen for the strikers in Montrose. Legal proceedings were also initiated on several levels, including actions challenging the legality of the governor's proclamation.

At Telluride, Major Zeph T. Hill assured all SMCCA businessmen, whom he called "the best people," they could conduct their trade without military interference. He ordered all saloons closed, except those whose clientele were mostly SMCCA members. All saloons at Ophir and Ames were shut down. No group was allowed to assemble on the street or in halls or even homes without written permission from the military command. The order included benevolent societies, church societies, public entertainment, and private socials and dances. A 9:00 p.m. curfew was imposed, with a more elaborate pass system put in place. The press, telephone, and telegraph came under military censorship. Since the mail was federal territory, Hill was careful not to interfere with that. He also avoided the courts. He ordered all firearms to be confiscated and registered, permitting those who were not known to be union members or supporters "to retain possession of their firearms, upon being identified by responsible parties acquainted with the military officials."[21] The "responsible parties" were the TMA and SMCCA members. Soldiers were dispatched to raid the homes of suspected agitators and to systematically search for and seize the "250 improved repeating rifles" supposedly purchased by St. John for the riot of 1901. No such firearms were discovered.[22]

Citizens were protected against unlawful search and seizure, unless they were pro-labor. Citizens were allowed to have guns in their homes, unless they were pro-labor. Citizens were given permission to congregate in groups, unless they were pro-labor. Citizens were allowed to loiter without proving they had employment, unless they were pro-labor. One need not commit any crime to be arrested, separated from family, jailed without due process, and deported. For this privilege one only needed membership in the Miners' Union, Federal Union, or Carpenters' Union. The reign of terror so often mentioned had finally begun.

Hill ordered raids and more arrests. Fourteen men were taken at Ophir. Others were taken in Telluride and the high basins. John E. Conn, the wealthy Finn of aristocratic lineage who had fled his native land due to dangerous politics, was again arrested. Monday's raids mainly targeted Italian-born miners; Conn, Hoskings and Murley were the only non-Italians arrested. A few men were allowed their liberty if they promised to leave town on their own accord. By the end of the week, another seventeen were deported to Ridgway and warned not to return.

An outcry against the unjustified declaration of martial law and the illegal arrests and deportations swept from coast to coast. Peabody dodged, claiming it was "necessary to prevent agitators...from entering San Miguel County."[23] The *Rocky Mountain News* condemned his action. The *Post* denounced it. Many asked how former Attorney General Eugene Engley fit into the category of agitator. Peabody's only move was to warn Hill to be careful how he dealt with the alleged troublemakers.

Dreading lawsuits for his actions, Hill informed Peabody he was sending two undercover military men to "mingle" with the growing congregation of union men at Montrose. "They will be conducted to Montrose by a deputy sheriff, who will abuse them quite roundly when they get off the train. My object is to give them some standing with the strikers and the agitators...." The spies were to report daily to Hill what was being said among the union men. "In this way I hope to forestall any chance for damage suits that might be brought by these people after the military occupation is over." Hill knew his actions were tainted; otherwise he would have nothing to worry over.[24]

A few union supporters, including Mary Agnes Mahoney, who had worked at the Miners' Hospital as a cook, began taking trips to Montrose to bring food and supplies and messages to the deported men. Mrs. Mahoney, wife of the supposedly murdered John Mahoney and mother of eleven-year-old Irene, had become fond of the union boys since the day St. John got her a job at the hospital when her husband walked out on her for the third or fourth time. John was presently in the Cripple Creek district, working as a strikebreaker. Mary offered the boys at Montrose her services as cook and lodging house manager. When she tried to return to Telluride, she was ordered off the train.[25]

Men who were being paid by the union to mine coal some miles below Telluride were also arrested, Tony Langeri among them. Other men took over the operation and soldiers forced them to flee for their lives over Hastings Mesa. The coal was being used for the benefit of the strikers and their families, and with the arrests, Hill effectively cut off their supply. No heat in a bitter winter was certain to drive the undesirables from the district.

On Monday, January 11, Bulkeley Wells's private cavalry was mustered in with high ceremony, with Wells elected as captain. "Troop A," First Squadron of Cavalry, Colorado National Guard was born with an initial membership of forty-one. Twenty-two were cowboys and a rancher from the west end of the county, two were mine managers (Wells and Charles Chase), several were TMA deputies (including

The second story of the Work Building was used as the armory. Deputies and Troop A men stand in front, armed and ready for action. The company clerk (first man on the left) was deputized and wore a revolver. Next to him is Harry P. Craver (in uniform), both a deputy and member of Troop A. Craver was a close comrade of Walt Kenley, also a Troop A deputy. The proprietor, F. D. Work, stands in the doorway (man with white shirt and folded arms), with a militia officer to the left. Payroll Deputy William Beattie leans against the window (second man from right in top photo).

Courtesy Homer Reid Papers, American Heritage Center, University of Wyoming.

Walt Kenley), and the rest were Smuggler-Union employees, SMCCA members, and a few young men barely out of high school.[26] The Miners' Union knew it was the Cavalry Troop from Hell. With the unleashing of Troop A, more of Hill's men were discharged from duty.

That same week Wells, now on the governing board of the SMCCA, drew up resolutions for the Citizens' Alliance to honor and endorse the actions of Major Zeph T. Hill under martial law. Happy days were here again.

Except... Except the men working the mines left in droves when they got their first month's paycheck. Few were coming in from outside the state. The WFM appeared to have the influence to stop men from seeking work were the Constitution didn't exist. Wells put large advertisements in the local and outside papers, but qualified miners were in short supply. Curry spread word the strike was over and the mines were all working at full capacity. Numerous company advertisements for more miners proved him false. Other editors said he was practicing "rank, rotten deception."[27] Press censorship may have buried the truth locally, but a number of reporters roving about town were slipping stories to the outside newspapers.

A union victory in the Clear Creek County courts swept fire and hope into every deported man: The thirteen union men charged with blowing up the Sun and Moon Mine at Idaho Springs were exonerated, and cheers rose loud enough from Montrose that Wells and Howe and Hill could hear it.[28] Wells called for more forceful action. Curry advocated hanging the agitators since insurrection was treason and treason was punishable by death. He complained loudly that Montrose was harboring the criminals and suggested they be run out. Both Captain Wells and Major Hill asked the governor to find a way to force Montrose authorities to make the strikers there move on.

In his own kingdom, Major Hill was exasperated by the theft of pass cards. He was irritated by Finns whispering to laundry girls in Finntown to quit work so soldiers wouldn't have clean underwear. He was incensed that several returned deportees continued eluding his "Argus-eyed sentries." Rumor cackled that some returnees had been back for more than a week without notice. A few came in simply to get their clothes and friends' belongings. One of these, H. C. Voss, was doing this on a large scale. He was caught, but after explaining his mission and promising to depart in two days, Hill allowed him to collect his comrades' trunks and luggage and take them via train to Montrose.

Joe Barnes, on the other hand, was a particularly irritating thorn in Hill's backside because he was becoming a local hero with his audacious feats. Joe and his brother Jerry were proprietors of a lodging house on South Spruce Street. Jerry had gone to Durango to stay clear of the military's claws. Hill knew Joe's mail was forwarded to Jerry in Durango, where the brothers owned other enterprises, but the rumors persisted that Joe was bringing cash to the deportees' wives, and his jovial moral support to the strikers in Telluride. Both Joe and Jerry were marked for committing the same horrendous crime: they had posted bond for union men arrested by Rutan. They were also Socialists, a felony in Telluride. Hill was determined to capture Joe, who had been deported with Carpenter before Christmas, but Joe slipped in and out and in again, his latest excursion wrinkling the strikers with glee. One of William Hutchinson's express wagons was called to the Telluride Brewery to deliver a large barrel to the Barnes' rooming house. The barrel was rolled inside the house into a back room and the lid removed. Joe popped out, undoubtedly grinning. Word reached Rutan, who raided the establishment, but could find neither Joe nor the barrel.[29] (Shortly after that Hutchinson sold his livery and left town — perhaps under pressure and threat of arrest.) Curry scoffed and scowled — strikers were laughing at both the civil and military authorities, so Curry condemned the tale as false.[30] Joe in turn thumbed his nose at Curry, assuring him via locally mailed letter that he had indeed slipped in again, in a barrel, and intended to stay. He still eluded Hill, despite boastfully announcing, "Here I am!"[31]

Hill needed to step up the "cleaning process," as Captain Harry Gamble called it.[32] Agitators such as Joe Barnes were making a mockery of him. He raided Finntown, finding no one to arrest. He hauled in six more Italians who were easier targets. Several of them were family men, and their wives and daughters rushed to the jail to plead for their release. Deputies sent the women away.

In the bitter cold hours the next morning (temperatures were below minus fifteen Fahrenheit most of January[33]), the young wife of one man hurried back to the jail. Unable to speak English, she tried to talk to him through the barred window, afraid she and her baby would starve without him. Wrapped in a few shawls and holding her baby against her heart, the thinly clad woman pleaded with Captain Sholtz, who finally allowed her to see her husband in jail. The meeting was brief.

The six prisoners were marched swiftly to the depot between twelve fully armed soldiers. The woman with the baby "attempted to fall in line with her husband, but was brutally prevented by the soldiers, who forced her back on the sidewalk."[34] Cold, terrified, and weak from poor health, the woman went after her husband, but the soldiers' pace was so fast she was unable to keep up. Tears flooded her desperate eyes. When she reached the depot she hastily purchased a ticket for Montrose.

The men were herded into a coach and, for a few stolen moments, said their goodbyes at the windows. The young wife tried to board with them, but soldiers forced her into a separate coach. She wrapped her baby in the shawl from her own back, kissing it gently to hush its cries, bending over it in broken sobs. Other women clung to their husbands' hands through the coach windows, bewildered, angry, frightened. Mabel, the seventeen-year-old daughter of a prisoner named Tony Munchiando, raced across the ice-cloaked platform, tears streaming down her face. She reached for her father who stood to meet her at the window, but he was "threatened by the bayonets of his captors."[35] He tried to reassure Mabel and she cried out, "Oh papa, what are they going to do with you?"[36]

His words could not comfort her. The windows were shut and the train belched steam. Mabel's heartbreak came out of her in tears and anguish as she whirled round to the gawkers on the platform and said, "I think it's a goddamn shame for men living in this country to be treated in such a goddamn manner!"[37] She collapsed against the depot wall in uncontrollable sobbing.

Several reporters captured the scene and when it hit the papers that evening in Denver and other cities the next day, the public became inflamed.[38] Yet a few editors reviled the girl as debauched because she swore. Celebrating the hurt, Curry snorted that the merciless editorial assessment "of the strumpet creating a scene at the Telluride depot" was correct and that even the worst of homes would be happier without such a girl.[39]

Finally sparked to life, Judge Wardlaw lashed Curry for his "scurrilous attack against this stricken, helpless girl," and for further assailing her with "vile, uncalled for epithets." The girl's language was strong because she was driven to desperation, Wardlaw growled in an editorial retelling the story; anyone's would be. But the innocent young woman was not a slut and Curry had no excuse for maligning her, no matter who or what her father was.[40]

The outcry put a few stoppers in Major Hill's rough handling of the deportees' families. Four days later, when Hill sent another six men out on a special train under bayonets, although a few were prodded into line away from the embraces of their wives, he gave his officers instructions to allow the kinfolk to say goodbye on the train. He didn't want any more bad press.

But bad press came by the wagonload. It boiled out of the *News*, *Times*, and *Post* in Denver, and every major paper across the country. Several small presses in Colorado stung relentlessly, particularly the *Durango Democrat*, now owned by Thomas H. Tulley. Unlike his predecessor Dave Day, Tulley was pro-labor and constantly poked at Hill, declaring the major started the insurrection in Telluride. He published amusing parodies of Hill's general orders. Quips, such as, "The citizens of Telluride are hereby authorized to breathe — but not out loud," peppered every page.[41] Hill, Adjutant General Bell, Peabody, and Polly Pry were the butt of his musings, and he proposed they create a joint monarchy and rule Colorado together. The arrogance of Bell was becoming alarming to even pro-Peabody citizens, and Hill's campaign of running men out of their homes without due process, or any process at all, had banged into the courts, giving newsmen more kindling for the growing fires of outrage.

Guy Miller, Carpenter, and Langeri filed for an injunction restraining Hill and Rutan from continued deportations and from interfering with the return of those deported. They had filed in Grand Junction where Judge Theron Stevens was holding district court. Stevens dismissed the suit. Curry's elation exploded in large headlines on February 2, 1904: "The Union Thrown Out of Court." Having an advantage in understanding the law, Wardlaw jumped off the pages of the *Examiner* with "NOT THROWN OUT," explaining the injunction wasn't granted due to incorrect technical points, and simply needed to be reapplied for without confusing "property rights" and "civil rights." Stevens implied the plaintiffs' civil rights "had been trespassed upon," but the suit was one of "equity, not of law." Giving Major Hill another shudder, the district judge also stated, "I know of no law which authorizes the seizure and deportation of persons in the manner" described.[42]

An attorney for Carpenter, Sumner, and a few other property owners visited Hill, asking if the major would allow the men to return to Telluride. Because Hill denied their return, he was certain legal action was coming his way. He wrote to Peabody, "I shall not pay attention to the court's orders unless you instruct me to the contrary."[43] Shaking off the last remnants of the Constitution, Peabody told Hill not to permit service on himself or "to obey the mandate of the court if such order is issued."[44] Thus Hill's mission of cleaning out the county under his own "laws" galloped forward, with eighty-three law-abiding citizens deported from Telluride by February 2.[45]

The program was advancing so well that a few days later the military forces were reduced to sixty-four. Captain Wells's cavalry troop of "excellent" and, as Curry labeled them, "noble"[46] men had been training with Troop B and was nearly ready to take over Hill's mission. These noble men had been demonstrating their competency throughout town: several recklessly waived their pistols around in a "threatening manner;" one abused his superior officer with obscene language; and one named Clyde Koehler attempted to kill Sergeant Linderfelt, barely missing him with a bullet. Fourteen members of Troop A nobly brimmed themselves with cheap whiskey and proceeded to nobly raid the cabins and houses of citizens in the lower part of town one Saturday night like a horde of crazed baboons. They broke down doors, shattered windows, shot off their weapons inside dwellings, chased men and women out of their homes, kicked, thrashed, shrieked, and whooped nobly until Major Hill's provost guard took them into custody.[47] They were thrown in jail. Wells simply recruited a few more like them. Koehler was found guilty, sentenced, pardoned by Peabody, and returned to duty.[48]

One of their number, H. P. Craver, who was also a deputy on the payroll of the TMA, wobbled around town under the influence, threatening to shoot strikers. Kinney McLean collared him and locked him in the city jail. Deputy Craver complained loudly that he was an officer of the law. The powers in the SMCCA warned Kinney, who had been a union committeeman, to take care who he arrested. City councilmen were already riding Marshal Joe Corey, and McLean easily could fall under their disapproval if he wasn't careful. Corey was elected city marshal when McIntosh was injured, but he displeased the aldermen because he refused to arrest strikers and was too soft on agitators. Corey stood firm, pointing out that men he was supposed to arrest had committed no crime. He was given no warrants. His one-time partner, Night Marshal Chris Geyer, had thrown in with the SMCCA early on and was enthusiastically banging union heads for the cause. While Geyer was lauded as a brave man, Corey was called a coward and several city councilmen demanded his resignation. Standing before them he said, "Until charges are preferred against me and proven, I will not resign."[49] The council chafed and burned over it for months before finally harassing Corey into departure in July. In the meantime, Mayor Rogers assigned gunman Deputy Willard Runnels to assist Corey in his duties, unleashing Runnels to crack union bones both in the county and the city. Corey's close friend, Kinney McLean, watched Corey's back and stepped lightly himself. With a young wife and baby daughter, Kinney was in no position to take on the power of the SMCCA. He had many friends among them, some asking him to join Troop A, but he simply smiled and kept himself close to his duties as a policeman. Major Hill and Captain Wells could carry on their roundups without his help. Jurisdiction was ambiguous at best, anyway; likewise was "crime."

Three nonunion miners working for Bulkeley Wells at the Bullion Tunnel got into a fight that ended in bullets and flesh wounds. A TMA deputy guard also was shot during the scrap, a bullet ripping into his calf. Major Hill and Captain Wells left the arrest to Sheriff Rutan, despite Hill's martial law prohibiting gunplay and limiting gun possession to the "best citizens." Had they been union men they would have been deported without trial and condemned as anarchists. As nonunion men, they simply were fined and sent back to work. Curry actually condoned the gunplay, saying it was a "satisfaction to know that the boys up at the Smuggler are prepared to shoot and will shoot if occasion requires."[50]

Hill had appointed a provost guard to police the town, yet the guard "policed" only those citizens belonging to the unions, their supporters, and soldiers gone astray. The city marshal's office policed the rest. Saloons were closed if frequented by union members, but those serving strikebreakers and cowboys remained opened. Initially Hill considered closing all saloons during martial law, but the mine managers requested certain establishments remain open since nonunion miners were unlikely to stay working in a town that impeded their weekend pleasures. These saloons were ordered to close by midnight, yet the curfew was 9:00, which meant if one drank until midnight, he or she would have to sleep at the bar. Or it meant those who were the "best citizens" were not subject to curfew, either.

Martial law was enforced arbitrarily, and therefore was not martial law at all. It was the caprice of Major Hill, and he was about to gift wrap it for Captain Bulkeley Wells. Troop A, in full battle array, was now on maneuvers in Telluride's streets under the drilling of Troop B's Lieutenant Strickland. While they were out terrorizing the citizenry with their skirmishes, Wells lavishly entertained Major Hill and his staff at the Smuggler complex and discussed plans for the imminent power transfer.[51]

Bulkeley Wells, Captain of Troop A.
Restored photo © M. J. Martin 2003.

Wells was the perfect replacement for Hill. He was born to privilege, confident his actions were always right, correct in his deportment, convinced laborers were beneath him; a thirty-second degree Mason, an Elk, a shining thing of the Mystic Shrine, willing to be lord and master. He had dazzled everyone at the military balls of recent days, his gracious arrogance charming both men and women. He paraded his beautiful wife, Grace Daniels Livermore Wells, and their four young children before the elite of Telluride. Grace's brother, Robert Livermore, and his new bride were introduced all around this "dangerous" town during their honeymoon stopover the second week of February. Called "Buck" by his intimates, Wells loved to hunt bear, deer, and fowl; loved to shoot in competition; loved to play polo and race fine horses; and loved to "rough it" — camping in splendor with silver service, fine china, and the best champagne.

The son of businessman Samuel Edgar Wells and Mary Agnes Bulkeley, Buck Wells was born in Chicago on March 10, 1872. Educated at Roxbury Latin School and Harvard, Buck's career ignited when he married Grace, the daughter of Colonel Thomas L. Livermore of Boston, on October 16, 1895. With the Livermore fortune to back him, he had nowhere to go but the stars. He had been in Telluride a little over a year and he was already on the executive board of the TMA and the SMCCA. He had a deputy sheriff's commission and was captain of Troop A. He had the ear of Governor Peabody and the confidence of James McParland, head of the Pinkerton National Detective Agency's western office. With Troop A at his command, the power of the sheriff's office, the military, the SMCCA, the TMA, and the spy network of the Pinkertons, thirty-one-year-old Buck believed he was about to have the Miners' Union on a silver platter, gutted and drained. Nothing would stand in his way. Everything existed to serve him.[52]

As soon as Wells's Troop A had been organized, he had turned it to personal use. Since there was no rioting or rebellion to quell, he suggested to Hill that the soldiers spend some time excavating sites where Crazy Miller had said Will Barney's corpse was buried. They moled into the back of a livery and along the banks of the

San Miguel. When no bones popped up in those places, they started digging out an old cesspool near the Denver and Rio Grande freight depot. Soldiers under disciplinary action pulled this last duty. The diversity of locations should have convinced Wells that Crazy Miller was indeed crazy, yet Wells was determined to find the remains of this "valued" employee, a man whose age, marital status, family, or where he came from was still unknown to the company. Wells never met the man, but Barney was now valued for other reasons, especially his corpse.

How Wells's obsessions, Barney's bones, and soldiers digging out cesspools fit into the plan for restoring law and order in San Miguel County baffled many citizens. Everyone knew the martial law drama was a farce in the first place. Instead of creating peaceful conditions, it inflicted hostility and fear, confusion and enmity. It gave men excuses to indulge their prejudices, even where none had previously existed. It dignified violence, crucified harmony. The *Journal*'s increasing viciousness reflected that of the law and order faction now in control. Curry and Painter belonged to the SMCCA. They were Masons, Elks, and benevolent proponents of "mob justice." In the early years, the *Journal* never promoted violence against any group or individual; now Curry was like a thing coughed up from a Gorgon's belly.

With Curry any union member or supporter, socialist or immigrant had become worthy of not only deportation but also of assault, humiliation, and ridicule. Men such as Andrew and Harry Floaten, who he had known for years as conscientious and kind businessmen, were now "vermin." Charley Sumner was an "exiled cuss" with "no brain," whom Curry threatened with injury should he return. Anyone who didn't own his or her own home and thus paid no property tax was "scum." Italians were "Dago anarchists," and "swine," whether they were men, women, or children. He rejoiced whenever a soldier "prodded" an "agitator" with a bayonet. He honored the escalating violence inflicted by the TMA's hired gunmen masquerading as deputies.[53]

Deputy Walt Kenley, now assigned to ride the train daily from Telluride to Montrose in order to point out anarchists to Hill's soldiers, encountered Mrs. Mary Mahoney en route to Telluride. Kenley informed the soldiers at Placerville and Mrs. Mahoney was removed. The next morning she was sent back to Montrose and, during the excursion, gave Kenley a piece of her mind. For this Curry labeled her a "she-male agitator." At the Montrose depot a deported union man laughed at Kenley for "letting a woman roast him." Calling Mrs. Mahoney vile names, Kenley told the union man, "If she was a man I'd have licked the stuffing out of her two weeks ago." This threat of violence angered H. C. Voss who confronted Kenley, asking him not to abuse Mrs. Mahoney. Kenley struck Voss, knocking him down, then beat him repeatedly with his fists. Curry gleefully reported the incident, suggesting Kenley should have used "his boot heel" instead of his fists. Others had tried to stop Kenley beating Voss, but Kenley's friend held them at bay with a pistol while Kenley thrashed the union man on the depot platform. Had a union man done the same to Kenley, he would have been shot on the spot. Because Voss had defended Mrs. Mahoney's name, Curry slandered the woman further, claiming Voss not only assumed the role of her protector, but also assumed "to her roles that are supposed only to follow the marriage rite." Anything vile or degrading Curry might use was served as front-page "news."[54] Wardlaw again renounced Curry's vulgarity and cheapness, suggesting Dave Day had taken over the *Journal*.

That afternoon, Kenley was on the Montrose train returning to Telluride when he encountered a number of the deported men who were traveling to Ouray. During

Kenley's seek-and-remove mission that took him through the coaches, he encountered a union man standing in the aisle. The man refused to step aside when ordered, calling Kenley a "son of a bitch." The words gave the deputy a clear right to use his fists, and he pummeled the man between two seats. Curry rejoiced. Union men got their "due," their "needings," the blood and bruises so well deserved, and Kenley, Runnels, Meldrum, Lute, Gray, and all the rest of the TMA deputies were noble men for delivering it.[55] But F. E. Curry wanted them to get more. Hatred had no bottom.

Because of the increasing number of beatings, abuse, and threats, Andrew Floaten's wife, Margaret, circulated a petition calling for the governor not to reduce Major Hill's troops and leave the citizens unprotected. She demanded protection for all the people, fearing that martial law placed into the hands of a mine manager, his cowboys, and deputies would result in despotism and bloodshed. Peabody ignored her.

Major Zeph Hill and the remaining troops were withdrawn from the district. On February 21, 1904, the command of "Camp Telluride" was placed into the eager hands of Captain Bulkeley Wells.

CHAPTER 20

MOB RULE

"Only the mob and the elite can be attracted by the momentum of totalitarianism itself."

— Hannah Arendt

"I'm no Pinkerton!" George Riddell growled at Guy Miller. "How dare you accuse me of being such a vile depraved specimen of humanity!"

Miller and WFM Vice President Williams felt certain Riddell was a Pinkerton spy. They not only had heard a scab laugh about the fact, but also suspected something was not quite right with Riddell. He had money out of thin air. He wrote letters nearly every day. He disappeared to Denver and returned, when he had no reason to. Miller disliked him. Williams found him too meddling, overly solicitous. Joe and Jerry Barnes liked him, but those two liked everyone.[1]

Rumor was rife among the exiles at Montrose that the military had spies among them, the Pinkertons had spies among them, and someone was "giving away" their private conversations. Another union man, Pat Harrigan, came under scrutiny as well, both he and Riddell swearing on their honor they weren't spies. Yet spies took oaths of allegiance to the WFM "on their honor" and still betrayed all who trusted them.

More despised than "tin soldiers" (as the National Guardsmen were called), Pinkerton labor spies were often roughly dealt with when discovered, and Riddell knew if the rumor caught fire he might be beaten. He stood his ground, even when Guy Miller accused him in front of Oscar Carpenter. He protested his innocence, damned the Pinkertons, cursed the militia, and swore against Governor Peabody. He pressed his case, bringing the "foul accusations" to WFM President Charles Moyer himself, saying he had to "endure a great deal more" abuse at the hands of the militia and SMCCA "than the majority" of union men. If he were a Pinkerton, he contended, he would have been "shown favors." Despite extreme tension and the bad feeling between Guy Miller and Riddell, the operative was unharmed. Carpenter chose to stand by Riddell, and likewise did President Schmelzer of the Silverton Miners' Union. This time Riddell would escape, with only sideways glances of distrust among his comrades and a two-year ban from WFM meetings.[2]

Strain, fear, and grinding weariness were cutting through the strikers, as hope of a settlement grew dimmer. Strike relief funds from the WFM were the only source of revenue sustaining the men, headquarters putting out thousands of dollars per month.[3] Union men twanged one another's nerves. Many longed for the leadership of Vincent St. John and pleaded for his return, still keeping his memory alive as president of Local 63 on paper, even though Guy Miller officially had that office.[4] At the recent meeting they replaced Carpenter as secretary with Englishman Stewart Forbes. Forbes was still in Telluride, whereas Carpenter could not return.

The deported union men continued their efforts to go back to their homes, most coming up against the fists of deputies and the bayonets of soldiers. Attorneys filed more petitions in the courts, but the legal process crawled.

Bulkeley Wells poses with arrogant grace in his Colorado National Guard uniform.
Courtesy Robert Livermore Papers, American Heritage Center, University of Wyoming.

At Montrose the union soup kitchen had closed. Mrs. Mahoney had been running it and WFM headquarters had been funding it. Not enough men remained to keep it open. Guy Miller had departed to solicit funds in Chicago and a number of union men abandoned the strike, seeking employment in other states. Even WFM President Moyer was unable to buoy them. All they had left was Oscar Carpenter; even Charley Sumner had returned to his mother's home in Grand Junction. They needed a new strategy or the strike would be lost. Despair gnawed at their souls.

The exiles joined their brothers in Ouray and Silverton, Schmelzer offering a plan for the men to return to Telluride in number, with an escort of Ouray and Silverton union members. While the plans were mapped out, Mrs. Kittie Heit, known as "Aunt Kittie" to miners in the region, took in numerous deported men at her St. Elmo Hotel in Ouray.[5]

In Telluride at least seventy-five strikers remained, keeping pickets on private ground at John Conn's stores near the Bullion Tunnel and Tomboy. Many of the men were housed by the Barnes brothers and fed twice a day at the Cosmopolitan Restaurant. Three days after taking the reins, Captain Bulkeley Wells, on Major Hill's advice, ordered the Cosmopolitan shut. He had no legal reason to do so except as a "measure to influence the men to go back to work"[6] by depriving them of meals provided by the WFM strike fund. C. W. Baldwin, the proprietor, believed he had a right to run his business without interference, reopening the restaurant a few days later. Wells ordered Baldwin arrested and deported.

Captain Bulkeley Wells had strutted into office at military headquarters in the First National Bank with ceremony and fanfare. Impeccably groomed and graciously smiling on his subordinates, Wells was of medium height, lean, tanned from hunting, with large dark brown eyes and thin brown hair slicked back and parted nearly in the middle. His crisp new uniform of broad-brimmed hat, dark blue tunic, jodhpurs, tall black boots, and gold insignia amplified his natural arrogance; a peacock all spangled, spurred, and armed with a pistol. He was in his element as glittering sovereign, now able to mold martial law to suit his whims. He had fifty-two men in Troop A at his command, but under martial law, he also had command of the sheriff and marshal's offices, considering himself "mayor of the city and governor of the county," as Curry extolled him.[7]

Wells continued censorship of the telegraph, putting his cyanide plant superintendent, Walter B. Reed, in charge of the operation. This gave him personal access to all telegrams sent by union members and supporters. Wells also censored the telephone and newspapers. He left the federal mail alone, as Hill had. At Placerville he stationed ten men under county commissioner John R. Galloway, now a lieutenant in Troop A, to kick incoming deportees off the train. Galloway was so fat he was unable to button his new uniform, being forced to wear civilian clothes while on duty.[8] Deputy Sheriff Beattie was assigned similar "deportee spotting" duties at the Ophir station.

Because Pinkerton Riddell reported faithfully on the plans of the union men at Ouray, Wells learned several deportees thought they might return to Telluride over Imogene Pass. The winter was bitter cold, but the snow was not heavy enough to make the high trails impassable. Determined to stop any man from coming home, Wells ordered his soldiers to build a stone redoubt on the peak overlooking Imogene. This small structure sheltered a couple of soldiers who were instructed to shoot anyone who refused to stop on command. The redoubt was called "Fort Peabody" and stood as a symbol of enmity to anyone who refused to yield to Wells's tyranny.

Troop A, sharpened by a few days' training with Troop B prior to that troop's departure, had another spree through town in celebration of the Cosmopolitan's closing. Their only discipline was building the redoubt and digging for Barney's bones in the cesspool. Most of the men and boys knew nothing of military order. They had signed on because they had guns, horses, and a mutual loathing. Just five days after Wells had taken command, six members of Troop A were "playing" with pistols in the armory on the corner of East Colorado and South Alder, on the second floor of the F. D. Work building. Ward Mooney, only eighteen-years-old, was swinging a pistol around his finger "after the manner of cowpunchers" when it fired. "The bullet plowed...deep into" Brick Drummond's head, a fellow soldier.[9] Drummond was just twenty-one. Both were close friends, having grown up in Telluride, and Mooney collapsed in frantic horror at his deed. Wells ordered Mooney thrown into jail. The town reeled. Although Curry made a general comment against the "promiscuous handling of firearms by inexperienced, careless amateurs,"[10] he attempted to blame the Miners' Union for the killing.[11]

Wardlaw exploded from dozy contentment once again. "Time to shut up," he screamed at Curry in bold type. "The limit of idiocy was attained by the *Journal*...when...it suggested that the 'elements of revolution and anarchy' were in a measure responsible for the death of Brick Drummond." Wardlaw had his fill. "The ugly ravings of the *Journal* give rise to no feeling but disgust." Wardlaw demanded Curry "shut up" and allow wounds to heal.[12]

Even if Curry did cork it for a month, with Wells in command, wounds ripped wide open. Wells's closest allies — Wheeler, Wrench, Herron, and Lavender — had

Only a shell of the tiny "Fort Peabody" remains on top of Imogene Pass. The stone redoubt, several yards above, is little more than a pile of rocks.

Photo by the author.

returned from their long vacations in sunnier climates, and pushed Wells for a swift end to affairs. Wells stepped up arrests and harassment, including that of a sixty-year-old miner who had been arrested five times under Hill and six under Wells. He was released when he asked what the charge was against him.[13] As long as any remnant of the Miners' Union was in Telluride, the SMCCA refused to be satisfied. Wells and several of the larger mine managers were still unable to get qualified workers to remain with them. Inexperience translated to weekly accidents and deaths, including the electrocution of Burtis M. Nutting, who had worked with Walt Kenley for the Club Cattle Company in Delta. Kenley had helped him get a job with the Smuggler, and the cowboy died.[14] Some soldiers also signed on as miners, staying only a short time. Finding the altitude too strenuous, strikebreakers frequently worked for one month's paycheck and scuttled. Yet at the Smuggler, even old employees preferred to work under someone other than Wells. Mel Robbins, now with a wife and baby, had resigned and went to work for the Alta. Kemp, Nicholson, and others were gone. Like the Tomboy, the Smuggler had few local miners left. Those brought in from elsewhere by the company's scab herder, Donald Cameron, balked when discovering conditions in the district, even though Wells lured them with a new reading room, complete with piano, and contracted hospital care with Dr. J. Q. Allen.

One man named A. A. Pratt, who had been brought in from Denver to work at the Smuggler, quit immediately after learning a strike was on. He stayed in town a day too long and Wells had him thrown in jail, charging him with "obtaining money under false pretenses." He was taken before Police Justice Holmes with Wells appearing as a witness against him, "although there had been no agreement made with" Wells.[15] Wells claimed Pratt had agreed to reimburse the cost of his train fare if he quit work before thirty days. Since no contract was produced and Pratt said the company's Denver officers lied to him about conditions, Holmes dismissed the case.

Within hours of Pratt's arrest, Wells ordered Sheriff Rutan to round up any man without visible means of support. This "vagging bee," as Wardlaw called it, was supposedly under the authority of the sheriff, and not a military proceeding.[16] Yet Wells had his soldiers accompany the deputies and some of the deputies, such as Kenley, were Troop A members. Thirty-four union men were dragged before Police Justice Holmes on February 29.

The courtroom was lined with armed soldiers from Troop A, although it wasn't a military matter. Captain Bulkeley Wells, making no distinction between his role as manager and as military commander, was on hand to direct the proceedings, while Deputy District Attorney Howe, who was also attorney for the Tomboy Company, questioned the men. Holmes was little more than a spectator. After Howe's "rigid examination" of each man, he determined the sentence, and Holmes merely parroted what he said. Seven men were dismissed, but the rest were fined twenty-five dollars and given until the next day to pay, leave town, or secure employment. Wells offered them employment at the Smuggler. All refused and Howe threatened that if they didn't comply with the sentence, they would be arrested and made to work on the streets at two dollars per day until the fine was discharged. Witnessing the proceedings, Andrew Floaten noted the court, military, city, and county officials "appeared to be a 'Miners' Compulsory Employment Agency,' for use and benefit of the mine managers and Citizens' Alliance."[17] Although several men had enough cash in their pockets to sustain them for months, they still were found guilty.

Henry Maki, a diligent thirty-two-year-old Finn who had arrived in Telluride in June 1903, had $304. He showed it to Howe and the police justice.

"Put the money in your pocket," Howe snapped.

Maki had a fair grasp of the English language, but was baffled by the proceedings. Unable to understand why anyone would fine him or jail him for vagrancy when he had money to maintain himself, he protested, refusing to pay the fine.[18]

None of the twenty-seven men paid fines. Several fled town before the Tuesday deadline. Thirteen of them, including Maki, voluntarily returned to the courthouse at 3:00 and others were forcefully arrested. All were marched to jail by Runnels and a squad from Troop A. Maki again pulled out his cash, insisting he was not a vagrant. This time Sheriff Rutan relieved him of it.[19] On Wednesday morning, Runnels ordered Maki and four others to the cesspool excavation near the freight depot. Soldiers had dug out the putrid muck down to bare rock, and since they failed to find Barney's corpse, the vagged men were to shovel the human waste back into the pit. The day was bleak, a piercing wind moaning over the peaks. Runnels gave each man a shovel and ordered them into the sickening, oozing mounds of muck while two soldiers with carbines stood guard.

Maki stood staring in disgust.

"Get busy, there!" snarled a soldier.

Maki refused to move. The four other miners straightened up, astonished, hope jolting through them. The soldier threatened Maki, but he dropped the shovel and flatly announced, "I cannot shovel that; that stink too bad."[20]

"Give him his shovel," Runnels told the soldier.

"The soldiers dug the hole," Maki said. "Let them fill it up."

The soldier slammed the shovel back into Maki's hand. The bold Finn hurled the shovel as far as he could and, turning to Runnels, he said, "I don't shovel this stuff."

"What! You won't work, eh?" said Runnels. "We'll fix you!"

Slapping a pair of handcuffs on Maki's left wrist, Runnels hauled him to the nearest telegraph pole, about a hundred feet away, and shoved him forward, his chest against the pole. He shackled the Finn's wrists together around the pole, snorting, "Now you'll stay there awhile," and walked away.

Henry Maki was clad in only a light jacket, his hands cuffed so tight that he was unable to move them. He could not move his body enough to keep warm in the frigid wind, yet he took the abuse in stride. Soldiers heckled him. Several citizens came to gawk and jeer. Miners' Union Secretary Stewart Forbes heard of the "barbarous act" and "raised such a protest" that Maki was finally released one hour and twenty-five minutes after being subjected to public ridicule and freezing temperatures.

While unlocking the handcuffs, Runnels instructed two soldiers, "Shoot him, if he moves," and had the soldiers keep their bayonets pointed at Maki. Vagrants were fearsome creatures. Runnels and the soldiers marched him back to jail. The man's hands were "chafed and swollen from the handcuffs." He was cold and Runnels could see it, but before shoving the Finn into the cell, Runnels removed all the blankets. For thirty-six hours Maki was denied food, water, and blankets. Strong, tough, and a symbol of heroic defiance to every man, woman, and child under local oppression, Maki walked tall the next day when Runnels hauled him to a mine to force him to work off his fine.[21] How mining figured in the city or county's public works was unexplained, although if Wells benefited from Maki's and four other vagrants forced labor that day, then perhaps Wells counted as public works since he was said to be "part of everything."[22]

Captain Wells denied having any knowledge of the pillory incident, yet most of the town knew of it within the first few minutes. Runnels bragged of it for several days. Troop A soldiers had participated. Stewart Forbes said he protested to both Sheriff Rutan and Captain Wells, and when he attempted to photograph Maki, soldiers shoved him away. The story exploded past the censors the day it occurred. John Wardlaw condemned it not only as "barbarous" but also as "illegal," pointing out how the "officials themselves realize their power and authority was exceeded."[23] Outrage brought one of the WFM's most celebrated allies to Telluride on the Thursday evening train.

Edmund F. Richardson, a distinguished Denver attorney with a reputation for being fiery, unfaltering, eloquent, and arrogant, came to defend sixteen common vagrants. His fierce tongue once caused an opposing attorney to scramble up a chair and onto the courtroom table, terrified Richardson would thrash him.[24] Fearless, immaculately dressed, his eye unwavering and his back straight, this man immediately appealed the vagrancy cases to the county court, where a blanket bond, with Andrew Floaten as surety, released Henry Maki and the others Friday morning.

The cases were tried Saturday before Judge Wardlaw. The courtroom was packed, everyone anxious to see Richardson eat a few men alive. Unlike Engley, Richardson was not a man anyone in Telluride dared to deport on a whim. This time Wells and his soldiers were conspicuously absent, the captain distancing himself from the entire affair, as if he had nothing to do with it.

"Harry Maki is the last straw," Richardson said in a tone that made Howe's lips quiver, and the celebrated attorney decried the brutality, promising to investigate the "right" that allowed civil authorities to arrest, detain, and enslave strikers as vagrants.[25] All sixteen vagged men testified, Maki once again showing his restored cash and recounting how he was shackled to a pole, the backs of his hands now bloody and peeling. Another man proved he had over $1,000. None had been without funds.

Howe's only witnesses were Deputies Runnels and Kenley and Night Marshal Chris Geyer. None could cite any incident where the men on trial had asked for a meal or money. Howe insisted the men were vagrants because they had been loitering and he launched into the "crimes" committed in the past three years, saying Richardson would not "defend these criminals with such touching eloquence" had he witnessed the reign of terror. Howe was mired up to his eyebrows. "It makes no difference whether a man has money in his pocket or not," Howe babbled, "whether he had gained that money by previous labor or other means...idleness constitutes a violation of the statutes."[26]

Richardson suggested Howe should be arrested the next time he was standing about on the street corner. Howe glowered. Razor-edged, the defense attorney then took Geyer, Runnels, and Kenley's flesh from their bones, forcing Kenley to admit he was on the payroll of the mine managers' association. As Richardson summed up his case, criticizing the civil authorities and particularly blasting the felon Kenley, he concluded no evidence supported conviction.[27]

The court concurred; state vagrancy statutes didn't allow conviction unless the "evidence is very complete and the offense most aggravating." Turning to the sixteen accused men, Judge Wardlaw said, "Gentlemen, you are discharged."[28]

Maki and his fellow union men were bewildered at first, having been convicted in the police court despite their same testimony. As their freedom sunk in and the courtroom emptied, the men thanked Richardson and Floaten and their supporters. Howe muttered that the militia should have solely handled the affair, then no WFM lawyers could interfere.[29] Richardson, Floaten, and Forbes left the courthouse and walked to the Sheridan Hotel, and irate Walt Kenley bolted after them.

"You used some rough language referring to me today," Kenley said to Richardson as he stepped in front of the attorney. "Just what did you mean by it?"

Richardson said nothing.

"I want an apology right now," Kenley snapped.

"I am making no apologies," Richardson replied coldly, attempting to pass the deputy.[30]

"You'll apologize to me or take the consequences!" Kenley snarled, his fist smashing into Richardson's face. The attorney crashed backward into the screen door at the hotel entrance.[31]

With his hand to his left eye, Richardson staggered to his feet, but Kenley struck him fiercely in the face again, knocking loose his teeth, blood spilling from his mouth. Andrew Floaten rushed to help Richardson and Kenley thrashed him across the face with an open hand, knocking his glasses askew. Men leapt forward to help Floaten and Richardson as Kenley attempted to get in a few more punches. Policeman Kinney McLean grabbed Kenley and the provost guard hauled the deputy to the county lockup.[32]

Walt Kenley was unrepentant, bellowing he'd do the same all over again. Curry applauded him, suggesting the deputy should have "bided his time until he got" Richardson "to one side where he could have done the job up right before the military could have got to him and stopped him."[33] The vicious proposal inspired Runnels and a number of other deputies to stalk Richardson as Floaten and Forbes escorted him to the train Sunday morning. Wells and a squad of soldiers appeared on the scene to see the attorney safely on board. Floaten filed a complaint of assault and battery on his and Richardson's behalf against Kenley. The deputy was out on bond that night.[34] A week later he pled guilty to the charge on Howe's recommendation; he was fined only twenty dollars because assault by an officer was less of an offense than vagrancy in Telluride.

With most of the strikers gone from the region and Governor Peabody convinced Captain Wells and Troop A had restored peace, Peabody revoked martial law in the district, the declaration being bugled in the streets at 10:30 p.m., Friday, March 11, by Wells. Troop A was relieved of duty, the provost guard discharged, and gambling reopened with a vengeance. The Citizens' Alliance toasted and praised Captain Wells and Peabody while the town partied until after midnight. The long awaited heavy snow started falling and Wardlaw twittered that it had stayed away due to martial law. Curry was not amused. The *Examiner* coolly stated that with martial law ended, "The eighty-six men who have been deported from this county are at liberty to return."[35] Curry was not amused. Joe Barnes was at liberty to come out of the woodwork, and he did, having eluded the militia of Hill and Wells with a grin.

At Silverton about thirty Telluride union men prepared to go home, and at Ouray about fifty packed their belongings to head for San Miguel County. News that a large number of imported scabs in the Cripple Creek district had joined the WFM and now were renewing the strike there sent up a cheer from the deportees. Neither Curry nor Wells were amused.

Charles Painter, on the other hand, gaily offered, "Let's forget everything," as if his *Journal* had never incited deep hatred against anyone.[36] Although advocating harmony in general, Wardlaw pointed out the *Journal* was rather inconsistent, since it promoted the beating of attorneys and other such anarchistic pleasantries. "You're a peach — simply a peach," said Wardlaw. "But the trouble is, you're a little too ripe!"[37]

In an editorial to Telluride's citizens, Wardlaw offered friendly advice to all parties in the conflict, contending all had suffered equally in the strike, but, he pointed out, "The strikers...have not exceeded their rights."[38] He believed the deportees had a right to return and get on with their lives, concluding the strike was over, although it was not. The lanky judge, who once announced, "I am against the Union," seemed to be climbing a different mountain. He dared to go a step further, drawing attention to the absence of the "old time prosperity" in the camp. The new miners were from out of state with no ties to Telluride, sending their money elsewhere while in the community only a month, then moving away. Several businessmen who were union supporters had also departed, unwilling to remain in a town that refused to compromise. The permanent population had dropped dramatically. Even Ernest Sumner had left the *Examiner*, he and his wife moving to Grand Junction. When pneumonia suddenly dragged Police Magistrate Warrington Robinson to the cemetery, his wife and children moved to Denver. Several doctors, merchants, and a number of others had recently pulled up stakes. Hatred was draining the town's soul.

The day after martial law was revoked, Tony Munchiando and Tony Langeri came back to their anxious families. Two more deportees came the next morning, and six came in the evening. They arrived from Silverton via Ophir. By Monday at least thirty arrived on the evening train, embraced by their wives and children in tearful reunions. None were armed. None made any threats or noise, simply going to their homes and rooming houses in quiet groups.

"Something must be done," was the cry from Citizens' Alliance members who worked their comrades into a frenzy with rumors that the returned union men would arm themselves and assassinate those who had abused them. Word spread through town that all SMCCA members should meet at Red Men's Hall, 8:00 p.m. sharp. Buck Wells, Charles Chase, Cooper Anderson, John Herron, Ed Lavender, Deputies Kenley and Runnels, Night Marshal Geyer, Wheeler, Painter, Van Atta, and most of Wells's office staff and Troop A members were in the crowd, Wells and Herron taking the lead. The meeting lasted an hour, after which the crowd dispersed, some going to the armory with Wells, who distributed government rifles to a number of men. Others went home to get rifles and pistols, and all met again in a half hour at the First National Bank building. About fifty men assembled, heavily armed, dressed in overcoats and slouch hats.[39]

Most of the mob was anxious to start, although some older men, such as ex-Mayor Hilgenhaus, were fearful of encountering armed union thugs. No thugs existed, and Wells knew it, assuring the crowd they would have the upper hand. The union men, and especially their supporters, suspected nothing. The order was given and the mob swarmed down Fir Street to the alley between Colorado and Pacific, banging in doors, ransacking houses, grabbing men from their beds. They searched for weapons, finding none.

The returned deportees were not the only targets of the mob. Anyone who had ever supported the Miners' Union or posted bond for union men, or were members of the Federal Union, were now subject to mob violence and arrest. Antone Matti, the local brewery agent, was ripped from his bed, his wife shoved and insulted when she protested. Jerry Barnes, who had come up from Durango to join Joe, was struck and dragged out of his home (Joe vanished into the woodwork). Stewart Forbes, a graduate of Queens College, Cambridge, England, was taken from the arms of his wife and three children and shoved out into the icy night without a coat.

Several dozen more men joined the mob, swelling the ranks to about a hundred. They tore through town in smaller groups, one consisting of the mayor and more "conservative" men, ransacking various parts of Telluride where they knew "agitators" and "anarchists" resided. Attorney Luther Kinikin, having been retained by the Miners' Union and recently involved in their efforts to shut down illegal gambling, feared he would be taken. He telephoned four friends to join him at his house with shotguns. He recalled, "We pledged ourselves to shoot simultaneously if the mob stepped on my premises." The "conservative" group approached and one among them yelled to take Luther Kinikin, the "baldheaded son of a bitch," to put an end to his "law enforcement crusade." Although swept along by the mob, Mayor Rogers objected; Kinikin was his friend.[40] The attorney was saved.

"The hum of voices, the shouts to open doors and banging on them produced a queer sensation, a feeling akin to that caused by the noise of an approaching tornado," the *Examiner* reported (March 19, 1904). Each time they grabbed a victim, they brought him to a vacant lot near city hall, where the captives were herded together in the cold and guarded until all the undesirables were collected. Six Italians were seized from a rooming house in lower town, while another portion of the mob launched an assault on Andrew and Margaret Floaten's house.

Andrew and his brother, Harry, had been living in Telluride for thirteen years and had been running the largest dry goods business, the Peoples' Supply Company, since 1893. The two Norwegians were exemplary citizens, both naturalized, docile, well-loved members of the community. Forty-seven-year-old Andrew was the more outspoken of the two. He had a seventeen-year-old son, Lewis, who had been injured in a hunting accident. Highly intelligent and compassionate, Andrew had given a series of travel lectures, scientific presentations, and wrote for various magazines. As a Socialist he advocated non-violence and Christian precepts always, but as had been demonstrated when Kenley attacked Richardson, Andrew was not afraid to interfere to stop violence. He preached at no one, although if the affluent scorned his politics, he offered the scriptural passage from James 5:1-6, which warned the rich against cheating those who labor for them.

This cold night, Andrew was half undressed, preparing for bed near the kitchen range, unaware of the terror raging through town. Margaret was already upstairs in bed. Hearing men in the alley behind his house, Andrew realized he was about to be invaded. Within seconds, someone smashed the butt of a rifle into his front door, demanding he open up. He rushed into the first floor bedroom as the mob shattered the glass panel in the front door. From the top of the stairs, Margaret demanded to know who was entering her home. They gave her no reply, destroying the lock and swarming in. Margaret screamed as the men ransacked the house, searching for Andrew, shouting, shoving, kicking things over. Andrew darted into the closet, but Walt Kenley shoved a pistol into the Norwegian's face.

"For God's sake, have you come to kill me?" Andrew gasped, knowing he had committed no crime.

Kenley ordered him out of the closet.

Regaining some composure, Andrew asked if the deputy had a warrant for him. Kenley said he did.

"Then please read it," said the merchant.

"I don't need any warrant for you," Kenley snarled, and he and the others shoved the barefoot, half-dressed man out of the room. Andrew asked to put on his shoes. The deputy swore at him, "You don't need any shoes," and smashed the pistol into Andrew's head, cutting a deep gash that spilled blood down his face and over his shirt.

Margaret rushed to her husband with his shoes, hat and coat in her hands, begging them to allow Andrew his clothing, but the men pushed her away and another of them struck him again in the head. Kenley yanked Andrew by the arm and the mob dragged him from the house, forcing him over ice and frozen mud down the alley. Taunting, laughing, insulting him, they came to a large puddle and one of them yelled, "Shove the son of a bitch through the water!" Kenley obliged. By the time they reached a sidewalk, Andrew pleaded to be allowed to use it; his feet were cut and bleeding. Kenley refused, striking him over the head again with the revolver. Someone screamed for Kenley to shoot the merchant, but Kenley shoved him forward until they reached the group at the vacant lot.[41]

The mob had gathered sixty-five captives, most wearing no coats or hats against the icy night. Several were without shoes or shirts. About fourteen were injured. None had been found with weapons. None resisted the onslaught, except one frantic man named W. A. Schiller. He decided to protect his home by threatening to shoot the invaders, although he had no gun. This fellow so enraged the mob that they fired a volley of shots at his house. He escaped through the rear window, only to be captured a day later and run out of town.[42] One man said Runnels robbed him of fifty dollars. Another man named Matthew Lingol was beaten into submission because he ran from the mob. He and Floaten were among the most seriously injured. Several men offered their handkerchiefs to bind up Andrew Floaten's head, but still his wounds bled profusely.

After freezing in the vacant lot until after midnight, the men were herded into an empty storehouse on Colorado Avenue that had been recently used for the military commissary. Again they were subjected to abuse, but a few of the wives and friends managed to get them shoes and coats. Above this store was the office of Dr. E. D. Allen, and a man rushed there to get help for a family member with pneumonia. The man was roughly seized by a storehouse guard and hauled down the street.[43]

The mob's original plan was to march the men to Keystone below town, then force them out of the county. Since the mob members were getting chilled in the frigid air, they decided to send the men out by rail. A special train was hastily chartered and the captives were herded into two coaches, a man yelling after Floaten, "If that fellow comes back to Telluride, he'll be hung!" As the train pulled out at 2:20 in the morning, Wells and his fellow mine managers, and Kenley, Runnels, Painter, and the "best citizens" in Telluride comprising the mob, whooped triumphantly, firing "volley after volley of shots as a parting salute." Sixteen armed guards rode the train to Ridgway where the sixty-five men were told to get out. They were warned, "Any son of a bitch who tries to come back to Telluride will be shot."[44]

Fifty-three of the terrorized, weary, and injured men walked the eleven miles to Ouray, reaching that town and the welcoming arms of Aunt Kittie and the Ouray Miners' Union just after dawn. Floaten was so weak they left him on the road and fetched him with a milk wagon, his wounds tended by a Ouray physician. Twelve of the men were barefoot or too exhausted to walk, waiting at Ridgway for the scheduled train. Andrew Floaten, Stewart Forbes, and Antone Matti took the first train out of Ouray to Denver to lay the outrage before Governor Peabody.

The governor dodged Floaten and ignored angry telegrams from around the state. A Durango telegram asked him, "Does not the situation at Telluride warrant and demand that you send troops there under command of an officer not identified with the mob who last night outraged law and constitution...?"

"The dispatch does not warrant a reply," said the governor, "and none will be made."[45]

Despite the Denver newspapers raging with the story and interviewing Peabody himself, for three days the governor claimed he had "no official information of the trouble at Telluride, and shall take no action unless the authorities make the demand."[46] Since the Telluride "authorities" were members of the mob and the governor was told such, his statement was worse than ludicrous. Finally Floaten and the others appealed to attorney John Murphy, since Peabody continued pretending to be unavailable. Murphy telephoned the governor without result, finally dispatching a letter, asking him to give audience to the three representatives from Telluride. Murphy explained numerous men were prevented from returning to their homes and were threatened "that if they do return their lives will be taken."[47] Peabody ignored Murphy as well.

Since the beginning of the strike in Telluride this was the first instance of actual brutal violence, and the armed mob that attacked citizens in their homes, destroyed their property, and flagrantly violated their civil rights, was now illegally patrolling the streets of the town. Members of the WFM executive board told the press the miners "have a perfect right" to return.[48] Although Bill Haywood suggested the deportees should return "armed simply for their own protection,"[49] President Moyer and other officials said the men would not be armed. The WFM pressed the governor to protect the union men and their sympathizers as he had protected the mine owners and theirs.

"If these people think they have been wronged, let them ask redress in the courts," Peabody said, making it clear the militia was reserved for the MOA.[50] "There is one thing, however, upon which I shall insist most firmly as long as I am governor of this state. This is that armed men will not be allowed to parade in this state unless authorized to do so by proper authority."[51] Such armed men were "parading" at that moment in Telluride without authority, yet the governor aimed the statement at the Telluride deportees, not Wells's mob. State newsmen condemned the governor's statements in scathing editorials.[52] Attorney Murphy began proceedings to apply for a restraining order against Wells and his associates.

By this time national newspapers were screaming outrage, denouncing the Telluride mob and Governor Peabody. In an effort to excuse the mob's brutality, both Curry and Painter hammered out editorials claiming the deported men were vile butchers and criminals who had done worse to strikebreakers in July 1901, marching them "barefoot" over the range — the barefoot part was to justify taking barefoot men through the ice and cold Monday night.[53] Barefoot also came from General Gardner of "Wrath of God" fame, and was entirely fabricated, but Curry knew repetition created fact. Even if it was merely an eye for an eye, the excuse didn't wash, since the 1901 strikebreakers had been armed and engaged in a shooting fracas; none were dragged from their homes in the night; and none were established merchants tending to their own business. The temperatures in July were no match for the below freezing temperatures of March, and nearly three years had lapsed since the riot, thus giving no one the pretext for acting in the "heat of passion." Curry knew this, submitting, "No good American will attempt to condone or justify mob action," then did his utmost to condone mob action until he shrieked at those who condemned Telluride that "no other course was left except a complete and abject surrender to anarchy...."[54] No other course left against sleeping, unarmed men?

"No violence was done" by Wells's "orderly" mob, Curry screamed, claiming the 1901 union men had robbed their victims of "everything of the slightest value," beating them, shooting them, leaving them for dead.[55] Each time he mentioned the incident it grew worse, but the record and testimony from scabs subjected to the July 1901 expulsion failed to support this. The extent of the violence against them was the one gunshot blast that struck three men with pellets, none seriously injured. A report that a Cornishman named Edwin Thomas was beaten by strikers and left for dead, Thomas himself refuted.[56] He came back within the first week and went back to work at the Smuggler. Jack Hyde had been untouched as well. No amount of justification from the *Journal* could wash Telluride's hands.

Even a weak assist from the *Examiner* failed to redeem the guilty, Wardlaw taking a sudden about-face as if fear rattled his bones. Had someone warned him to take care? He had defied the SMCCA by discharging the vagrants, and many citizens declared this action signaled the deportees to return. Wardlaw at first described the cruelty and violence of the March 14 rampage in clear terms, had dispatched all to the Denver newspapers, yet later noted it was "one of the most practical and impressive object lessons ever administered." He launched into an editorial claiming the mob had just provocation because Vincent St. John had ruled the town two years before, during which time Barney, Smith, and Collins were murdered. "St. John...was like an octopus with tentacles extending in every direction."[57]

Such absurdity suggested Wardlaw had gone on a drinking binge after the mob outrage and was still teetering. The *Journal* gushed how good it was to see him "back on the side of decency."[58] Whiskey was the refuge of the frightened. Rumor claimed Sheriff Rutan was hiding inside a bottle as well. As Rutan's drinking habits trickled into public notice and his apathy became apparent, it was said the TMA trusted him only with simple tasks, as he "might forget himself some day and get lost in the courthouse."[59] Wells and the Citizens' Alliance were running the town, with more than a dozen armed civilians illegally patrolling the streets and depot. These men personally informed any remaining undesirable citizens they had to leave, quietly threatening them with violence. Twenty strikers departed Thursday, six Friday, and more than thirty Finns, their wives and families, fled together, terrified the mob would come for them next. They took all they owned, having no desire to ever return. Other citizens fled as well, letting it be known they were afraid of the mob and would not return until Telluride was restored to order.[60]

On behalf of Andrew Floaten, Matt Lingol, and fifty other plaintiffs, John Murphy filed a petition for an injunction with District Judge Theron Stevens at Montrose. On March 22, Stevens granted a temporary restraining order and injunction against the defendants — Wells, Runnels, Kenley, Herron, and thirty others — barring them and any members of the Citizens' Alliance, the MOA, their agents, and "all persons acting under their direction and control" from interfering with the lawful return of the plaintiffs to San Miguel County. The order also enjoined the defendants not to "molest, injure, or deprive the plaintiffs...from enjoying all their natural and legal rights."[61] On behalf of the plaintiffs, Ouray property owners Florence Ziller and Thomas Malesardi signed the $2,000 bond required by law for an injunction, and the additional $1,000 emergency bond.

Most of the deported men were still at Ouray where Forbes, Floaten, and several others joined them. Frustrated and angry men in the Silverton and Ouray unions suggested having an armed escort of fifty or more go with the deportees to protect them, but the idea was scrapped early on. Some union men had written to St. John, asking him to lead them home, and they "expected him at any time."[62] St. John's spirit was with them, but he, Jardine, and Carpenter were presently committed to work in Utah (after Carp's wife and children left Telluride, the WFM sent them to join St. John and Jardine).

The thought of St. John leading a large body of men into Telluride made his enemies shudder. Rumors that he and several 1901 strike leaders were returning with the deportees reached town via Riddell, who prepared to join the throng. In an effort to scare away or allow the arrest of any of those involved in the 1901 strike, Howe hastily concocted charges against these men, despite the cases being dismissed in July 1903.

News of the restraining order hit even harder. Wells and the SMCCA needed to find a way to defy the court. Because of the large number of defendants, writs would take a while to serve. In the meantime, Wells and Howe whipped out a petition to Governor Peabody the day the injunction was signed, asking for troops to protect them from the mob that was assembling with the "purpose of destroying property and inflicting personal injuries."[63] No proof of such a mob with this intent was necessary. Runnels signed as proxy for Rutan, making most of the thirty-one signatures from members of the armed mob who ransacked the town March 14. Neither the elected sheriff nor Marshal Corey signed this petition that claimed they were unable to maintain control. Alderman Fred Pheasey refused to sign it, yet the SMCCA left him alone.

Unlike the evasive delays Peabody gave to the union men when they asked for military protection, he swiftly ordered troops to Telluride under Adjutant General Sherman Bell on March 23 and declared the district in a "state of insurrection."[64] All military orders were kept secret, sealed from reporters. The same day he also ordered troops to Trinidad to suppress the coal strike there, the coal companies and railroads putting up $200,000 to pay for the expedition.[65] Attorney General Miller cautioned Peabody to have the troops "act in conjunction with the civil authorities...." Arrested persons were at some point to "be dealt with by the courts," however, Miller advised, it was up to Peabody's discretion as to how long the military detained these prisoners before they were given due process.[66]

Since at least thirty-six hours were needed to mobilize, General Bell called out Captain Wells and Troop A, instructing Wells to implement martial law without delay. This manifestly disregarded the district court's restraining order against Wells. Bell told Wells to "recruit some 100 men if possible" into Troop A in order to control the district until Bell's arrival.[67] Wells immediately signed up twenty-one more men, all of whom had taken part in the mob violence on March 14, and were also under the restraining order. They included Manager Herron of the Tomboy, William E. Wheeler, Ed Lavender, and W. B. Van Atta, Floaten's business competition. A telegram to Norwood brought a "detail of cowboy cavalrymen" from that area, "tearing down the street" on their horses in enthusiastic whooping,[68] giving Wells eighty men at his command. He placed censors at the telephone and telegraph offices and established headquarters in the First National Bank building once again. He personally warned Judge John Wardlaw, editor of the *Examiner* and correspondent for the Denver newspapers, "It will not be necessary for you to send out any news concerning the Telluride situation."[69] Then at 4:40 that afternoon, Captain Wells read the proclamation declaring San Miguel County under martial law.

General Bell had boasted on March 17 that if he had been present when the Citizens' Alliance mob invaded union homes, "I would have shot every man in Telluride before I would have permitted these men to be deported."[70] He now gave Wells orders to block the deportees' return, defying the court and civil law. Immediately after martial law was proclaimed, Wells declared any deportees attempting to return "will be regarded in the same light as they had been heretofore."[71]

Since Wells was informed of the injunction by telephone only and the court papers had yet to be placed in Sheriff Rutan's hands, the deported men remained in Ouray, consulting with WFM President Charles Moyer. Moyer had joined them on Thursday, March 24. He wired the governor to ask if returning

A 1904 political cartoon of Sherman Bell, first published in the Victor Record.

Author's collection.

miners would be protected. Peabody replied, "I have no disposition to interfere with or intercept the movements of unarmed citizens…."[72] Moyer took that to mean that once the court papers were served, the miners could return unmolested. On Moyer's advice, the men decided not to return all at once, selecting just twelve of those with families or property to go first. Depending on how the militia in Telluride received them, then the rest could follow. Moyer let the authorities know none of the men would be armed.

The following day, March 25, while a severe snowstorm hammered the region, General Sherman Bell rolled into Telluride on the train with 276 soldiers and a Gatling gun. Having been uninformed, the residents were taken aback by the display, assuming Captain Wells and his men would maintain the district. No one actually believed the miners planned an assault on the town. Bell set up headquarters in the more spacious Sheridan Hotel and ordered Troop A to patrol the town as provost guard, while his soldiers protected mines, mills, power, and water supplies. Just prior to the troops' arrival Sheriff Rutan received the restraining order documents from Montrose, but waited until he consulted Bell before serving the papers. That afternoon Peabody wired General Bell with an urgent message:

> *You will permit unarmed, law abiding citizens to return to Telluride unmolested, and if they conduct themselves in a lawful, orderly manner make no interference, but incendiary talk or actions, or arming themselves*

The fashionable hotspot in Telluride was the Sheridan Hotel, which General Bell used as his headquarters.

Denver Public Library, Western History & Genealogy (#X94).

> *with deadly weapons, or having same in their possession, will not be permitted or tolerated and you will promptly arrest all such persons and hold them for trial and punishment by the civil courts.*[73]

Peabody was afraid to openly defy the court, or at least wanted it on record that he had done his part, for he and his MOA advisers were plucking the Federation's goose that very moment.

CHAPTER

21

THE GOOSE IS COOKED

"Scorpio blanditur vultu, pars postera pungit."

"A scorpion presents an attractive face but its rear stings."

— Old Latin proverb

Plump and stumpy Van Atta stood on the platform of the railroad depot Saturday evening at 6:45 with three other members of Troop A. These fully armed mob members, now legalized as soldiers of the Colorado National Guard, suddenly came under fire. In a panic they darted into the depot and had the station agent telephone General Bell. Six or seven bullets "passed near" them, they said, coming from the direction of town where many Italians resided.[1]

Bell ordered Captain H. D. Smith to take sixty men to scour the area. Smith discovered only one shot had been fired and that it was possible the shot wasn't from a gun, but a firecracker. Nevertheless, they surrounded East Telluride and systematically searched it, particularly the dwellings of Italians, because these were said to be the most rabid of agitators. The search continued straight through the night. No guns were found.[2]

The military and Governor Peabody were obsessed with finding the cache of repeating rifles St. John allegedly had purchased with union funds in 1901. The original numbers quoted were 250, but now Peabody said, "If these men will dispose of these 1,600 guns and prove good, law-abiding citizens, that is all we ask."[3] Rumor

The Telluride depot around 1902.

Denver Public Library, Western History & Genealogy (#X91).

whispered these 250 or 1,600 guns had been purchased prior to the strike of 1901 and cached in some abandoned adit in Marshall or Savage Basin, or were brought in a few at a time and hidden away beneath the floorboards of a house in the Italian neighborhood. Pinkertons fed this rumor, as illogical as it was. If the miners had the rifles, they certainly would have them in hand, not in storage. St. John himself had repeatedly informed authorities the guns the men used July 3, 1901, were "the personal property of the miners" and they each owned or bought their own to defend themselves as the Constitution allowed. "One could not pass over the public roads without being shot at by the mine guards," he said.[4] Neither he nor the union had bought the "latest" rifles; the union financial records confirmed this. The miners' guns were everything from single-shot rifles to shotguns, repeating rifles to handguns. The shooting fracas that broke out was no organized affair, the men not armed uniformly. Had they been, many more than two Smuggler employees would have died.

But the governor was losing ground: his administration was being condemned from all sides, in the press, the courts, and by his own party. A number of soldiers in the field even took the side of miners, especially in Cripple Creek, stealing food and blankets for miners' families in dire need.[5] Finding the cache of rifles would redeem executive action and prove its righteousness. Somehow Peabody had to balance his public "law and order" image with his desire to wring the life from the WFM. It was looking as if he were in collusion with the railroad magnates, coal barons, and the MOA.

The MOA believed if the executive officers of the WFM could be held in custody long enough, it would throw the Federation into disorder and collapse. Executive President Charles Moyer was unprotected in the outpost of Ouray. The WFM had recently distributed posters with a drawing of the American flag with the inscription, "Is Colorado in America?" above it, and statements of constitutional violations on each stripe. At the poster's bottom left corner was a halftone of a man handcuffed to a pole near a train. "The picture represents Henry Maki," the caption pointed out, describing why Maki had been shackled to a pole. The poster was a plea for donations to help the miners in their battle against tyranny. Moyer's and Haywood's signatures were at the bottom. The posters appeared in Telluride on March 25.

Although for many years pictures of the American flag were used in advertisements and covered with words, the WFM poster was the excuse for the MOA to act. In Telluride, Alderman M. A. Wood, an SMCCA member and new recruit in Troop A, filed a complaint against Moyer for "desecration of the flag."[6] Since the poster was merely a picture of a flag, "desecration" could not legally apply, but prosecution of the charge was not the aim: arrest was. Police Justice P. A. Lilley issued a warrant and Rutan rang up Ouray County Sheriff Maurice Corbett to arrest Charles Moyer and take him to Ridgway, where Rutan would take him into custody.

Corbett politely requested Moyer accompany him to the Ouray County Courthouse, not to Ridgway, since Corbett had no physical warrant. The local union lawyer was at hand to advise Moyer to refuse to go to San Miguel County. Rutan and Deputy Harry P. Craver, a Troop A member, took the train to Ridgway and a wagon to Ouray. En route they passed another wagon heading for the Ridgway depot. In it were deportees Stewart Forbes, Tony Langeri, Jerry Barnes, Matt Lingol, and five others, those chosen to test the injunction. At Ouray, Rutan quickly took Moyer into custody and telephoned General Bell to inform him nine deported men were heading to

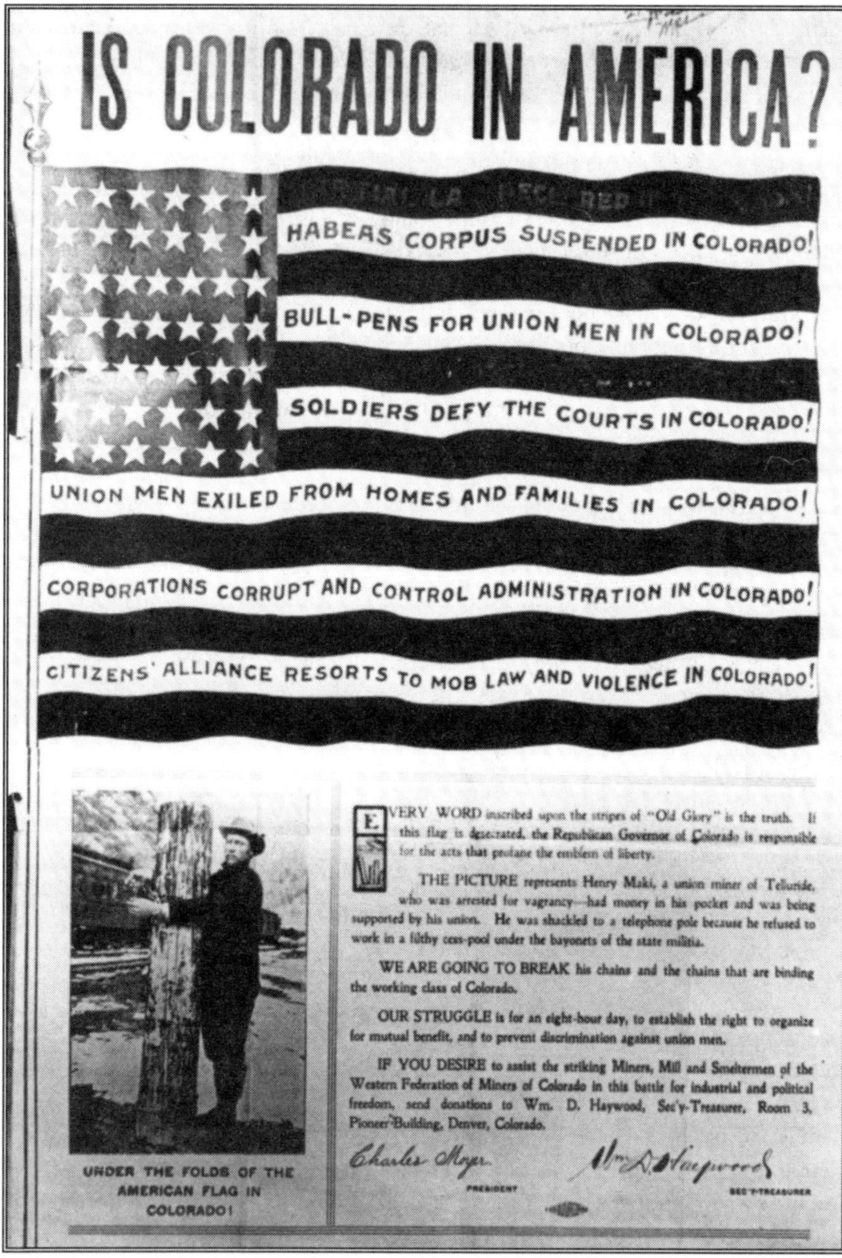

The WFM used this poster to solicit funds for the union. The first posters distributed did not include the photograph of the man shackled to the pole. On the posters that included the image, the WFM stated the photo "represents Henry Maki," without claiming it was Maki. This is the "flag" that generated the charges of desecration brought against Moyer and Haywood.

Courtesy Haywood Trial Papers 60-1.1.246, Idaho State Historical Society.

Telluride. Bell ordered Captain Wells and thirty men from Troop A out on a special to intercept the men.

Forbes and his comrades, all unarmed, law-abiding, and with no warrants against them, got aboard the same passenger train to Telluride as Rutan, Craver, and Moyer. The train was stopped at Vance Junction, where Wells and his soldiers rushed aboard. Without explanation, Wells ordered all but Forbes, Barnes, and an Italian to board the special with Troop A, the special taking them back toward Ridgway immediately. It stopped at the county line, at the wye on Dallas Divide at nearly 9,000 feet elevation. The night was dark, icy, and silent, and the snow that had fallen for two days lay in deep drifts across the hills. Wells ordered the six men into the snow, warning them never to return. They struggled on foot all night to return to Ridgway, a distance of more than ten miles.[7]

Meanwhile, Forbes, Barnes, and the Italian were arrested by the military at Telluride, taken before General Bell and interrogated. Barnes and the Italian were released to tend to their affairs; Forbes was thrown in jail. The civil authorities tried to concoct charges against Forbes, but finding nothing, he was released and allowed to leave town on March 29, his wife and children remaining.[8]

"The deported men may return if they behave themselves," Governor Peabody declared to the press yet again.[9] "No man who is law-abiding and attending to his business will be interfered with," General Bell announced publicly.[10] Yet, despite verbal and written promises of non-interference and protection, despite the injunction writs being duly served on all the defendants, no deported person was able to return unmolested.

President Charles Moyer had to spend a few nights in Telluride's jail before attorneys Murphy and Richardson reached his side. Justice Lilley fixed his bail at $500, even though the maximum fine for desecration of the flag was $250. Harry Floaten posted bond, and Moyer walked out of the jail at 10:30 Tuesday morning. Under orders from Peabody, executed by General Bell, the military immediately rearrested Moyer, taking him to the barracks in Red Men's Hall. He was not charged with any crime; the arrest was termed a "military necessity." Peabody informed reporters he intended to have Moyer rearrested "on new warrants every time he secures his release on bond." To "suppress insurrection," Moyer was to be kept confined indefinitely. A writ of habeas corpus was sought at once.[11]

A warrant was issued in the Telluride court for William D. Haywood's arrest on the same charge. Reporters grilled Peabody, who "refused to discuss the proposed arrest of Secretary Haywood," but the governor announced, "Moyer and Haywood will be prosecuted to the limit of the law."[12] (The flag desecration charge was never pursued.) Hearing of the plot, Haywood turned himself in to the Denver police, thus avoiding being hauled over to Telluride and kidnapped by the militia. He was released on bond and went about his business.

The WFM recognized the threads of conspiracy between the governor and the MOA and let it be known the plot would fail. Even if censors intercepted every telegram from President Moyer, the union would endure. Vice President Williams would act in Moyer's place; Big Bill Haywood would pour his strength into the miners, bellow to the press, call down the wrath of the nation.

Yet in Telluride the legalized mob and the military had the Miners' Union in an iron grip. More men were arrested in town and at Ophir, roughly interrogated before General Bell, and thrown in jail. Others were forced to leave the county and warned not to return. When district Judge Stevens granted a writ of habeas corpus in the

Moyer case, returnable on April 11, and Sheriff Rutan certified in writing he had delivered the writs to Bell and Wells, Governor Peabody ordered them to ignore the court.[13] Consequently, Judge Stevens adjudged Bell and Wells guilty of contempt, ordering Sheriff Corbett to arrest them. Since the officers claimed they were "exempt and free from arrest and interference and process of the civil courts" because of martial law, Corbett was unable to carry out his orders. Angered by the total disregard for the law, Stevens declared no insurrection was manifest in San Miguel County "unless such condition existed by virtue of the unlawful usurpation of power by the governor and militia."[14]

The WFM attorneys appealed to the State Supreme Court, while Moyer continued to be held without charge in Telluride's jail under military custody. In the meantime, the militia read his every letter and telegram, even those to his wife. In order to impress upon him the power and might that wrapped itself around him, Moyer was made the "guest of honor" at military demonstrations. The soldiers were inspected, put through drills, and exhibited the killing force of the Gatling gun. Curry gloated, "There is reason to believe that the object lesson had a beneficial effect on the socialistic generalissimo."[15]

That first week in April, Bell reduced the troops to about ninety-eight men, leaving Troops A and D on active duty. General Bell and the Gatling gun remained. During the entire occupation nothing disturbed the peace, except the arbitrary arrests and deportations, twelve more men being shipped out with the troops. No clear line between the civil and the military authorities was evident, since numerous soldiers on active duty had been deputized and were also guarding or working as miners in the mines.

Battery A demonstrates the power of the Gatling gun in the field below Telluride.
Courtesy American Heritage Center, University of Wyoming.

While General Bell and his staff lavishly wined and dined at Wells's expense at the manager's house, Deputy Bob Meldrum was working at the Tomboy as a hired company guard. Meldrum heard a miner was drinking liquor on the premises and confronted the man, Olaf Tissel (also spelled "Thissel"). The man was drunk and unarmed, and when Meldrum opened Tissel's coat to search him, the miner bunted Meldrum in the face with his head, loosening the gunman's teeth. A scuffle ensued, and Meldrum shot the miner in the belly, Tissel taking twenty agonizing minutes to die. Meldrum was arrested. A coroner's jury ruled the "shot was fired in the heat of passion with felonious intent."[16] Although bond was $10,000, Meldrum was released, his sureties being Manager Herron, Wheeler, and Lavender, the men who had hired the gunman as a deputy. Meldrum was sent to Denver to get his grin fixed, ending up with a mouth full of gold-capped teeth. "He looked like a million dollars every time he smiled," miners said after that.[17] On his return, the gold-grinned killer went back on duty, General Bell ignoring his violence.[18]

Any violence perpetrated by the military or the deputies or the roving armed citizens Bell ignored, but the least word or criticism against the military by a union member or Socialist became a hanging matter, although very few who were arrested were ever charged with any crime.

In preparation for the transfer of military command, Bell took Troop A through maneuvers around town at a gallop, trundling the Gatling gun along and out to the ballpark. Sergeant Jack Bowman put the men through training and target practice, cutting down trees at 800 yards, according to Wardlaw. Bowman allowed Sheriff Rutan to have a "crank" and the sheriff was aghast at the firing power. "That is the most wicked machine I ever witnessed in operation," he said.[19] And once again, Bell impressed upon the town he meant to use the wicked machine if necessary. Intimidation at its finest was meant to scare off the remaining union men and Socialists.

On April 5, Captain-Manager-Deputy Wells and the Telluride Mine Operators' Association attempted to screw the lid on the district by issuing the following statement:

> *We do not propose to enter into negotiations with the Western Federation of Miners. We do not recognize a union in Telluride. There is no strike in Telluride. All our mines are working with a full force of men, and we do not know what kind of settlement can be made. With us there is absolutely nothing to settle.*[20]

None of the large mines had a full force. Scabs still worked for one paycheck and departed. Wells actually complained he could not get qualified miners.[21] Three days after the announcement, seventy-four of the previously deported men informed their family and friends in Telluride they were returning on the evening train. They made no secret of it. The governor and General Bell had said if they were law-abiding, they could return. None were under outstanding warrants, so they boarded the train at Ridgway and headed home.

Before the train arrived, Bell ordered Troops A and D and the Gatling gun squad to the depot. He also ordered out the high school cadets, who had rifles but no ammunition.[22] Bell's intention was another object lesson. He awaited the train astride his horse, dressed in full regalia, coiled with gold braid, his spurs polished, and his jaw set. About a hundred armed Citizens' Alliance men joined the soldiers, taking up positions on both sides of the street just north of the troops. Several hundred family members and friends of the deportees hurried to the depot as well. Although the train was late and the night raw and cold, wives and mothers, children, friends, and sisters waited for their men to return, their hearts full of welcome.[23]

At the first faint sound of the whistle and the sighting of the engine's headlamp, the crowd pressed forward, but the soldiers ordered them back. The train steamed in, slowed, halted, and Bell ordered the Gatling gun squad to fire the weapon over the top of the coaches into the mountainside, a warning that anyone defying his golden power would be cut down instantly.[24] Silenced in terror, the anxious crowd of wives and friends could only wait, tears falling down their cheeks.

Troop A lined the platform and as the passengers stepped off, they were shoved into the depot's waiting room, their bodies, clothing, and luggage searched for weapons. "Not so much as a jack knife was found upon these anarchists," Curry said, attempting to validate the treatment with the derogatory label.[25] The men had upheld their end of the agreement, yet they were then hustled into a line and marched between the soldiers to Red Men's Hall and held under bayonets. Although they were given a meal, the men "were subjected to ridicule and other humiliation" and not allowed access to their families.[26]

Bell believed violence would result if the men were allowed to go home, but the violence feared was not at the hands of the union men: he feared the Citizens' Alliance planned to treat them harshly or even lynch them. Yet instead of taking the SMCCA members to task, removing their guns, and ordering them to obey the law, he ordered the deportees to be marched back onto the train at 9:30. They were escorted by Captain Wells and a detail of thirty men to Dallas Divide. The men were ordered off the coaches into the icy darkness about midnight and warned never to return. A volley of shots drove the warning home. Like those before them, the men were forced to make their way on foot through the snow to Ridgway.

Six of their number, whom Curry called the "more worthless and persistent of the leaders,"[27] Stewart Forbes, Tony Langeri, Pinkerton Riddell, H. C. Voss, Swan (Sven) Gook, and Joe Barnes, who had been popping in and out without notice until now, were arrested by Bell and thrown into the city jail. After interrogating Gook and Langeri, both of whom had families, Bell released them upon their promise to leave as soon as they disposed of their property. Within the week all but Forbes were released. His wife was cautioned to mind her tongue, because she was speaking too loudly against injustice. Finally Bell allowed Forbes to dispose of his personal property and depart with his wife and children, only on the promise he would never return. The Forbes family left on April 14.

The wives of many deportees petitioned General Bell to allow their husbands to return. He refused them, claiming that if any were permitted to remain another large group would soon follow. Several of the wives packed their belongings and left, heartbroken, unable to believe in American justice for all.

During his escapades sneaking in and out of town, Joe Barnes had so injured the pride of Major Hill and Wells that General Bell now considered him a dangerous man. Finding Joe's most dangerous asset was his grin, Bell turned him over to the city authorities. They cooked up a charge of distributing handbills without license and fined him twenty-five dollars. The twenty-five-year-old Joe, who had survived the ravages of smallpox and stood tall against anything that assailed him, refused to pay the fine. He was detailed for duty on the streets. With a broad grin he refused to work, tossing the hoe away, daring his captors to harm him. A soldier threatened to shoot him. Joe defiantly said, "Pull the trigger!" Dragged back to jail, he was given "the bread and water treatment."[28]

Incorrigibles were best shipped out and the machinery of deportation swallowed dozens more each week, leaving Thomas Nelson acting as both president and secretary-treasurer of the Miners' Union. He and his wife had charge of the union's holdings, including the hospital building. They ran it as a rooming house, quartering between forty and sixty of the remaining strikers. The men were marked for exile, yet as long as they didn't open their mouths, Bell only stalked them.

At this point Bell was busy with habeas corpus proceedings in the Moyer case, the State Supreme Court having ordered him to deliver Moyer to the bench.[29] He complied, bringing Moyer to Denver encased in the iron power of Captain Wells and Troop A. The WFM and President Moyer found no relief from the court: Of the three judges, two were Peabody's friends and one of those, Judge Gabbert, had lived in Telluride, still held mining interests there, and was a director of the Bank of Telluride. Gabbert's ruling in favor of Peabody and the militia appalled the dissenting judge, Robert W. Steele. Steele wrote, "The court has not construed the Constitution, it has ignored it.... If one may be restrained of his liberty without charge being preferred against him, every other guarantee of the Constitution may be denied him.... If the law is as this court has declared, then our vaunted priceless heritage is a sham...."[30] Moyer appealed his case to the federal courts.

He was returned to military custody at Telluride, the train stopping at Ridgway where the troops and prisoner had to transfer to another train. Numerous deportees stood on the platform at the depot. One, Gus Mohrhardt, asked Captain Wells for permission to come aboard. He said, "I can show you I'm all right," and stepped into the coach. Wells shoved him off the train, snapping, "You can show me nothing." As big Walt Kenley swung aboard the rear coach and the train pulled out, deportees heckled him. Austrian Floro Flor, who had been taken from the arms of his wife by Kenley, shouted, "Go on off with you, you scab-herding son of a bitch!"

With his beans steaming, Walt yelled to a soldier, "I'm jumping off — tell the Captain I'll be up tomorrow."

The soldier informed General Bell, who immediately ordered the conductor to back up the train the thousand or so yards it had advanced. He ordered the soldiers to "take Flor."

Already on the run, the Austrian darted into the town, soldiers shouting at him, "Halt!" His failure to obey brought a hail of gunfire from their revolvers, bullets whizzing down the village streets, people diving for cover. Flor suddenly tumbled and the soldiers thought they had their quarry, but he bolted again, disappearing into a building.[31] Some reports said the soldiers winged Flor and three men and killed a horse in their efforts to avenge a mere insult. A few days later Ridgway residents signed a condemnation of Bell and his troops for "shooting up [the] town while its streets were occupied by men, women and children."[32]

Failing to capture Flor, the indignant soldiers and deputies got back on board at General Bell's command. The train continued to Telluride, arriving at 8:30 that evening with Charles Moyer calm and resigned. As a greeting, Sergeant Jack Bowman and his squad "turned loose a cannonade from the Gatling gun" over the top of the locomotive and the soldiers and Citizens' Alliance mob at the depot let out a wild cheer.[33] Moyer was marched back to his cell through a heavy snowstorm that muffled mob revelries that night. General Bell declared, "No one but God Almighty himself or Governor Peabody...will release Mr. Moyer."[34]

On April 25, the machinery of deportation was transferred to Captain Buck Wells when the general turned the district over to his command and departed. Fortunately the Gatling gun went with Bell. A number of "rapid-firing" guns, however, were left in the hands of the untrained soldiers formed out of a mob. Wells perched one gun in a guard tower atop the roof of the Smuggler Mill. Others were used in a similar manner around the district.

Relaxing some of the regulations of martial law, Wells lifted press censorship and allowed fraternity meetings. He called upon those citizens he didn't want cluttering up his town and told them to depart. He was master of the polite threat and gracious intimidation, for instead of flatly telling them they must go or be shot, he said, "I cannot protect you from a mob if they decide to come for you with a rope."[35] Tony Ralla, a property owner with a pregnant wife, refused to leave, but the remaining strikers at the Miners' Union Hall were waited on by Wells on April 28. Wells informed Sullivan of the Federal Union that these men were a menace to the peace of the community and that they must either go to work or quit the district.[36] They failed to obey his orders.

With all but thirty men relieved of military duty, Mayor Rogers swore in most of the SMCCA as "special police," the armed citizens patrolling the streets in shifts at night. Afraid to protest too loudly, Judge Wardlaw did the next best thing: he resigned the Republican Party and became a Democrat and was elected as a delegate to the convention, giving Curry fodder for a tirade.[37] Curry shredded the judge as irresponsible and disloyal to those who elected him to his Republican position. Yet Wardlaw was hardly the judge Curry needed to worry about. Judge Theron Stevens, wise and slow to anger, was seething over the extreme disregard for the Constitution in San Miguel County.

District court was about to convene in Telluride, and on the evening of May 9, Judge Stevens was due in on the train from Ouray. Captain Wells had gone to Glenwood Springs for the Republican convention, leaving the young Englishman, Lieutenant Louis Lomax, in command of Troop A. As usual, the town was quiet. Then that evening's issue of the *Journal* hit the streets.

Finding gossip better news than fact, Curry had published a front-page story under the ominous headline, "Bringing Back the Agitators," insinuating the distinguished district judge was bringing a dozen or more of the deported men to Telluride with him.[38] Curry claimed this information was from a reliable source who had telephoned him from Ouray; however, many astute people believed Curry invented the "slush" because he was already sickened by the judge's restraining order and contempt charges against Bell and Wells. The story promptly threw Lomax into a panic and the SMCCA into a rush for their revolvers.

Lomax rang Wells in Glenwood who instructed him to arrest the men as soon as they stepped off the train. Thus Lomax ordered Troop A to the depot with rifles and bayonets. The Citizens' Alliance was bristling and armed as well, lining the street near the depot.

Just as the evening train puffed to a standstill, "the soldiers sprang upon the platforms of the coaches ready to arrest" the agitators. Only a few salesmen, regular passengers, and the judge were on board. Judge Stevens was forced to walk between armed soldiers and armed citizens as he made his way to the hotel omnibus. He was "very much displeased and heartily disgusted at the war-like demonstration" and became more so when he reached the hotel and was given a copy of the *Journal*. He "denounced [the story] as a personal insult to himself as well as to the district judiciary."[39]

Disappointed to have no one to arrest and humiliate, the soldiers and the Citizens' Alliance mob dispersed. When Sheriff Rutan opened district court the following morning, a stern judge glowered from the bench. Stevens read a statement to the assembled crowd of lawyers, prosecutors, and spectators:

> *Gentlemen of the bar, I came here yesterday for the purpose of opening court and transacting such business as I felt the conditions would justify.*
>
> *I find a different condition here than what I had expected. The demonstration at the depot last night upon the arrival of the train could only have been planned and executed for the purpose of showing the contempt of the militia and a certain portion of this community for the civil authority of the state and the civil authority of this district.*
>
> *I had always been led to suppose...that in a republic like ours the people were supreme; that the people had expressed their will in a constitution which was enacted for the government of all authority in this state. That constitution provides that the military shall always be in strict subordination to the civil authorities.*
>
> *It is doubtless construed differently, however, by the executive, who has declared this county to be in a state of insurrection and has declared martial law within its limit. In effect, therefore, the executive has said that there is no law in this county, except the military commander. I can only believe from the indications...that the executive and the militia...are willing that this court should be opened and such business transacted and such orders of the court executed as meet the approval of the military commander...but that such portion of the orders of this court, or the decrees of this court as do not meet with the approval of this militia and the people of this county, may not be executed. Under such circumstances the court would not be in a position to enforce its lawful orders....*
>
> *Such being the case, it would simply be a farce to attempt to enforce the civil law in this county. It seems to the court, further, that the members of the bar of this county, with a few exceptions, have become imbued with the military spirit to such an extent that they would not feel right assisting this court in the proper transaction of the business of the term. Under such circumstances the court will be greatly hampered should it attempt to do business. For that reason I have decided that, until a different condition exists, until the supremacy of the civil authorities is acknowledged in this county, I shall not attempt to transact any business within its limits.*
>
> *It is, therefore, ordered that all matters pending and undetermined in this court be continued until the next term.*
>
> *It is further ordered that court be adjourned* sine die.[40]

As if a thunderbolt had struck from a clear blue sky, the courtroom reeled in numbed silence. The game the town was playing was no game to Judge Stevens. He walked off the bench, instructing the attorneys to put their protests in writing. That evening they presented a statement regretting the action of the court, assuring the judge they were not "imbued with the military spirit," and wished to conduct the numerous cases on the docket with a willingness to "assist in the enforcement of all lawful orders of the Court."[41] Stevens ordered their statement written into the record, and court was adjourned.

Judge Wardlaw felt a certain vindication, his spirit shaking off the darkness of suppression, and he blasted Curry for the "unwarranted publication of a mere rumor" that brought about a cessation of the May term of district court. The *Journal*'s "apparent policy," he said, "is to foment trouble and keep alive existing animosities."[42]

District Court Judge Theron Stevens.
Restored portrait © M. J. Martin 2003.

At first Curry tried to duck out by claiming others had duped his informant in Ouray, and that the demonstration at the depot was the ordinary reception for every train. He then muttered that maybe a few more armed people were on hand, maybe thirty more, or sixty more. Suddenly he lashed out, striking with poisonous condemnation at Judge Stevens, the Miners' Union, and John Wardlaw. He insinuated Stevens merely had set out to exact revenge on the community because it was combating the element he had identified with. Stevens's only purpose "was to humiliate the people of this community; to accentuate and strengthen the impression...that this is a lawless community." Curry accused the judge of protecting "men charged with the most diabolical of murders, [men] still fugitives from justice." And the worst thing the honorable judge ever did was to appoint St. John, an "illiterate...utterly incompetent and unfit for the work," to the committee to examine the county's books in 1901. How he could bestow a position of "honor and responsibility" on "one of the most despicable and dangerous characters that has ever infested this community" was beyond the pious Curry, who declared the act "a gross and flagrant affront to the property owning, tax paying, reputable elements of society." Curry conveniently forgot Harry Servis had pled guilty to embezzlement. The "willful adjournment by the judge," Curry foamed, was a "violation of the rights of all persons accused of crime, to say nothing of an undignified display of a petulant temper." Henceforth and forever, the editor proclaimed the district judge an anarchist, and Curry knew time would prove him right and the judge wrong.[43]

Wardlaw shook his head. "The desire to create excitement and shoulder the blame upon Judge Stevens"[44] was so far out of line, the county judge could only gasp in disbelief.

After the shock subsided and all cases were indeed held over until the fall term, San Miguel County went right on trampling the Constitution. On May 11 Captain Wells returned from Denver, his wife, children, and mother accompanying him. Grace had been in Boston a few months, enjoying brighter and freer society.

The wife of Charles Moyer arrived as well, determined to be near her husband until God or Peabody released him. Although confined, Moyer wrote his president's report with the heading, "Bullpen at Telluride, Colo., May 24, 1904," for the WFM's twelfth annual convention. Determined and in good spirits, Moyer rallied the miners not to weary in their efforts to gain an eight-hour workday and the right to organize, despite the MOA's iniquitous tactics. The only hint at animosity toward those who illegally incarcerated him was his identifying Peabody as a species of his own and urging the brothers, should they discover others like him anywhere on the globe, to "relegate them to the cage in which they belong." He explicitly warned the members against using violence of any sort, assuring them the remedy was in the strike, the boycott, and above all, the ballot.[45]

Moyer's wife remained in Telluride until his release on June 15. She was a comfort to him and a guest of Margaret Floaten, the Wisconsin-born thorn in Wells's jodhpurs. As a Socialist activist, Margaret locally kept the spirit alive behind the scenes. Her husband, Andrew, was allowed to return, as long as he kept his mouth shut. Word was the Socialists planned to nominate him for governor in the upcoming election.

Harry, his younger brother and partner in the People's Supply Company, landed on Wells's offensive list. Although in the same Masonic fraternity as Wells, Harry was an undesirable, a Socialist, a thing without rights. Wells questioned him about his connection with the Miners' Union, discovering Harry "received on deposit funds of the local union" and transacted other financial business for them — no crime.[46] Wells informed him he "was too friendly to the union and he must quit the district." Harry's friends protested to Wells personally, but Harry was forced to leave the first week of June. He had business in Denver — but he said he would be back.[47]

Many Telluride citizens were tiring of the farce and even Congressman Hogg, once a pro-militia man, wired the governor in frustration, "I guess you had better call off this foolishness here."[48] But on that very day, June 6, an explosion in the Cripple Creek district at the Independence depot killed thirteen men and wounded several others. The WFM was immediately blamed. In that district, wholesale rioting on the part of the Citizens' Alliance and the MOA broke out, the MOA using a noose to convince the pro-union sheriff to resign. Union men were beaten and robbed, homes and businesses destroyed, and hundreds of men deported with the assistance of the military, including Pinkerton Manager McParland's brother, a shoemaker. The WFM condemned the hideous dynamiting and the rioting, promising to investigate.[49]

The bombing threw Wells into an arresting frenzy in Telluride. He hauled more than seventy men into court to account for themselves as if they had something to do with the outrage. Some were released, the rest deported. It was more an excuse for raids rather than an actual fear, for on June 15, 1904, Governor Peabody revoked martial law in San Miguel County and relieved the soldiers of duty. The *Examiner* called for an end of the "gun packers" reign, an end to the SMCCA.[50] The SMCCA was unwilling to let go, continuing deportations on the whim of Wells, Wheeler, Painter, and Herron, most without provocation. A reason appeared in only one case: a drunk Swedish miner yowled around the saloons that a far worse explosion than the Independence dynamiting was planned for Telluride on July 3, the anniversary of the Smuggler riot. Wells had him arrested and Kenley invented a scene where Moyer told him personally that such was in the works.[51] Although both the miner and Kenley were blowing gas, when pranksters set off a giant firecracker in an alley and caused the alarmists to scream "Revolution!" the mayor jumped. As if a simple proclamation would stop a war, on June 30 Mayor Rogers prohibited the use of any gun, firecracker, or explosive for the entire first week of July, and forbade anyone who wasn't a duly sworn law officer from carrying any concealed gun, knife, razor, or slingshot.[52] The SMCCA said this didn't include them. People hungered for peace, but the leading antagonists in the Citizens' Alliance refused to allow harmony to blossom.

Believing the end of military presence meant he could return home without harassment, Harry Floaten consulted with Lieutenant Governor Warren W. Haggott in Denver, and Haggott assured him if he had committed no crime, no reason existed to prevent his return. Haggott promised him protection of his rights.

Having transacted his business in Denver, Harry returned to Telluride on the evening train June 23. Night Marshal Geyer met him at the depot and told him he was under arrest.

"I asked what the charge was," Harry related in a public statement. "He said, 'You will find out.' I was taken to the sheriff's office and told to remain there. After a while I requested the sheriff to show cause for my detention. He told me to wait a few minutes, and when the minutes were up told me I could go. At my home in the evening later I was waited on by a committee of five. They stated that but for their interference a mob would have had me before this time, and advised me not to stay, but if I would leave on the morning train I could stay over night. I took their advice."

In dismay, Harry continued, "What are things coming to in this state? I have lived in Telluride over thirteen years, and there isn't a person that can point out that I ever disobeyed any law. All I have on this earth is there — my home, my wife, child and business. They say I have been doing business with the union. It is true, but is that a crime?"[53]

The Citizens' Alliance boycotted the People's Supply store, harassing Andrew, Margaret, and Harry's wife, Mary "Mamie" E. (Jackson), until they felt compelled to sell the company and all its stock. Andrew and Margaret, "high-principled, superbly true" citizens, moved to Fort Collins: Telluride's loss.[54] Harry left, but returned later, living in the town for several years, still harassed by Wells.[55]

As if to signal the Citizens' Alliance their reign should end, District Attorney Sam McMullin of Grand Junction requested Deputy District Attorney Edward Howe to turn in his resignation. Howe refused, forcing McMullin to remove him. McMullin said Howe had been using the office to manipulate the law and conduct legal harassment. He replaced him with E. L. Oakes, a local attorney. During that fall's term of district court, McMullin quashed all the latest charges Howe had filed against St. John, Jardine, Washburn, Brennan, and the rest; he would "no further prosecute the pleas of the people" in these cases.[56]

The majority of the people worked toward peace and healing, yet a handful of malcontents under Wheeler, Wells, Painter, Herron, and a few others seemed unable to let go of their supremacy. Wells and Herron were particularly fond of having things run their way, although this meant the mines dropped in production. The Smuggler especially was losing money, its workforce at only 200 men, when it once had more than 500. Because he could not keep or even find qualified miners, Wells discontinued operations at the Smuggler on July 1, the properties being worked exclusively by lessees. He publicly blamed the WFM for the shutdown, claiming miners were afraid of reprisals from that organization. The WFM said experienced miners simply refused to work for the Smuggler-Union Company, and since other mines in the area were working without a hitch, the WFM appeared to be correct. Some, such as Senator Buckley's Ophir Consolidated, continued to employ union miners and had not lost one day of work since the millmen struck. Employing his usual blustery threats, Wells once again said the mine would remain closed indefinitely unless the town stood behind him.[57] He felt the pull of the tide and disliked losing his absolute power.

The Thiel Detective Service in Denver sparked Wells to action when they informed him they had absolute proof a man named Jesse W. Shields had assisted in burying Will Barney. Wells sent Runnels to Silverton to arrest the man. Either incompetent or just plain stupid, the detectives had wrong dates, hearsay for facts, and forgot to check the suspect's whereabouts. Shields had been in the Durango home of his parents with a broken arm during the months around Barney's disappearance. His parents and his doctor verified the fact. The case evaporated. Wells's disappointment ruptured into a mission to find Barney's body so he could prove once and for all the Miners' Union was everything he said it was. He refused to let them win.[58]

The mouthpiece of the malcontents, F. E. Curry, could not find it in himself to back off either. A new boy on the block, Austin E. Benson, joined the *Examiner* to help the timid Wardlaw keep the screamers at bay. Benson signed on as manager and assistant editor. Straightaway, he editorialized that the only way for Telluride to make progress was to banish the offensive terms "agitators," "anarchists," "trouble-breeders," and all the rest from the local newspapers. He wanted an end to petty malice and the cry of "Wolf! Wolf!"[59]

Believing their judgment correct, Curry and the malcontents said peace would come only when the last agitator was gone and the leaders were brought to justice. The group finally ran Marshal Corey out, replacing him until the elections with Chris Geyer. Like Kenley, Marshal Geyer thought fists and guns superior to the courts, and he carried out the bidding of the Citizens' Alliance with gusto. He was turned loose on nineteen-year-old Grover Skelton, merely because Skelton's brother, Pearl, had been arrested for complicity in the Independence depot bombing. Pearl was in Pueblo with his wife at the time of the explosion and was released. But he was a union man, so Grover had to go. Geyer took him to Keystone and told him to "Git!" The boy ran back to his mother's house in Pandora. Geyer heard of it, barreled into the house, striking the widowed mother over the head with his pistol, cursing, and bellowing. He grabbed the boy, but Grover's sister tried to stop him; Geyer kicked her down and shoved her off the porch. Grover was handcuffed and dragged out of town.[60] He returned and his brother Pearl also came home. This time they were left alone.

Bored by the slumbering peace that edged in, Manager Herron, his brother David, the Tomboy bookkeeper, and faithful Runnels went on a spree on August 21, hurling rocks down the airshaft of the Black Bear Mining Company. The four men working inside rushed out. They were lined up and driven at gunpoint over to Silverton. The four were stockholders in the Black Bear.[61] They were also WFM members: undesirable citizens.

Having dug themselves into a pit, the mine managers found nowhere else to go. They left the deportation game to the Citizens' Alliance, who discretely carried it on right through December. The managers needed to make the mines profitable and to do so meant fair wages and fair hours. Even Buck Wells gave up his pout and reopened the Smuggler at the end of August, hiring 150 miners. In November the managers agreed to an eight-hour day for all employees, millmen included, and set a standard scale of wages that was almost identical to that proposed by the millmen in the first place. All the mines and mills in the area adopted it.[62] Wells was one of the first to implement it, admitting he did so in order to get the pick of the very best men in camp. Union men hired on, but he refused to recognize their affiliation. Moyer assumed this meant the mine managers had no objection to hiring WFM men, commenting to the press that it indicated, "they will ask no questions." Moyer said, "Neither will we."[63] He was mistaken.

Anyone with a scrap of sense asked why the hour and wage standard hadn't been accepted before the strike. It would have been quite simple to listen to the millmen, negotiate, and settle without putting an entire community through the nightmare of military occupation and mob rule. Arrogance got in the way. Arrogance still greased the managers' wheels, for Wells made it clear they had implemented the eight-hour day "not in recognition of the demands of the Federation," but because they knew the legislature was about to enact such a law. Wells asserted the TMA "had no negotiations whatever with the" WFM and would never knowingly employ any WFM members.[64]

In an effort to build the bridge of sanity and peace, the San Juan District Union declared the strike off on November 29. President Frank Schmelzer stated,

> *We would infer from the action taken by the mine managers in granting the eight-hour day and a satisfactory wage scale that it is their desire that peace and normal conditions shall again prevail in San Miguel County.... We will expect the mine owners to co-operate with us to prevent the Citizens' Alliance from continuing assaults upon the rights of men 'to work when, where and for whom they please' and reside in any community which they may select.... We demand that the Citizens' Alliance shall be prohibited from employing armed forces and murderous thugs, contrary to the laws of the state. We shall expect that the laws shall be made effective and that the restraining order of the district court shall be obeyed.*
>
> *As members of organized labor, we have endured untold sufferings through unlawful imprisonment, through invasion of our homes, through confiscation of our property and deportation by Citizens' Alliance mobs. We cannot forget the brutal and barbarous indignities that have been heaped upon us; but we will endeavor to endure these cruel memories now that our demands have been granted and we have the opportunity of resuming our vocation on an eight-hour basis.*[65]

In open defiance, the Citizens' Alliance sent their hired gunmen, Meldrum and Runnels, now deputy town marshals, and Marshal Geyer to round up union men who came back to town that week. Some of the men were visiting with Antone Matti when the gunmen invaded his home through his front and back doors. Others were in homes around town. In arresting three, Runnels explained their crime: "No one who speaks against Peabody can remain in" the county.[66] Two were released, but five were forced on the train and told not to return.

As news correspondent, Wardlaw sent the story out to the Denver and Pueblo newspapers. Curry attacked him as a liar, claiming the men were arrested for a row in the red-light district on Wednesday and left of their own free will. In a brilliant investigative report, Wardlaw and Benson asked, "Who is the real liar?" and said if the men were arrested for disturbing the peace, then why were their cases not processed through the court? The *Examiner* verified the facts of its story to the smallest detail, proving the men had only arrived on Thursday, and were forcibly deported merely for being union men. "The *Journal* maliciously lied," was the conclusion.

"Businessmen have implored [Painter] to put some restraint on the editor's inflammatory pen," but Painter was Curry's co-conspirator and gladly added to the inflammation. "Many residents of San Miguel County hold the Telluride *Journal* directly and primarily responsible for the trouble, discord and dissention which has overwhelmed this place for the past three years, and regard it as the worst and most venomous agitator and trouble breeder the city has ever been afflicted with."[67] Wardlaw

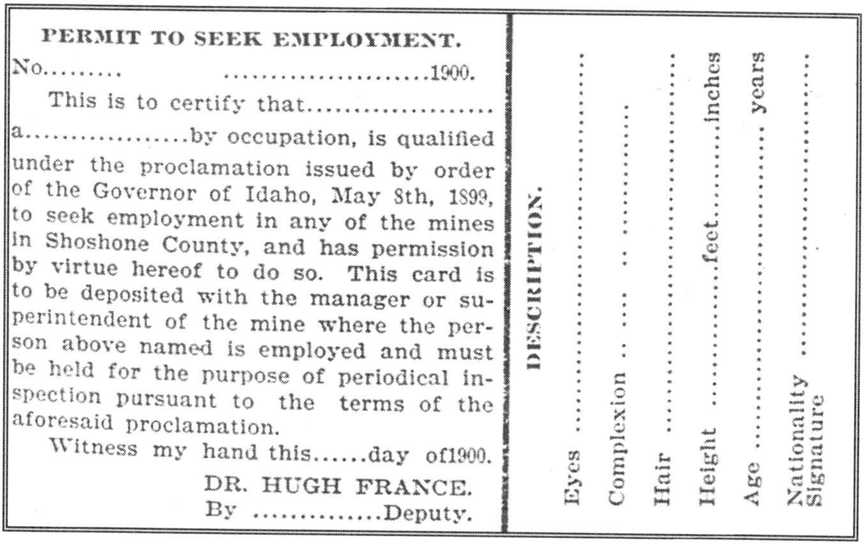

PERMIT TO SEEK EMPLOYMENT.

No........ ....1900.

This is to certify that....................
a..................by occupation, is qualified under the proclamation issued by order of the Governor of Idaho, May 8th, 1899, to seek employment in any of the mines in Shoshone County, and has permission by virtue hereof to do so. This card is to be deposited with the manager or superintendent of the mine where the person above named is employed and must be held for the purpose of periodical inspection pursuant to the terms of the aforesaid proclamation.

Witness my hand this......day of1900.

DR. HUGH FRANCE.
By ....Deputy.

DESCRIPTION.

Eyes ....
Complexion .. ..
Hair ....
Height ....feet...........inches
Age ....years
Nationality ....
Signature

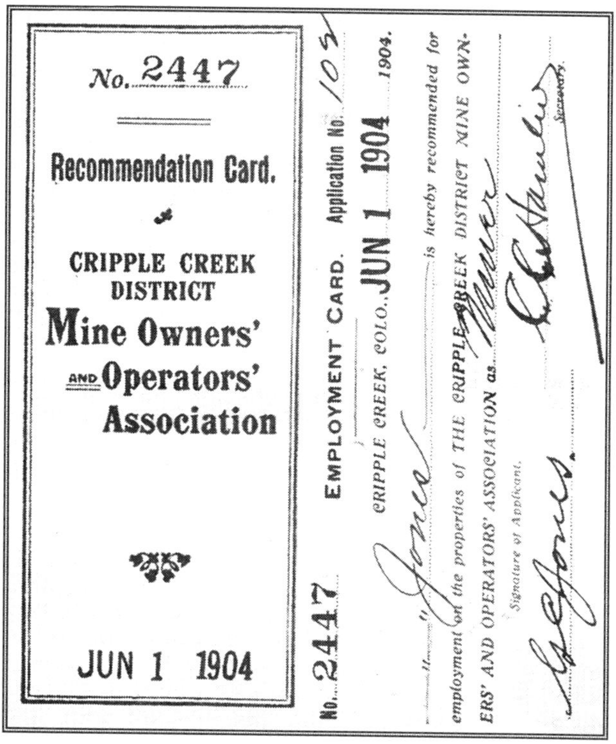

No. 2447

Recommendation Card.

CRIPPLE CREEK DISTRICT
Mine Owners' and Operators' Association

JUN 1 1904

No. 2447 EMPLOYMENT CARD. Application No. 105

CRIPPLE CREEK, COLO., JUN 1 1904 1904.

" Jones is hereby recommended for employment on the properties of THE CRIPPLE CREEK DISTRICT MINE OWNERS' AND OPERATORS' ASSOCIATION as Miner

Signature of Applicant. Secretary.

The "card system," designed to exclude known union men from employment, was initiated as a means to destroy the WFM. These are two examples, one from Idaho, the other from Cripple Creek District. Several San Juan district mine owners, including Wells, used similar cards. They also hired employment agents with the sole purpose of filtering out union miners.
Author's collection.

and Benson put into print what the frustrated majority of citizens were too timorous to say. Did Painter, Curry, Wheeler, or Wells listen?

Almost immediately the mine operators adopted the card system, whereby only miners who were approved or recommended by the MOA and had no known affiliation with the WFM would be employed.[68] This effectively squelched union power in the district, and Curry smirked. Painter beamed. Wheeler turned his frown a wee bit clockwise. Strutting Buck Wells felt he had won.

Yet his victory left suppurating wounds across the landscape and dark holes where bitterness festered. What Wells gained, Telluride lost, never again to have vitality and booming exuberance. As if sucked into the darkness, the beloved, progressive, wise, and benevolent Senator William S. Buckley died that last week in November at the age of forty-four. He had been the voice of reason in the nightmare of suppression; he had brought truth amidst the lies; he had brought peace, justice, and forgiveness. This man of "perfect integrity," as the *Denver Times* wrote, who was "Plain in manners, direct in speech, acting from the promptings of a mind that knew no devious ways, tender of others' feelings, generous and courageous," was gone forever.[69] The miners had always called him their friend. His absence left an enormous void in San Miguel County, and a dreary and silent winter slipped over the region.

CHAPTER

22

THE FIRE-EATER

"Always forgive your enemies; nothing annoys them so much."

— Oscar Wilde

Election year had caught Governor Peabody with his pants down. His entire two-year term had been snarled in war on organized labor, racking up a war debt of over a million dollars. The Telluride campaign alone cost $199,292.[1] The MOA hadn't yet paid one cent of their promised funds for their extensive use of the militia. Peabody tried to defend himself as he sought a second term.

In a statement to the voters, he excused his bad management by blaming the WFM for nearly every crime in the state. He even dragged in the Pinkerton assertion that Vincent St. John had "bought and paid for 250 rifles, with which he armed his Federation followers" to mount an assault on the Smuggler-Union Mining Company in 1901 — a year he wasn't even in office. He offered no evidence for his accusations.[2]

The general response was, "While Peabody's 'explanation and defense' was long on showing up the misdeeds of the other fellow, it was almighty short on reasonable vindication of his own trampling under foot of the Constitution and the laws of the commonwealth."[3] Even within his own party he was denounced for his unconstitutional assumption of power overriding all civil laws. Yet the powerful MOA and the smelter trusts insisted he be nominated for reelection.

The Democrats nominated former Governor Alva Adams, a man of sincerity and vision. Peabody and the MOA had reason to fear. But the contest turned vicious, with evidence of election fraud on both sides, although Adams was unaware members of his party were engaging in illegal tactics to get him elected. "Repeaters" voted early and often in the Democrat precincts in Denver, one showing over 700 returns where only 100 voters registered. The MOA forced employees in the mining camps to vote Republican or lose their jobs. A Republican officer of the Denver Union Water Company openly bragged, "We rule...the people have nothing to do with nominations and elections."[4] A more sinister picture emerged years later when Pinkerton Operative Charles A. Siringo, who had worked for the agency for more than twenty years, wrote of his activities under James McParland. In some cases Pinkerton agents gathered men from saloons and paid them to vote a certain way. Siringo himself said, "I voted eight times, as per McParland's orders — three times before the same election judges."[5]

On November 8, 1904, Alva Adams was elected by a large majority. Peabody screamed, "Fraud!" The State Supreme Court investigated and tossed out the returns from many Democratic precincts in Denver. This still left Adams with a 9,764 lead. He was inaugurated on January 10, 1905. Peabody and the MOA brought their protest into the chambers of the Republican legislature. The Hydra of conspiracy thrashed in the background, the MOA, coal corporations, smelter trusts, and railroads having made agreements with Peabody prior to the election. On January 21 Governor Adams filed counter charges against Peabody. In the end no one won.

A Republican committee, which was drenched in suspicion, ousted Adams and the citizens of the state protested the "deliberate theft of the governorship." A joint session of the General Assembly determined the only way Peabody could be declared the winner was if he agreed to resign within twenty-four hours after inauguration. When the corporations offered an inducement of his full two-year salary, Peabody finally agreed. He was inaugurated and on March 17 resigned, his Lieutenant Governor Jesse F. McDonald taking office that same day. Colorado citizens raged.[6]

Governor Alva Adams.
Author's collection.

Alva Adams didn't exit quietly. "Force has triumphed," he said. "A brutal minority placed in the governor's chair a man the people repudiated. Ninety-five percent of Colorado citizens know that Peabody was not elected. All of those connected with the crime know it.... A stolen governorship will bring only reproach and disaster to Colorado and Republicanism."[7]

The MOA kingpins were undaunted. They were without Peabody, but McDonald was a Republican and a Lake County mine owner and banker, sympathetic to their position. If they needed him, they could count on him. Now so well organized throughout the West, the MOA concentrated the power of the Pinkerton Agency in their battle to destroy the WFM. Better than the militia, Pinkertons infiltrated nearly every local in the West, a veritable army of operatives choreographed by James McParland from the Denver office. McParland's bible was the MOA publication, "Criminal Record of the Western Federation of Miners," also known as "The Red Book."

The MOA published the thirty-two-page Red Book in July 1904. It listed crimes committed from 1894 to 1904, from the Coeur D'Alene region to Telluride, condemning the Federation as responsible for every one of them. This booklet passed itself off as a reputable, factual document, giving dates, names, and gruesome details. As propaganda designed to make the public shudder, it shrieked in bold type and italic whenever a dramatic scene was presented. Evidence, conviction, and truth were absent, these not being necessities for either the MOA or the Great Detective. McParland fondly quoted it. Curry did likewise. Just because it was in print, they decreed it unimpeachable.

The list of crimes for Telluride featured Vincent St. John as a fugitive from justice and the leader of the terrorists. The Red Book was the first to claim John Barthell was shot by *mistake* by the strikers, ignoring all eyewitness accounts. It also claimed Bill Jordan was crippled for life, ignoring medical reports that said he had no more than a flesh wound. It described Thomas Ballard as an old man, "shot through both arms and crippled for life," who walked "half naked" to Silverton.[8] Again the truth escaped the Red Book's claim: At the time, Ballard was forty-nine, and his own testimony said he received flesh wounds in one arm and a hip. He was fully clothed and never crippled, and he had gone back to work for the Smuggler.[9]

The MOA's Red Book also included among the Telluride crimes the "Torture of Edwin Thomas," a Cornishman.[10] Again, no such torture had occurred: Thomas's testimony in a 1906 hearing indicated he wasn't even beaten.[11] The Red Book blamed Judge Stevens for allowing the "human wolves" to escape indictments and the prosecution of the murder of Arthur Collins. As if the MOA had an actual document in hand, the booklet said 250 rifles were ordered "on union letterhead" and paid for by St. John.[12] Again, that was fiction: The Pinkertons craved such a document but never found one.[13] Had they found one, St. John would have been scratching lines in a cell wall.

Continuing with its indictment, the Red Book presented the Will Barney murder in uppercase boldface that nearly flung itself off the paper. Barney was now officially a "shift boss," and was heard to cry out, "DON'T MURDER ME!" To make the case more disturbing, the "YOUTH WHO HEARD HIM CALL OUT ALSO MYSTERIOUSLY DISAPPEARED" and "HIS REMAINS HAVE NEVER BEEN FOUND."[14] The MOA's source for this fabrication was F. E. Curry, for no law officer in San Miguel County ever questioned the whereabouts of Sam Servis, the "youth." Sam was a married man of twenty-three whose gallivanting wife, Elizabeth H. Servis, had run off to Denver without him. Curry's latest tale said Sam was "done away with" by St. John a few days after Barney vanished. If this was true, it was an amazing thing that yet another dead man filed for divorce. Sam was supposed to be killed in June 1901, yet court records showed Sam filed for a summons against his wife on August 6, 1901. When he discovered her in Denver, he then filed to dismiss the cause on September 28. He even paid the court costs and service fees that month, proving he was very much alive.[15] But Curry never mentioned anything about Sam's disappearance until 1904, at which time Sam was not around to refute it. The young man had moved to Denver in early 1902, before his cousin Harry Servis went off to the state lockup. Gus Roth was informed where Sam went, and had an address for him up until November 1902. Sam went his own way after that, and so did Harry when he was paroled in May 1903. Harry was in Denver in October 1903 when he was served with a court order in a case related to the old embezzlement charges. He refused to appear in district court at Telluride and San Miguel County won judgment by default, the court ordering Harry to pay the county $745.[16] Neither man returned to Telluride ever again. Curry simply took advantage of their absence, knowing he could say anything about them and it would stand undisputed.

The MOA's Red Book also listed Wesley Smith as a Telluride Miners' Union murder victim, although no body or evidence was ever found. It also resurrected an old tale that at one time even Curry had put to bed: the "MURDER OF JOHN MAHONEY."[17] There just weren't enough bodies in Telluride to make a decent list, so once again, the MOA boys created them out of what was beginning to appear to be their favorite source: divorce. The Red Book killed John Mahoney on October 10, 1901, and claimed "rewards brought no results."[18]

Mahoney, of course, was never dead and no rewards were ever offered for him in any form. This was settled and silenced in 1901. He and his wife Mary popped from mining town to mining town, having married in Leadville in 1891. They had one child, Irene. John had family in Salida. He had a nasty habit of getting bored and running off without Mary, at which times she fended for herself, often running small lodging houses. She went to Cripple Creek and back to Telluride so many times she lost count, sometimes with John, sometimes without. Sometimes he helped support her, many times not. In early 1902, he went to Teller County where they both had lived in 1900. In Telluride Mary opened a boarding house on Colorado Avenue, had a falling out with the owner, and moved to another house on Columbia Avenue at the end of October 1902.[19] To make ends meet, feisty, determined Mary hired on as a cook at the Miners' Union Hospital. With John refusing to support her and Irene, Mary was granted a divorce on June 8, 1903, John being found guilty of "willfully" deserting her. The lengthy divorce document said John was personally served the summons where he lived in Teller County and was "a man in good bodily health." It proved he was alive in June 1903, a full year and eight months since his "murder."[20]

Was it possible divorce cases were used with the intent to create murder victims? Stacking coincidences suggested a modus operandi on the part of the Pinkertons, who claimed to have investigated the vanished men. If they were unable to find one living man, that wouldn't reflect badly on them, but the Pinkertons declared all of the men could not be found — was it plain incompetence or their M.O.?

Mary Mahoney herself cozied up a few times to the Pinkertons because she thought she might get something out of it. When she felt the union snubbed her, she courted their enemies. More often than not, this fractious woman remained devoted to the WFM, even when McParland tried to put the fear of hell into her during the Idaho trials in 1906-07.[21]

In answer to the MOA's Red Book, the WFM published a thirty-one-page booklet shortly after the Red Book hit the streets. The WFM titled theirs, "Category of Crime of the Mine Operators' Association, A Partial List Showing 851 Men Murdered in Less Than Four Years."[22] The murders included the men who died by the hundreds in mines due to company negligence, the victims of the MOA riot at Cripple Creek, and of the militia in the coalfields, and all those shot, beaten, deported, and robbed by mine managers and their guards. The WFM indictment also included MOA bribery of government officials, conspiracy, and a condemnation of Peabody's use of the militia to crush organized labor. Unlike the MOA, the WFM presented their indictment in normal type, factually refuting the list of crimes in the Red Book, pointing out some of the men charged with crimes in the MOA book weren't even union men. Then the WFM went one better than the MOA: They called for a congressional investigation, which U.S. Senator Patterson supported. The investigation was begun straightaway and, by January 1905, documents were presented to the Committee on the Judiciary in Washington, D.C.[23]

Although the Peabody wars had drained the Western Federation of Miners, the organization still stood strong. Because the WFM had called for a government investigation numerous citizens swung back to its side, believing any organization open to an investigation was not as guilty as the MOA painted it. Donations poured in. Curry, Painter, Howe, Wells, and Wheeler, on the other hand, believed the MOA's Red Book was as accurate as if God had written it. Justice would hang Fire-eater St. John, and Wells intended to supply the rope. The Pinkertons still labored over the case on the

Smuggler-Union Company's behalf. It was just a matter of finding something that would assure a hanging.

In Telluride enmity had solidified, no one giving an inch to those with conflicting views. It was impossible for arrogance to lower itself and admit its wrongs, and equally impossible for justice to accept invalid assertions. The majority disliked Wells, Wheeler, Painter, and Curry's fatal resolve against union men, yet the Wells faction was the more powerful, and Wells seemed to carry tremendous influence in the town. Because of him, the TMA, and Citizens' Alliance, more than 200 people had been exiled from town from autumn 1903 to the end of 1904. Another 200 or more had left during that time, sickened by oppression. More than 700 miners had left the region when the strike was first called, believing they had no future where arrogance smothered negotiation. The exodus dominoed through several businesses; without customers, merchants generated no profit, and thus numerous businessmen and their families left town as well. Others left simply because the "Curried" atmosphere was unbearable. Among these was Alderman Fred Pheasey, who took his family to Alaska in 1905. For the same reason neither Charlie Sumner nor his brother returned. Charlie found himself a new wife, Lulu Belle Currier, and had taken employment with the Grand Junction *Daily Sentinel.* Telluride had left both these men with a sour taste on their tongues.

A number of the more determined deportees, such as John E. Conn, Matt Lingol, Tony Ralla, and other real estate owners, brought lawsuits against the county, the state, and several individuals, including Cal Rutan. Many of the men now resided in Ouray and nearby counties, waiting until reason and justice returned to San Miguel.

Telluride citizens certainly wanted to shake off the torpor of antagonism and the influence of the Citizens' Alliance. In the spring city elections of 1905, they demonstrated their disapproval of illegal force by electing Democrat Kinney McLean as city marshal, deposing Geyer. They had also had their fill of Walt Kenley's two big fists, and he thundered off to Cripple Creek, taking employment as a guard for the MOA. In that district he was arrested for cracking craniums with too much force, but was later released. The Teller County sheriff had relieved him of a special revolver given to him by the Telluride mine managers. When he was discharged, he requested the revolver back by pointing a gun in the deputy sheriff's face and grabbing the revolver. He and several cohorts raced off on horses, a posse in pursuit. The ensuing gun battle ended with Kenley back behind bars.[24] Gunmen Meldrum and Runnels, on the other hand, seemed content to mellow out in Telluride, Runnels cheerfully shooting stray dogs as a deputy city marshal. Meldrum had been acquitted in the Tissel murder — his excuse for killing an unarmed drunken miner was self-defense — and he continued as full-time day guard at the Tomboy. (Later he would shoot another drunk miner at the Tomboy, with a similar outcome.) In October of 1905, Runnels was elevated to city marshal when Kinney McLean died suddenly of acute peritonitis. Kinney was thirty-three. His wife, Alice, followed him to the grave just six months later, leaving their two babies, Thelma and Frederick (named after Uncle Fred Pheasey), orphaned. Kinney's sister in Boston took the children.[25]

Several other well-loved citizens died that year, adding to the heavy sense that good times would never be recaptured. In January of 1905, longtime resident Charles Peterson died mysteriously, some thinking he took poison. He and Barney Borg, friends of the Miners' Union, had bought the Clipper Saloon from Pheasey and McLean. On the last day of May, former Sheriff Oscar D. Downtain took his own life, putting a pistol in his mouth and pulling the trigger. He had been in Telluride nearly a quarter century, raising his family and taking part in city and county politics. He left a widow and a son, Clarence, seventeen, and a daughter, Florence, fifteen. The fifty-six-year-old Downtain had gone down to tend his horses just a half hour before he died. The *Journal* gave "rheumatism" as the reason Downtain shot himself, a strange motive for a man devoted to his family. Downtain had taken abuse from Wrench and Curry over the affair at the Smuggler in 1901, although Curry praised him in the obituary.[26] No one would ever know how much he had suffered.

A few weeks later, on July 17, Mel Robbins tumbled out of a fishing skiff on Trout Lake, his friends in the boat unable to save him from the icy water. His body was recovered three days later. Robbins had grown up in Telluride, passed through his ugly stage as a Smuggler thug, and softened into a kind family man. He left a widow and baby to mourn him.

Despite the shadows prowling at Telluride's edges, tragedy and loss seemed strangers to Buck Wells, who indulged his circle of admirers with lavish dinner parties and offered short-term leases on the Smuggler to his Harvard pals when they came for the summer holidays. Although Grace and their children were spending more time at their Colorado Springs home and in Boston with her family, Buck pranced and danced, wined and dined, sometimes even having Judge Wardlaw and his wife in his dinner crowd. He enjoyed being the center of the party, the golden boy, the man above men.[27] As he was the manager of numerous properties in various locations, he contracted attorney Edward L. Oakes to act as his personal employment agent at Telluride, a post designed for the express purpose of excluding WFM men from employment at the Smuggler. With a salary of $110 per month, Oakes was to make background checks on applicants and to hire only those approved by the MOA.[28] Wells didn't want to waste time doing this himself — he had greater duties to tend. He did want to maintain his Troop A, but the cowboys tired of the game with no one to rope, and drifted back down county. The "cow punchers...cannot be induced or expected to attend drills," Buck wrote General Bell, advising the adjutant that Troop A "be mustered out of service."[29] With the demise of his personal cavalry, Captain Wells had to hang up his spurs. Yet in April he was given higher honor and enough gold braid to make his jodhpurs bulge.

Dogged by bad press and faced with more than two million dollars worth of litigation stemming from his war on labor, Sherman Bell resigned as adjutant general of the Colorado National Guard, his letter dated April 5, 1905. Governor McDonald appointed Bulkeley Wells to take Bell's place at the urging of his MOA advisers. Although unqualified as a military leader, Wells had trampled the Constitution with such aplomb that he seemed the perfect man for the office, one who would use his position to further the MOA's war on the WFM.[30] Polly Pry gushed over the appointment, whipping up a fairy tale about General Wells for the *Denver Post*. He loved the new title as much as she did. Her salivating prose described him as a shining god, and she gave this Chicagoan an accent that "would accelerate the pulse of an Oxford

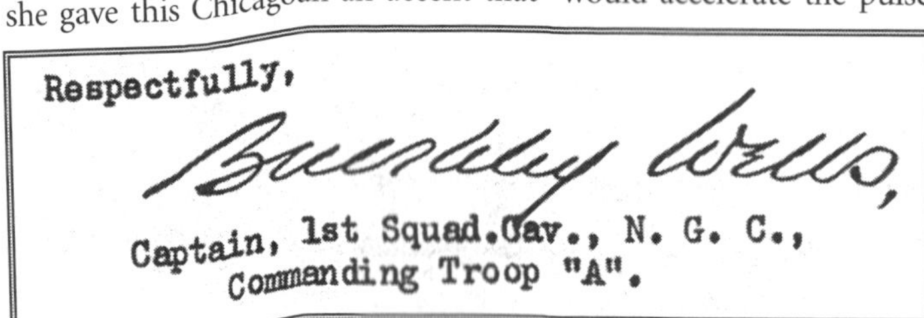
Respectfully,
Bulkeley Wells,
Captain, 1st Squad. Cav., N. G. C.,
Commanding Troop "A".

Captain Wells's signature.

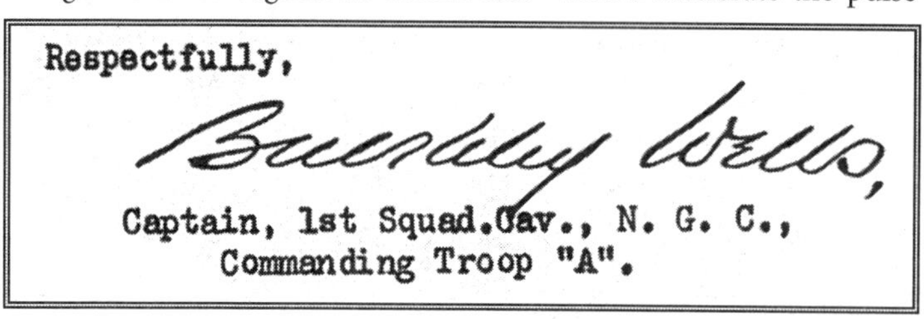don."[31] He told her he would run the Guard like a polo club, and build it into the best state militia on earth. Amen.

Buck then murmured how much he loved polo and tennis, riding and fishing, but above all, his favorite avocation was hunting. "There is a certain class of men who are definitely impressed by Wells's hunting skill," Polly burbled. "They don't trot in his class, to be sure." These were the anarchistic, illiterate agitators, of course; those Wells hunted down and deported in lieu of having them stuffed and hung on his study wall. Polly invented a quote for the worst of them: "What does that la-de-da Boston dude think he is going to do down here [in Telluride]," she had St. John saying when Wells arrived to take over management of the Smuggler.[32]

Little did poor Polly realize that no member of the Miners' Union ever cared what hobby the mostly British and eastern seaboard mine managers followed. Miners were accustomed to the privileged and moneyed class being in Telluride and Bulkeley Wells's advent was of no consequence; he was merely another like the rest. Her attempt was to set Wells apart from all the other hunting men and women of Telluride — she didn't know many followed this pastime, including the sisters known as Feather Legs and Bad Eye, not just mine managers. Her primary aim was to illustrate the superiority of Wells over St. John in all things. Indeed, Wells was a good hunter. He went off to shoot bear and brought back deer, or went for grouse and brought back trout; not that he misidentified his game, rather, he was always willing to kill whatever he found at hand.[33]

Now with the strike out of his craw, General Wells had more time to pursue hunting, but by the end of 1905, it wasn't bear he had in his sights; it was the elusive Fire-eater. On December 30, a monstrous crime launched the final chase, and Wells's Pinkerton shrubbery shakers rushed out ahead of him.

On that date in Caldwell, Idaho, former Governor Frank Steunenberg opened his gate, triggering a bomb that tore through his body. He died in hideous agony. Shortly, the alias-laden Harry Orchard was arrested for the crime. He had loitered around town as if wanting to get caught, and once caught, it was only a matter of days before the Great Detective, Pinkerton James McParland, manipulated the State of Idaho into letting him take control of the case.[34] McParland gave the serial bomber Orchard religion in exchange for his confession. The detective habitually extracted confessions from killers, yet the amazing thing about Orchard's confession was how conveniently it wrapped up nearly every unsolved crime the MOA had ascribed to the WFM in Colorado. More conveniently, it did what Peabody's militia failed to do: it delivered the WFM executives in one gift-wrapped box for the hanging. Orchard claimed he went on his crime spree at the command of President Moyer, Secretary Haywood, and a one-time officer, George Pettibone. He said the men (McParland's "Inner Circle") paid him to kill dozens of men, to blow up mines, assassinate managers, blow up the Independence depot, and murder Frank Steunenberg. He also said that just before the bombing, he had perfected the Steunenberg plot with Vincent St. John, now Miners' Union president in Burke, Idaho. The MOA in several states had been cringing for the last two years as St. John organized men in Utah, Nevada, California, and Idaho, building the WFM's strength and taking part in creating a new union that would include all wageworkers. Orchard's confession came right in time.

Although the trophy confession was blubbered out over a space of a few days, it took McParland coaching Orchard for nearly a year to polish the document with previously published or known details, and for Orchard to memorize it so it slipped off his tongue without the slightest variation. In the subtlest manner, McParland promised immunity, and less subtly, promised redemption. Orchard soaked up so much religion during his tutelage with McParland that the Great Detective had none left for the WFM executives. McParland spoke for God, and God didn't want the WFM executives redeemed. He wanted then hanged.[35]

With the first rough scraps of Orchard's confession in hand in February 1906, McParland plotted with the officials of Idaho and Colorado, the MOA, and General Bulkeley Wells to kidnap Moyer, Haywood, and Pettibone and bring them to Idaho where they could be hung for conspiring to murder Steunenberg. The kidnap plot was to prevent extradition snags and they found judges to sanction it. While Wells diligently carried it out, McParland instructed Shoshone County Sheriff Angus Sutherland, who had a long hatred for the WFM in northern Idaho, to collect the Fire-eater from Burke.

In addition to claiming Vincent St. John had his fingers in the Steunenberg plot, Orchard told McParland Vint had paid Steve Adams to assassinate Arthur Collins at Telluride. If McParland could have Adams corroborate this, then St. John could definitely be executed. Thus McParland ordered his operatives to assist with the arrest of Adams, who was residing near Baker City, Oregon, at that time.[36]

Sunday night, February 18, St. John and his wife retired in their small rented house in Burke, knowing nothing of the deep plots McParland spun. Late in the night they were startled awake when their door suddenly burst open and Sheriff Sutherland and several deputies rushed into their bedroom. Clara clung to Vint as they yanked him from the bed and ordered him to dress. He demanded to know why they had come for him with guns drawn. They refused to say. He asked where they were taking him. They refused to tell him. They roughly cuffed him and dragged him into the darkness, Clara running after them in her nightgown. As they disappeared into the icy night her breathless terror convulsed her with sobs.[37]

Taken to the Boise penitentiary, St. John was charged with conspiracy in the murder of Steunenberg on nothing more than the word of a confessed killer. For a few days he wasn't allowed to see family, friends, reporters, or attorneys. When McParland snapped him into his pincers, the detective didn't bother to offer him religion for a confession. St. John's reputation for integrity and fidelity prevented McParland from trying to get him to sell his brothers for thirty pieces of silver and a cigar. Both McParland and St. John were Catholics, yet the detective's religion made him self-righteous; St. John's made him a humble servant. McParland had only disdain for this quiet man and judged him so evil a creature that he was determined to see the Fire-eater's body writhing on a gallows.[38] McParland also had a client who wanted St. John hanged, and that client's desire was more important than Vint's redemption. The client was General Wells, who, with several state officers and Pinkertons, had brought Moyer, Haywood, and Pettibone to Boise. Bob Meldrum tagged along as an assistant to Wells.

So certain these WFM men were guilty simply because Orchard said so, McParland and Governor Gooding boasted to the press none would be leaving Idaho alive. Wells told St. John, "Now you will hang," before any formal charges were filed.[39] When the news of the arrests hit Telluride, Curry was the happiest man on earth. If ink could gloat it did that day, with triumph glistening in every black letter of the issue. The front-page headline led with, "VINCENT ST. JOHN ARRESTED IN IDAHO."[40]

Suddenly the gloaters took a dive. Orchard was unable to cough up a particular date for the week he supposedly spent plotting with St. John only three months earlier. No witnesses could connect the Saint with the serial bomber. Orchard's friend, Steve Adams, had been arrested and, contrary to McParland's instructions to have him sweated in solitary, was locked in with Orchard. Orchard told him what McParland wanted from him, supplying Adams with information and details. Adams's wife, Annie, and her children were brought to the prison as well, as a means to "sweat" him. It worked. Thinking he had a promise of immunity, Adams folded. Yet Steve Adams, now under McParland's control, failed to back up what Orchard said about St. John. He had visited St. John in Burke in the first week of July 1904 — miners frequently sought out Vint for assistance, employment, or advice. McParland asked Adams, "Did you have any talk with [him] about the Steunenberg matter?" Adams flatly replied, "No, sir."[41] No matter how much the Great Detective tried to manipulate the situation, no evidence even hinted at St. John's involvement in the Steunenberg assassination. Mine owners in Idaho had retained the services of Captain Wilson S. Swain of the Thiel Detective Service in Spokane, and Swain had shadowed St. John from January through December of 1905 without ever seeing Orchard. Swain testified at the grand jury investigation that St. John hadn't engaged in anything even suspicious. McParland declared this competing detective a liar and Orchard the only man who spoke the truth.[42] If anyone contradicted Orchard's confession, that person became the liar in McParland's view, not Orchard, even though Orchard admitted to lying to McParland.[43]

The Great Detective had made it clear to Orchard in the beginning that he wanted to be able to charge St. John and Oscar Carpenter with a capital crime for his Telluride client's satisfaction. McParland now included these two as members of the Inner Circle. To please McParland, Orchard attached St. John to his own bombing plot, his first rambling accusations not carefully thought out. It is important to note that Orchard gave McParland the information that Adams had carried out crimes on behalf of St. John, that Adams had buried Barney, and that Adams had shot Collins. Orchard said all this *before* Adams was even arrested, giving McParland the potential and much needed corroborator of Orchard's confession.[44] This strongly suggested Adams's confession was manipulated, if not entirely created by McParland from previously published information and Orchard's contributions. While Adams's confession was merely twenty-four pages, Orchard's kept expanding under the tutelage of McParland, until it became book-length, and Orchard decided he wanted it published as a book, his mentor encouraging him to work toward that goal because it would "refresh his memory on the events...and make him still a better witness."[45] Unlike Orchard, Adams was incapable of churning out details of events, even when McParland coached him with leading questions. When Adams was pressed to single out St. John as the person who paid him to shoot Collins, Adams failed to acquiesce: One time he said his pals Art Baston and Ed Minster said Moyer and Haywood wanted the job done; another time he said Sherman Parker, a union leader in the Cripple Creek district, wanted the job done; and another time he said he had to do the job himself because Minster threatened to kill him if he didn't.[46]

Working hand-in-glove with General Wells, McParland informed him that it was beginning to look like Orchard and Adams wouldn't be able to conclusively connect St. John to any murder. Wells had a backup plan, should the Idaho plan fail: In the second week after St. John's arrest, Wells had his faithful shift boss Bob Wright get something on file at Telluride in order to make extradition of St. John possible. To assure nothing interfered, the county commissioners appointed M. L. Brown as justice of the peace to carry out the plot. On February 27, Wright brought charges against St. John of conspiracy to murder, and of the murder of Ben Burnham, the same old nullified charges stemming from the 1901 riot. Brown issued the needed documents, placing the warrant in Meldrum's hands.[47] Wells wanted more, declaring to the press, "A clear case against St. John for murder could be made out if the body of a missing Telluride miner could be found."[48] This was Will Barney, the stonemason, memory of him so clouded now that even Wells forgot what the man did for the company. Wells had word from McParland that Adams knew where the body was, thus hope for a good hanging still beat in the general's chest. But at least they had Wright's charges to start the fish jumping.

As expected, Idaho declared they had no case against St. John, but, McParland made certain to have the other warrant in place before he was released. Wells arranged with the Colorado MOA secretary Floyd Thompson to get extradition papers from Governor McDonald. Knowing no valid charges could be filed against St. John in Colorado, his attorneys, family, and friends feared extradition meant certain death, very likely at the hands of a mob. Terrified by the whispers and the old promises of the Citizens' Alliance to hang Vint high, Clara sought strength and help from his mother in San Jose, California. The fifty-eight-year-old Irish widow immediately wrote to WFM Vice President Williams in Denver. She informed him her daughter-in-law had been told that once Vint was in Telluride, the sheriff "would turn over the keys to the jail" to the Citizens' Alliance.[49] Since Vint and Clara left Telluride they had been threatened several times, via anonymous letters and shadowy persons, that the MOA had a price on St. John's head. A few times men pulled guns on him, and shots were fired at him in Utah. For this reason he was listed as "J. W. Vincent" at the thirteenth WFM convention in Salt Lake in 1905.[50]

Since the charges against her son had already been thrown out of court, Mary St. John knew it was only a ploy. She wrote to Williams, "If it be in your power to protect this brave, innocent man, I hope you will leave no stone unturned in order to do so. I am alone. He is my only son, and, while I know and we all know, he is innocent of any crime, still we must use all necessary precautions against treachery."[51]

Williams replied, "Rest assured that your son will receive every protection from this organization." He also wrote to Colorado Governor McDonald, requesting the State "place around St. John the necessary safeguards" to protect him from the "fury of outlaws" and the "vengeance of a mob." McDonald gave his promise.[52]

On March 16, Idaho Governor Gooding honored the requisition papers and turned Vincent St. John over to Sheriff Cal Rutan, Deputy Meldrum, and a Pinkerton. "The promptness with which Sheriff Rutan secured possession of his man and got away from Idaho without encountering annoying legal delays was due to some very skillful manipulating on the part of" Idaho's prosecutor, Curry gleefully reported (March 19, 1906). St. John's attorney was told his habeas corpus hearing was postponed until the next day, and when his attorney had left late in the afternoon, Gooding let Rutan take St. John, processed through a justice of the peace. Rutan put him on a train immediately, WFM attorneys hearing of it and rushing to board with them. But just outside of Boise at Nampa station, where the train stopped, Rutan had Meldrum slip St. John onto a different train while Rutan stood about the platform as a decoy. At the last minute, Rutan boarded the train St. John now was on, and whisked the union leader away from his attorneys.[53]

The attorneys wired WFM headquarters in Denver. Immediately two other attorneys were dispatched to meet St. John's train, which was delayed by heavy snow through Utah. At Grand Junction Rutan, Meldrum, and their prisoner came in so late they had to stop overnight. In the morning, on March 17, a *Sentinel* reporter interviewed Rutan, who said neither the prisoner nor anyone else gave him any trouble en route. St. John was not handcuffed at any time during the trip, simply required to sit with Rutan or Meldrum beside him. Rutan said the prisoner "chatted very pleasantly at times...." But for the most part, St. John had nothing to say. He was present during this interview, but "seemed to be in deep study" and all the reporter squeezed out of him was, "I am entirely innocent of the murder charges preferred against me."

St. John sat smoking a cigar, silent and patient, missing his wife, seemingly disheartened. Dressed in the "garb of a miner," he wore a corduroy coat with a WFM lapel pin and a slouch hat. The reporter was surprised to find him "rather small in stature, erect in bearing, and of respectable appearance..." since rumor had painted him as something with thorny hide and fiery eyes. St. John's reputation and "remarkable career as an organizer" made him a large figure in the West, the reporter said, "fearful and all powerful" in some circles, yet "...truly a leader and a hero among certain large classes of men."[54] Rutan and Meldrum escorted Vint to the train, St. John walking between them, carrying his suitcase as if the three were businessmen.

At Montrose the WFM attorneys from Denver caught up to Rutan and his prisoner, riding the rest of the way to Telluride with them. Wells and the MOA would have been better off leaving St. John in the hands of the attorneys they ditched in Idaho, for the WFM sent Judge Orrin N. Hilton, a man "regarded as perhaps the most capable, resourceful and eminent criminal lawyer of the state."[55] Hilton was a retired judge from Michigan who had been in private practice in Colorado for more than twenty years. Tall, lean, sharp-witted, and white-haired, with steady steel-blue eyes and a frog-like smile, the fifty-seven-year-old Hilton was highly esteemed, even by his enemies.

Judge Orrin N. Hilton, Attorney for the WFM.
Author's collection.

At Telluride, St. John was arraigned before Justice M. L. Brown. Instantly Hilton questioned the jurisdiction of the illegally appointed justice and made a motion for dismissal on the grounds that Brown was not qualified to sit as an examining magistrate. Brown, who had been a Citizens' Alliance member and a member of Troop A, refused to recuse himself. He said he would hear the arguments, which he did, overruling the pleas of jurisdiction, and committed St. John for trial. Criticizing this unprecedented ruling later in district court, Hilton said Brown "heard it as examining magistrate himself, before himself, argued it to himself, through himself, and decided it as a *de facto* and *de jure* Justice of the Peace, by himself; signed the commitment, and the illustrious deed that made the name of Brown immortal was finished."[56]

In St. John's preliminary hearing, witnesses recounted the day of the riot and the death of Ben Burnham, several of the prosecution's stars making it clear St. John arrived on the mountain more than an hour after the riot commenced and Burnham had been shot. None was able to say they saw him at the Sheridan dump where Burnham was shot. Yet he was denied bail. Hilton immediately made application for bail to district court and it was granted, bond set at $10,000, despite no evidence to support the charges. The WFM was St. John's surety. After spending three weeks in the foul county jail, upon his release St. John took Sheriff Rutan's hand in a gesture of friendship. Rutan had endeavored to treat him with respect, to see that his time in the cell was not unnecessarily harsh. St. John could see Rutan was in poor health, the sheriff's face puffy and ashen, his robust shoulders stooped. Some said Rutan's rheumatism and some said his drinking habits sucked out the lawman's energy.[57] St. John parted without enmity and went to Denver at once.

A few days later a disastrous earthquake toppled San Francisco and a number of newsmen quipped, "It must have been St. John and the Western Federation of Miners!"[58] An article in the *Miners' Magazine*, smelling of Harry Jardine's sulfurous humor, poked fun back with, "We might as well confess." The WFM had decided to take out San Francisco with a "carload of McParland's best...XXXX strength...." Corroboration could be found in Harry Orchard's confession, "Volume 1631, page 25,489."[59] St. John and the WFM deeply needed the comic relief.

While the nation turned its eyes to the terrible tragedy of the quake, Hilton quietly pursued St. John's case, easily winning a change of venue. Even Curry admitted a "strong and very bitter prejudice" existed against St. John among a certain element in San Miguel County, and a trial there would only result in a mistrial; but, he said, it was the "disciples of Socialism and anarchy who have made affidavit in support of the contention of their patron saint for a change of venue."[60]

Once in the jurisdiction of Mesa County District Court, Hilton presented a motion to dismiss. St. John was charged with eight counts, a confusing glut all pointing to the "murder" of Ben Burnham. Hilton shredded the information in a brilliant and witty attack on the folly of it. He pointed out the law didn't allow a person to be prosecuted "for any offense denominated by the common law felony (murder, arson and forgery excepted) unless the indictment for the same shall be found by a grand jury within three years next after the offense shall have been...committed." The district attorney protested, saying St. John was a fugitive, but Hilton declared he was at no time a fugitive, and the present complaint was filed on February 27, 1906, two years too late, invalidating several counts.

St. John was charged with conspiracy to commit murder, his confederates being Jardine, Brennan, Conn, Macari, et al, yet he was the only person charged. "Conspiracy cannot be committed by one person," said Hilton.

In the same complaint, St. John "is absolutely omitted as a participant charged with having committed the crime," Hilton explained to the court. "In the first count he is already charged with having committed the murder, in the second count the same pursuant to the conspiracy, and in the third count he is charged — or attempted to be charged with conspiracy to do it, and in the latter part of that count his associates kindly take the job off his hands.... Attention will be later called as to this form of pleading at once forbidden and unauthorized."

While the district attorney squirmed, Hilton established before the court that nothing actually showed Ben Burnham was murdered. "He could have killed himself," Hilton quipped. The prosecutor could not charge murder simply because someone was killed.

Hilton had enough ammunition to have six of the counts quashed without a fight; but he pressed on, bringing before the court the prejudice and conspiracy of the Citizens' Alliance and the mine operators under the leadership of "that redoubtable swash-buckler Bulkeley Wells." Hilton pointed out this information was filed originally before the illegally appointed M. L. Brown, who, in his "blind devotion to a divine cause" and the MOA, placed the warrant in "the hands of the most celebrated gun fighter [Bob Meldrum] that Telluride ever knew, and he was sent after the defendant." Hilton put it on record that Brown was a tool of the Citizens' Alliance. He also mentioned how swashbuckler Wells "valiantly" drove the union men from their homes and usurped the courts in Telluride, letting it be known that the charges against St. John were actually nothing more than a continuation of illegal manipulation and oppression.[61]

This declaration so rattled District Attorney Hugo Selig and his cohorts in Telluride (Congressman Hogg among them) that after the district court judge quashed six of the eight counts on Hilton's motion, Selig declared the remaining two dismissed. It was election time and none of the MOA candidates wanted Hilton exposing their dirty linen in a trial that promised to be the biggest in western Colorado's history. Said Selig, "There is not enough evidence to convict the defendant on the charge of murdering Benjamin Burnham." In fact, there was no evidence.[62]

Capital charges were not brought against Vincent St. John because he committed crimes: Charges were assembled out of anything possible for the sole purpose of confining this man in a cell or a grave. St. John wasn't deemed "dangerous" because he shot people — on the contrary, he was known to be opposed to violence and never resisted arrest. It took extreme provocation to rile his temper and still, words were his only weapon. Even Sheriff Rutan felt no risk traveling several days with the man. Since the warrant for St. John was given to Meldrum, it was quite likely Rutan's presence on that trip thwarted a resurrection of the old plan for Meldrum to find some excuse to shoot the prisoner. The MOA gave warrants to men such as Meldrum just for that purpose.

An MOA detective in Idaho at the time of St. John's arrest inadvertently told the press why the MOA feared this man. He didn't mention anything about murders, bombs, or bullets making the Saint dangerous, but he did point out the deadliest thing about him.

"St. John has given the mine owners of the district more trouble in the past year than any twenty men up here," the detective said. "If left undisturbed he would have the entire district organized in another year. As it is, there are hundreds of miners added to the union membership through his efforts."[63]

And now Vint was free again, and the mine owners choked on their failure.

CHAPTER 23

GENERAL WELLS WANTS A HANGING

"Many a good hanging prevents a bad marriage."

— William Shakespeare

Vincent St. John escaped the conspiracy-soaked noose of the MOA again and took a nearly shattered Clara with him to Chicago, where he was a delegate at the second annual convention of the Industrial Workers of the World (IWW) in autumn 1906. This was the "One Big Union" that surged from the loins of the Western Federation of Miners and other labor unions, the WFM becoming the IWW's Mining Department, although run separately with Moyer still as president. The IWW was a union that included miners, factory workers, waiters, cooks, and all wageworkers, men and women, American and foreign born. It was a glorious dream for Isaac Cowen, William Trautmann, Eugene Debs, Haywood, Moyer, St. John, Guy Miller, and other labor leaders, yet in its early stages it was convulsed by what author Patrick Renshaw termed "infantile disorders." Politics and personality clashes, the unseating of delegates, allegations of fraud against the first and last president, Charles O. Sherman, and violations of the IWW constitution nearly destroyed it before it had begun.[1]

Moyer and Haywood, yet imprisoned in Idaho awaiting trial, kept abreast of the rabid events at the convention. St. John was elected to the executive board, but Moyer was incensed with him. During that turbulent second convention certain actions St. John believed were necessary to salvage the organization brought a harsh denunciation from Moyer. He accused St. John of trampling the IWW constitution to depose Sherman for his crimes. At one point Sherman's gang actually took over the IWW headquarters and his thugs whacked St. John in the back of the head with a blackjack when he arrived at the door. Sherman was only part of the problem. The Socialist Labor Party[2] leader, Daniel DeLeon, an inflexible intellectual, appeared to some to be trying to take over the IWW. Although St. John, "earnest, conscientious, loyal" as Haywood saw him, was willing to compromise and make things work, when he threw in with DeLeon on certain issues, he was branded a traitor by other factions.[3] Unfortunately, the hostilities of the convention leapt into the labor press afterward, articles that St. John saw as "willful and malicious misrepresentations."[4] St. John actually didn't care for DeLeon and disliked the Socialist Labor Party (which would eventually split with the IWW). The political, ideological, and larval organizational difficulties were complex and nearly crippling.

"Realizing that no good can come from a discussion carried on in the heat of passion," St. John wrote, "I have refrained from making any statement...until calmness and reason had taken the place of the feeling ('of anger, indignation and disgust') these articles and comment caused."[5] Vint wrote a long letter of explanation to all the WFM via the *Miners' Magazine*, refuting false reports, defending General

Secretary Trautmann against accusations of fraud, and demonstrating he only had done what he believed was right in the struggle against "grafters and traitors" within the IWW. He admitted to having been blinded by prejudice at one point, but had corrected his wrong assumptions once he found the truth. St. John's humble apology and clear explanation bolstered his stand among rank and file WFM members who loved him. Not once did he utter a derogatory word against Moyer. Moyer still condemned him.

Earlier that year St. John had been elected to the WFM's executive board, but the rift between him and Moyer widened into an ideological abyss after the second IWW convention. The same abyss pulled apart the anti-political direct actionists and the conservative political actionists in the WFM, splitting the "compromising" WFM and the "revolutionary" IWW, and yanking apart Moyer and Haywood permanently.[6]

Believing body and soul in the idea of the One Big Union and the overthrow of the wage system that subjugated the worker to the employer, St. John became general organizer for the IWW. In this capacity, although still an executive board member of the WFM at this time, he and Clara moved to the booming, raucous mining camp of Goldfield, Nevada, at the end of 1906. Goldfield barely had civilized edges, the hot-blooded population and turbulent labor atmosphere boiling into a war that pitted Citizens Alliance, MOA, Samuel Gompers' American Federation of Labor (AFL) and others in a peculiar alliance against the WFM, IWW, and their allies. Racked with MOA lockouts and scrip forced on miners in lieu of wages, WFM strikes and boycotts, flocks of spies and counter spies, Citizens' Alliance and MOA thugs, and the AFL riling moderate WFM members to go against the radical IWW, the camp became volcanic. Adding St. John's energy to this magma only intensified the mine operators' hatred for him and they concentrated their efforts to hammer him into the dirt.

He roused the MOA to rage when he delivered what they termed "inflammatory speeches" at rallies from Goldfield to San Francisco as Haywood's trial for the murder of Steunenberg approached. In part, St. John was soliciting defense funds for Haywood, Moyer, and Pettibone, yet his impetus was a mounting anger against the "capitalists [who] use the machinery of the law to clear [labor leaders] from their path." His own appalling experience at their hands crawled up inside him like a molten worm, eating at his normally quiet soul until it bolted out of him in fiery words. "The only way to destroy the Western Federation of Miners," he said, "is to hang every man who has a union card in that organization. If they start the ball rolling by attempting to execute Haywood, Moyer, and Pettibone, we will see that the ball rolls to the bitter end. They have shown us that they will stop at nothing. Well, we will meet them at their own game."[7] St. John's warning to the MOA brought thunderous applause from his audiences. The MOA condemned him.

In Goldfield St. John was unstoppable. He brought together skilled and unskilled workers, agitated for a minimum wage in the entire camp, inspired and drew in hundreds of men and women to the WFM and IWW. For one fleeting glorious moment in history, it seemed the goals of the One Big Union were possible and St. John wrote of that hour with elation: "Under the IWW sway in Goldfield, the minimum wage for all kinds of labor was $4.50 per day and the eight-hour day was universal.... No committees were ever sent to any employers. The unions adopted wage scales and regulated hours. The secretary posted the same on a bulletin board outside of the union hall, and it was the LAW. The employers were forced to come and see the union's committees."[8]

Such triumphs on the part of the already despised labor organizer brought the MOA's absolute hatred of St. John to near detonation in Goldfield. They had to get rid of him before they could control the WFM and use the AFL to weaken the WFM's marriage to the radical IWW. Yet they feared openly taking him out of the picture because his exceptional organizational talents were coupled with the adoration of the rank and file across the West. They could not risk an uprising in this volatile camp. In March 1907, an opportunity to use the courts to clear St. John from their path presented itself in bullets and blood. The MOA acted swiftly.

With a strike in progress, tempers flared and a WFM man named M. R. Preston shot and killed Goldfield restaurant owner John Silva on March 10. Preston claimed self-defense. He and a friend were arrested. Just before their trial a mysterious witness came out of the woodwork, claiming he heard St. John and his comrades plotting to have Silva killed. The men he named were the exact same men the MOA wanted out of the path. The witness's testimony before a grand jury secured the arrest of St. John, Harry Jardine (now a WFM local officer), and several others the last week of April, the news plastered in the *Journal* by an elated Curry. St. John was taken to Hawthorne and remained in jail until June, when O. N. Hilton came to his rescue again. During that time he "cheerily" conducted union business from jail, but the "moderate" union men now in control of the Goldfield WFM in his absence, took the opportunity to thrust the miners into opposition with the IWW, and signed an agreement with the MOA to end the strike. This caused numerous IWW members to accuse the WFM of scheming with the AFL and mine owners.[9]

The Goldfield MOA conspiracy was so obvious even to its supporters that the *Goldfield News* made no secret of the MOA's offer to drop all charges against St. John if he left town.[10] Hilton forced their hand instead: The district attorney dismissed the indictments for lack of evidence.

Much to the distress of mine owners across the West, this round with law and order only made St. John more of a hero in the hearts of the IWW members. As soon as he had posted bond, but wasn't quite free of the false indictment, they invited him to speak across the nation. He had engagements from July 23 in Salt Lake City until August 5 in Paterson, New Jersey, with every major city in between. It was a grueling, exhausting trip for a man who had been confined for weeks. Nor was he fond of speaking before crowds. Yet the flagrant violations of law against him and so many of his comrades poured out of him in passionate words, holding his listeners spellbound. He was no orator, as so many remarked. It was his "plain, unassuming way," and his "deep integrity" that held his audiences. He was more than an inspirational leader; he had become "beloved."[11]

By the time he reached Cincinnati he was hoarse, yet still they stood cheering him. After the talking was over, they had to "tear St. John away from the capitalist reporters" who were enthralled by him. The IWW local members "hiked him over to our headquarters, where we had him to ourselves until about two o'clock in the morning." They didn't want to let him go. Finally someone suggested "it was just possible that he was the kind of animal that sometimes slept, and with a tired smile, Vincent agreed that he was one of that kind."[12]

St. John was still on tour when astounding news flashed across the country. At Chicago he announced to the crowd, "Haywood is free!" and the thunderous cheers bounced into a joy-fest, where they nearly carried St. John off.

The Idaho jury had declared Big Bill Haywood, "Not guilty," despite the memorized confession of Orchard, despite the canvassing of potential jurors by the Pinkertons, despite the incredible manipulations by James McParland. When Haywood went home to Denver, a massive crowd greeted the irascible radical. His triumph distressed the prosecution and, although they whacked Pettibone over the head with Orchard's confession, he, too, was acquitted shortly thereafter. Seeing failure on the wall, Idaho released Moyer, the charges dismissed.

After being robed in certainty, fed on certainty, bathed in certainty, James McParland, the Great Detective, the mouthpiece of Divine Providence, was crushed by the verdicts. He had one man to blame: Stephen W. Adams.

William D. Haywood, Secretary of the WFM.
Author's collection.

On September 8, 1906, after being fed fine food and all the cigars he wanted since he had signed his confession in February, after being called "even a more important witness" than Orchard, Steve Adams delivered a lightning bolt to the prosecution. His wife, Annie, slipped it past the guards at the Idaho penitentiary into the hands of a WFM attorney:

> *This is to certify that the statement that I signed was made up by James McParland, detective, and Harry Orchard, alias Tom Hogan. I signed it because I was threatened by Governor Gooding, saying I would be hanged if I did not corroborate Orchard's story against the officers of the federation union of miners. Stephen Adams.*
>
> *Witness: Annie Adams.*[13]

McParland cajoled, threatened, quoted scripture, turned Gooding and Prosecutor James H. Hawley loose on him, and even put him in the cell with Harry Orchard again. McParland told Adams he was locking him in with his pal to calm Orchard's unraveling nerves. Orchard wailed he would kill himself if Adams didn't back up his confession and Harry showed Steve a suicide letter, telling him how he'd do himself in.

Steve Adams had had more than enough. He told Warden Whitney, "This case smells of hell."[14] Through his wealthy uncle, Warren Lillard, Adams finally came into the hands of WFM attorneys and they filed for a writ of habeas corpus. Up to this time, Adams was held in the penitentiary on no charge whatsoever, his wife and children kept in the women's ward. Annie's brother, Adams's brother, Annie's friend Mary Mahoney, and Lillard had been working diligently to free Adams from McParland's and Orchard's spell. Lillard was Steve's mother's seventy-year-old brother, a robust man who "worked like a Trojan, sparing neither time nor his own money," on behalf of those in whom he believed.[15] Lillard forced Idaho to allow lawyers to speak with his nephew, but after they left, the deputy warden took Adams to a cell previously occupied by a man who had been executed. Adams was "stripped and searched, and locked" in the death cell, although still not charged with any crime. This terrifying treatment only convinced him not to give in. "I was glad I had given my wife the statement to take away with her," Adams said, "...for when I was taken to a murderer's cell I was afraid something would happen to me."[16] Although fearful, Adams now had the support of family and friends and much-needed attorneys, enabling him to stand against the Great Detective, despite McParland telling him "that if he persisted in his course that he must prepare to meet his God."[17] McParland warned him he would face trial in Idaho and Colorado on charges of murder.

In retaliation for leaving the prosecution without a corroborating witness to Orchard's confession, McParland turned Steve Adams over for the hanging. Buck Wells, obsessed with hanging St. John, was currently digging up Boomerang Hill for Barney's remains. But he jumped at the chance to hang Adams instead, and put into motion extradition proceedings. A reporter snorted that Wells would "hang Jesus Christ if it would serve his...interests!"[18] Wells and the Colorado MOA were actually helping to finance the Idaho prosecution of the WFM, and paying huge Pinkerton bills, which included drinks and cigars.[19] In a letter to a potential witness, Wells declared, "I am actuated only by a desire and a determination to bring to a just punishment the men who are responsible."[20] If Wells couldn't get a WFM officer hanged, any hanging would do. Steve Adams was the ram in the bush. When the extradition papers from Colorado were in hand, McParland threatened Adams again, letting him know he would be executed for the murder of Arthur Collins. Steve still refused to withdraw his recantation.

Since this was prior to Haywood's trial, McParland thought it best to keep Adams in Idaho, believing a chance still existed to bring him back into the fold. If he could be tried for the murder of two timber claim jumpers in northern Idaho and be found guilty, he could then be induced to corroborate Orchard's confession in the Haywood trial. McParland talked this over with Wells. They hatched a plot to have Adams charged with the claim jumpers' murders and snatched to Wallace by Sheriff Angus Sutherland before a writ of habeas corpus could be issued. "It was agreed that the Colorado deputies should act as though I had deceived them and helped out Sheriff Sutherland," McParland wrote in his report.[21] This was carried through, and Adams went on trial in Wallace, Idaho, in February 1907.

Although Adams's confession was admitted in this trial and McParland and Buck Wells testified against him, the jury failed to convict, seven favoring acquittal. A mistrial was declared. Hastily, Idaho shifted the case to a different court, hoping to get a better jury, but once again, Adams was not convicted, the jury similarly split.

On the stand, Steve Adams detailed McParland's pernicious plotting from the time of his arrest in February 1905:

> *On the sixth day after I was taken to Boise and put in the cell with Harry Orchard, I was taken to the office of the penitentiary and introduced to Detective McPartland [sic].[22] He told me about 'Kelly the Bum' and other men who had turned state's evidence and had been set free. He told me some bible stories, too, but I am not familiar with the bible. He kept me until 4 or 5 o'clock in the morning trying to make me confess.*
>
> *McPartland told me that he wanted to convict Moyer, Haywood, Pettibone, St. John, and Simpkins, whom he called 'cutthroats.' If I would*

> *not help to convict them, he said, I would be taken back to Colorado and either hanged or mobbed. If I did help I would only be taken to Colorado as a witness. When we parted McPartland told me he was my friend. They put me back in the cell with Harry Orchard, who talked to me about the need of backing up his story. I was frightened. The next day McPartland called again. I said I would do what he wanted me to do. He told me to think well what I was doing, to think of my family and save myself. He said he would see that my family was cared for.*
>
> *When the confession was made, McPartland led me on step by step and showed me all that he wanted me to say. He told me that what I said about the Tyler and Boule murders* [the Idaho claim jumpers] *was only taken with the idea of making a strong chain of evidence to convict the officers of the Western Federation of Miners. He wanted the names of the officers of the federation used as much as possible all through the confession. Two or three days later Warden Whitney brought the confession to me to sign.*[23]

McParland's own reports to Gooding confirmed Steve's courtroom statement. The confession, which McParland had taken down originally as questions and answers, was transcribed into narrative form. The narrative was what Adams signed, as if he had related it himself. The original question and answer form was read in court, revealing McParland indeed guided Steve, the Great Detective having provided details from "notes in his pocket." The narrative document, inconsistent with Steve's own manner of speech, concluded, "I hope that the reign of terror inaugurated by Moyer, Haywood and Pettibone...will cease."[24]

With Idaho failing twice to convict Adams, McParland gave up hope of bringing him back into the plot. By 1908 Adams was to be tried in Colorado, giving Bulkeley Wells a crack at hanging him. No longer adjutant general, Wells was a colonel in the Colorado National Guard and, in this capacity, on the governor's staff. He still held a deputy's commission in San Miguel County and was managing the Smuggler-Union. He and his wife were living apart, but her younger brother, Robert Livermore, was working a lease on the mine and lived in a house on the property near Wells. Wells traveled often, being at the Smuggler intermittently.

On August 7, 1907, just a few months prior to Steve Adams's return to Telluride, Buck Wells had finally exhumed the bones he had been seeking. Wells proclaimed them Will Barney's bones, bones that would bring at least one good hanging. At the time of discovery he thought the man for the gallows would be St. John, yet the district attorney advised, despite the presence of the bones, that no evidence existed to connect them to St. John. Nothing existed to show they were even Will Barney's bones, except Steve Adams's claim, but he supposedly came into the picture after the victim was long dead. The jailhouse ravings of convicted murderer Crazy Miller had not led to the discovery of the bones, making his testimony invalid as well. Steve Adams's confession did produce the bones, yet that very confession failed to connect St. John to them outside of hearsay. Wells was forced to concede he had no case against St. John, a galling blow that inspired Curry to note "that a proper application of the unwritten law [lynching] at the proper time...would have been the means of saving several valuable lives... but it is too late now to apply this remedy."[25] Wells did have a case against Steve Adams, however, and that, he was certain, would finally give him a tidy legal hanging.

The Turkey Basin region supposedly was the burial place of Will Barney.
Photo by the author.

Stephen W. Adams arrived in Telluride on Tuesday, December 31, 1907, in the custody of Wells's mine guard and gunman C. C. Hicks and Sheriff Charles A. Fitzpatrick. He was locked in the county jail, the sheriff listing his occupation as "Socialist and Anarchist and Dinamiter [*sic*]."[26] Fitzpatrick had replaced Cal Rutan only a few months before, when Rutan died on September 25, a "physical wreck." Rutan had been taken to a Denver hospital, broken by alcohol and tormented by hallucinations, his body bent by rheumatism.[27] His final year was marked by further lies on behalf of the MOA and Citizens' Alliance, which only heaped more torment on him, for he knew the truth. He had even lied under oath in the Haywood trial, trying to dance around the fabrications with sudden memory lapses whenever the defense attorneys cornered him. One such lapse came when he was questioned about illegally deputizing gunmen Meldrum and Runnels on behalf of the mine owners who paid their salary. Rutan suddenly couldn't remember who brought the gunmen to Telluride, or who paid their wages, yet he had initially had the two men in his house until they found beds of their own. The county never paid them one thin dime.[28] The lies twisted Rutan, and, disgraced in his own mind, he slid inside a bottle and died. Like him, Sheriff Fitzpatrick was willing to please Buck Wells; he was dispatched to Idaho to grab Steve Adams the instant that state dismissed all charges against him. The mission was accomplished with such alacrity that no WFM attorney or any of Adams's family were on the train with him.

Warren Lillard wanted the best defender possible for his nephew, and he and the WFM quickly brought Orrin N. Hilton into the picture, along with Leon O. Whitsell. Because of Adams's confession, Hilton was unable to convince the court to release him on bail. Trial was set for the May term of district court.

As the weeks turtled by and Hilton and Whitsell gathered witnesses from around the country who had lived in Telluride when Arthur Collins was murdered, Wells and the remnants of the Citizens' Alliance grew anxious. Rumor hinted Hilton could find the means to clear Adams. The prosecution's scant material evidence and few witnesses were not enough to prove the WFM or St. John had anything to do with the murder. Steve Adams's confession was all they had to point to his guilt, and nothing corroborated the confession. General Wells (he still used the title) called himself the "star witness" for the prosecution, since McParland had given him a copy of the confession before Wells had met Adams and taken him to Telluride in 1906. Wells was certain Adams would hang for the murder, especially since Judge Theron Stevens was retired; but he wanted the WFM held responsible, at least before the public. He needed something to condemn the union, something that would swing a jury toward his view — something explosively convincing.

At 2:03 a.m. on March 28, 1908, while Telluride slept and a light snow drifted over the peaks, C. C. Hicks was operating the big searchlight in the Smuggler Mill tower. Suddenly he heard a blast, but was unable to tell where it came from or what it was. Before he had as much as considered leaving the tower to investigate, Buck Wells came out of the manager's house in his pajamas. He called to Hicks in the tower. Night watchman Richard Kermode had heard the noise during his rounds, but he, too, was unable to tell where it came from and, seeing Wells outside in his nightclothes, went to him. With a few scratches and bruises on the right side of his face, Wells explained to them that a bomb had gone off beneath his bed. He said he was "blown to the ceiling" of his second-story porch bedroom and knocked unconscious, finding himself in the opposite corner from the bed in a "heap of bedding and debris, and some of the bedding smoldering with fire."[29]

Showing none of the typical symptoms caused by close proximity to a blast, such as damaged eardrums, or any sign of having been unconscious or even being burned by the "smoldering bedding," Buck Wells personally telephoned friends in town to come see the damage. The mill foreman awakened Buck's brother-in-law, Robert Livermore, and told him the general had been blown up. Thinking Wells was in fragments, Livermore rushed to the office, only to find him sitting at his desk. Livermore and others were convinced the WFM had sent an assassin to kill Wells because of Adams. A motley band rushed off to the jail to lynch Steve Adams, but Wells allegedly rode up and stopped them, saying, "If we get one legal conviction, our troubles with the union will be over. If we make one illegal move, our name is mud!"[30] The odd thing was, the Miners' Union was so small in the district at this time that it was practically feeble and was by no means causing even the remotest "trouble" for Wells. In August 1907, sixty Liberty Bell miners went on strike to end the card system at that mine and within hours Manager Chase settled. Union men were to be employed, no questions asked. Wells followed suit, abolishing the card system, however, he stated no WFM men would be employed. A brief shoot-out between Marshal Runnels and some Greek nonunion miners accused of highgrading woke up the town a week later and that had been the extent of trouble; not a wink until — the bomb.

Curry flew multi-tiered banner headlines that afternoon about the "fiendish attempt" to end General Wells's life. He filled pages with details from Wells's own mouth, described the events and the scene meticulously, and proclaimed it to be the work of the anarchistic WFM. "How long shall the anarchists rule?" Curry roared.[31]

Huge rewards for the apprehension of the assassin were offered, eventually climbing as high as $25,000.

In following issues, Curry published photos taken of the bomb's damage. The rooms were "just as the explosion left them," nothing having been disturbed. The photos contradicted his original report, revealing far less destruction than he described. His sensational headlines said the house was "badly wrecked," yet no structural damage actually occurred. The worst damage was downward through the flimsy porch bedroom floor, and through the dining room ceiling below it, where two joists, wainscoting, and ceiling plaster were damaged. Curry said, "Every bit of glass in the dining room windows was blown out," yet the photo clearly showed nearly every pane still in place, some cracked, only a few shattered. Curry said the porch bedroom was a "complete wreck," yet the photos showed otherwise. Although Wells claimed he and the bedding were blown to the ceiling and landed in the opposite corner of the twelve-by-eight-foot room, the photos showed the mattress and bedding just heaped over the foot of the metal cot, and no damage to the cot frame, only the wooden slats scattered. Wells supposedly was beneath the bedding, yet had he been, he would have been draped over the footboard. He kept his loaded Colt .45 under his pillow, and he said the gun was blown into the wall with such force that "a piece of wood was driven into the muzzle, entirely closing it."[32] A number of people questioned why the gun wasn't driven into Buck's head, if he was sleeping on the pillow; or why his head wasn't smashed into the wall or ceiling, his ribs crushed, his bones fractured. Unanswerable questions began to flutter from many lips.

A time bomb supposedly had been set between three and five hours before the blast by someone who knew where the guards were, when the maids left, where Wells was, and where he slept. The assassin was said to have broken into the house via a rear, second-story bathroom window just before the snow started falling. No boot prints, not even slight traces, were in the snow. Wells slept in the second-story enclosed porch room and not in any of the main bedrooms. He was in town playing cards and the maids had already gone when the assassin slipped in and placed the bomb. Yet Sheriff Fitzpatrick found not "a single scrap of material of any kind" to support the theory of a timed device. Such a bomb, whether triggered by a wind-up clock or acid vile, would have left material evidence.[33]

People flocked to the scene to see the wreckage and "congratulate [Wells] upon his most miraculous escape."[34] It seemed all too miraculous to many, even his friends. "Some believed it to be a hoax on the part of Wells," was the whispered conclusion of the day. "From that time on the popularity of Bulkeley Wells declined in Telluride," according to a contemporary who knew Wells and liked him.[35]

The WFM response to the "attempted assassination" was incredulity. Miners suggested Wells had "chucked a giant firecracker" under his bed. If they wanted to blow up a manager, they would use enough dynamite to shift the entire clapboard house to a neighboring county. Wells replied, "Yes, I know the miners are saying that it was a put up job and that I did it myself. Well, the only answer I have to make to that is that I promise not to do it again." He pointed out how other witnesses had been bombed over the years and gave as the motive for the attempt on his life his role in the upcoming Steve Adams trial. "I am the most essential witness," he said.[36]

Wells may have believed he was the star witness, but Marshal Runnels knew everything Wells knew regarding Adams. No one attempted to bomb Runnels. The bomb exploded just in time to color Adams's preliminary hearing the first week of April. Hilton petitioned for a change of venue. Curry fumed, "The statement that

Adams cannot secure a fair trial here is preposterous and without foundation."[37] The petition was readily granted. Trial was set for June 22, 1908, in Mesa County District Court at Grand Junction. Adams was required to remain in jail at Telluride until then.

On June 22, the Grand Junction courtroom teemed with spectators; Arthur Collins's widow, her sister, his brother George, WFM President Moyer, and McParland among them. Miners, Pinkertons, Socialists, reporters, and Telluride citizens jostled for seats. On the bench roosted Judge Sprigg Shackleford, long-necked, lean, and stern, his white mustache hiding his mouth, his icy eyes watching the bevy of attorneys before him. Adams's team consisted of Hilton, Whitsell, Ralph Esteb, and Samuel Wheeler. Deputy District Attorney John Fry, James S. Carnahan, and ex-Congressman Herschel M. Hogg of Pueblo, previously a Telluride resident, buttressed the prosecution under District Attorney Hugo Selig. The defense intimated Carnahan and Hogg were "employed by private interests" to assist Selig, but they declared the San Miguel County commissioners had retained them. The judge allowed them to stay aboard. San Miguel County wanted to make certain Adams went to the gallows. The change of venue threatened that objective in some minds, thus Hogg's muscle was solicited.[38]

Jury selection battles chewed up the first two weeks, an entire second venire needing to be called after the first was exhausted. By July 6 a jury of ten farmers, one Fruita merchant, and one Grand Junction automobile repairman was selected. They would be carefully sequestered; no newspapers or communication with family were allowed. Both prosecution and defense were "well pleased." Adams seemed satisfied, but spent most of that morning gazing at his wife, Annie, knowing the moment that was to determine his fate had finally arrived.

Having spent the last three years behind bars, forty-one-year-old Steve Adams was beginning to look like a cave cricket, his powder-blue eyes watery, his pasty face stressed by unending fear. His thinning brown hair was neatly combed forward and he was dressed in a suit and tie, a carnation in his buttonhole and WFM pin on his lapel. As he glanced around the courtroom, reporters noted apprehension in his eyes, yet he favored Annie with an elfin smile.

Younger than her husband, Annie was pretty and vivacious. With plucky determination she had stood by him through the dark days of the Boise incarceration, the Idaho trials, and the confession, which she never believed. She had broken down a few times, but Mrs. Mary Mahoney "put a good deal of nerve" back into her.[39] Annie was Mrs. Anna G. (Diamond) Dickson, the widow of a miner and mother of two children when she first met Steve in Telluride. Mary Mahoney had introduced them, Mary knowing Steve from Cripple Creek when she ran a rooming house there in 1900. Annie and Steve married on November 15, 1903, at the First M. E. Church in Telluride.[40] They had added a third child to the family during their incarceration. Now sitting directly behind her husband, Annie watched the court proceedings with unflinching concentration as the corpulent Hogg pumped up his lungs for the prosecution's opening statements.

Steve Adams stared straight ahead, sometimes rocking back and forth, while Hogg described the agonizing last hours of Arthur Collins for the jury. Hogg launched into the confession that would prove Adams killed Collins, and Hilton objected. The court had not decided the question of admissibility of the alleged confession. The confession "was not free and voluntary," Hilton said, and its introduction "violated the defendant's constitutional rights."[41] Judge Shackleford overruled the objection, allowing Hogg to proceed.

Hogg said Adams had given his confession while being held at the penitentiary in Boise. He said the document described how Adams shot Collins in the back on November 19, 1902. It accurately detailed the Collins card game, the house, and the surroundings. When Adams was later brought to Telluride in May 1906, Hogg said, he showed Marshal Runnels and General Wells precisely where he stood to fire the shot, where he went afterward, where he hid the shotgun at the highgraders' shack, and where he helped bury the body of W. J. Barney. Since that time, Barney's body was found in the precise spot indicated, "with an empty purse on the breast," confirming the confession as factual in every word.[42] The empty purse was a figment of Hogg's imagination tossed to the jury for effect, but Barney's bones were rattled as incontrovertible evidence. Thus began the "greatest criminal trial in western Colorado,"[43] its nucleus Steve Adams's confession, its sinew Will Barney's bones.

Stephen W. Adams, from the Daily Sentinel.
Restored photo © M. J. Martin 2003.

Since the confession was ruled admissible in Adams's two Idaho trials, Selig and Hogg were confident Shackleford's decree would be the same. While Shackleford addressed the question, the jury was removed. Contending the document was illegally obtained by the "grossest intimidation," Hilton called James McParland to the stand to explain his unique method of extracting confessions. Hilton had a transcript of McParland's testimony from Adams's last Idaho trial and, for the most part, simply followed it. The sixty-four-year-old, white haired, paunchy, and soft-spoken detective had boasted he "never failed to get" a confession "from a guilty man."[44] He calmly answered Hilton's questions in the same manner as he had in Idaho. He admitted he had not advised Adams of his rights, had not allowed him access to an attorney or clergyman, had offered immunity through stories of Kelly the Bum and other confessed criminals, had threatened with tales of hangings and lynchings, and had suggested that once Adams confessed, forgiveness was his.[45]

Hilton pointedly asked, "Did you tell the defendant expressly by whom he would be forgiven?"

"Yes," said McParland, "by God Almighty."

"Did you have any assurances as to that?" Hilton asked, as sniggering skipped around the room.

Hogg exploded, "I object to that question! A good many people would call that blasphemy."

"And others would call it absolution," Hilton retorted. He returned to McParland, dragging him through the disappointment and wrath caused by Adams's recantation and McParland's strenuous efforts to bring Steve back.

Hogg objected so frequently to this line of questioning that he finally ceased to arise from his place at the table, simply repeating, "I object," without stating why.

Finally Hilton pointed out a reason was needed to make a valid objection, otherwise "a phonograph could just as well be set here grinding out" the protest.

Rolling to his feet, Hogg spat out his reasons for his objections and slung a few jibes at Hilton. Hilton slung back. Hogg huffed, "I object to his comparing me to a phonograph, and I won't stand for it either."

With a stern rebuke, Shackleford silenced the contenders and the examination continued, McParland's testimony being corroborated by the stenographer who originally took down Adams's confession. The stenographer admitted McParland "directed" him as to what to take down and what to leave out. Sometimes "the substance of the conversation" was written in at a later time. He said Adams had acknowledged the confession as "free and voluntary."[46]

In conclusion, the defense said the confession was inadmissible because it was obtained while the defendant was illegally incarcerated, deprived of his rights, intimidated, as well as offered inducements. The defense had shown the document wasn't even in the language of the defendant. The stenographer confirmed this. The prosecution begged to differ, Selig presenting the State's rebuttal since Hogg wore himself out with objections. The State maintained the confession was voluntary, and because Adams took several months before recanting it, the time lapsed proved it. Judge Shackleford took the question under consideration overnight.

The next day Shackleford dropped a bomb on the prosecution: The confession was barred. Citing previous similar cases and McParland's own testimony, where he admitted using stories that allowed the defendant to conclude he would either be hanged if he failed to confess or set free if he did confess, the judge ruled the Adams confession had been obtained illicitly. McParland didn't tuck tail and scamper away; he simply noted the WFM had corrupted this judge as well.

Selig and Hogg thought their case was sinking. Shackleford tossed them a lifeboat. The confession was out, but, said the judge, he would allow General Wells and Marshal Runnels to repeat whatever the defendant had said in their presence. Wells was ready, having read the confession more than once; he nearly had it memorized. He had Barney's bones in a box. Wells had personally worked up the evidence in the Collins case himself, and it perfectly matched Adams's confession. Adams would hang.

Oddly, what Wells said on the stand conflicted with the confession, and what Runnels said contradicted Wells. Nor was the confession as factual as Hogg had presented. The confession put two women and one man playing cards with Arthur Collins on the fatal night, when the party actually consisted of three men and one woman. If Adams had watched the card game for as long as the confession claimed, he would have certainly noticed only one woman was in the party. The confession made Adams an employee in the cyanide plant before working in freight, when the company records showed him working in the freight department only. He was under Foreman John R. Hogan, who said Steve Adams worked from October 16 until November 11, 1902. He recalled Steve said he was quitting to go prospecting. The confession killed off Will Barney in 1902, when Barney's disappearance was in 1901. It placed Vincent St. John in Telluride in September 1903, when St. John had left May 18, 1903, and did not return until his extradition in 1906.

Where it concerned the murder of Arthur Collins, the confession appeared to be created from the raw material of the original newspaper reports, with details added to make the killing a plot of the sinister Inner Circle. Neither the confession nor testimony from Wells, Runnels, or McParland established just which of the WFM officials had ordered the Collins assassination. The confession said Arthur Baston and Ed Minster were in Telluride when Adams arrived on the scene, and they told Adams they wanted Collins killed. They said Haywood and Moyer wanted the job done. McParland's report to Hawley, the Idaho prosecutor, said Sherman Parker and W. F. Davis, WFM officers in the Cripple Creek district, wanted the job done.[47] Runnels said Adams told him "headquarters" wanted it done, and Wells said St. John wanted it done.

The confession said Baston gave Adams the shotgun used to kill Collins. Yet Runnels said the defendant informed him that St. John instructed Adams to get the shotgun from a man named Ryan. The confession and testimony in the Idaho trials put the payment for the assassination at $1,000. Wells put it at $2,000. The confession said Adams and Baston remained in Telluride three days after the murder, then they went to Altman (near Cripple Creek) where Baston paid Adams $250 for the Collins killing, saying he would get the rest from Haywood. It said Baston later brought $750 to Adams at his Altman cabin. "Baston told me Vincent St. John knew all about this murder," the confession stated, "but I did not talk to St. John about it."[48]

Wells had read that line more than once and had heard it read in the trials, yet on the stand he painted the picture he preferred, since the confession was out and he was the star. He said St. John himself paid Adams and Baston $500. Wells spoke with the confidence of a man who knew his own importance, adding, "Later in Salt Lake, St. John gave them $1,500, which they divided."[49] Had Adams ever actually told Wells or McParland such a thing, St. John would have been charged the minute the statement fell from Adams's lips, just as Haywood was charged on the word of Harry Orchard.

His testimony demonstrated how intensely Wells still hated St. John. He would connect him to the Collins murder, even if it meant lying under oath. Since Collins's death, Wells had been building a case against St. John, yet he kept hitting legal snags and unalterable facts. Adams's trial was his last chance. He wanted Adams to hang as St. John's proxy.

CHAPTER

24

IF BONES COULD SPEAK

"As for believing things, I can believe anything, provided that it is quite incredible."

— Oscar Wilde

"I killed Arthur Collins with a shotgun," Steve Adams had stated in spring 1906. At least that was how Marshal Willard Runnels recalled Adams saying it during the trip from Boise to Telluride. Runnels said Adams told him the Winchester shotgun was the best for the job. It was a sporty conversation, as if the men were comparing fishing rods. He said Adams then went down on one knee as if holding such a weapon and pretended to fire it.

Weaving a tale of multiple murders, beatings, and bombings in which Adams was a participant, Runnels finally blurted, "Oh, he talked about so many killings he was in that I didn't pay much attention."[1] The statement sent a murmur through the courtroom.

Dr. Ochsner's testimony was as stirring as the marshal's, the prosecution drawing out painful details while Arthur Collins's widow and her sister winced. One reporter noted how Marguerite Collins glanced with disdain and disgust at Annie Adams whenever the prosecution scored a point, and Annie flashed sharp looks in return, the reporter tagging it, "The battle of eyes."[2] During Ochsner's testimony, the battle waned. The doctor, who had been barred from practicing medicine in San Miguel County in November 1903, described the wounds Collins received from seven pellets of buckshot. In a hushed courtroom, he softly unfolded the man's agony over the spectators.

A. B. Blainey recalled the manager's dying words: "I did not think those people would shoot me in the back." This statement was never included in original testimony, the initial reports saying Collins asked "why anyone would wish to shoot" him. He was sedated with opiates after that and the attending doctors had not allowed him to speak, yet Blainey's statement hung before the court like a bloody wraith pointing at the defendant, that willing tool of Vincent St. John: Steve Adams.

Stephen W. Adams was a simple miner, born in Missouri, son of the now-deceased David Adams and Amanda Lillard. Amanda's brothers, Warren and Ephraim Lillard, were well-known, wealthy businessmen, Ephraim serving as warden of the Kentucky penitentiary for seven years. From an old and respected southern family, the Lillards had served in the Confederate Army, but Adams had little interest in his uncles' politics.

Adams's two brothers, Joseph and William, were now mining in Creede, Colorado. He had followed them to Cripple Creek around 1898 (he couldn't recall the year) because Joe wrote to him of the "easy living" in the gold camp.[3] He joined the WFM at Cripple Creek, never much of a vigorous member. He was neither a Socialist nor took part in political action. When the strikes were called, often times he hoofed off to other camps. Just six days after he married Annie in Telluride, November 15, 1903, Steve was arrested in Cripple Creek by the militia because he had "made threats against workmen" during the strike there.[4] Numerous union men were arrested that day because the Vindicator Mine was blown up. Adams was detained for weeks. On December 2 he and several others were charged in the explosion, but later the charges were dropped. Orchard later confessed to bombing the mine and the Independence depot, naming Adams as his accomplice. For a man who purportedly received more than a year's wages each time he killed or beat enemies of the WFM or blew up mines and railroads, Adams had nothing to show for it. He lived in near poverty most of the time, engaging in petty theft on occasion. Having lived with Orchard sporadically, Steve and Annie finally moved to Oregon August 1, 1905, to make a "fresh start," Adams filing on a 160-acre homestead in Baker County near his uncle's ranch.[5] After his arrest in February 1906, Adams believed if he signed a confession for McParland he would be allowed to go back to his homestead. Annie thought the same, McParland telling her that if Adams corroborated Orchard's confession, "he would be able to go" home "in a short time, and there would be no prosecution."[6] Instead, they were both locked in the penitentiary. Some weeks after Adams had signed his confession, he was allowed to live with his wife and children in a cottage inside the prison, but he was not free to leave, see anyone, or consult with an attorney. Once he recanted, he was again taken from his family and thrown in a "death cell." His uncle, Warren Lillard, told him he believed he was innocent. Adams had some greasy stains on his record wholly unrelated to the Federation, Lillard knew, but he believed Orchard and McParland had cooked up the heinous crimes in the confession. General Wells firmly believed otherwise and intended to prove to the jury Adams was a cold-blooded hired assassin for St. John and the WFM.

When Wells took the stand his entrance caused a hum of admiration. Reporters noted his immaculate grooming, his "suave, calm manner," his elegant suit, his "steady brown eyes" set in "a handsome face," and the "tone of finality in which he answered all questions." The thirty-six-year-old mine manager knew he was "born to lead," and leading this trial to the conclusion he desired was his goal.[7]

Wells testified before the court how he had carefully preserved material evidence from November 1902: the shotgun shell found sixty-two feet from the house, the window sash with the pellet hole in it, and the fragments of window glass glued together on cardboard so one could see the eight-to-ten-inch wide pattern of the shotgun blast. He counted each hole. He also introduced the badly rusted, sawed-off Winchester shotgun found in 1903. The items were entered into evidence.

The defense objected to the shotgun as evidence, since the prosecution failed to show it was the weapon used or had ever been in the hands of the defendant. Judge Shackleford allowed its entry, because it matched the gauge of the shell found near the house. As mentioned earlier, the Winchester Model 1897 twelve-gauge, pump-action shotgun was the most popular in the West. This one easily could have belonged to the highgraders who operated out of the shack. Wells was convinced it was the very weapon used to kill Collins. He testified that Adams had asked him during the trip if he had "found the Winchester shotgun" that Adams had hidden below Pandora. This question didn't prove the shotgun found was the weapon used; it hinted that McParland, who knew the shotgun had been found, had given the information to Adams. Wells claimed Adams had also asked if they found the ejected cartridge on the trail and if they knew he met two men on the track that night. Again, questions of this nature coming from Adams too conveniently covered information previously published in newspapers. McParland quite likely instructed Adams what to say to Wells, since it was important Wells believed the confession. The Great Detective spelled out entire conversations for others to use: His reports to Gooding showed he told Orchard what to say to WFM attorneys, and told Baker County Sheriff H. K. Brown what to say (a two-page conversation was scripted) to manipulate Lillard so Adams would withdraw his recantation.[8] Since Wells had taken over building up the Collins case, and was both McParland's ally and paying client, they frequently discussed the details, which Wells admitted under oath.

Although the sheriff's investigation immediately following the murder of Collins put the shooter fifteen to twenty feet from the window, hidden in the shadow of a towering spruce, Wells said he had determined the shooter stood where the empty shell had been found up on the trail, and he marked it so on a plat. Wells had instructed J. S. James, a surveyor, to make a plat of the Smuggler buildings so the jury could see the forty-five-by-sixty-foot, two-story house in relationship to the Smuggler and Pandora Mills and the cyanide plant. Wells had James place four figures at the card table to represent the night Collins was shot. The table was eight feet from the window. The window opened onto a roofed porch eight feet wide. The plat showed the tower with the arc light, but did not show the spruce tree. Wells had ordered the tree cut down shortly after his arrival in 1902. At the

This view of the Collins house at the Smuggler Mills complex was taken by Joseph Byers a few days after Arthur Collins died. The buckshot passed through the porch window to the right of the door. The shooter could not have stood seventy feet away and up the trail, but had to be about where the photographer stood to take this shot. From the Denver Post *1902.*

Author's collection.

time it was generally believed the offending tree had given the assassin a place to hide. According to the confession, Adams had waited quite some time in the spruce shadow watching the card game, waiting for the "lady beside Collins" to lean over toward the lady across from Collins, the assassin not wishing to shoot "a lady."[9] This was an odd notion when the killer had a sawed-off shotgun, used for its ability to spread pellets broadly.

Actual facts showed Maude Blainey was seated directly in the line of fire across the table from Collins. Her husband was on her left, Kemp on her right. Collins had his back to the window. The card table was quite small.

Despite the original reports, Wells insisted the shooter had stood where the shell was found, not merely because it was found there, but because Wells fancied himself a master sleuth. He explained to the jury he personally calculated the position by "sticking a pencil in a" pellet hole and "sighting along the pencil."[10] The idea was good; the method was inaccurate. It put the shooter seventy feet from the window, and seventy-eight feet from Collins, at an angle above his target, and on the far side of the spruce. It meant he had to sight his target through the boughs of the tree and through a window with the shade half down. In contemporary photographs the single tree was dense and broad. Almost every tree in the area had been cut down; this one had been preserved for shade at the front (west) of the house. Stumps peppered the hillside. The arc light tower was on the uphill side of the tree and, at night, would have created a denser shadow in the boughs. It also meant the killer was in the light in full view of anyone either on the trail, at the small house just above, or at the Pandora Mill 400 feet to the west. Yet Wells insisted his calculations were correct.

Wells explained to the jury how Adams had told him he watched Collins for several nights (Runnels said ten nights), waiting for an opportunity to shoot. He supposedly spent ten nights on the treeless hill watching the back of the house, and no one noticed him. This was in the heart of the Smuggler Mill complex, where the mills and plants operated twenty-four hours a day, with men coming and going on the trails. At the time Collins resided in the house, several other persons did as well. One thing Wells and Runnels didn't know was that Collins was gone for some weeks and had returned to Telluride just five days before the assassination.[11] Perhaps Adams

The spruce and arc light tower are clearly visible to the left of the Collins house (center), in this 1902 photo of the Smuggler Mills complex.

Courtesy M. L. Shaffer.

watched only those five days. Wells said on November 19, 1902, Adams stalked at the front of the house. The window shade was partially down, giving him a clear shot at Collins — "clear" if one failed to consider the spruce, the shadows, and the overhanging porch roof.

"As soon as he fired," said Wells, "he jumped back, ejected the empty shell and threw in another and stood for a minute waiting, because he had heard that Collins was protected by a bodyguard and he meant to kill anyone who attempted pursuit."[12]

Despite the "waiting," Wells had Adams meeting Theodore Becker and Harold Smith on the narrow-gauge railroad track in the moonless night, Adams having gone a half-mile in the time it took Becker and Smith to walk 200 feet. Adams didn't know the men and they didn't know him. Becker recounted his story for the jury, but under Hilton's cross-examination, he could neither describe the man he saw nor say for certain what the man was carrying. The night was too dark, he said, admitting he couldn't even tell if the man was carrying the object over his shoulder or just in his hand. In a night that dark, Adams would have set a world record for the half-mile stroll on a railroad track, for Becker had described the man he saw as "walking."[13] Hogg had asked a Smuggler employee, Walter Reed, to test how long it took him to walk the same distance on the track, which he said he did in five minutes, in the daylight. Hilton said the testimony was irrelevant, since no one ever proved the defendant had taken the route in question.

Wells said Adams told him that after he passed the men, he then "hid the gun in a clump of trees at Royer Creek." This was at the highgraders' shack where young men found the rusted shotgun that originally Wells and Rutan had decided was not in working order when abandoned. That detail was forgotten, although when Wells was asked to extract a cartridge from the shotgun to show the jury, he failed.

Confirming Wells's testimony, Runnels said when he and Wells brought Adams to Telluride to show them Barney's grave, Adams also showed them the spot where he stood to shoot Collins. It was near a stump, about seventy feet from the window, up on the trail. Stumps were obviously plentiful, since the one fifteen feet from the house was the original stump in question. The closer stump figured in the first news reports, yet somehow now it was adjusted to fit Wells's position of the shooter.

Under Hilton's thorny cross-examination, gunman Runnels, now the city marshal of Telluride, became "nervous," keeping "his eyes on the ceiling," according to the *Journal*'s reporter.[14] He was frequently directed to turn to the jury so they might hear him. Runnels suddenly suffered a series of memory lapses, unable to recall if Wells was armed or wore a deputy's badge during their clandestine trip with Adams from Idaho to Telluride in 1906. Wells had admitted he was armed with two revolvers on his belt and one in his grip. He also wore a badge, but not in the capacity of an officer of the law, he claimed. Was the badge just jewelry? Runnels was also armed and wearing his silver star, posing as a marshal but not "in the capacity" of a lawman.[15] The jury was left to figure that out.

James C. Mills, Jr., deputy warden from the Idaho penitentiary, had accompanied Adams, Wells, and Runnels to Telluride during the Barney bone hunt, and his testimony confirmed neither man had a warrant or any legal documents to take Adams as a prisoner. The trip was done via a secret arrangement with McParland, Hawley, and Warden Whitney in Idaho. Mills said he couldn't recall what Adams said during the trip or what he said once at Telluride. Back on the stand the following day, Mills suddenly recalled entire conversations, confirming Adams had said he shot Collins, where he stood, and where he hid the shotgun. Hilton knew from the testimony Mills was merely repeating what Runnels had said. Mills admitted his memory was refreshed when he read the marshal's testimony in the newspaper. With new questions unrelated to Runnels's testimony, Mills returned to, "I don't know" and "I don't remember." He did recall that Adams was under heavy guard at all times, yet he had not been charged with any crime in either state. Adams was not free to leave or go where he wanted, and it was understood he would be returned to the custody of Idaho when Wells was finished with him. The guards and guns were explained as "protection" for Adams. Runnels, Mills, and Wells claimed the WFM would have killed Adams to stop him from testifying.

"Did any members of that organization attempt any violence?" Hilton had asked Wells.

Wells said no one attempted anything. The WFM had no access to Adams in Boise and was unaware Wells took him to Telluride to locate the remains of Barney.

"And the finding of the body of Barney was not your only purpose?" Hilton said.

"No, not my sole purpose," Wells replied. "I wanted to secure information [from Adams] that would lead to the conviction of other parties."[16]

Initially Wells and McParland had planned to validate Adams's confession by his pointing out correct locations; this material could then be used to corroborate Orchard's assertion that the WFM executives hired assassins to put men out of the way. Since that failed, the present purpose was to put Adams out of the way. The prosecution felt their case was going well, perhaps even better than if they had used the confession.[17]

Buck Wells's testimony "proved by far the strongest that the state has introduced," the *Sentinel* reported (July 10, 1908). "He stood Judge Hilton's cross-fire without once losing his self possession, and only once his pleasant, well-modulated voice raised above its customary pitch, and that was when Judge Hilton asked him if he had had charge of the Telluride bull-pen. Wells flashed up then, and his words were sharp and crisp, 'We never had a bull-pen in Telluride.'"

It was a mere matter of semantics — the men in the crammed county jail at the time called it the "bull-pen."[18] Wells refused to allow Hilton to drag him through the mud of what Curry had recently termed "humane" deportations designed to save on lynchings.[19] Wells was here to convict Adams — the *Sentinel* said precisely that — and not to have the WFM "buzzard" lawyers convicting him for his benevolent dictatorship.[20]

Wells *did* want to give the jury the Barney tale; he wanted them to know how the bones were found where Adams said they would be, thereby confirming the Collins murder scenario. But Hilton blocked him, and Shackleford sustained. At least Runnels had managed to cough out most of the Barney saga before it was prohibited. Runnels said Adams moved the body in September 1903 from a spot just fifteen feet off Boomerang Road near Alta. This was a corpse that had been lying on the surface for over two years, barely hidden behind a log, astonishingly attracting no notice except from one or two flies. Would the jury believe the tale? Runnels said his story came from Adams, who heard it from Carpenter, making it stratified hearsay. Why did the prosecution allow Runnels to sputter hearsay until adjournment that night? The tale had nothing to do with the Collins murder. The answer grinned from the name used most abundantly in the story: Vincent St. John.

Runnels cranked up like a cowboy spinning windies at a campfire. Perhaps he forgot he was under oath. "Adams told me that Barney was gotten into a stable where somebody hit him on the head with a gun but the gun went off. They carried Barney's body out of the stable, down the tracks to the foot of Boomerang Hill and left it there with a Dago and an Irishman who took it up the road. The next morning he told St. John that the body was gone. St. John said there was a committee appointed to see that the remains were buried."[21]

St. John paid the Irishman to leave town, yet he wanted to know where the corpse was, so paid Adams and Baston to go find it in September 1903. Apparently Runnels wasn't expected to make sense. He said when the news spread that the "soldiers were coming," which was late November 1903, St. John "sent Steve and Carpenter up after the body again." Runnels knew St. John left in May of that year, but he had the full attention of the courtroom, and he spun the windiest windy any Wyoming range rover ever spun. He told about an Ophir man named Kelly who had hidden the body. Steve and Carp got the man drunk, had him lead them back up Boomerang Hill, and the man ran off. They found the corpse despite him, just fifteen feet off the road, less than a mile from Alta. It was amazing St. John failed to find it in those previous twenty-seven months. They took out their sacks and commenced stripping the ripe corpse, but Carpenter "got deadly sick." Adams put the clothing in a sack and merrily scooped up the entrails and put them in with the clothes. The body parts were folded into the other sack. They hauled the sacks a few hundred feet downhill, buried one and hauled the clothing and entrails sack clean down to the river, a distance of six miles. "They threw the shovel away and the sack with the entrails and clothing was chucked into the river."

When Runnels finished that windy he tossed in another: "Steve then said that St. John told him that a man named [Wesley] Smith had been murdered in the Union hall and that he had been put in a grave with another man who had just been buried."[22]

It was a brilliant idea: burying the murdered with the recently expired. The entire town knew the day after Wes Smith disappeared, four Liberty Bell avalanche victims were buried: Gus Kroul, Andrew Aho, Olaf Swanson, and Louis Lundborg, the latter three being union men. A joint funeral occurred at the Methodist church, under the auspices of the Miners' Union and Odd Fellows; half the town walked to the cemetery in the heavy snow. Had Adams told Wells that Smith was buried in one of those graves, Wells would have dug up all four without hesitation. It seemed Runnels had created the scenario on the stand.

As sinister and thrilling as the tales were, the jury had to take them as hearsay. The prosecution knew Runnels was for effect, but Wells's "damaging evidence" would win the case. When the prosecution rested, Curry grinned on the sidelines, proclaiming Wells and Runnels "good strong witnesses." The defense was "unable to shake their testimony," he gushed, calling Orrin Hilton "very caustic and bitter."[23]

The *Sentinel* reported July 11, 1908: "Mr. Wells made a most excellent witness." The reporter also praised Judge Hilton as a powerful defense attorney, "superior" to the famous Clarence Darrow and Edmund Richardson, who conducted the defense of Haywood, Pettibone, and Adams in the Idaho trials. In the end, the fight would be between Wells and Hilton, the reporter said.

Judge Orrin N. Hilton presented a thin case for the defense, based solely on eight witnesses who put Steve Adams at Mrs. Mary Mahoney's boarding house at the time Collins was murdered. Hilton's opening statement was straightforward, without a hint of fire. He said he would disclose the "true animus and spirit" that drove the prosecution to target Adams as "an object for slaughter by the Pinkertons and Bulkeley Wells."[24]

The defense's first move was to establish that, at the time of the murder, Steve Adams had not resided at the Colorado House as the prosecution presented, but at the rooming house run by Mary Mahoney on Columbia Avenue. Wells had earlier testified Adams had shown him the very room in the Colorado House where he had lived at the time of the Collins murder. Wells had no ledger or other witness to corroborate this. Mary Mahoney, now remarried as Mrs. Meenan, said she previously had run the Colorado House, but moved to the other house three weeks before Collins was shot. Adams had arrived in Telluride in late August or early September that year, residing first at the Colorado House with Mary and moving when she moved.

At the grand jury inquest into Collins's death in 1902, Mary Mahoney had testified and, at that time, had consulted with Wells and then-Deputy District Attorney Howe. Because Adams had picked up his paycheck the day before Collins was shot, then left Telluride on November 23, he came to the attention of Wells. At that time, Mary had testified the boarder had been at supper and then at breakfast the following day, but she didn't know where he was in between. Hogg brought this out on cross-examination, since Mary now said Adams was playing cards with three other boarders when news came that Collins had been shot. Mary originally had been brought to the trial as a witness for the prosecution, but having recalled the card game, she was now part of the defense. She named all fifteen of her former boarders, including Baston and Minster.

Hilton didn't rely heavily on her testimony. Of the fifteen boarders in the house in November 1902, seven of them were now in court. These were two women, Theresa Woodford and Catherine Cassidy; and five men, William Miller, Jack Griffin, Andrew J. Tyner, and Frank Price, all miners; and Walter Hull, a painter and paperhanger. None of them still lived in Telluride; most lived in other states. Although some were weak in delivery, most were compelling in their testimony, all corroborating the contention that Steve Adams was at the rooming house playing cards with William Miller, Frank Price, and a man now dead, William Noyer, when Collins was shot. The two women and Griffin and Tyner were in the sitting room as well that night, the men discussing the mining contract they had at the Smuggler. All of the witnesses said they had never seen Adams or anyone else bring a shotgun or any gun into the house. The men who knew Adams well, such as Tyner, said Steve never owned a shotgun.

The testimony established that Adams and the three card players had started the game around 7:30 after supper and were still playing when Hull came back from the Wharton Saloon around 10:00 and announced, "Manager Collins has been shot." Hull heard the news in the saloon. He clearly recalled Adams and the others playing cards. He didn't want to be a witness but had been subpoenaed because others remembered him bringing the dark news into the house.

On cross, Hogg gouged into each witness, suggesting they had all conspired together to create this scenario, attempting to impeach them by demanding other details. Only Mary Mahoney Meenan and Miller had a few memory freezes. The others were unshakable — the date, the night, the circumstances all immortalized by the reality of murder.

Catherine Cassidy had been visiting Theresa Woodford and other friends in Telluride, Cassidy having lost a brother in the Smuggler fire the year before. The morning after Collins was shot, the women went to the cemetery to visit the graves of the miners who died in that disaster. The two days were fixed by tragedy in their memories. Woodford recalled tiny details, explaining how she had broken a lamp chimney curling her hair the night Collins was shot. She had also asked the men at the table for a box of candy, something her meek soul remembered because it was the first time she ever dared to do such a thing.

Jack Griffin and his mining partner, Andrew Tyner, had worked in mining camps throughout the West. Griffin was a good friend of St. John (he didn't bring this out at the trial) and had been in Burke with him. He was in that camp when subpoenaed as a witness. The man had a memory for every shift boss and manager he ever worked under, and answered every question, even under Hogg's battering, "with scarcely a pause for reflection." He and Tyner had worked in Cripple Creek after the Smuggler was shut down on account of the murder. They returned to Telluride a few months later, hiring on at the Tomboy until the strike of September 1903 occurred. Adams was back in town with them, and Tyner "stood up for him" as best man at his wedding. Tyner had been a member of the Miners' Union and recalled being deported.

Trying to shake him, Hogg said, "Do you remember the exact time?"

"No, not exactly," Tyner replied. "We had Thanksgiving dinner in the Telluride jail and Christmas dinner in the Montrose jail." He laughed. He had quit the WFM since then — the adventures of deportation were too trying on him.

Hilton spotted a bell rope dangling in Tyner's testimony and pulled it on re-cross. "Do you know who had charge of this deportation?" he asked as soon as Hogg released the witness.

"I think Mr. Wells had charge of it," Tyner replied with a smile.

A jubilant "didn't-we-say-it-was-so" chortle broke from the audience and Hogg boiled to his feet, growling at Hilton, "Your people ought to be excluded from this courtroom!"[25]

Before Hilton crossed swords with Hogg, Judge Shackleford stamped out the battle, disappointing the audience who relished the lively skirmishing between the two attorneys. Hogg blustered at Tyner so he might prove the deportations were due to the deadly threat of the WFM, causing Shackleford to cork the ex-Congressman's questions.

The smiling, charming Tyner left the stand like a hero and Frank Price took his place. Price recalled playing cards with Adams and the others the night Hull came in and announced Collins had been shot. It was fixed solidly in his head, and Hogg failed to dislodge it.

Price was the final witness. On the word of Mary Meenan and the seven witnesses, the defense rested its case. Hogg was pleased. The defense based its case entirely on establishing an alibi. District Attorney Selig proposed that Judge Shackleford instruct the jury that "an alibi is easy to fabricate and hard to disprove."[26] Shackleford agreed to give the instruction, despite Hilton's strenuous objections. Hogg was very pleased.

Selig ground up the defense in the first round of closing arguments. "Why," he said, "if there was no charge against him, did Adams confess to such a terrible crime?" The confession, as recited by Wells, Runnels, and Mills, made it clear Adams had murdered Collins. He said the motive "existed solely in the $2,000 offered by Vincent St. John."[27]

Selig attacked the eight-member alibi, saying the defense could not find one decent person in Telluride to testify and had to bring in these witnesses, all under the control of Mrs. Meenan, who was running the show. Her testimony to the grand jury in 1902, Selig pointed out, had caused the indictment against Adams then, yet now she repudiated it. "A woman of that character" was not to be believed.

It was Selig who was not to be believed: Adams had not been indicted in 1902 at Telluride. All the grand jury indictments stemmed from the 1901 riot and Adams was not in Telluride in 1901.[28]

The district attorney explained to the jury that Wells was only doing what "he felt was his duty to make an honest inquiry" into the death of Collins when he brought Adams from Idaho. "General Wells...is a man of such courage and prowess that he actually dared to go to sleep" while he had this horrifying confessed killer, this "tool" of the Inner Circle, Steve Adams, in his custody.

In closing Selig asked the jury to do "your duty without allowing your judgment to be swayed by the eloquence of the famous" Judge Hilton.[29]

With a froggy smile and a polite word of thanks to the jurors, Hilton told them this was the third in "a series of prosecutions" of this man. "What a strange delight there must be in hunting this simple fellow.... All the machinery of Idaho and Colorado, and all the devotion of Bulkeley Wells, and all the learning of a dozen lawyers, and all the cunning and trickery of the Pinkertons have united to hunt down and destroy this man and break his wife's heart.... General Wells has journeyed...with sublime valor to bring this defendant to Colorado, but he has told you he did it for the purpose of getting bigger game."

Because Wells had "vile political differences" with organized labor, Hilton said, he brought the Pinkertons, those "human jackals," into the hunt, but "perjury and false witness always" followed in the Pinkerton wake. Several times Hogg leapt to his feet to protest, and Judge Shackleford told the jury to "take no notice" of anything concerning the Pinkertons, since they weren't involved in the case. Hilton said indeed they were.

Turning back to the jury, he said, "No man does anything without a motive. What motive is shown here? All labor troubles had been settled in Telluride; there were no differences of opinion between Collins and this defendant. And yet the state has mapped out for this man a career in crime actually unbelievable."[30]

Not only had Adams's and Orchard's confessions mopped up every conceivable crime that could be laid at the WFM's door, but both documents showed a pattern of glaring incompetence on the part of the two men that made anyone question why they would be assigned to do any job. Both failed at their missions, refused to carry them out, or bungled them so frequently, according to the confessions, that they would have been viewed as appalling high risks. McParland and the prosecutors always claimed the motive for such crimes was revenge for events that occurred two to six years earlier. Hilton said it made no sense.

He said the confession "purporting to have been made to the redoubtable General Wells" had nothing to corroborate it, not even the shotgun. "If there is one single iota of testimony connecting this defendant with this gun — " Hogg jumped up again, but before he said anything, Hilton continued, "I ask this distinguished jack-in-the-box to show you just what it is."[31]

Hogg was incensed. Unruffled by the sniggering in the courtroom, Hilton slashed Wells's "outrages" against law and decency, recounting how he was "armed to the teeth" when he took Adams out of Idaho on no legal authority, just so Wells might hang the WFM leaders. "He was willing to do anything to get them," Hilton said.

He attacked the testimony of Wells, Runnels, and Mills, showing they had contradicted one another in several instances, one or the other of them lying. "You decide which," Hilton told the jury. Wells and his co-conspirators brought Adams into the state illegally, Hilton continued, which "entitled them to a term in the penitentiary. Bulkeley Wells had no more right to kidnap this man than I have to kidnap any one of you."

Herschel M. Hogg, Congressman and attorney, as he looked in 1902.
Restored portrait © M. J. Martin 2003.

Hilton concluded by telling the jury that the State never proved Adams was at the scene of the crime, but the defense proved, with eight witnesses, that the defendant was at a card game when the crime was committed. "Not one of our witnesses has been impeached," he said, "even in the slightest degree."

In rebuttal Hogg ripped loose. "With his bushy white hair bristling aggressively, perspiration streaming from his face, his arms waving with the energy of his argument, [he] shouted out in his clear, ringing tones,"[32] his denunciation of Steve Adams, Annie, Mary, and especially Hilton. Hilton had shredded Hogg's pride and Hogg lashed back, calling Hilton "a vulture, that, after feeding on carrion and rottenness, spews it out on whatever happens to be below." Hogg attacked the integrity of Mary Meenan and Annie Adams, implying Annie was an adulteress because she had a third child born while Steve was incarcerated and the other two weren't his either. "Whose children are they?" he yelled at the jury. If Hogg was unaware Annie had the first two with her first husband and created the third with Steve when they were allowed to stay in the prison cottage in Idaho, then he showed signs of incompetence. If he was informed, then he was vicious in his thrust for conviction. A case could be made for incompetence, since Hogg also said Adams got the shotgun from a man named O'Brien.

In attacking the alibi, Hogg asserted it was impossible for people to recall what they were doing on a specific night six years past. It didn't matter to him that most people recalled in detail where they were and what they were doing when horrific news arrived — Collins's assassination shocked the entire town, including those who disliked him. Even forty years after Lincoln was shot, people recalled precisely where they were when the news hit them. Hogg accused Hilton of rehearsing the witnesses.

He blasted Hilton for "vilifying" Runnels, Mills, McParland, and especially Wells. "Mr. Wells went to Idaho on a lawful purpose in a lawful manner," he said and ran the jury through the entire story, even how Wells had consulted with McParland and Hawley. Nothing was wrong with Wells having a copy of Adams's confession or going over the details with McParland. Wells wasn't out to get the WFM; he only wanted to bring heinous criminals to justice. "Wells said he was anxious to bring Adams to Colorado to learn what had become of Barney," Hogg pointed out. "Barney had been a faithful employee of the Smuggler-Union Company, and is there any reason why Wells should be vilified by such a man as Hilton, simply because he wished to learn the fate of a man who had served him well?"[33]

Again Hogg appeared uninformed. Or did he say this just for effect? Wells never knew Barney. He worked for the company only three months. When Wells found the bones purported to be his remains, did he give this "faithful employee" a decent burial or pay the Pinkertons to try to find his sisters, his daughter, or any relatives? No. Wells put the bones in a box and used the man's skull as a shop ornament to condemn St. John and the union. How did the bones become so important? Did Wells know the truth about them? Did he know the truth about Barney? If only the bones had a voice, they might stand up and say where they came from, for they most definitely were *not* the remains of William Julius Barney. This alone could prove the confession was created. Sheriff Rutan, who had been present at Barney's divorce, could have corroborated the facts. Did he ever tell Wells the truth? Or did he take his secret to the grave?

Wells needed the bones for one purpose, but Hogg failed to note that Barney, the valued employee who served Buck so well, was really valued only as a dead man and served him best in that capacity. Hogg roared at the jury that Wells did nothing more than his "duty" in the battle against the immoral, criminal, abhorrent Western Federation of Miners and its evil Inner Circle. "I tell you," he panted from his exertion and the stifling heat in the courtroom, "so long as we are confronted with the awful conditions brought about by these men, we will stand and defend our rights. So long as they go about saying that we must divide with them whatever we have, these troubles will continue."[34]

Shaking his finger at the jurors, Hogg warned them that if they shirked their duty and failed to convict Adams, the same hideous conditions would soon spread right into their own communities. Giving farmhands living wages was worse than death.

The speech took up most of the morning, and in the end it was declared "exceedingly strong and brilliant" by the *Sentinel* reporter (July 15, 1908). With the weight of the world on their backs, the twelve men were removed at 11:00 a.m. to debate Steve Adams's fate. They elected J. P. Talley, the Fruita merchant, as foreman. Because of Judge Shackleford's instructions, they agreed they would not even consider the alibi testimony or the political underpinnings of capital versus labor. Barney's bones said nothing to them, thus that saga was out as well. The judge had also instructed, "Proof of motive is not required to convict, provided it is shown that the defendant, beyond a reasonable doubt, committed the crime," and, "If a confession is corroborated by the facts in the case, the fact that the confession itself was made while the defendant was in custody without process...does not necessarily invalidate the confession." For the twelve men it came down to deciding if the confession as Wells, Runnels, and Mills presented it, was fact. They called for the plats and the window glass glued to the cardboard, and after studying them, discussed their common knowledge of shotguns. They decided to take an informal ballot. They found they had a unanimous verdict on the first round.

Because they had been out only an hour, when court reconvened, Hilton and Hogg had not yet returned; nor had Annie and numerous others. For the first time since the trial began Annie was absent from her place behind the railing at the defense table. Steve Adams had only two of his lawyers to stand by him and his face was ashen. The jury wasn't expected to return so soon. He sat "nervously wetting his lips" as Shackleford turned to the jury.[35]

"Have you reached a verdict, gentlemen?" said the judge.

"We have," replied the foreman. He handed the sealed document to the clerk who passed it to Shackleford, who read it and silently handed it back to the clerk. The clerk announced to the court, "The defendant has been found not guilty."

Adams was at a loss. He sucked in a breath. The district attorney moved to discharge the defendant.

"You may go," Shackleford told Adams.

As the bailiff stifled budding applause, the courtroom swirled into activity around Adams. His friends pressed in on him, and he hurried to the clerk's room to thank the jurors. He scarcely could speak, tears filling his eyes, as he took each man's hand in gratitude. The jury, when it was selected, was believed to be more inclined toward the State than not. The swiftness of the decision had astounded the prosecution. Why, several reporters asked, did they not convict?

Talley explained that most of them were quite familiar with shotguns of all sorts, ten of them being farmers who had orchards, chickens, and other temptations that brought in foxes, coyotes, and raiders of all variety. The shotgun was the farmer's best friend. Thus the scenario with the sawed-off shotgun did not fit their experience. From the position of the wounds Collins received, from the window glass Wells had preserved that showed the tight pattern of buckshot, the shooter had to be standing no more than twenty feet from the window. If he had been uphill on the trail and seventy feet from the window, the buckshot would have scattered so broadly as to cover most of the window or more; it would have gone downward at that angle into the floor or, in the very least, into Collins's posterior. They were correct. A sawed-off shotgun at twenty feet made an eight-to-ten inch pattern in window glass, and the wad punched the thirteenth hole. At seventy feet the wad would have had no power to reach the glass and the twelve pellets of buckshot would have spread much too wide.[36] Wells's own carefully preserved evidence had freed Steve Adams. The jury didn't even know about the spruce blocking the line of fire from that seventy-foot location, but their decision was sound.

Adams walked out into the brilliant July sunlight, a crowd of well-wishers around him. A reporter yelled, "How does it feel, Steve?"

"I can't tell you," he said. "It's too much for me."[37]

CHAPTER

25

SUNDOWN

"When we remember we are all mad, the mysteries of life disappear and life stands explained."

— Mark Twain

"I have no comment whatever," General Wells flatly told reporters when they asked what he thought of the verdict.[1] His silence spoke volumes. And the verdict spoke volumes. By setting Adams free, the jury declared Wells's testimony false. They did not believe him.

The *Denver Republican* was incensed, claiming the WFM's "alibi mill held out" against justice, but the trial "enlightened the nation regarding" the evil nature of the "Inner Circle." The paper said the people of Telluride owed Wells their thanks because he had exposed this "enemy of society" and eradicated it from San Miguel County. Seeking the "punishment of the guilty" was his "sacred duty," a duty he "would not renounce under threats and attempts upon his life."[2]

His "sacred duty," or more accurately, his obsession, had actually freed Steve Adams. But Adams had no intention of thanking him. Adams wanted to be as far from Wells as the world would allow. Although the WFM had defended him, Adams wanted to be away from them as well, wishing only for a quiet place free from reporters and fear. Although it was possible the WFM paid eight witnesses for an alibi, it was highly unlikely, and in the end, the jury said the alibi hadn't influenced their decision: It came down to physics alone. Adams was still in shock when he was escorted from the courthouse on July 15, 1908, to O. N. Hilton's temporary offices in the hotel. Among his lawyers and friends, he finally "broke down and wept like a child." Years of uncertainty and anguish convulsed out of him and "it was some minutes before he was able to regain his composure."[3] That evening he, Annie, and Lillard boarded the train for Oregon, free of officers and Pinkertons, free at last.

In exonerating Steve Adams, the jury exonerated the Western Federation of Miners, Hilton believed. Judge Hilton had been an attorney for the WFM for many years; he remained so because he firmly believed in the organization. The consistent failure to gain convictions against the WFM, even in those counties friendly to the mine owners associations, cast much doubt on the MOA's charges that the organization was criminal. Nor did the assertion that the WFM was "militant and dangerous" hold, for, as Suggs put it, "At no time did the WFM engage in armed resistance against the constituted authorities, even when their extreme harassment and provocation might have justified it."[4] After the Adams trial, Hilton told the newsmen emphatically, "There never has been an inner circle in the Federation.... I have never heard either in private or in public any advice or counsel given by any officer of the Federation to any person to violate any law.... We [he spoke for the WFM] never have wanted anything but peace. But if we must fight to assert our rights, we will fight, and as vigorously and with as much success as we have displayed in this case."[5]

Although Hilton's words sounded militant, he was not speaking of bullets, but legal battles. He was the unflinching descendant of an illustrious line of English barons and knights; a man highly respected, a stalwart Republican, a Mason, and loyal counselor to the WFM, a position he would hold for twenty years. He was not fond of Big Bill Haywood on a personal level, but he had deep respect for the WFM officers and the organization's fight for justice in the workplace. He did have a special affection for St. John, one that would last through St. John's departure from the WFM, one that would last until death. He knew St. John had never ordered anyone to kill Arthur Collins or had been involved in anything to do with Barney's disappearance. Orrin Hilton believed so deeply in the man he called "the Saint"[6] that St. John's declaration that Barney was never killed convinced him Wells invented the entire incident of finding the remains. That was a possibility, since only a select few ever saw the bones.

Hilton was unaware Will Barney was alive and still involved in the dispute over his grandmother's estate in Boulder County. The property dispute was finally settled

This photo of William Julius Barney on his homestead in Montana (or Idaho) is dated August 4, 1908.

Courtesy C. H. Lamb.

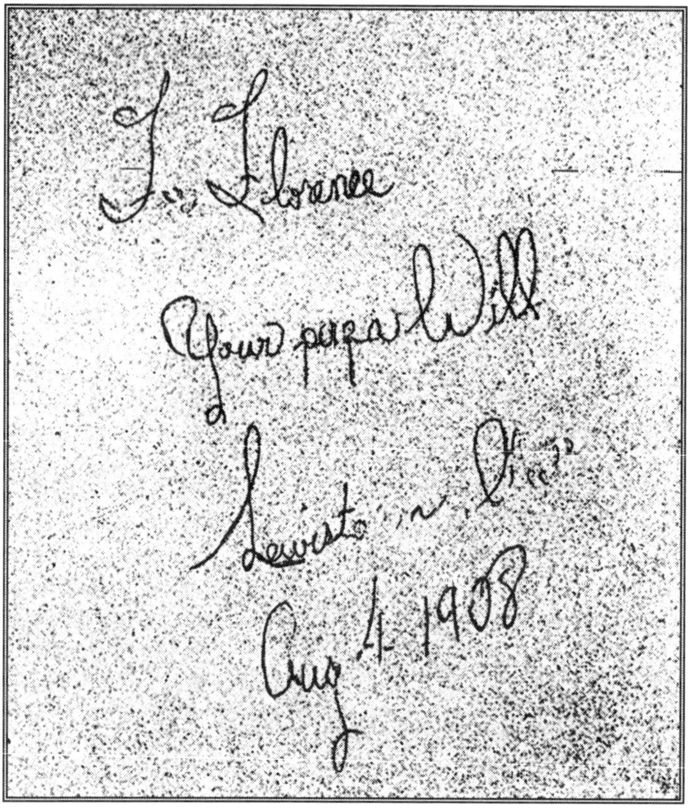

This inscription appears on the back of the photo of Will Barney. It was written in pencil on a very dark cardboard photo-mounting (enhanced). Courtesy C. H. Lamb.

in 1911.[7] Will's father, William Miles Barney, had died in St. Louis while visiting his daughters, Katherine and Emma, in October 1905; the body was returned to Longmont, Colorado, and interred in the family plot. The father left what he had to his second son, Ralph, half-brother of Will. Ralph and Will were too far apart in age, separated by different mothers and the fracture in the family, for them to care about one another. Ralph knew Will had married, but he didn't know when or where, or even Sarah's surname; nor did he know Will's daughter. Ralph had written Will off as dead in 1898.[8] That was the year Will and Sarah brought Florence into the world, and for the child's tenth birthday, Will sent her a photograph, via his sister Ella Johnson. Florence was still living with her mother and stepfather near Timnath (the Sylvesters vanished sometime after 1910). On the back of the photograph of himself with two horses, Barney had written, "For Florence, Your papa Will, Lewistown [or Lewiston], Mont. [or Ida.]."[9] The penciled inscription was too faded to discern which place; however, a William Barney had filed for a 160-acre homestead near Lewistown, Montana, August 8, 1912.[10] Will was once again pursuing what he loved most on a homestead somewhere far from the noise of intrigue. Wherever William Julius Barney went in the end, his divorce record was enough to prove he was never a murder victim in Telluride. It proved Vincent St. John had spoken the truth about him, and it condemned the Adams confession as a Pinkerton invention.

Since Will wasn't dead, the bones on Boomerang Hill remained an enigma. If Buck Wells actually found an "entire skeleton" in the spot McParland and Adams described, how did it get there? If Steve Adams's confession was a setup from the start, and Adams had never had anything to do with the "reburial," then who buried this corpse?

After recanting his confession, Adams himself told the court in Idaho, "Last June [1906] I was asked if I knew of a certain location near Telluride, being a spot designated by Orchard as the burying place of a man named Barnum [Barney], said to have been killed there. I said I did and was then informed that I was to go down and pick out the place." Adams said that during a three-day secret sojourn in Telluride in the custody of Wells and Runnels, "I found the place described, but there was no grave. I think they expected to find some evidence against Vincent St. John."[11]

Orchard knew about the bones before Adams was ever arrested; McParland's private reports to Gooding proved that. Orchard was the person who led McParland to Adams. McParland had a list from his MOA clients of all the "murders" he wanted solved, a list that included only murders of nonunion men. It should be noted the Pinkertons never took up the investigation of the mysterious murder of August "Mexican Sam" Baral, the Liberty Bell union millman shot in September 1903. Nor did the Mine Operators' Association list this genuine murder in its Red Book. McParland's Telluride list contained the names of Collins, Barney, Smith, and Mahoney. The only bona fide murder victim was Collins. Mahoney was never investigated (they knew he was alive), and Smith was oddly sidelined, with Collins and Barney becoming the focus. Collins and Barney were Smuggler-Union men — that company paid the Pinkertons well and Buck Wells kept the investigation in motion. Smith was a Liberty Bell man — that company preferred to settle with the union rather than prolong disputes (once it had disengaged from the MOA), which meant they were likely not paying the Pinkertons for any further Smith investigation. Wells pursued the Collins and Barney cases because he wanted St. John dead. That was the bottom line.

After Orchard led McParland to Adams, and Adams was put in the cell with Orchard to discuss McParland's needs for a Telluride wrap-up, it was glaringly convenient for Adams to confess to precisely what was needed. Adams, the sluggish union member known for his occasional drunkenness, suddenly became so tight with St. John and Carpenter that he popped off Collins for them, shuffled Barney's bones a few yards for and with them, was told all their bloodiest secrets, and trusted to keep his mouth shut. Unlikely.

Harry Orchard had some sort of hold over Steve Adams, a glimpse of it evident in Orchard's suicide play, which he performed just to get Steve to withdraw his recantation. Prior to that and the saving influence of Annie, Steve did whatever Orchard and McParland wanted, because he had been threatened as well as "promised to go free if I followed directions."[12] McParland's report of June 4, 1906, stated, "Orchard informed me that in talking with Adams since the latter's return from Telluride he is satisfied that Adams has located the exact place where the body of Barney is buried."[13] Why would Orchard be satisfied, unless he was the one who had put the bones in the spot and had left the hatchet mark on the tree to designate his cache? Orchard obviously knew the spot well and had described it to Adams. Most Telluride residents knew where the old sawmill was on Boomerang Hill; directing someone to a spot so many yards east of the mill would be no difficulty.

If Adams had buried bones here for someone else, no reason existed for him to mark this tree after he moved the bones a mere one hundred yards or less from the original point. No reason existed to separate the bones and clothing, especially if one were shifting them so short a distance. No reason existed for anyone to move this body in the first place, if it had been rotting here fifteen feet or even one hundred yards off the road for twenty-seven months without notice. No reason existed for murderers to carry a dead man two miles out of town and then six miles up a mountain. No reason existed for "hobnailed boots" to suddenly become "congress shoes cut through the soles." Wells supposedly found the boots in 1906, yet no clothing, no coat, belt, trousers, or any other items were with these boots — it was quite improbable such items rotted completely in just five years.[14]

Above all, Adams's description of the body was all too obviously invented. A bullet-riddled body lying on the ground for two and a quarter years not only would have had greater predation and insect activity, but would have been skeletonized and scattered in that length of time. Bears, wildcats, coyotes, and stray dogs were not the only San Juan animals fond of a femur. Marmots had an uncontrollable need to gnaw, and small rodents, including mountain packrats, scurried off with prized ribs and metacarpals. Jays, crows, ravens, and other birds, which were plentiful in the mountains, would have claimed their share of the spoils. The hot fierce sun of three summers and heavy snows of two winters would have finished what blowflies had started. Yet when Steve Adams found his man, Barney's body was intact, but for the hand. The state of decomposition was consistent with a body exposed to the elements for only a few months, not years.[15]

The story was preposterous. It had to have some other explanation.

Orchard, the alias-burdened, confessed-serial-bomber bigamist, who had deserted his first wife and child, was a queer fish, something odder than the fur-bearing trout. Although he credited all his crimes to WFM executive orders, he also admitted, and McParland confirmed, that he had been affiliated with the railroad detective D. C. Scott and MOA detective K. C. Sterling, discussing things with them before doing a few Cripple Creek "jobs."[16] Scott gave Orchard cash. The notion Orchard was an MOA operative was strong in his day; McParland spent a lot of energy denying it. Orchard also was associated with spy Gratias.[17] Yet, it was more likely Orchard was nothing more than a sociopathic sycophant, dogging the WFM leaders so they'd take notice of him, and killing for the thrill of it. On the stand during the Haywood trial, the defense attorney asked Orchard, "It was your habit to lie about everything, wasn't it?" Orchard replied, "Yes sir, whenever it suited my purpose."[18] If that was so, then when did he tell the truth? His "confession," wherein the details were all from previously published reports, obviously "suited" his purpose to save his hide and to please McParland who worked him like soft dough with God talk, a false friendship, and false praise. McParland found him disgusting, but said so only in his reports.[19]

Since Orchard told McParland about Barney before Adams's arrest, Orchard clearly knew the story. The anti-labor press had satiated the entire state with murder tales of Barney. Orchard had been in Telluride in autumn 1903, where he purchased a revolver from San Juan Hardware,[20] which he claimed he used to rig the Vindicator bomb in Cripple Creek. Thus it was very possible Orchard placed the "Barney" bones here to create a murder for his own purposes — perhaps with a view to blackmailing a WFM officer — a good reason to mark the spot. Orchard was always looking for ways to get money out of people. He even posed as a hail insurance salesman for that purpose. If he wasn't in the mood for a killing, he could have taken bones from a cemetery, which would not create a second murder and additional suspicion. If homicide was more exciting, the victim more than likely was killed near Alta, because the bones were a short distance from that mine and mill complex — it was very rare in the days of foot and horse travel for a killer to haul a corpse eight miles away. If the bones did exist and belonged to a genuine victim, McParland's reports point to Harry Orchard as the most likely killer.

For his horrendous confession, Orchard received a death sentence, commuted to life imprisonment. McParland had promised him the State would "do good" by him, and Idaho did. He lived out his long life in comfort, had his own cabin, his own garden and a dog, and was allowed to raise poultry for the prison and build furniture. He became something of a celebrity. He wrote and rewrote his confession, had it published, started a prison bible study group, and enjoyed visits from many people. Visitors included state officials, the family of his last victim, Frank Steunenberg, and representatives of the Seventh-day Adventist Church, who had witnessed his baptism and helped with his books. On a few occasions, Bulkeley Wells dropped in to see Orchard, providing him with a typewriter and other gifts.[21] Orchard's last book was titled, *Harry Orchard, The Man God Made Again*, which was his confession presented as religious testimonial, and woven with condemnation of the WFM and the "atheist" Haywood. He bequeathed the rights to his church. Orchard had kept his last alias, the one that gave him fame, since his real name, Albert Horsley, had been forgotten. He died on April 13, 1954, at the age of eighty-eight.

James McParland, the man who gave Harry Orchard absolution and the recognition he so long desired, limped away from his battles with the WFM in poor health. He clung to his belief in a depraved Inner Circle until death forced him to let go. Orchard had been the confirmation. Yet Morris Friedman, who had worked for McParland in the Denver office of the Pinkerton Detective Agency, had testified at Haywood's trial that "Manager McParland" was the "inventor of the world famous Inner Circle."[22] None of the Pinkerton operatives — not even George Riddell or A. W. Gratias who had become union officials before their exposure, or even the spy embedded deep in the Haywood defense team — could supply proof the Inner Circle existed. It was McParland's figment and because he believed it was real, he felt he had to destroy it.

Once he developed the conviction that he and the Pinkertons were on God's side, he didn't care what laws were bent or morals swept away in his quest to eradicate the enemies of the realm. Since Pinkerton clients were the business elite and corporations opposed to organized labor, unions became the "enemy," the WFM being the worst. For McParland, the WFM was identified as an anarchistic secret society like those of Ireland, even though the WFM was neither secret nor a society, but a legally established organization, not opposed to government or business, but merely working in the interest of gaining just wages and safe working conditions. McParland would stop at nothing to take it down because he believed his authority came from "Divine Providence." To carry out God's will meant he was free to break laws and lie until every man he judged evil was hanging on the gallows. Since his days in Pennsylvania he was comfortable lying under oath. In the Haywood trial and the Adams trials, he lied frequently, even claiming he never joined the Ancient Order of Hibernians. Documents showed he had. Some might excuse this portly old man as senile, but Charles A. Siringo, a Pinkerton who had worked for more than twenty years as an operative, detective, and spy, and McParland's personal bodyguard in Idaho, declared the agency "corrupt." In 1915 he wrote a book charging the Pinkertons with election fraud, jury tampering, fabricated confessions, false witnesses, bribery, intimidation, and hiring killers for its clients, men such as Tom Horn and Bob Meldrum. Documents and time sustained many of his assertions, but the agency trounced him with lawsuits and made his small book nearly vanish from the planet.[23]

The Pinkertons were powerful and their clients were powerful. Anyone who dared to cross them soon found he or she had committed a terrible mistake. James McParland went to his grave as their shining Great Detective, the man who destroyed the Molly Maguires and brought the Western Federation of Miners to its knees. To the last he had believed he solved the murder of Steunenberg single handedly, and had exposed the Inner Circle's heinous crimes to the world. Only because of that organization's vast evil powers did Haywood, Moyer, and St. John escape the gallows. McParland died believing the men were unworthy of forgiveness.

On May 18, 1919, McParland choked out his last breath in Denver's Mercy Hospital at the age of seventy-five. He left a widow and no children. The *Denver Post*, *Rocky Mountain News*, and the *Denver Catholic Register* filled columns in tribute, recounting his Molly Maguire tales and Harry Orchard triumph, tucking in fiction and numerous lies along the way. It mattered little, since the man had become a legend and the legend was six feet beneath the sod in Mt. Olivet Catholic Cemetery. The *Post* said he was "a savior" to many, "respected" and "feared, even by his enemies."[24]

The Western Federation of Miners never respected him. The members and officers saw him as a cunning liar and manipulator bent on destroying their union. He and the mine owners' associations had so thoroughly blighted the WFM from 1904 on, using lock-outs, blacklisting, relentless prosecutions, and Pinkerton spies, that the Federation's membership had dropped dramatically. Other factors contributed, such as the complete ideological disconnection between Haywood and Moyer and the IWW and WFM, the latter organization severing all ties with the IWW in 1908. But the rank and file departed from the WFM in weariness and dread of further persecution. In the end, men wanted to work, to feed their families, and to live without strife. In Telluride, Bulkeley Wells saw to it WFM members were kept from working in the Smuggler.

C. R. Waters of the Silverton Miners' Union was sent to investigate Telluride's Local 63 in 1909, because membership there had dropped suddenly from 402 in May of 1908 to 200 by October. Union miners told Waters that as soon as they paid their monthly fee they were discovered and discharged. A new spy was obviously in the woodwork. Their names were entered on the local blacklist, making them unable to get work in San Miguel County where Wells was still the "dictator," as Moyer had called him. Some mine managers in the county preferred union men, but they were few. Wells pushed his views on most. He had slacked off in the year prior to the Adams trial, but after losing to Hilton, Wells took out his revenge on the union miners in Telluride, pursuing his blacklisting with renewed vigor after July 1908. Worse still, the Local 63 Secretary-Treasurer Andrew Nylund disappeared with $329.95 in union assessments, and Waters had to take the books and install a new lock on the union hall door. He surveyed the once-magnificent hospital, and found that portion of the building in a "deplorable condition" due to neglect. He advised they rent the building to save it. He also advised they elect all new officers, which they did, but it was of no use. Local 63 was collapsing. By October 1909, only 108 members were left. The Telluride Miners' Union was reduced to little more than a "coffin society," those unions whose sole purpose was to tend to the sick and bury the dead. Without any funds coming in, it wasn't even good for that. Suffocating under a $17,000 debt to WFM headquarters, Local 63 turned the deed of the hospital over to Charles Moyer at the end of that year.[25] Wells was finally winning his war.

Although now serving as apartments and a laundry, the Miners' Union Hospital building still stands.

Photo by the author.

The Federation was losing members throughout the West as well, especially since the divorce between Moyer and Haywood. Haywood now put all his revolutionary energy into the IWW, and Moyer led the WFM into the grave of conservatism, the latter union reorganizing as the International Union of Mine, Mill, and Smelter Workers. It tottered as a mere skeleton until Moyer was forced to resign in 1926, and a few years later no breath remained. Since 1907, Moyer had also become embittered with St. John, and St. John lost all interest in trying to reconcile their differences, eventually working directly against Moyer and his compromising conservatism that St. John saw as surrendering to employers. After the long incarceration in Idaho, Moyer came out a changed man: subdued, cautious, and believing compromise was the only way to achieve the WFM's goals. Yet too much compromising with the mine owners' associations led to the loss of miners' rights and weakened union power. Managers such as Wells set up changing rooms, where miners had to strip after work and submit to inspection to prevent highgrading or theft of tools. St. John called these compromises slavery.

Conservatism was anathema to him. He saw it as death for the wageworker. Any step forward toward justice in the workplace was gained by demand-*demand* ended child labor, *demand* secured a living wage, *demand* brought an eight-hour day. St. John's own false imprisonments and abuse at the hands of the MOA only made him more determined such things didn't happen to others. He denounced conservatism and Moyer's guidance of the WFM and resigned his executive board position in that union in 1908. The final blow was delivered by one of the WFM's most "conservative" members in Goldfield, Nevada, in 1907.

On occasion mine management had employees sit for the photographer. Although foremen, boarding house staff, and sometimes office staff were included, these photographs were used to identify miners during the time the discriminatory card system was in place.

© Main Street Photography, Walker Studio Collection; used with permission (26T).

With the Goldfield false indictment against St. John quashed and his speaking tour at an end, he had returned to that town in autumn 1907 as IWW organizer for the WFM. This feral and tumultuous camp vented turmoil, the Miners' Union steaming because mining companies were paying in scrip. Although the WFM debated calling a strike, St. John's involvement was minimal, his business in the camp related to organizing. He incurred the wrath of the WFM "conservatives who were against strife," merely because he spoke against compromises that benefited only the MOA. He had a legion of enemies among the MOA and Citizens' Alliance who despised his return to Goldfield, but he told them, "Do your worst."[26]

On Tuesday evening, November 5, just after 5:00, St. John and a friend, John Gilbert, had left the Stope Saloon on Main Street and were chatting with one another as they walked along the crowded street toward home. As they neared the Palm Restaurant, next to the entrance of the Miners' Union Hall, Vint glanced up and saw WFM conservative P. H. "Paddy" Mullaney coming at him, a .45 caliber pistol in his hand. Just six feet away, Mullaney cocked the gun and leveled it at St. John, firing two shots rapidly, putting both bullets through St. John's right wrist while St. John was still moving toward his assailant. In an instant St. John was shot a third time, the bullet striking his left arm as he turned on his heel and fled, people on the sidewalk and street scattering in panic. Mullaney continued firing at him, one bullet striking a miner named John Tennant in the leg, and another grazing Bob Kelley.

Deputy Sheriff Call grabbed St. John as he ran down the street with blood gushing from his wrist. He was taken to the nearby Palace Saloon and someone bound his wrist with a towel. They took him out to the "bus and told the driver to take" him to the Miners' Union Hospital, but Undersheriff Bart Knight ordered the driver to take him to jail. St. John told Knight his wrist was broken and he was bleeding profusely and he wanted to get to the hospital.

"You go to the jail," Knight ordered, "and you'll get a doctor down there."

They took St. John to the jail, and after some argument and exasperating delays, Deputy Jack Grant, a friend of St. John, arrived. Jack finally took him to the Miners' Hospital, where the doctor found he had lost so much blood that Vint was in a state of near collapse. One bullet had blown a hole clean through his wrist, the second had shattered the ulna and ripped tendons, causing damage that permanently crippled his right hand. The third bullet, in his left arm, was less serious.

The bullet in Tennant's thigh had so badly shattered the bone that the man's leg had to be amputated. Kelley was patched up and sent home.

The sheriff arrested Paddy Mullaney. He yowled how the revolutionary St. John was about to shoot him and he just beat him to the draw. "I am not at all sorry for shooting St. John," Mullaney said. He was sorry about the other victims, but he was surprised he hit them since he thought he fired only two shots. He whimpered, "St. John had made threats that he would kill me and also that he always went armed."[27] A few newspapers claimed St. John had a "big revolver."[28]

Neither the witnesses nor the deputies saw St. John with a gun. The sheriff did not find one on him. St. John said he was not armed, nor had he had any words with Mullaney. The MOA press cooked up a tale about a feud between him and Mullaney that had supposedly erupted due to the question of a strike, or over a fight for the leadership of the WFM Local 220. Suddenly witnesses claimed they heard an entire conversation between St. John and Mullaney on the crowded street just seconds before the shooting, backing up the gaseous stories. None was even remotely true. Union officers said St. John had nothing to do with Local 220's affairs, had not been personally involved with Mullaney, and "took no part whatever in the discussion on this subject [the strike], either at the regular or special meetings."[29]

Another tale said St. John had accused Mullaney of being a spy, and yet another babbled about the WFM men accusing St. John of being a Pinkerton. The *Goldfield Daily Tribune* claimed St. John wanted the accusers expelled and Mullaney did not. The absurd tales were promoted to discredit the labor leader and prejudice the minds of the miners. They failed.

Union men of various WFM Nevada locals rallied to St. John and denounced the false press reports, including the report that J. M. O'Neill, editor of the *Miners' Magazine*, reprinted from the wire service. O'Neill had followed Moyer's lead in his falling out with St. John. The McCabe Miners' Union jumped O'Neill, calling on the WFM executive board "to reprimand the editor...for his unwarranted abuse in assailing the integrity of our brother, Vincent St. John, whose loyalty to the working class is so well known." The men called Vint "one of the most valued members of the" WFM and "one of the bravest advocates of the cause of labor."[30] Most miners and even MOA men knew St. John was not one for carrying a gun despite endless threats against him. After this, it was said, his wife urged him to do so.

Confined to the Miners' Hospital with fever and infection, St. John was unaware of the battle on his behalf. His friends and numerous union men insisted Paddy Mullaney shot St. John only because Mullaney was staggering drunk. Mullaney's failure to plant a bullet in a vital spot at six feet supported the statement. His notion that he had pulled the trigger twice, when he had done so five times, also gave it support. Although Mullaney had been a Cripple Creek deportee and victim of General Bell, he was now a "stanch conservative" who claimed he hated strife. If this was the action of a conservative who detested conflict, St. John had good reason to dislike conservatives. When the "conservatives" carried guns and the "revolutionaries" went unarmed, it was cause for reflection.[31]

The local mine operators rejoiced in the shooting, publishing in their periodical, *Goldfield Gossip*, what the IWW termed "an incitement to murder." The MOA blamed St. John for their stocks plummeting during the spring 1907 strike, and said, "Wherever he is, there is trouble." Then they plunged their bayonet:

> *Mr. St. John will clearly understand that we have no personal grievance against him when we say that it was a great pity that the man holding the gun in the recent shooting did not have sense enough to hold it straight. Speaking as one who puts the public good foremost in his argument always, Mr. St. John will pardon us for remarking that he would look much better dead. We will...be more than pleased to attend his funeral...in the hoped-for event of anything sudden in the shape of a bullet overtaking him, we will cheerfully subscribe for a wreath for the coffin and count it money well spent.*[32]

The week after the shooting, Big Bill Haywood came to his comrade's side. St. John was too weak to leave the hospital. By the end of the month the local union had voted to strike against the use of scrip for pay and the MOA called on President Theodore Roosevelt to send federal troops. The camp was about to be engulfed in the nightmare of military occupation. The Citizens' Alliance called for St. John's lynching.[33] By December 16, Vint's wife and friends managed to get him out of town, arriving in Chicago three days later. He thought he was well enough to attend to his duties as IWW assistant secretary, but he collapsed a few days later. He was confined to a Chicago hospital until February.[34]

Whatever misfortunes befell St. John, Curry was certain to note them in the Telluride *Journal*, always with a slant to make St. John appear deserving of the worst. Even small scraps were turned into monsters: when the agreement St. John had signed with Edgar Collins (during the 1901 riot at the Smuggler) was presented as evidence in a trial, Curry headlined it as the paper signed by "the Czar [where he] agreed to call off his Mob of Murderers."[35] No other labor leader did Curry so carefully pursue, although the editor failed to note the achievements of this dedicated man. Curry seemed incapable of leaving his bitterness behind. Perhaps he thought he was still serving Buck Wells and the "best citizens" in Telluride. Wells himself still fought to discredit the union whenever opportunity offered, writing to *Harper's Weekly* in August 1908 to demolish the Maki story as fabrication, claiming Maki was merely a vagrant, chained on a warm day for less than thirty minutes so he wouldn't run away. Yet for most, the war was over, the hurt a thing to leave out of conversations, and the union a shadow. Many of those left in Telluride acknowledged the town lost much more than she won, if she won anything at all; some thought she had lost her soul.

Telluride never regained the rambunctious vivacity she had at her peak, with miners streaming down the mountainsides like ants to honey on Saturday night. Even Judge Wardlaw departed in 1908, opening a law practice in Denver and remaining an active Democrat until he died. The wounds of deportation and prejudice, of Wells's arrogant power and Curry's blatant lies, eventually closed, but they left rough scars that still caused an aching deep in the secret places of memory. Some families would never forget. Will Barney's story rose in the night as a ghost tale and finally faded into the wind. Curry, too, eventually sank into the dirt, like all editors do, leaving absolutely nothing behind but the spoor of his bitterness.

The dashing Bulkeley Wells gradually lost his power to influence Telluride, spending less time there. Although he had acquired an interest in the First National Bank and became president, as well as president of the Western Power Company, he traveled extensively, looking after various other mining interests for his father-in-law's company. He belonged to numerous country clubs and men's clubs in nearly every major city in the nation, spending his leisure time in high-stakes card games, playing polo, and enjoying the glamorous life. His wife, Grace, and he lived apart and he indulged himself in liaisons with other women. One of these was Mrs. Louise (Sneed) Crawford Hill, a Denver socialite whose wealthy husband was an invalid. She kept a full-length portrait of Wells "on the first stairway landing of her palatial home"[36] where she entertained the elite of the nation, including Harry Payne Whitney. E. B. Adams, who succeeded to the post of local Smuggler company attorney when Howe died in 1908, spent much time with Wells and knew him intimately. He excused Buck's infidelity with the question, "Was he to blame because Nature had generously endowed him with such overpowering masculine charm?"[37] On a trip with Wells to San Francisco in 1912, Adams thought it humorous how Wells furtively introduced his current "lady" as "Mrs. She." Grace Livermore Wells finally brought suit for divorce in 1918, and Wells made no protest.

Wells lost what few friends he had in Telluride when he became embroiled in a bitter lawsuit with Charles Chase and the Liberty Bell Mining Company in 1911. The case involved mining infringement, Wells accusing Chase of developing ground that belonged to the Smuggler. Wells eventually won the case and over $400,000 in damages from the Liberty Bell. Chase was so incensed that he "declined to speak" to E. B. Adams for "nine years thereafter," although Adams played a minor role.[38]

Always ready to entertain old friends and charm new prospects, Wells was generous with the Livermore wealth, allowing Harvard friends to work leases on the Smuggler and walk away with as much as $50,000, which did the company no good. The Smuggler continuously failed to pay dividends to the stockholders. Sometimes Wells's yearly expenses soared as high as $75,000. He dropped $3,000 on E. B. Adams as a bonus after winning the suit against Chase. Wells lived the good life, swaggering through other men's fortunes, including some of Whitney's, having met Harry at a New York card game. Wells invested Whitney money in mining ventures in the West, including a revival of the old Comstock in Nevada, which failed miserably.

"It would be unfair to say that Bulkeley Wells gambled the Livermore money and the Whitney money and the money of others that I knew of," E. B. Adams wrote, "but certainly he hazarded their money in some highly dubious mining and other ventures."[39]

Wells left Telluride permanently in 1923, the Livermores sending Robert to deal with the "distasteful and seemingly impossible task" of saving the Smuggler-Union from bankruptcy.[40] Livermore shut the Smuggler down in 1928 and after that sold it to the Denver Metal and Machinery Company, which planned to salvage equipment and rails. The property changed hands a few more times, and was eventually purchased by the Idarado Mining Company. Buck Wells went on speculating and charming his way across the West, remarrying in January 1923. The woman was Virginia L. Schmidt, a "beautiful platinum blonde" from Battle Mountain, Nevada, twenty-five years younger than Wells. They married in Salt Lake City, surprising all their friends, especially Buck's, who thought he was still

The Smuggler-Union reduction plants about 1928. Fire had destroyed the old Pandora Mill in 1919 and the flotation plant and several other buildings in 1920.
© Main Street Photography, Walker Studio Collection; used with permission (315T).

connected with Louise Hill. It was rumored the marriage surprised Louise so much that she was instrumental in having the Whitney money cut off around this time. The couple settled in San Francisco where Virginia was well known in social circles. She bore him two children.[41]

As investments failed and his youth, glamour, and power waned, Wells withdrew bit by bit from society. He became ashamed of his losses, forced to downsize his standard of living. Only A. D. Snodgrass, who had been the chief clerk in the Smuggler company's Denver office and had followed Wells to San Francisco, remained at his side. Yet the bare threads of friendship and the love of a young wife and small children were not enough to sustain him. Virginia and the children were living in Nogales, Arizona, when he went to his small office in the Hobart Building in San Francisco and penned Snodgrass a note on May 26, 1931.

"Dear A.D.," Wells wrote, "As the result of all my difficulties and worries, I feel my mind is bound to go — either that or a stroke, and I will not be a charge upon anyone. Nothing but bankruptcy is possible so far as my 'estate' is concerned. Do what you can for Mrs. Wells. B.W."[42]

Wells defined his soul by wealth and power and when he found himself without either, life was no longer worth living. He had no final words for his wife, no blessing for his children, Virginia's two or Grace's four. He took his revolver from his desk drawer, put it to his head, and pulled the trigger. His life was over at fifty-nine. His friends were shocked, explaining to reporters he was too proud to come to them; had he done so they "gladly would have helped him out of his financial difficulties."

He had lost "his Midas touch," one reporter wrote, which to Wells was the only thing that gave him meaning.[43]

The man Wells had hated for so long and with so much vigor, Vincent St. John, had died in the same city only two years before. Wells probably never noticed his passing. Mainstream newspapers barely noted it. Like Wells, St. John died financially broken, yet Vint didn't die in bitterness, fear, or shame. St. John had never sought wealth or power, having always lived simply, oftentimes fighting poverty at the corners of life, content that Clara was the most important wealth of his heart. He had defined life as service to others, and his passing left a legacy so broad and far-reaching that it was taken for granted by those generations that came after him. Along with other early day revolutionaries — Moyer, Haywood, Debs, Mother Jones, Elizabeth Gurley Flynn, and the rest who believed only with personal sacrifice could they achieve a better world — Vincent St. John and his comrades gave the nation child labor laws, minimum wage, humane hours and working conditions; they rallied for freedom of speech and industrial justice, and stood strong against unjust war. St. John died content that he had lived what he believed, his heart faithful to his vision of a just nation, faithful to his fellow workers, his wife, his family, his church, and his God.

Bulkeley Wells, age 45.
Author's collection.

This quiet, unpretentious miner from Telluride had been chosen by the convention delegates to lead the IWW out of its birthing agonies and, once rival factions had been removed, with St. John at the helm, members flocked in by the thousands. He and Clara moved to Chicago to a humble flat on North Clark Street and, from 1908 to 1914, St. John guided the organization as general secretary (the position of president had been abolished). It wasn't a position that offered riches, for the IWW neither glorified leadership nor created administrative hierarchies, paying salaries commensurate with that of the wageworker. The man Curry had labeled illiterate and incompetent ran a nationwide organization of nearly 100,000 members. He guided the Wobblies, as members were called, into direct action and away from politics, the IWW refusing to adopt the views of any political party. Some said this steadfast dedication to the worker's freedom of choice would eventually lead to the IWW's undoing. In St. John's hands it was the only means of keeping the organization focused on industrial justice. Although still adhering to socialism, St. John was no longer affiliated with any Socialist party. He was adamantly opposed to anarchy and decentralization, keeping guard against "anarchist freaks."[44]

The Wobblies went down into the gutter, as Haywood termed it, to bring the enslaved of the sweatshop up into the sunlight. They organized everyone, skilled and unskilled, native and alien, and saw to it the Chinese, Japanese, and blacks were not excluded as they were from other labor organizations. Employing some of the most original strategies the labor movement had yet seen, the IWW invented the sitdown strike, the union winning many of its contests, including the triumphant textile workers strike at Lawrence, Massachusetts, in 1912. It was the IWW's heyday and St. John left his imprint upon the union like no other leader, his leadership engendering "universal trust and respect."[45]

"The Saint...was one of the best leaders ever produced by the IWW," organizer William Z. Foster wrote of him.[46] "I never met a man I admired more," said Elizabeth Gurley Flynn, the "rebel girl" who was an IWW dynamo with charm and missionary zeal.[47] Flynn led free speech rallies, strikes, and organized workers in Lawrence, Paterson, and other cities. She would hold Vint as her model and inspiration even long after his death. Rank and filer Henry McGuckin listed St. John among his unsung heroes.[48] Despite being well loved by most, St. John had a handful of detractors and even enemies within the union. The ideological clashes between him and William Trautmann turned the latter against him so completely that, years later, the embittered Trautmann labeled him a "psychopathic case" akin to Hitler.[49] Trautmann, who was "expelled [from the IWW] for insubordination," claimed St. John's mind was "unbalanced by the tortures inflicted in Goldfield," making him dangerously delusional.[50] Nothing supported Trautmann's claims, and his extreme animosity made them suspect. In 1913, Trautmann joined DeLeon's splinter group in Detroit. He preferred political action and condemned St. John's direct action philosophy. For St. John, Trautmann's and DeLeon's Socialist Labor Party was too "ultra-legalistic, narrowly restricted, and doctrinaire."[51]

James P. Cannon, another IWW organizer and later leader of the American Trotskyites, was an admirer of DeLeon's dedication, although he thought "his conception of political action was...rendered sterile by legalistic fetishism."[52] Cannon was not generally free with praise, yet personally recalled St. John in a way that set Vint apart as a labor leader:

> *St. John's positive qualities as a man of decision and action were contagious.... He was not a back-slapper but a leader, with the reserve that befits a leader, and he didn't win men by argument alone. In fact, he was a man of few words. The Saint lived his ideas and methods. He radiated sincerity and integrity, and unselfishness free from taint or ostentation. The air was clean in his presence.*
>
> *The young men who fought under his command...swore by The Saint. They trusted him. They felt that he was their friend, that he cared for them and that they could always get a square deal from him.... John S. Gambs remark[ed]: 'I have heard it said that St. John, among outstanding leaders, was the best loved and most completely trusted official the IWW have ever had.' He heard it right.*[53]

St. John's selflessness built an organization that inspired its members to self-sacrifice and solidarity, yet he suddenly resigned as general secretary at the end of 1914 and disappeared into the hills of New Mexico. Big Bill Haywood took his place. The rank and file wanted their Saint back.

Rumor said St. John simply grew tired of the sodden skies of Chicago and "had a sudden urge to return to the wide open spaces" of the West.[54] But it would take more than that to pull the dedicated revolutionary out of the struggle for industrial justice. Something overwhelmed him. The bottom had fallen out of his heart. That year his beloved Clara had died, that fragile shy woman that meant more than the world to him. Years ago he had vowed to protect her from the terrors of his enemies, but the rabid persecution and violence against him took its toll on her health. Without her, his heart was dark. He had his sister and aging mother, who would live until 1916, yet only the mountains of the West could fill up his longing, for his soul was bound to the rock and trees and sky where he had labored as a miner for so long.

Vincent St. John, General Secretary of the IWW, as he appeared about 1914, age 38.
Restored portrait © M. J. Martin 2003.

He and a few friends bought a copper claim in the dusty hills near Jicarilla, New Mexico, and organized a small mining company so they might sell shares to finance the project. In July 1915, Vint disappeared into the pines to develop the claim himself. He was thirty-nine, glad to have the sun on his back, a pick in his hands, and the company of millions of silent stars by night. Hard work soothed his longing heart. When he left the IWW on January 1, 1915, he became self-employed, which made him ineligible for membership in the organization. Membership was limited to wageworkers. Thus he was no longer affiliated with the Wobblies, except through personal friendships with several members and a brief project in 1917, when he revised a pamphlet he had first written in 1912 on IWW history. He remained working in the Jicarilla Mountains, far from the growing madness of the world.[55]

In Europe the "War to End All Wars" was raging and America was about to leap into it. The IWW's anti-war stance came to the attention of those who hated organized labor, and employers, eager to end Wobbly power, pushed the idea that the IWW was a subversive organization. The federal government went after the Wobblies with a grappling hook, despite General Secretary Haywood's invitation to inspect the union's files. On September 5, 1917, the Department of Justice simultaneously raided forty-eight IWW local headquarters across the nation, seizing every inch of material from personal letters to ink pens — anything that might prove the organization was subversive. A federal grand jury in Illinois passed down indictments against 165 IWW leaders.[56]

On October 31, 1917, Vincent St. John was arrested on a federal warrant while he was working inside his copper mine. He was taken to Carrizozo and lodged in the

Lincoln County jail until December 6. He was then arraigned in federal court in Albuquerque and held until December 13, after which time he was transported to Chicago and locked in the Cook County jail. On March 23, 1918, he was finally released on $10,000 bond.

The "Great Trial," as labor organizations called it across the nation, began in April 1918. It was wholesale persecution on a grand scale, and blanket convictions were expected. Whatever the federal government thought it had to convict the Wobblies, it certainly had nothing to convict St. John, who was not even a member of the organization. Since July 1915, he had left his dusty hills only twice — once in May 1916 to see to his mother's burial in San Jose, of which he told no one, and on that trip a return via Boulder County to look into tungsten claims, which he stated for the record; and once to attend a meeting of the stockholders in his copper mine in Chicago in July 1917. At that time, since he was in the city, he visited a few old IWW friends. "There was no other significance to the visit and the government did not introduce any evidence to prove the contrary," St. John wrote in his statement to his attorney.[57]

On the stand, the federal prosecutor grilled St. John on the IWW history pamphlet he had written. It proved nothing. St. John was charged with conspiracy to violate the Selective Service Act and the Espionage Act, but the prosecutor had no evidence to support the indictment. At one point Vint was asked if he would have used "open warfare against the government" if he thought it would achieve his goals in the IWW.

"If I had any such idea as that," he replied, "I would have been out organizing a military organization instead of an industrial one."[58]

The government failed to prove St. John had ever attempted to influence men not to register for the draft or had actually violated any laws at any time, yet he was found guilty with all the rest — a blanket verdict for a blanket indictment. He received a sentence of ten years imprisonment and a $30,000 fine. It was an ugly day in an ugly time in the nation's history.

Judge K. M. Landis declined to permit St. John to file a "pauper petition" for appeal because, Landis said, the trial evidence showed St. John "owned a gold mine."[59] His little copper claim was more rock than copper and had been barely developed, but the judge insisted it was a gold mine.

Vincent St. John was transported in the custody of U.S. Marshals to the federal prison at Leavenworth, Kansas, being registered as Prisoner Number 13144-L on September 7, 1918.[60] He wrote, "The pleasure of a trip to a penitentiary had long been promised me by my most solicitous enemies." Here it was at last — what his enemies had always wanted, although a few preferred he was in a coffin, not a cell. He said at least now he had free room and board and could "dispense with the bother of a name." He wasn't going to let a miscarriage of justice reduce him to rage and hatred, although one of the deputy marshals told him, "I'm sorry — sorry that you didn't get twice as much." Vint called the federal penitentiary "the government's home for unlucky malefactors and holders of unpopular opinions," himself being the latter. It was a cozy place, where "the guards are organized as a part of the American Federation of Labor. This no doubt is a great source of comfort to the members of organized labor who are confined for union activities. Every beating a union beating." St. John noted how one could get alcohol and drugs as easily as bibles in this formidable bastille. "It dispels the notion that law breakers never get into prison. Some of them do, and draw a salary from the government."[61]

Although St. John made the best of a nightmare, he fought for freedom from the moment he was sentenced, appealing his case to the Circuit Court of Appeals. Elizabeth Gurley Flynn, who had been expelled from the IWW in 1916, started an all-out campaign for the Saint's release. On April 25, 1919, while his case was under appeal, he was released on bond again, living in the Peoples' House in New York. Haywood was out on bond during his appeal as well, but he was unwilling to face another battle; in poor health and disillusionment he fled the country, never to return. He died in Russia on May 12, 1928.

Ever the idealist, ever believing truth and justice would see him through, Vincent St. John stuck out the appeal process, even though the trial had reduced him to rags and he lost his copper claim. He remained amazingly cheerful, all the anger of his 1906-07 storm having long since blown out to sea. On his behalf, Elizabeth Gurley Flynn, who had founded the Workers' Defense Union, hammered at every resource and person she could, including Alexander S. Lanier, former Captain in the Military Intelligence Division. He studied the case and wrote to President Wilson, "There can be no question, in my opinion, of the absolute innocence and wrongful conviction of...Vincent St. John."[62]

Flynn also solicited help from the famous attorney Clarence Darrow and Judge Orrin Hilton. The still-vigorous sixty-nine-year-old Hilton jumped in immediately, but said there was little the law could do, pointing out, "It is particularly unfortunate that the Saint failed to file a motion for severance."[63] St. John could have walked free had he demanded a severance of his case on the grounds of non-membership and inactivity with the IWW during the period covered by the indictment. Others had done so, but he stood with his old comrades, confident his case would be dismissed without trial, believing in justice. It seemed so odd to him that justice was absent from federal court in the great republic he so loved.

"Correspondence between myself and members of the IWW during this period [1915-1917] was purely of a personal nature as the record will show," he stated in his appeal. "I was offered employment as an organizer among the miners but refused to undertake the work. This the record also shows.... My name appeared among others on a list of persons to take the office of Secretary of the IWW in case the incumbents of that office were arrested. I was not consulted in this matter, nor did I agree to accept. The government introduced no evidence to prove that I was aware of, or had agreed to, this tender of office.... The only explanation that I have as to the government's connecting me with the case is that it was the policy to prevent anyone who might assume the handling of the IWW's affairs from doing so. There is not a single line of proof adduced by the government to show that I was a party to any conspiracy or agreement as charged in the indictment."[64]

On April 19, 1921, St. John received a telegram at the Peoples' House in New York: His appeal failed. He and forty-six others were asked to surrender to federal authorities by midnight, April 25, or be considered fugitives. Although he lost his appeal, the circuit court did reverse two of the counts against him and modified the sentence to ten years and a $20,000 fine.

Vincent St. John freely gave himself into the custody of U.S. Marshals and was taken back to Leavenworth. Flynn, who had helped found the American Civil Liberties Union in 1920, put all her energy into fighting for his release and that of the other political prisoners. She solicited funds, appealed to government officials, and sent friends to visit the men and bring them hope in the darkness. In August 1921, her friend Caroline went to Leavenworth on one of these missions. "To my astonishment," she wrote to Elizabeth, "I found the Saint in the Hole." Some sort of clash had occurred, and the participants refused to make an explanation to the guards, and thus were thrown into solitary. Caroline badgered the warden, who checked St. John's record, and, finding it "absolutely clear," promised the man would be taken out.[65]

In autumn of that year St. John's attorney, whom he addressed as "Friend Weinberger," proposed a parole package that included a small pension. Vint utterly refused it. He wanted a pardon, not parole. "My only reason for applying for a pardon was to give that outfit a chance to remedy the injustice of the Chicago trial as it applied to me personally," he replied. "I'll not consider a parole under any circumstances. As my attorney I expect you to stand on the record and to insist that I get the benefit of what the record shows. Drag them out into the light."[66] He underlined this last sentence, his mind set on the government proving to him it had not usurped truth and justice. He had to believe justice was still a reality, or there was nothing left to believe.

Because several of the political prisoners were soon released, Flynn found, through her investigations and correspondences, that "even more sinister prejudices than operate against the group operate to keep" St. John in prison. She discovered some powerful mine owners in the West, "who hate him implacably" were seeing to it the government didn't release him.[67] Despite this, she refused to give up. Numerous unions joined her, writing resolutions for St. John's immediate release to the President, senators, congressmen, and the press. Finally her tireless dedication paid off.

The war was over and President Wilson was out of office. The prejudice and panic that had swept the country sank with the Wilson administration's assault on the nation's soul. On June 23, 1923, President Warren G. Harding signed a document stating, "Vincent St. John is a fit object of executive clemency," and the Saint walked free.[68] He had served thirteen months of a sentence that was designed to leave him a broken old man.

He had nothing left but the love of his gentle sister, Helen, and numerous friends. That was all he needed. In quiet obscurity he went to live with an old friend, Oscar Witthoft, in San Francisco. Helen lived in nearby San Rafael. Oscar was a small humble man, and member of the Lithographer's Union. He loved the Saint like a brother, feeling privileged to share his little bit of nothing with the man who had led the IWW toward a beautiful dream. Vint had touched that dream for one sweet moment in Goldfield when the entire camp operated on living wages and humane hours...one moment he saw it was possible.

Now he left it to younger revolutionaries. He had come out of prison gaunt and infected with tuberculosis. The hardships of prison, the years of toil, the loss of Clara, and his old injuries had all driven deep into his vitality and pulled the strength from him. His last years of freedom were spent in poor health. He had spells of weakness and severe dizziness, but never let it stop him from contributing to Elizabeth Gurley Flynn's work, writing letters, offering suggestions, cheering her with his humor. In his last letter to her, dated June 6, 1929, he was preparing to visit some workers. He was upbeat but noted he was "pretty nervous yet" after he had collapsed while visiting at Alameda. "As soon as I get a bit of strength back," he wrote, "we'll hit the trail."[69] A few days later he lay in the Catholic St. Francis Hospital, his life slipping out of him.

Oscar wrote to Elizabeth in Portland, begging for help. Neither he, Helen, nor Vint had enough funds to cover medical care, and Oscar wanted the Saint to have the best doctor possible. He desperately wanted him to live. Elizabeth jumped immediately. Hardly anyone in the various unions had heard of Vincent St. John for the last six years, yet as each heard the call from the devoted Rebel Girl, they gathered their quarters and dollars and sent them to her. The Saint was not forgotten. Even old Judge Hilton sent a large donation. Individuals, unions, the ACLU, old friends and young admirers, doctors and lawyers all reached out to the Saint in his helplessness.[70] "Elizabeth, you are a darling!" little Oscar wrote, overwhelmed with gratitude.[71]

Helen and Oscar stayed at Vint's side in the hospital, unable to leave him, unwilling to let him go. "The poor Saint is in a bad way," Oscar wrote to Elizabeth, telling her Vint asked for nothing, but that he was concerned about their losing sleep over him. "Always the same, thinking of others and never of himself," Oscar said.[72]

It was that selflessness that was taking the fifty-two-year-old St. John home. He had neglected his health while giving his time to others. From x-rays, Dr. J. Underwood Hall, who had attended Vint's mother when she died, found the tuberculosis from the days he had spent in prison had suddenly spread.[73] His lungs were failing; his blood pressure went off the charts. Nothing could be done. As the sun was sinking over the gold-splashed bay on June 21, 1929, Vincent St. John gave his sister and Oscar a gentle smile, and ever so silently slipped into the golden light of unknown tomorrows.

The funds gathered from friends were used to bury St. John. Oscar paid his hospital expenses since his stay was brief. Oscar and Helen picked out a plot in Oakland's Mountain View Cemetery, ironically among the wealthy and famous. They chose the spot because it was in view of the mountains and the bay, "a place of beauty for a soul of beauty." They "made the most simple burial" as he had wished it, but they hadn't enough funds for a tombstone.[74] More than seven decades later Bay area labor advocates furnished a memorial stone.

In the days after the Saint died, Oscar was adrift. Elizabeth had asked him to send her a "small memento" from Vint's things, and he confessed to her, "So far I have had to keep away from our home so as not to face the loneliness produced by his absence."[75]

St. John's absence was barely noted in the mainstream press, but in the IWW newspapers he was given a huge front-page tribute.[76] He was recalled with deep affection as a leader, a hero, a man of selfless dedication and unassuming manner. One friend wrote, "I shall always cherish the memories of the times we had together. They were not only happy but inspiring days."

Another said, "He remains a living inspiration to those of us who were privileged to know and love him. We carry on his name. We are stronger, wiser and better fighters for knowing The Saint."[77]

Twenty-five years after his death, Elizabeth Gurley Flynn still felt his absence. She always would. On the anniversary of his death in 1954, she wrote:

A nebulous cloud passes suddenly over the sunshine of my mind,
a stabbing awareness of loss pierces its calm,
a memory of you, my lifelong and truest friend,
grips at my heart;
Twenty-five years dead, dear Saint —
I miss you as if it were yesterday,
I'll miss you forever.
I'll never live so long that you are dead to me,
Nor will you ever be dead so long, you do not live on in me.[78]

NOTES

CHAPTER 1: **RATTLING THE BONES** (Notes for pages 2-8)

1. Jensen, *Heritage of Conflict*, 25-37.
2. *Solid Muldoon*, July 8, 1887.
3. *A Report on Labor Disturbances in the State of Colorado*, United States Senate, Document 122 (hereafter cited as *Labor Disturbances*), 35-37.
4. Lukas, *Big Trouble*, 73.
5. Siringo wrote two books detailing his experiences as a Pinkerton operative: *A Cowboy Detective* and *Two Evil Isms*. The Pinkerton National Detective Agency brought a lawsuit against Siringo regarding the publication of both these books.
6. For a clear presentation of the strike, see *Heritage of Conflict*, 72-87.
7. *Labor Disturbances*, 36.
8. Ibid, 37-38.
9. Western Federation of Miners, *Category of Crime of the Mine Operators' Association*, 26.
10. *Labor Disturbances*, 85.
11. *San Miguel Examiner* (hereafter cited as *Examiner*), April 26, 1902.
12. Pinkerton Operative No. 42, A. W. Gratias: report to his superiors, dated June 9, 1904, Clinton Jencks Collection, Box 13, University of Colorado at Boulder, Archives (hereafter cited as Jencks, CU Archives). Other examples of such manipulation, disruption, and instigation, too numerous to cite here, are in various Pinkerton operative reports available in the Pinkerton's National Detective Agency Collection MSS #1037 (hereafter cited as PNDAC) at Colorado Historical Society (CHS). These are also available at Idaho State Historical Society, as the operative reports were used in *State of Idaho v. William D. Haywood.*
13. *Labor Disturbances*, 78-79.
14. *Labor Disturbances*, 79.
15. *Western Federation of Miners, Proceedings*, Fifth Annual Convention, (hereafter cited as *WFM Proceedings*) Salt Lake City, May 8, 1897, as quoted in the MOA's *Criminal Record of the Western Federation of Miners* in 1904, 8.
16. *WFM Proceedings*, Ninth Annual Convention, Denver, May 27-June 6, 1901.
17. *Statement of the Western Federation of Miners*, presented by Senator Thomas M. Patterson, United States Congress, Document 163, 5.
18. Undated clipping in author's possession.
19. Guy E. Miller, "The Telluride Strike," January 1904, later published in Langdon's book, *The Cripple Creek Strike 1903-1904*, 210. This book preceded Langdon's *The Cripple Creek Strike, a History of the Industrial Wars in Colorado 1903-4-5*, which incorporates material from the first and expands it, since the strikes continued. The latter book will be cited as Langdon, *The Cripple Creek Strike, Industrial Wars.*
20. Miller, "The Telluride Strike"; Langdon, *The Cripple Creek Strike 1903-1904*, 210.
21. Curry used this term frequently, from 1901 on, throughout the Telluride *Daily Journal* (hereafter cited as *Journal*).

CHAPTER 2: **A CORPSE NICELY DISMANTLED** (Notes for pages 10-24)

1. McParland reporting his conversation with Adams to Gov. F. R. Gooding, June 9, 1906 (PNDAC, CHS).
2. The Ancient Order of Hibernians has disputed the Molly Maguire label. When McParland joined the fraternal order, he joined the chartered AOH, not any group called Molly Maguires. For detailed studies of the Pennsylvania cases, see Broehl, *The Molly Maguires*; Kenny, *Making Sense of the Molly Maguires*; Crown, *A Molly Maguire on Trial*; Crown and Major, *A Guide to the Mollie Maguires.*
3. *Shenandoah Herald* (Pennsylvania), May 6, 1876.
4. Orchard's real name was Albert E. Horsley. He was arrested in Idaho under the name Thomas Hogan and had several other documented aliases.
5. For the full details of Orchard's confession see *Big Trouble*; Orchard, *The Confessions and Autobiography of Harry Orchard* and *The Man God Made Again.*
6. McParland to Gooding, February 5, 1906 (PNDAC, CHS).
7. "Conspirators on the Run," *Industrial Union Bulletin*, March 9, 1907.
8. This is a portion of Adams's confession as summarized April 13, 1906, by James H. Hawley, prosecuting attorney in the case against William D. Haywood for conspiracy to murder ex-Governor Steunenberg of Idaho. Hawley, "Steve Adams' Confession and the State's Case Against Bill Haywood," *Idaho Yesterdays*, Winter Issue 1963-1964 (Idaho Historical Society), 22.
9. McParland to Gooding, March 8, 1906 (PNDAC, CHS).

10. McParland to Gooding, March 14, 1906 (PNDAC, CHS).
11. Kenny, *Making Sense of the Molly Maguires* , 109-111, 228-230, 292 on Daniel Kelly. Several of the Molly Maguire (AOH) members were quite probably innocent, hanged because of perjured testimony from McParland, who was an agent provocateur within their ranks. Confessed killers such as Kelly the Bum, who were promised freedom if they testified against the leaders of the AOH, were also likely perjurers. For detailed accounts of these cases, see books listed above in Note 2 and Lukas, *Big Trouble*, 195.
12. McParland to Gooding, April 17, 1906 (PNDAC, CHS).
13. McParland to Gooding, April 4, 17, and 20, 1906 (PNDAC, CHS).
14. McParland to Gooding, April 17, 1906 (PNDAC, CHS).
15. McParland to Gooding, May 2, 1906 (PNDAC, CHS).
16. From Wells's and Runnels's testimony during the trial of Steve Adams, Grand Junction *Daily Sentinel*, July 10, 1908, and *Journal*, July 11, 1908.
17. McParland to Gooding, June 4, 1906 (PNDAC, CHS).
18. Ibid.
19. McParland to Gooding, April 9, 1906 (PNDAC, CHS).
20. McParland to Gooding, June 9, 1906 (PNDAC, CHS).
21. Steve Adams trial testimony, *Journal*, July 11, 1908.
22. McParland to Gooding, June 9, 1906 (PNDAC, CHS).
23. Ibid.
24. *Journal*, July 11, 1908.
25. McParland to Gooding, June 9, 1906 (PNDAC, CHS).
26. Ibid.
27. Ibid.
28. McParland to Gooding, June 30, 1906 (PNDAC, CHS).
29. Friedman, *The Pinkerton Labor Spy*, 203.
30. *Heritage of Conflict*, 197-218. For detailed analysis of the plot to kidnap these men, see *Big Trouble*.
31. *The Pinkerton Labor Spy*, 203.
32. *Journal*, June 29, 1901.
33. Gressley, *Bostonians and Bullion*, 149. Robert Livermore recalled, "The pitiful remains were exhibited...to convince the union sympathizers that the murder had been committed, but I believe they only commented that the 'mine owners had done it to discredit the Union,'—the usual union explanation."
34. Photograph of placard in possession of Albert Fuentes in 1976. In a conversation with the author, Fuentes said union leaders chopped up Barney's body in the forest and the head rolled down the hill, thus Barney's ghost was always looking for his missing head.

CHAPTER 3: **WILLIAM JULIUS BARNEY** (Notes for pages 25-29)

1. *White Pine News*, Ely, Nevada 1896-1900.
2. Ralph B. Barney, unpublished manuscript in possession of Ralph C. Barney.
3. County records have not verified this claim; however, William Miles Barney homesteaded 160 acres near Valmont (1869 land patent): S1/2NE & N1/2SE Sec. 13, T1N and R70W. Henry Ludlow homesteaded 160 acres immediately south of Barney's. Savory's land was immediately north.
4. Ralph B. Barney.
5. Military records, Civil War Book O, p. 32; Transcript of Records Third Regiment Cavalry, Colorado State Archives (CSA).
6. Family information from various family friends and descendents of William Miles Barney and his first wife Melissa Rannells, and descendents of him and his second wife Helen Barclay.
7. Colorado State Census 1885, Schedule 2: Productions of Agriculture in the County of Boulder.
8. *White Pine News*, May 2, 1896.
9. *White Pine News*, August 9 and October 18, 1900.
10. Polly Pry interview with O. B. Kemp, Denver Post, November 30, 1902.

CHAPTER 4: **TELLURIDE STARTS A WAR** (Notes for pages 31-47)

1. Articles of Incorporation, Smuggler-Union Mining and Milling Company, recorded in Book 27, pp. 269-272, San Miguel County Courthouse (SMCC).
2. *Journal*, December 30, 1899. *Denver Times*, April 23, 1899, said the purchase price was $1,500,000 for a 51 percent controlling interest. *Denver Post*, July 8, 1901, said the purchase price was $2,500,000.
3. J. A. Porter's *Eighth Annual Report of the Smuggler-Union Mining Company* showed a profit of $76,000 for 1898, even with massive expansion of the plant and improvements in the mills (Smuggler-Union Mining Company file, American Heritage Center, University of Wyoming Center, hereafter cited as AHC-UW); *Journal*, June 3, 1899.
4. *Journal*, Special Edition, December 30, 1899.

5. Ibid.
6. Purington, *Preliminary Report on the Mining Industries of the Telluride Quadrangle.*
7. Hall, *History of Colorado*, "San Miguel County," Vol. 4, 313-320.
8. WFM Local 63 Union Charter (Jencks, CU Archives).
9. Charles Lewis, telephone conversation with author, May 1978. Lewis said his father, John, was a union man who fully supported St. John. In 1904 or 1905, Wells ordered the Lewis family to leave the district. A John Lewis appears in the San Juan District Union minutes in 1899 as a committeeman (Jencks, CU Archives).
10. San Juan District Union minutes (Jencks, CU Archives).
11. Charles Lewis, in conversation with author, May 1978, recalled St. John at his father's house saying this many times. Also Elizabeth Gurley Flynn, personal notes on Vincent St. John, Industrial Workers of the World Collection, I.D. No. 130 (hereafter cited as IWW 130), Archives of Labor and Urban Affairs, Wayne State University (hereafter cited as Wayne State).
12. San Juan District Union minutes (Jencks, CU Archives).
13. This and the preceding two quotes are from Charles Lewis, May 1978.
14. *Journal*, April 25, 1901 and *Examiner*, April 27, 1901. Also *Labor Disturbances*, 106.
15. Collins's statement to the *Journal*, April 26, 1901.
16. *Examiner*, April 27, 1901.
17. Ibid.
18. *Examiner*, May 4, 1901.
19. All quotes in this and the preceding paragraph from *Examiner*, April 27, 1901.
20. St. John's letter dated, April 23, 1901, as printed in *Examiner*, April 27, 1901.
21. *Denver Post*, November 30, 1902.
22. *Examiner*, April 27, 1901.
23. Charles Lewis, May 1978; *Examiner*, April 27, 1901.
24. *Examiner*, April 27, 1901.
25. *Examiner*, May 11, 1901; Charles Lewis, May 1978.
26. *Journal*, April 27, 1901.
27. Wells to the *Journal*, May 1, 1901 and *Examiner*, May 4, 1901.
28. *Examiner*, May 4, 1901.
29. *Miners' Magazine*, August 1901.
30. St. John's letter printed in the *Examiner*, May 4, 1901.
31. *Examiner*, October 19, 1901; testimony of J. W. Elliott at the fire inquest, San Miguel County Coroner's Inquest record, November 25-26, 1901 (Idarado Papers); boxes of ammunition were stored in the tram house and during a later fire, exploded, injuring two men; *Journal*, November 26, 1901.
32. Charles Lewis, May 1978.
33. *Examiner*, May 4, 1901.
34. Ibid.
35. Ibid.
36. Louis B. Macari, affidavit, April 23, 1901; *Examiner*, May 4, 1901.
37. Mike Musso, affidavit, May 3, 1901; *Examiner*, May 4, 1901.
38. *Examiner*, May 11, 1901.
39. Ibid.
40. *Examiner*, May 18, 1901.
41. *Examiner*, May 4, 1901.
42. *Examiner*, May 4, 11, and 18, 1901.
43. Ibid. Also, *Miners' Magazine*, August 1901; *Denver Post*, May 20 and 30, 1901.
44. This nickname is spelled variously as Shadagee, Shadigi, and Chatagee, all phonetic spellings derived from Chateaugay, a town in upper state New York.
45. San Miguel County Clerk and Recorder, Book 70, p. 561 (SMCC); Smuggler-Union, records, Idarado Papers, Ouray (hereafter cited as Idarado Papers).
46. This and the preceding two quotes from M. L. Shaffer, unpublished notes; *Denver Post*, December 14, 1902.
47. *Denver Post*, December 19, 1902.
48. Ibid.
49. M. L. Shaffer, unpublished notes.
50. Ibid. *Denver Post*, December 19, 1902, has similar phrasing.

CHAPTER 5: **DEPUTIES, SCABS, AND A DASH OF CURRY** (Notes for pages 48-62)

1. *Miners' Magazine*, July 1901.
2. *Journal*, April 23, 1901.
3. *Journal*, May 11, 1901.

4. E. O. Anderson, personal communication.
5. *Journal*, May 22, 1901.
6. This quote and all those in the preceding four paragraphs are from *Examiner*, May 25, 1901.
7. This and the preceding three quotes are from *Examiner*, May 25, 1901.
8. *The People of the State of Colorado v. William Jordan*, Case #384, District Court Record, May 23-June 6, 1901, (SMCC).
9. Ibid.
10. All quotes in this and the preceding three paragraphs are from *Examiner*, June 8, 1901.
11. *Journal*, June 6, 1901.
12. This and all quotes in the preceding paragraph are from *Examiner*, June 8, 1901.
13. Quote is from *Miners' Magazine*, July 1901. The assault information from *Examiner*, June 15, 1901; Miller, "The Telluride Strike; *Miners' Magazine*, July 1901.
14. *Examiner*, June 15, 1901.
15. *Examiner* named the Metropole, another union restaurant and saloon, as the site of this gunplay; *Miners' Magazine* named the St. Louis, which was the actual place where it occurred. William Hudson who signed a complaint for the night marshal owned the St. Louis, not the Metropole. City magistrate records, June 14, 1901, (SMCC).
16. *Miners' Magazine*, July 1901; City magistrate records, June 14, 1901, (SMCC).
17. Ibid.
18. *Denver Post*, July 8, 1901.
19. *Denver Post*, December 14, 1902. Thieman interviewed St. John several times for the *Denver Post* and spent some weeks in Telluride in 1902.
20. Telluride Miners Union No. 63, itemized list of expenditures (Jencks, CU Archives).
21. *Denver Post*, May 30, 1901.
22. *Examiner*, May 11, 1901.
23. *Labor Disturbances*, 107.
24. Ibid.
25. *Denver Post*, June 18, 1901.
26. Thiel spy reports, John F. Campion papers (Jencks, CU Archives).
27. Several Western newspapers repeated this tale when referring to W. J. Barney's mysterious disappearance, but nothing in the union records or any St. John interviews indicates it was true. The *Denver Post* of November 22, 1902 claimed Barney had been asked to join the union but replied he "had no use for unions."
28. *Journal*, September 6 and 9, 1901.
29. Per Gus E. Noraling, Finlanders of Telluride, letter to the *Journal*, September 9, 1901.
30. *Journal*, June 29, 1901.
31. All quotes in this and the preceding paragraph are from the *Journal*, June 29, 1901.
32. *Journal*, June 29, 1901.
33. Ibid.
34. *Journal*, July 1, 1901.
35. *Examiner*, June 29, 1901.
36. *Journal*, June 6, 1901 and *Examiner*, June 8, 1901.
37. *Examiner*, June 29, 1901.
38. *Journal*, June 29, 1901.

CHAPTER 6: **DEATH STALKS THE HOLIDAY** (Notes for pages 63-75)

1. *Journal*, June 6, 1901.
2. This and the preceding quote are from *Examiner*, June 29, 1901.
3. *Denver Post*, December 15, 1902.
4. Telluride Board of Trade, *Resources and Mineral Wealth of San Miguel County, Colorado*; *Journal*, Special Annual Edition, December 30, 1899; *Journal* and *Examiner*, various dates; U.S. Federal Census, 1880, 1900; Miller, "The Telluride Strike," January 1904, updated 1905 as published in Langdon, *The Cripple Creek Strike, Industrial Wars*; Telluride Board of Trade, records.
5. *People of Colorado v. Vincent St. John*, Case #5686, Mesa County District Court, "Argument of O. N. Hilton on defendant's motion to quash the information" (hereafter cited as Hilton, Motion to Quash), (Jencks, CU Archives); *People of Colorado v. Vincent St. John*, Case #686, District Court Record (SMCC), preliminary hearing, March 21, 1906.
6. *People of Colorado v. Vincent St. John*; Karl G. Olin (translated by S. Forsman), "Vörå Man Martyred at Telluride," Jakobstads Tidning, Jakobstad, Finland, November 6, 1993.
7. *Labor Disturbances*, 107; *Denver Post*, July 3, 1901.
8. Olin, "Vörå Man Martyred at Telluride."
9. A few accounts, including the first stories in the news, had Ben Burnham shot at the Bullion Tunnel, but subsequent investigations showed Burnham worked at the Sheridan. Later testimony by

witnesses all put Burnham at the Sheridan dump. The testimony in the preliminary hearing of the *People of Colorado v. Vincent St. John*, March 21, 1906, including that of R. M. Wright and William Hay, put Burnham at the Sheridan. *Hilton, Motion to Quash* (Jencks, CU Archives); *Journal*, March 22, 1906.
10. *Durango Democrat*, July 5, 1901.
11. *People of Colorado v. Vincent St. John*, preliminary hearing; *Hilton, Motion to Quash* (Jencks, CU Archives); *Journal*, March 22, 1906.
12. *Miners' Magazine*, August 1901; Journal, July 3-6, 1901; *Examiner*, July 6, 1901; *People of Colorado v. Vincent St. John*, preliminary hearing; *Hilton, Motion to Quash* (Jencks, CU Archives); *Journal*, March 22, 1906.
13. *Journal*, July 5, 1901.
14. Charles Lewis, May 1978; *People of Colorado v. Vincent St. John*, preliminary hearing; *Hilton, Motion to Quash* (Jencks, CU Archives); *Journal*, March 22, 1906; *Labor Disturbances*, 108.
15. All quotes in this and the preceding four paragraphs from *People of Colorado v. Vincent St. John*, preliminary hearing; *Hilton, Motion to Quash*; *Journal*, March 22, 1906.
16. Ibid.
17. *Denver Post*, July 4, 1901; *Miners' Magazine*, August 1901. McMullin's telegram was the first Orman received, yet McMullin was 127 miles away from the scene.
18. Ibid.
19. *Denver Post*, July 4, 1901; *Miners' Magazine*, August 1901, as quoted from original telegram.
20. *Miners' Magazine*, August 1901.
21. Ibid.
22. Ibid.
23. *People of Colorado v. Vincent St. John*, preliminary hearing; *Hilton, Motion to Quash* (Jencks, CU Archives); *Journal*, March 22, 1906.
24. *Denver Post*, July 4-8, 1901; *Journal*, July 5, 1901; *Examiner*, July 6, 1901; *Miners' Magazine*, August 1901; *WFM Proceedings*, Tenth Annual Convention, Denver, May 26-June 7, 1902; John H. Murphy's report; *People of Colorado v. Vincent St. John*, preliminary hearing; *Hilton, Motion to Quash* (Jencks, CU Archives); *Journal*, March 22, 1906.
25. *Denver Post*, July 8, 1901.
26. *Labor Disturbances*, 108.
27. On maps of 1898, 1900, and 1904 this is called Royal Gulch/Creek. On modern maps it is Royer Gulch/Creek. During testimony in several cases it was called "Roger" Gulch. Although the oldest materials call it Royal, the Royer name will be used, since evidence exists that it was named for the Royer family members who pioneered here.
28. *Examiner*, July 6, 1901.
29. Ibid.
30. *Examiner*, July 6, 1901; *People of Colorado v. Vincent St. John*, preliminary hearing; *Hilton, Motion to Quash* (Jencks, CU Archives); *Journal*, March 22, 1906.
31. *Examiner*, July 6, 1901.
32. This quote and the one preceding from *People of Colorado v. Vincent St. John*, preliminary hearing; *Hilton, Motion to Quash* (Jencks, CU Archives); *Journal*, March 22, 1906.
33. Ballard's account, *Durango Democrat*, July 5, 1901.
34. Ibid.
35. Ibid.
36. Ibid.
37. Charles Lewis, May 1978; *Miners' Magazine*, September 1901, echoes his words in the statement from Press Committee No. 63: "Let every true man...leave the next generation something more in the way of better conditions."
38. Charles Lewis, May 1978.
39. *Denver Post*, July 4, 1901.
40. Ibid.

CHAPTER 7: **A BIRD WITH BROKEN WINGS** (Notes for pages 77-88)

1. All quotes in this paragraph from *Denver Post*, July 4, 1901.
2. *Denver Post*, July 4, 1901.
3. *Denver Post*, July 5, 1901.
4. Ibid. Also in *Miners' Magazine*, August 1901.
5. *Miners' Magazine*, August 1901.
6. *Denver Post*, July 4, 1901.
7. Ibid.
8. Figures furnished by John A. Holmberg, State Auditor, November 11, 1904, *Labor Disturbances*, 360.
9. *Big Trouble*, 299.

10. *Denver Post*, July 5, 1901.
11. *Journal*, December 25, 1902.
12. *Denver Post*, July 4, 1901.
13. Ibid.
14. Ibid.
15. Ibid.
16. All quotes in this and the preceding two paragraphs are from *Denver Post*, December 18, 1902.
17. *Denver Post*, July 5, 1901.
18. *Examiner*, July 13, 1901; inquest information also in *Journal*, July 5, 1901.
19. Sixteen to One Miners' Union, resolution for John Barthell (Jencks, CU Archives).
20. *Journal*, July 6, 1901.
21. *Journal*, July 6, 1901; *Denver Post*, July 6, 1901.
22. *Denver Post*, July 6, 1901.
23. Charles Lewis, May 1978; *Denver Post*, July 6, 1901.
24. All quotes in this paragraph from *Denver Post*, October 17, 1901.
25. *Miners' Magazine*, November 1901.
26. *Denver Post*, July 5, 1901.
27. Ibid.
28. *Denver Post*, July 6, 1901.
29. *Denver Post*, July 5, 1901.
30. *Denver Post*, July 4, 1901.
31. *Denver Post*, July 6, 1901.
32. Ibid.
33. Ibid.
34. *Miners' Magazine*, August 1901.
35. Ibid.
36. Ibid.
37. *Denver Post*, July 6, 1901.
38. Ibid.
39. All quotes in this and the preceding paragraph are from *Denver Post*, July 6, 1901.
40. Ibid.
41. *Denver Post*, July 7, 1901.
42. *Miners' Magazine*, August 1901.
43. From the original copy of the agreement, with hand-written corrections and signatures of each participant (Idarado Papers).
44. *Miners' Magazine*, August 1901.
45. Ibid.
46. *Denver Post*, July 6, 1901.
47. *Denver Post*, July 7 and 8, 1901;
48. *U.S. Dept. of Commerce and Labor, Bureau of the Census, Special Reports, Religious Bodies: 1906.*
49. All quotes in this paragraph from Charles Lewis, May 1978; *Miners' Magazine*, August 1901, has similar statements.

CHAPTER 8: **WALTZING ON THIN ICE** (Notes for pages 89-95)

1. *Journal*, July 8, 1901.
2. *Miners' Magazine*, September 1901.
3. *Colorado, Bureau of Labor Statistics, Eighth Biennial Report*, 1901-1902.
4. *Miners' Magazine*, August 1901.
5. *Denver Post*, July 7, 1901.
6. *Examiner*, October 19, November 2 and 9, 1901.
7. Charles Lewis, May 1978.
8. *Journal*, July 8, 1901.
9. *Examiner*, July 13, 1901.
10. All quotes in this paragraph from *Examiner*, July 13, 1901.
11. *Examiner*, November 9, 1901.
12. *Denver Post*, July 7, 1901.
13. *Plaindealer*, August 30, 1901.
14. *Denver Post*, July 8, 1901.
15. Ibid.
16. *Plaindealer*, July 12, 1901.
17. *Denver Post*, July 8, 1901.
18. This was a paid advertisement first placed in the *Examiner*, July 13, 1901, and the *Journal*, and repeated several weeks.

19. *Denver Post*, July 5, 1901.
20. *Denver Post*, July 7, 1901.
21. *WFM Proceedings*, Tenth Annual Convention, 1902; John Murphy's report.
22. Ibid.
23. *Examiner*, July 13, 1901.
24. Ibid.
25. San Juan District Union minutes (Jencks, CU Archives); *Denver Post* December 6, 1902.
26. *Denver Post*, December 14, 1902.
27. Telluride Miners' Union #63, committee reports, August-September 1901 (Jencks, CU Archives).
28. *Journal*, July 9, 1901.
29. *Examiner*, July 13, 1901; *Journal*, July 17, 1901.
30. *Examiner*, July 13, 1901.

CHAPTER 9: **SEND IN THE SPIES** (Notes for pages 96-103)

1. *Examiner*, October 18, 1902.
2. *Miners' Magazine*, September 1901.
3. Telluride Miners' Union #63, quarterly financial reports, receipts, disbursements and itemized expenditures; during O. M. Carpenter's terms of office the trustees without fail reported Local 63's books "correct and in good order" (Jencks, CU Archives).
4. *Miners' Magazine*, September 1901.
5. *Journal*, August 24, 1901.
6. *Journal*, September 4, 1901.
7. Ibid.
8. Ibid.
9. Ibid.
10. Ibid.
11. *Journal*, September 5, 1901.
12. *Examiner*, October 12, 1901.
13. Goldman's speech was printed in full in the *Chicago Daily Tribune*, Sept. 8, 1901.
14. *The New York Times*, September 7, 1901.
15. All quotes in this and the preceding two paragraphs are from Creelman, *On The Great Highway.*
16. U.S. Presidents: United in Service, "William McKinley," from website, http://www.whitehouse.gov/history/presidents/wm25.html; also, Morgan, *William McKinley and His America.*
17. *Journal*, October 1, 1901.
18. *Examiner*, October 5, 1901.
19. *Journal*, October 1, 1901.
20. U.S. Federal Census, San Miguel County 1900; Edward S. Pierce, "Telluride in 1900," 1952 memoir published in *Pioneers of the San Juan Country.*
21. *Denver Post*, April 23-25, 1901.
22. *Journal*, October 14, 1901.
23. Friedman, *The Pinkerton Labor Spy*, 118. Friedman had worked in the Denver Pinkerton office as McParland's stenographer until mid 1905.
24. McParland's personal reports and instructions to operatives in the Cripple Creek area unions and other WFM unions in the western states are available at the Colorado Historical Society, the Idaho State Historical Society, and CU Boulder Archives. Friedman also reproduces many of these documents verbatim in his book, *The Pinkerton Labor Spy*.
25. *The Pinkerton Labor Spy*, 9-11.
26. *The Pinkerton Labor Spy*, 7-11.
27. McParland to Luther M. Goddard, February 9, 1906 (PNDAC, CHS).
28. *Biennial Report of the Adjutant General of Colorado, 1903-1904* (hereafter cited as *Biennial Report*), "Report of Telluride Expedition No. 2," 215.
29. *Journal*, October 18, 1901.

CHAPTER 10: **THE LITTLE NAPOLEON OF LOCAL POLITICS** (Notes for pages 104-110)

1. *Examiner*, October 26, 1901.
2. Ibid.
3. Ibid.
4. *Denver Post*, October 17, 1901.
5. *Examiner*, November 2, 1901.
6. *Examiner*, October 12, 1901.
7. *Examiner*, October 19, 1901.
8. Ibid.

9. Ibid.
10. Ibid.
11. Ibid.
12. *Journal*, October 29, 1901.
13. This quote and the one preceding are from *Examiner*, November 9, 1901.
14. *Examiner*, October 26, 1901.
15. *Examiner*, November 2, 1901.
16. *Examiner*, October 26, 1901.
17. *Examiner*, October 19, 1901.
18. *Examiner*, November 9, 1901.
19. *Examiner*, October 26, 1901.
20. *Journal*, November 1, 1901.
21. *Journal*, October 26, 1901.
22. All quotes in this paragraph from *Examiner*, November 2, 1901. Lydia E. Pinkham (1819-1883) was a patent medicine inventor who purported to have the cure for "all those painful Complaints and Weaknesses so common to our best female population." Her 18% alcohol potion also helped a few men in extraordinary ways, if the ditties of the era were true:

    And Michael Kline, oh Michael Kline, oh Michael Kline,
    His tears were many for he had one nut, just one,
    Till he drank nine bottles of Pinkham's, just nine,
    And now they grow in clusters round his bum,
    in clusters, in clusters, in clusters round his bum.

    At her place of business she also gave out advice to women on emotional ills, sex, and family difficulties. Sumner viewed it as quackery and the advice as the "trash department."
23. Women gained the right to vote through a constitutional amendment passed by the people of Colorado during a general election on November 7, 1893. Colorado male voters passed the women's suffrage amendment by several thousand votes.
24. *Examiner*, November 2, 1901.
25. *Examiner*, October 19, 1901.
26. Ibid.
27. *Examiner*, November 2, 1901.
28. *Examiner*, November 9, 1901.
29. San Miguel County Clerk, Abstract of Votes, November 5, 1901, reproduced in the *Journal*, November 21, 1901.
30. *Examiner*, November 9, 1901.
31. This name is spelled in several ways in the local newspapers of the time. The spelling in the District Court Record for the November 1901 term, Case #395, is "Mengoni," which is used here.
32. *People of Colorado v. Joe Mengoni*, Case #395, District Court Record (SMCC); *Journal*, November 7, 9, and 13, 1901; *Examiner*, November 9, 1901. An attorney was assigned to represent Mengoni, but withdrew, and the court assigned E. L. Oakes to represent him. Oakes told him to change his plea to "not guilty," but the jury found him guilty and he was sentenced to 20-25 years hard labor in the state penitentiary at Cañon City.

CHAPTER 11: **HEROES IN THE SMOKE** (Notes for pages 111-123)

1. Every detail of the fatal fire at the Smuggler is from several primary sources, including the San Miguel County Coroner's Inquest record, November 25, 1901 (Idarado Papers); *Examiner*, November 23, 30, and December 7, 1901 and March 8, 1902; *Journal*, November 20-22, 25-27, 1901; Fire Inquest testimony; *Denver Post*, November 21-28, 1901; *Miners' Magazine*, December 1901; various personal memoirs. To preserve continuity of the tragic events, which are assembled from so many various reports and documents, individual quotes and details are not footnoted.
2. *Journal*, November 25, 1901.
3. Ibid.
4. San Miguel County Court records, fee book, December 7, 1901 (SMCC).
5. San Miguel County Coroner's Inquest record, November 25, 1901, over the bodies of 24 victims (named in the document) of the Smuggler-Union fire on November 20, 1901; I. L. Glenn, Coroner of San Miguel County (Idarado Papers). Also recorded in the *Denver Post*, November 24; *Journal*, November 25; *Examiner*, November 30, 1901.
6. Ibid.
7. San Miguel County Coroner's Inquest record, November 25, 1901, jurors J. P. Redick, C. C. Van Lanningham, I. J. McDonald, Charles DeWitt, E. L. Sleeper, L. L. Knudtson; verdict of the Fire Inquest over the bodies of 24 victims (named in the document) of the Smuggler-Union fire on November 20,1901, attested to by I. L. Glenn, Coroner of San Miguel County (Idarado Papers).
8. San Miguel County Coroner's Inquest record, November 25, 1901 (Idarado Papers).

9. *Miners' Magazine*, December 1901.
10. Ibid.
11. Collins's letter was printed in *Examiner*, November 30, 1901.
12. Interview with St. John in the *Denver Post*, December 6, 1902.
13. The Wage Scale contract included underground miners, machine men, trammers, drivers, timbermen, hoist engineers, station and cage tenders, outside engineers, firemen, blacksmiths, day laborers, and tramway men. All ten-hour men working on the hill were given one hour on company time for dinner. The boarding house staff had a separate contract under the Federal Union.

CHAPTER 12: **WINTER SPORTS** (Notes for pages 124-136)

1. *Examiner*, October 5, 12, 19, and 26, 1901.
2. *Examiner*, December 7, 1901.
3. *Journal*, January 3, 1902.
4. *Examiner*, November 2, 9, and 16, 1901; January 4, 1901; July 11, 1902.
5. *Examiner*, February 8, 1902.
6. Charles Lewis, May 1978.
7. Miller, "The Telluride Strike", Langdon, *The Cripple Creek Strike 1903-1904*, 213.
8. San Juan District Union meeting May 17, 1901, record of the Auditing Committee consisting of V. St. John, F. Wood, A. J. Horn (Jencks, CU Archives).
9. *Journal*, May 12, 1904.
10. *People of Colorado v. Harry L. Servis*, Case #396, November 1901, District Court Record (SMCC).
11. Kinikin, "Early Days in Telluride," *Colorado Magazine*, June 1949.
12. *Denver Post*, November 19-20, 1901.
13. *Mills' Annotated Statutes of the State of Colorado*, (hereafter cited as *Mills Statutes*).
14. *Examiner*, December 21, 1901.
15. Ibid.
16. *Journal*, January 3, 1902.
17. All quotes in this paragraph from *Journal*, January 3, 1902.
18. All quotes in this paragraph from St. John's letter in the *Examiner*, January 4, 1902.
19. *Examiner*, January 4, 1902.
20. All quotes in this and the preceding paragraph from *Examiner*, January 4, 1902.
21. *Examiner*, January 4, 1902.
22. *Journal*, January 7, 1902.
23. *Examiner*, January 11, 1902.
24. All quotes in this paragraph are from *Examiner*, January 11, 1902.
25. In original documents this is spelled the old way, "Telluride Business Men's Association." For clarity, the spelling has been modernized throughout to "Businessmen's."
26. *Examiner*, February 1, 1902.
27. Charles Lewis, May 1978.
28. St. John's letter to the public, printed in *Examiner*, January 25, 1902.
29. *Journal*, January 31, 1902.
30. All quotes in this and the preceding three paragraphs are from *Examiner*, February 1, 1902.
31. *Examiner*, February 8, 1902.
32. *Telluride Miner*, February 6, 1902 (Jencks, CU Archives).
33. All quotes in this and the preceding two paragraphs are from *Telluride Miner*, February 6, 1902 (Jencks, CU Archives).
34. *James J. Blazer v. 16 to 1 Miners Union No. 63 Case #569*, County Court Record; District Court Record, Book 19, (SMCC).
35. *Miners' Magazine*, November 1902.
36. *Examiner*, February 8, 1902.
37. *Examiner*, March 1, 8, 15, 1902; *Journal*, March 6-21, 1902. Eyewitnesses reported the harrowing details of these slides as well as miraculous escapes, heroic efforts to save those buried, and the long dark days of sorrow. Many of the victims had wives and children and a fund was established for the benefit of these survivors.
38. *Journal*, March 13, 1902.
39. *Telluride Miner*, March 13, 1902.
40. *Examiner*, March 29, 1902.
41. *Journal*, March 29, 1902.
42. *Examiner*, April 5, 1902.
43. *Journal*, March 13, 1902.
44. Ibid.
45. *Examiner*, March 15, 1902.
46. Ibid.

47. *WFM Proceedings*, Tenth Annual Convention, 1902; resolutions, 90.
48. San Juan District Union resolution, minutes (Jencks, CU Archives).

CHAPTER 13: **DEAD MEN DON'T DIVORCE** (Notes for pages 137-147)

1. *Examiner*, March 29, 1902.
2. *Denver Post*, December 6, 1902.
3. Flynn, personal notes on Vincent St. John (IWW 130, Wayne State).
4. *Examiner*, January 11,1902.
5. *Examiner*, March 22, 1902.
6. Frederick N. MacLean, unpublished notes, (Frederick N. MacLean file, AHC-UW). MacLean was K. A. McLean's son (the different spelling of their surnames is unexplained).
7. *Journal*, March 13, 21, 24, 1902 and August 8, 1907; *Examiner*, March 15, 1902.
8. *Examiner*, March 8, 1902.
9. *Denver Republican*, March 20, 1902.
10. For further information on the political plotting of the trusts and the motives they had for destroying the Bunker Hill, see *Heritage of Conflict*, 72-87.
11. *Journal* April 3, 1902 and March 24, 1902; *Examiner*, March 15, 1902.
12. *Examiner*, March 8 & 15, 1902; *Journal*, March 21, 1902.
13. Colorado Mine Operators' Association, record book, March 27, 1902-April 18, 1903, George E. Collins Collection, Denver Public Library, Western History/Genealogy (DPL/WH).
14. Reprinted in the *Journal*, March 10, 1902.
15. *Journal*, April 24, 1902.
16. *W. J. Barney v. Sarah E. Barney*, Case #975, April 4, 1902, County Court Record, Book 8, pp. 44-46 (SMCC).
17. *Mills Statutes.*
18. Ibid.
19. Ibid.
20. County sheriffs by law were required to summon the venire, which unfortunately allowed an official to actually personally select individuals who might favor his interests. If he wished these men to reveal nothing of a case to the public, he could select those in whom his trust was secure and give them instructions as to his wishes.
21. *Mills Statutes.*
22. Marriage record of Winthrop D. Sylvester and Sarah E. Barney.
23. *Denver Post*, November 27, 1902.
24. *Examiner*, December 27, 1901, from the Denver press.
25. Kinikin, "Early Days in Telluride."
26. Ralph B. Barney, unpublished family history, March 1, 1962.
27. *Examiner*, April 12, 1902.
28. McParland to Luther Goddard, Associate Justice of the Colorado Supreme Court, Feb. 9, 1906, marked, "Strictly Confidential." This is only one example of pages of McParland's expressed hatred and condemnation of the WFM. He had Secret Operatives in nearly every WFM local in the West, yet never uncovered actual criminal activity, other than that which he ordered his operatives to perpetrate (PNDAC, CHS).

CHAPTER 14: **BUCKSHOT IN THE BACK** (Notes for pages 148-162)

1. Even in 1907 McParland publicly claimed John Mahoney was murdered by the Miners' Union although everyone knew this was untrue (see Journal, April 22, 1907).
2. *Examiner*, April 12, 1902.
3. Ibid.
4. *Examiner*, October 18, 1902.
5. Ibid.
6. In a speech before the Tenth Annual Convention in 1902, President Boyce warned the officers against the rampant bribery practiced by those who opposed the WFM.
7. *Examiner*, October 18, 1902.
8. Smuggler-Union Mining Company records (Idarado Papers).
9. *WFM Proceedings*, Tenth Annual Convention, 1902.
10. San Miguel County and District Court cases against the saloon proprietors numbered in the dozens, all pleading the same, all settled the same; *People v. Fred Pheasey and K. A. McLean*, Case #429, May 1902, County Court Record (SMCC).
11. *Denver Post*, December 8, 1902; *Examiner*, May 24, 1902; *People of Colorado v. John Miller*, June 5, 1902, District Court Record, Book 18 (SMCC).
12. Colorado State Penitentiary Record for Prisoner #5487, John Miller (CSA).
13. *Examiner*, May 24, 1902.

14. *Journal*, May 22, 1902; *Examiner*, May 24, 1902; *People of Colorado v. John Miller* (SMCC).
15. *Examiner*, May 24, 1902.
16. Ibid.
17. Ibid.
18. *Journal*, May 22, 1902.
19. *Examiner*, May 24, 1902.
20. *Journal*, June 12, 1902.
21. The details of the John Miller shooting of John Touhy and the consequent trial are from *Examiner*, May 24 and June 7, 1902; *Journal*, May 22, 29 and June 12, 1902; *People of Colorado v. John Miller*; and the prisoner's record (CSA).
22. *People of Colorado v. Harry L. Servis*, Case #396, District Court Record, Book 18 (SMCC); *Journal*, June 5, 1902; *Examiner*, May 30 and June 14, 1903.
23. Colorado State Penitentiary Record for Prisoner #5487, John Miller, and Prisoner #5488, Harry L. Servis (CSA).
24. *Examiner*, February 21, 1903.
25. *Samuel Servis v. Elizabeth H. Servis* Case #987, County Court Record, Book 7, 501 and 517 (SMCC).
26. The year-end auditing committee report for 1901, District Court Record, Book 18 (SMCC), February 7, 1902, suggests the deputy county treasurer was collecting funds but not depositing them to the county's account, "due probably to dishonesty." The treasurer and the deputy were implicated, but no other record exists to indicate the names of the two county officials Sumner continuously referred to as embezzlers who escaped the law.
27. *Examiner*, June 7, 1902.
28. All quotes in this paragraph from *Examiner*, June 21, 1902.
29. *Journal* as quoted in *Examiner*, August 9, 1902.
30. *Examiner*, August 9, 1902.
31. *Examiner*, June 21, 1902.
32. Sumner used these terms throughout the *Examiner* in 1901-1903.
33. Details for the Ike Kling case from Colorado State Penitentiary Record for Prisoner #5611, Ike Kling (CSA); *WFM Proceedings*, Tenth Annual Convention, 1902; *Journal*, December 4, 1902; *Examiner*, December 6 and 13, 1902; *Denver Post*, December 12, 1902.
34. *WFM Proceedings*, Tenth Annual Convention, 1902.
35. *Examiner*, June 14, 1902.
36. *Examiner*, June 7, 1902.
37. Preston, *Aliens and Dissenters*; Constantine (Editor), *Letters of Eugene V. Debs, 1874-1912*; *Debs: His Life, Writings and Speeches*, Girard, Kansas: the *Appeal to Reason*, 1908.
38. Ibid.
39. Inscription on the Barthell monument.
40. *Engineering and Mining Journal*, July 25, 1903, article draft (George Collins Collection, DPL/WH).
41. *Miners' Magazine*, August 1902.
42. *Miners' Magazine*, September 1902.
43. Debs, *Writings and Speeches of Eugene V. Debs.* Speech titled, "Outlook for Socialism in the United States" (first printed in the *International Socialist Review*, September, 1900).
44. *Journal*, July 31, 1902.
45. Ibid.
46. All quotes in this and the preceding two paragraphs from St. John's letter in the *Examiner*, August 2, 1902.
47. *Examiner*, August 2, 1902.
48. *Miners' Magazine*, September 1902.
49. *Examiner*, September 20, 1902.
50. *Examiner*, October 11, 1902.
51. The *Denver Republican* promoted Peabody and his "conservatism" in editorials throughout August-November 1902. For a detailed analysis of the Peabody campaign, see Suggs, *Colorado's War on Militant Unionism.*
52. *Journal*, October 16, 1902.
53. *Examiner*, November 8, 1902.
54. *Examiner*, October 18, 1902.
55. Ibid.
56. *Journal*, October 16, 1902.
57. *Examiner*, November 1, 1902.
58. *Denver Post*, December 23, 1902.
59. James H. Peabody, inaugural address to the Fourteenth General Assembly, *House Journal*, 1903 (CSA).
60. *Journal*, November 6, 1902.

CHAPTER 15: **A NOOSE FOR THE SAINT** (Notes for pages 163-178)

1. Accurate distances are taken from the insurance maps of the Smuggler-Union Mining Company plant made by the Sanborn Map Company, 1904.
2. *Denver Republican*, interview with O. B. Kemp, November 20, 1902. In his own words, Kemp said he telephoned Collins's brother-in-law, Becker, and the deputy at the Bullion Tunnel before calling for a doctor.
3. *Journal*, November 20, 1902.
4. *Journal*, November 20, 1902; *Examiner*, November 22, 1902.
5. *Journal*, November 20, 1902.
6. *Denver Republican*, November 20, 1902.
7. *Denver Post*, November 20, 1902.
8. *Denver Republican*, November 21, 1902.
9. *Denver Post*, November 22, 1902.
10. *Denver Post*, November 20, 1902; *Journal*, November 20, 1902; *Denver Republican*, November 20, 1902; *Examiner*, November 22, 1902.
11. *Denver Post*, November 20, 1902.
12. This and the preceding quote from *Journal*, November 22, 1902.
13. Some reports said he died at 1:30. Details taken from the *Denver Post*, November 21, 1902; *Denver Republican*, November 22, 1902; *Examiner*, November 22, 1902; *Journal*, November 22, 1902.
14. *Denver Post*, November 30, 1902.
15. The Winchester Model 1897 pump-action shotgun came in many configurations. It was the first Winchester shotgun chambered for 2-3/4-inch smokeless powder ammunition. Produced in standard models starting in 1897, over 1,024,700 were manufactured until production ceased in 1957. Configurations included 26- to 32-inch barrels, various chokes, plain stocks, and both solid and takedown types of frames. Various Trap, Black Diamond, Pigeon, military, riot, and trench versions were also available. The M1897 is still highly prized today.
16. *Examiner*, November 22, 1902.
17. *Journal*, November 20, 1902.
18. *Denver Post*, November 22, 1902.
19. *Denver Post* interview with Kemp, November 30, 1902.
20. Benjamin B. Lawrence to the Colorado Mine Operators' Association, December 20, 1902 (George Collins Collection, DPL/WH).
21. Arthur Winslow's address at the first annual dinner of the Colorado Mine Operators' Association, December 13, 1902 (George Collins Collection, DPL/WH).
22. This is from a magazine clipping of an article in George Collins's scrapbook. The page simply has a penciled "MW" at the top and the date November 29, 1902 (George Collins Collection, DPL/WH).
23. *Examiner*, November 22, 1902.
24. *Journal*, November 20, 1902.
25. *Journal*, December 4, 1902.
26. Broehl, *The Molly Maguires*.
27. Magazine clipping marked "MW," and dated November 29, 1902 (George Collins Collection, DPL/WH).
28. *Journal*, November 27, 1902.
29. From the *Durango Democrat*, as printed in the *Journal*, November 27, 1902.
30. From the *Ouray Herald*, as printed in the *Journal*, November 27, 1902.
31. *Examiner*, November 22, 1902.
32. *Denver Post*, November 22-26, 1902; *Denver Republican*, November 22, 1902; *Journal*, November 22-December 4, 1902; *Examiner*, November 29, 1902.
33. *Denver Post*, November 22, 1902.
34. *Journal*, November 27, 1902.
35. *Denver Post*, November 21, 1902.
36. *Denver Post*, November 22, 1902.
37. *Examiner*, November 22, 1902.
38. *Examiner*, November 22, 1902; Sumner reported a long conversation between himself and Collins at the Maher resort in Denver.
39. This and the preceding quote from *Denver Post*, November 25, 1902.
40. *Journal*, December 4, 1902.
41. *Examiner*, November 29, 1902. This was also published in several Denver newspapers.
42. All quotes in this paragraph from *Denver Post*, November 25, 1902.
43. *Denver Post*, December 2, 1902.
44. *Denver Post*, November 27, 1902.
45. All quotes in this and the preceding paragraph from *Denver Post*, November 27, 1902.
46. *Miners' Magazine*, January 1903.

47. All quotes in this and the preceding four paragraphs from *Denver Post*, November 27, 1902.
48. *Examiner*, November 29, 1902.
49. All quotes in this paragraph from *Examiner*, November 29, 1902.
50. *Miners' Magazine*, January 1903.
51. Affidavits were sworn to and signed before Justice of the Peace W. Robinson, November 29, and published in the *Examiner*, November 29, 1902.
52. *Denver Post*, December 2, 1902.
53. *Denver Post*, November 28, 1902.
54. Salida Record as quoted in the *Journal*, December 11, 1902.
55. *Rico News* as quoted in the *Journal*, December 4, 1902.
56. *Rocky Mountain News*, November 28, 1902.
57. *Journal*, December 4, 1902.
58. *Rocky Mountain News*, August 8, 1895. The article used a fairly accurate drawing of Clark taken from a Telluride photo of the man, making it appear the story was factual. The story was fiction from start to finish, depicting Clark as being Jim Cummins (Cummings) because both men were from Missouri and both had been in Quantrell's (Quantrill) Raiders. Cummins was on the rosters; Clark's name has never been on any Missouri Partisan rosters. The year before he died, Clark told a tale of being with Quantrell's men in 1863, which he was, briefly. He deserted after the Lawrence Massacre and became a cowboy, riding herd from Texas to the Canadian border. He was in Ft. Lyon, Colorado in 1870 and in Leadville in 1886. He arrived in Telluride in 1888. He served several terms as town marshal, beginning in July 1889.
59. Cummins, *Jim Cummins' Book, Written by Himself*. Born in Clay County, Missouri, Cummins had been a member of Quantrell's Guerrillas during the Civil War, and ran with the James and Younger gangs for several years after the war. He died many years after Clark, and kept in touch with the Missouri Partisans-Quantrell's Raiders. His memoirs never mentioned Clark.
60. *Telluride Republican*, August 9, 1895.
61. The author has various materials that prove Jim Clark was not an outlaw. He was doing his job in Telluride and his movements can be traced enough to show he was elsewhere on dates during the crimes attributed to him *after* his death. C. W. "Doc" Shores's memoirs (housed at DPL/WHG) include a tale about him meeting Clark and his friendship with him, painting him as an outlaw with a good heart. The Clark tale was written in 1928 when Shores was in his eighties. Instead of relying on his memory, Shores relied on the *News* clipping of August 1895 that depicted Clark as the outlaw Jim Cummins. In an attempt to reconcile what he did recall of his friend with the clipping in his scrapbook, Shores made Clark kind and brave, and gave him excuses for his outlawry. Shores knew Clark for only a short time and had to fit the so-called outlaw incidents of the News into his story by mangling dates and places and making up impossible scenes. Records and original news stories show Shores was wrong and Clark was innocent. Matt Warner, one of the cowboys who robbed the San Miguel Valley Bank with Cassidy, never mentioned Clark as an accomplice in his memoirs. The marshal was J. F. Marion Dunlap during the bank holdup. Dunlap was still living in Telluride in 1900. James A. Beattie was sheriff at the time of the robbery and he and his posse pursued the robbers immediately.
62. *Examiner*, November 29,1902.
63. *Journal*, December 4, 1902.
64. *Examiner*, December 13 and 20, 1902.
65. *WFM Proceedings*, Eleventh Annual Convention, Denver, May 25-June 10, 1903; John Murphy's report, 162; *Miners' Magazine*, January 1903.
66. *Journal*, December 4, 1902.
67. Telluride Miners' Union #63, resolution to accept Barnes's loan, November 29, 1902 (Jencks, CU Archives).
68. *Examiner*, May 23, 1903.
69. *Examiner*, December 6, 1902.
70. All quotes in this and the preceding two paragraphs from *Denver Post*, December 6, 1902.
71. All quotes in this and the preceding three paragraphs from *Denver Post*, December 8, 1902.

CHAPTER 16: **A PINKERTON ON EVERY CORNER** (Notes for pages 179-193)

1. *People of Colorado v. Walt Kenley*, Case #516, December 4, 1902, District Court Record, Book 19 (SMCC).
2. *Denver Post*, December 14, 1902; Examiner, December 20, 1902.
3. *Denver Post*, December 8, 1902.
4. *Examiner*, December 20, 1902.
5. *Denver Post*, December 6, 1902.
6. *Denver Post*, December 7, 1902.
7. *Denver Post*, December 10, 1902.

8. *Examiner*, December 20, 1902.
9. *Examiner*, December 13, 1902.
10. *Denver Post*, December 10, 1902.
11. Ibid.
12. Wells's interview as reprinted in the *Durango Democrat*, December 16, 1902.
13. Colorado Mine Operators' Association, record book, March 27, 1902-April 18, 1903, George E. Collins Collection (DPL/WH).
14. *Durango Democrat*, December 16, 1902.
15. *Denver Post*, December 13, 1902.
16. Grand Jury Report December 15, 1902, District Court Record, Book 19, pp. 61-63 (SMCC).
17. *Denver Post*, December 17, 1902.
18. San Miguel County Sheriff's Office, Jail Register.
19. *Journal*, December 18, 1902.
20. *Examiner*, December 20, 1902.
21. *Denver Post*, December 17, 1902.
22. *Examiner*, December 20, 1902.
23. Ibid.
24. Ibid.
25. *Denver Post*, December 23, 1902.
26. *Miners' Magazine*, January 1903.
27. Information from Clarence E. Hagar report to Governor Peabody, October 1903, Peabody Papers Box 26716 (CSA); and San Miguel County Sheriff Bill Masters, interview, February 11, 2002.
28. San Miguel County Sheriff's Office, Jail Register records.
29. *Durango Democrat*, December 19, 1902; repeated in the *Journal*, December 25, 1902.
30. *Journal*, December 25, 1902.
31. *Examiner*, December 27, 1902.
32. Ibid.
33. Quoted from the Denver papers in *Examiner*, December 27, 1902.
34. Sarah Rannells died March 17, 1896. Leaving no will, her lands became tangled in a bitter and lengthy dispute between her heirs, which included Will Barney and his sisters, and the adult children of Sarah and their children. The estate was eventually settled in 1911; Decree of Final Settlement, Decrees, Book 337, p. 326-327, Boulder County Clerk and Recorder.
35. *Journal*, December 25, 1902.
36. *Denver Post*, January 14, 1903; *Journal*, January 15, 1903; *Examiner*, January 17, 1903.
37. *Ouray Herald*, January 2, 1903.
38. *Examiner*, January 24, 1903.
39. Ibid.
40. *Examiner*, January 10, 1903.
41. The official hospital agreement was signed by Charles A. Chase, Manager of the Liberty Bell and secretary of the MOA, and Vincent St. John, president of Sixteen to One Miners' Union, on January 22, 1903. Negotiations had gone on since the hospital's opening; Telluride Miners' Union #63, committee reports, November 1902 (Jencks, CU Archives); *Examiner*, January 10, 24, 1903.
42. *Examiner*, January 17, 1903.
43. *Journal*, January 29, 1903.
44. *Examiner*, February 21, 1903.
45. *Durango Herald*, as reprinted in the *Examiner*, March 21, 1903.
46. *Durango Telegraph*, as reprinted in the *Examiner*, March 21, 1903.
47. *Durango Journal*, as reprinted in the *Examiner*, March 21, 1903.
48. *Examiner*, February 28, 1903.
49. Barnes's letter as printed in the *Examiner*, January 10, 1903.
50. *Examiner*, February 14, 1903.
51. *Examiner*, April 18, 1903; *Ouray Herald*, April 24, 1903.
52. *Examiner*, May 2, 1903.
53. *Miners' Magazine*, May 1903.
54. *Examiner*, May 9, 1903.
55. *Examiner*, May 16, 1903.
56. *WFM Proceedings*, Eleventh Annual Convention, 1903; Murphy's report, 162.
57. *People of Colorado v. James Roner*, May 16, 1903, District Court Record, (SMCC). The ruling of Judge Theron Stevens in the Roner case applied to all cases arising from the grand jury indictments wherein pleas in abatement were filed, as Stevens had stipulated only the first case (Roner) would be argued and all others would abide by the decision of the first. Of the original twenty-two indicted, only eleven had been arrested, the others never found; however, all those indicted were released by this ruling and all fifty-seven indictments of the grand jury were quashed.

58. *Journal*, May 21, 1903.
59. Ibid.
60. *Examiner*, June 6, 1903.
61. *People of Colorado v. Vincent St. John*, Case #586, 587, 603, May 18, 1903-July 1903; District Court Record, Book 19 (SMCC).
62. *Denver Republican*, June 3, 1903; *Denver Times*, *Cripple Creek Press*, June 4, 1903; *Examiner*, June 6, 1903.
63. *Miners' Magazine*, July 1903.
64. *Examiner*, May 23, 1903.
65. Ibid.
66. All quotes in this and the preceding paragraph from *Examiner*, May 23, 1903.
67. *Examiner*, May 30, 1903.
68. Ibid.
69. Wenger, "My Recollections of Robert Livermore and events in his Journal," handwritten MS, December 1968, Robert Livermore Papers (AHC-UW). Also printed as *Recollections of Telluride, 1895-1920*. Wenger grew up in Telluride and knew Meldrum personally.
70. *WFM Proceedings*, Eleventh Annual Convention, 1903; John Murphy's report, 162-163.
71. *Miners' Magazine*, July 1903.
72. *People of Colorado v. Vincent St. John*, Case #586, 587, 603, motion entered in the district court record, July 7, 1903.

CHAPTER 17: **PEABODY SELLS THE MILITIA** (Notes for pages 194-206)

1. *Examiner*, January 17, 1903.
2. For a detailed analysis of Peabody's term in office (1903-1905) see Suggs, *Colorado's War on Militant Unionism*.
3. *Denver Post*, January 19-25, 1903; *Journal*, January 22, 1903.
4. Suggs, *Colorado's War on Militant Unionism*, 192.
5. *Labor Disturbances*, 58.
6. *Colorado's War on Militant Unionism*, 74-75.
7. This and the preceding quote from *Examiner*, March 14, 1903.
8. *Colorado's War on Militant Unionism*, 192.
9. Ibid.
10. Bishop Matz to J. P. Morgan, undated; Nicholas C. Matz Collection, Catholic Archdiocese of Denver Archives.
11. Catholics made up more than 16 percent of the population; *U.S. Dept. of Commerce and Labor, Bureau of the Census, Special Reports, Religious Bodies: 1906*.
12. *Miners' Magazine*, July 1903.
13. Ibid.
14. *Denver Catholic Register*, December 5, 1903.
15. *Rocky Mountain News*, June 21, 1903.
16. Charles Lewis, May 1978.
17. St. John's letter in the *Industrial Union Bulletin*, August 3, 1907.
18. *Examiner*, June 13, 1903.
19. All quotes in this paragraph are from *Examiner*, June 13, 1903.
20. All quotes in this and the preceding three paragraphs are from *Examiner*, June 13, 1903.
21. Telluride Miners Union No. 63, itemized list of expenditures for August 1903; minutes (Jencks, CU Archives).
22. San Juan District Union, minutes, August 6, 1903 (Jencks, CU Archives).
23. *Miners' Magazine*, November 1903; February 18, 1904.
24. *Journal*, December 3, 1903.
25. *Journal*, July 23, 1903.
26. *Examiner*, June 13, 1903.
27. *Examiner*, May 16, 1903.
28. *Examiner*, June 6, 1903.
29. *Examiner*, June 13, 1903.
30. *Labor Disturbances*, 149-150.
31. *Examiner*, July 11, 1903.
32. San Juan District Union, resolution, August 6, 1903 (Jencks, CU Archives).
33. *Miners' Magazine*, April and May 1903.
34. *WFM Proceedings*, Eleventh Annual Convention, 1903; Moyer's report, 208.
35. *Labor Disturbances*, 151-159.
36. Moyer to Carpenter, August 6, 1903 (Jencks, CU Archives).
37. *Examiner*, August 29, 1903.

38. Miller, "The Telluride Strike," in Langdon, *The Cripple Creek Strike, Industrial Wars*, 206.
39. *Journal*, September 3, 1903.
40. Miller, "The Telluride Strike," in Langdon, *The Cripple Creek Strike, Industrial Wars*, 206.
41. *Miners' Magazine*, November 1903; *Journal*, September 3, 1903; *Examiner*, September 5, 1903.
42. Joint resolutions, Miners' Union, Federal Union, Carpenters' Union, September 1, 1903 (Jencks, CU Archives).
43. *Labor Disturbances*, 167-169.
44. Ibid.
45. *Examiner*, September 26, 1903.
46. Langdon, *Labor's Greatest Conflicts*, 31.
47. *Journal*, September 24, 1903.
48. *Examiner*, December 26, 1903.
49. Miller, "The Telluride Strike," in Langdon, *The Cripple Creek Strike, Industrial Wars*, 206.
50. *Examiner*, November 7, 1903.
51. Miller, "The Telluride Strike," in Langdon, *The Cripple Creek Strike, Industrial Wars*, 207.
52. *Journal*, October 1-December 31, 1903.
53. Trustees report and account books of the Telluride Miners' Union, No. 63, O. M. Carpenter, Secretary-Treasurer, October 6, 1903 showed all moneys received, all disbursements and expenditures accounted for and the books in perfect order; at no time during Carpenter's tenure were there any discrepancies (Jencks, CU Archives).
54. Petition (Filed October 20) from Officers and Citizens of San Miguel County, October 19, 1903 (Peabody Papers, CSA).
55. Ibid.
56. Telluride's population had risen to 4,500 by the end of 1903 (Recorded by Sanborn Map Company July 1904).
57. Miller, "The Telluride Strike"; "Military Rule in Colorado," *Army and Navy Journal*, October 17, 1903; state auditor's report of November 1904, *Labor Disturbances*, 360. State Treasurer Whitney Newton, in his annual report to the governor on January 2, 1905, gave the total amount of certificates of indebtedness issued in payment of military expenses during 1903 and 1904 as $776,464, and recommended that the general assembly authorize the issuance of "insurrection bonds" to the amount of $800,000, payable in 25 years, and to draw not more than 3.5 percent interest annually.

    In 1907 several bills were introduced in the state legislature to reimburse the Mine Operators' Association for the "funds they expended in calling out the state militia and carrying on the war against the miners at Cripple Creek, Telluride, and other camps," *Daily Sentinel*, January 31, 1907. One bill provided for a payment to the MOA of $800,000; another, $900,000. In the opinion of many, most of this money would merely "cover the expense of useless extravagances and petty grafts."

CHAPTER 18: **TO HELL WITH THE CONSTITUTION** (Notes for pages 207-220)

1. This was a nickname used throughout the state, but was first assigned to Bell in the Cripple Creek campaign. Langdon, *The Cripple Creek Strike 1903-1904*, 147.
2. Langdon, *The Cripple Creek Strike, Industrial Wars*, 153.
3. Langdon, *The Cripple Creek Strike, Industrial Wars*, 5.
4. *Army and Navy Journal* as quoted in *Statement of the Western Federation of Miners*, U.S. Senate, Document 163, 38.
5. Painter to Hagar, October 27, 1903 (Peabody Papers, CSA).
6. Curry to Peabody, October 27, 1903 (Peabody Papers, CSA).
7. Howe to Hagar, October 26, 1903 (Peabody Papers, CSA).
8. Wrench to Peabody, October 23, 1903 (Peabody Papers, CSA).
9. Rutan, affidavit taken by Howe, November 13, 1903 (Peabody Papers, CSA).
10. Petitions, affidavits, letters detailing the situation and need for troops were all from the TMA and SMCCA members, hardly representative of the general population (Peabody Papers, CSA).
11. Miller, "The Telluride Strike," in Langdon, *The Cripple Creek Strike, Industrial Wars*, 279.
12. *Examiner*, November 14, 1903.
13. *Journal*, December 3, 1903.
14. A telegram signed by Wells, Wheeler, Herron, Howe, and Rutan made it clear they had no interest in settlement-they fully planned to open the mines and mills with nonunion labor; dated November 18, 1903 (Peabody Papers, CSA).
15. Affidavits (Peabody Papers, CSA).
16. *Examiner*, November 7, 14, and December 12, 1903; *Journal*, December 10, 1903.
17. Citizens' Alliance telegrams came at odd hours, one on November 19 at 11:50 p.m. in commemoration of the shooting of Arthur Collins (Peabody Papers, CSA).
18. *Examiner*, November 28, 1903.

19. Miller, "The Telluride Strike," in Langdon, *The Cripple Creek Strike 1903-1904*, 212.
20. *Journal*, November 23, 1903.
21. *Labor Disturbances*, 195; *Journal*, November 26, 1903; *Examiner*, November 28, 1903.
22. *Labor Disturbances*, 195.
23. *Journal*, November 26, 1903; *Examiner*, November 28, 1903; *Labor Disturbances*, 195.
24. Letters from Zeph Hill and Colorado National Guard documents clearly illustrate the exclusion of business from any union sympathizers. See Colorado National Guard Papers (CNG Papers) Box 10643, (CSA).
25. General Orders No. 67, from Adjutant General Sherman Bell, November 20, 1903 (CNG Papers, CSA).
26. Peabody to Hill, November 23, 1903 (Peabody Papers, CSA).
27. Hill to Peabody, January 3, 1904 (CNG Papers, CSA).
28. Hill to Peabody, November 26, 1903 (CNG Papers, CSA).
29. *State of Idaho v. William D. Haywood*, 1907, transcripts, testimony of Cal Rutan, Vol. 13, 4561-4581 (Jencks, CU Archives).
30. *Journal*, November 26, 1903.
31. Affidavit, Sheriff Rutan, November 26, 1903, wherein he furnishes a copy of the torn up letter from Haywood to Carpenter dated November 20, 1903 (CNG Papers, CSA).
32. *Journal*, November 26, 1903.
33. *Journal*, November 26, 1903; *Examiner*, November 28, 1903; Miller, "The Telluride Strike," in Langdon, *The Cripple Creek Strike 1903-1904*, 214.
34. Hill's military Circular No. 1, November 26, and Circular No. 3, November 30, 1903 (CNG Papers, CSA).
35. Hill to Peabody, December 10, 1903 (CNG Papers, CSA); and Peabody to Hill, December 12, 1903 (Peabody Papers, CSA).
36. Andrew H. Floaten, "Reign of Terror in Telluride," *Coming Nation*, December 12, 1903, as reprinted in the *Journal*, December 31, 1903.
37. *Journal*, December 3, 1903.
38. Floaten, "Reign of Terror in Telluride," *Coming Nation*, December 12, 1903.
39. *Examiner*, December 5, 1903.
40. Ibid.
41. *Journal*, December 10, 1903; *Examiner*, December 12, 1903.
42. *Norman Young v. J. C. Rutan, Sheriff*, Case #1146; *Tony Langeri v. J. C. Rutan, Sheriff*, Case #1147; County Court Record, Book 8, pp. 377-378 (SMCC); *Guy Miller v. J. C. Rutan, Sheriff*, Case #655; Marchiano #656, Langeri, #657, Macari, #658, all filed in District Court December 19, 1903; District Court Record, Book 19, pp. 321-323 (SMCC).
43. Major Hill, Field Reports, Camp Telluride, National Guard Reports; Court Martial records, Camp Telluride, 1903-04 (CNG Papers, CSA).
44. *Examiner*, December 19, 1903.
45. Sherman Bell's own words in an interview with the *St. Louis Star*, June 11, 1904 (CNG Papers, CSA).
46. The Citizens' Alliance placed ads in papers from Leadville to Denver, and in mining regions in Missouri, Michigan, and the Dakotas, in an attempt to attract miners who knew nothing of the strike; *Denver Republican*, December 1903; *Journal*, December 3 and 12, 1903; *Ouray Herald*, December 11, 1903.
47. *Journal*, December 10, 1903.
48. Ibid.
49. Polly Pry as reprinted in the *Journal*, December 10, 1903.
50. Hill to Peabody, December 8, 1903 (CNG Papers, CSA).
51. Hill to Peabody, December 20, 1903 (CNG Papers, CSA).
52. Ibid.
53. Hill to Peabody, December 21, 1903 (CNG Papers, CSA).
54. Ibid.
55. Hill to Peabody, December 20, 1903, including Howe's recommendations; Hill to Peabody, December 23, 1903 (CNG Papers, CSA); *Mills' Statutes*.
56. Miller, "The Telluride Strike," in Langdon, *The Cripple Creek Strike 1903-1904*, 214.
57. Miller, "The Telluride Strike"; *Miners' Magazine*, January 1904; *Journal*, December 24, 1903; *Examiner*, December 26, 1903.
58. *Journal*, December 24, 1903; *Examiner*, December 26, 1903.
59. *Examiner*, December 26, 1903.
60. *WFM Proceedings*, Twelfth Annual Convention, Denver, May 23-June 8, 1904; Hill to Peabody, December 23, 1903 (Peabody Papers, CSA). *Examiner*, December 26, 1903 and January 2, 1904; *Journal*, 24, 1903; Miller, "The Telluride Strike," in Langdon, *The Cripple Creek Strike 1903-1904*, 214.
61. *Examiner*, January 2, 1904.

62. *Examiner*, January 2, 1904.
63. Hill to Peabody, January 3, 1904 (CNG Papers, CSA).
64. Hill to Peabody, December 2 and 10, 1903 (CNG Papers, CSA); Peabody to Hill, December 12, 1903 (Peabody Papers, CSA).
65. Nathan C. Miller, *Biennial Report of the Attorney General of Colorado, 1903-1904*, 182-84.

CHAPTER 19: **REIGN OF TERROR** (Notes for pages 221-233)

1. *Labor Disturbances*, 198; Rutan to Peabody, December 28, 1903; Hill to Peabody, December 28, 1903 (Peabody Papers, CSA).
2. Governor James H. Peabody's executive proclamation declaring martial law in San Miguel County, January 3, 1904; for full text see *Labor Disturbances*, 198.
3. Nathan Miller to Peabody, January 4, 1904 (Peabody Papers, CSA).
4. Peabody to Hill, January 4, 1904 (Peabody Papers, CSA).
5. Hill to Peabody, January 5, 1904 (CNG Papers, CSA).
6. Bartholomew, *Anarchy in Colorado*, 92.
7. *Rocky Mountain News*, January 13, 1904.
8. Hill to Peabody, December 27, 1903 and January 5, 1904 (Peabody Papers, CSA).
9. *Biennial Report of the Adjutant General of Colorado, 1905-1907*, 46.
10. Hill to Peabody, January 3, 1904 (CNG Papers, CSA).
11. Hill to Peabody, January 5, 1904 (CNG Papers, CSA).
12. *Examiner*, January 9, 1904.
13. Miller, "The Telluride Strike" in Langdon, *The Cripple Creek Strike 1903-1904*, 215.
14. *Examiner*, January 9, 1904.
15. *Journal*, January 7, 1904; *Miners' Magazine*, January 14, 1904; *Biennial Report* "Report of Telluride Expedition No. 2"; Hill to Peabody, January 3, 1904, and National Guard records of arrests, Camp Telluride, 1903-04 (CNG Papers, CSA).
16. *Examiner*, January 9, 1904.
17. *Examiner*, January 9, 1904; Hill to Peabody, January 4, 1904 (CNG Papers, CSA).
18. Hill to Peabody, January 4, 1904 (CNG Papers, CSA).
19. *Journal*, January 4, 1904; *Examiner*, January 9, 1904; Hill to Peabody, January 4, 1904 (CNG Papers, CSA).
20. *Examiner*, January 9, 1904; Miller, "The Telluride Strike", in Langdon, *The Cripple Creek Strike 1903-1904*, 215.
21. *Examiner*, January 9, 1904.
22. Hill to Peabody, January 4 and 5, 1904 (CNG Papers, CSA); *Examiner*, January 9 and March 12, 1904.
23. *Labor Disturbances*, 198.
24. Hill to Peabody, January 6, 1904 (CNG Papers, CSA).
25. Ibid.
26. Hill to Peabody, January 11, 1904 (CNG Papers, CSA); Examiner, January 16, 1904.
27. *Durango Democrat*, January 7, 1904. The *Democrat* was no longer run by Dave Day. Thomas H. Tulley had replaced him.
28. *Examiner*, January 16, 1904.
29. Hill to Peabody, January 11, 1904 (CNG Papers, CSA); *Journal*, January 7, 1904.
30. *Journal*, January 7, 1904.
31. *Journal*, January 12, 1904.
32. Gamble to Peabody, January 21, 1904 (CNG Papers, CSA).
33. Both the *Examiner* and *Journal* reported in several issues these unusually cold temperatures for January.
34. *Denver Post*, January 15, 1904.
35. *Examiner*, January 23, 1904.
36. *Denver Post*, January 15, 1904.
37. *Pueblo Chieftain*, January 15, 1904.
38. Since a number of correspondents were in Telluride, several newspapers printed this story throughout the state, capturing the injustice and agony that was being repeated weekly. See *Denver Post*, January 15, 1904; *Rocky Mountain News*, January 15, 1904; *Pueblo Chieftain*, January 15, 1904; *Durango Democrat*, January 17, 1904; *Examiner*, January 16 and 23, 1904.
39. *Journal*, January 19, 1904.
40. *Examiner*, January 23, 1904.
41. *Durango Democrat*, January 7, 1904.
42. *Examiner*, February 6, 1904.
43. Hill to Peabody, January 8, 1904 (CNG Papers, CSA).
44. Peabody to Hill, January 11, 1904 (Peabody Papers, CSA).

45. *Labor Disturbances*, 199.
46. *Journal*, January 14, 1904.
47. National Guard Reports, Court Martial records, Camp Telluride, 1903-04 (CNG Papers, CSA); *Journal*, January 25, 1904; *Examiner*, February 13, 1904.
48. Court Martial Records, Telluride campaign 1904 (CNG Papers, CSA).
49. *Examiner*, February 13, 20, and July 2, 1904.
50. *Journal*, February 11, 1904.
51. *Journal*, February 11, 1904; *Examiner*, February 18, 1904.
52. Biographical information on Bulkeley Wells comes from Box 10691, FF5 (CNG Papers, CSA); Stone, *History of Colorado*, Vol. 2, 112-114; E. B. Adams, "My Association With a Glamorous Man, Bulkeley Wells," privately published MS 1961, from Alta Cassietto (this is also at AHC-UW); Gressley (Editor), *Bostonians and Bullion, the Journal of Robert Livermore 1892-1915*; Wenger, "My Recollections of Robert Livermore and events in his Journal," Robert Livermore Papers (AHC-UW); Marianne Wells Glaser (Wells's daughter from second marriage).
53. These epithets and incidents were recurrent throughout the *Journal*, February-July, 1904. The threat against Sumner was repeated February 25, 1904.
54. All quotes in this paragraph from *Journal*, February 14, 1904.
55. *Journal*, February 18, 1904.

CHAPTER 20: **MOB RULE** (Notes for pages 234-249)

1. This paragraph and the one preceding from a lengthy letter Riddell wrote to James Kirwan, April 28, 1907 (Jencks, CU Archives).
2. Ibid. Riddell was accused of being a Pinkerton spy several times, managing to escape the charges after an investigation by Moyer in February 1904, but was not allowed to attend WFM meetings for two years afterward. He continued "faithful" to the WFM, working his way into the Eureka, Utah local, where he became president. He was Pinkerton Manager James McParland's Operative No. 36 and was not kicked out of the union until 1907 when he was exposed. Union men never harmed him.
3. In December 1903 strike relief funds for Telluride were $5,000, but as the men abandoned the cause, funds were reduced to $1,000 by the end of February 1904; *WFM Proceedings*, Twelfth Annual Convention, 1904; financial report.
4. Vincent St. John was kept in name only as president of Local 63 through spring of 1904, as a symbol of hope. Few members actually expected him to return.
5. *Rocky Mountain News*, March 18, 1904.
6. *Examiner*, February 27, 1904; *Journal* February 25, 1904; *Miners' Magazine*, March 10, 1904.
7. *Journal*, February 25, 1904.
8. Ibid.
9. *Examiner*, February 27, 1904.
10. *Journal*, March 3, 1904.
11. *Journal*, February 26, 1904.
12. *Examiner*, February 27, 1904.
13. Fred Chaffin, age 60, a miner who once served in the Union Army, wrote to Bell after being repeatedly deported for no crime; Chaffin to Bell, April 4, 1904 (CNG Papers, CSA).
14. Burtis M. Nutting inquest record, 1904 (Idarado Papers).
15. A. A. Pratt, affidavit, March 3, 1904, as published in Miller, "The Telluride Strike," in Langdon, *The Cripple Creek Strike, Industrial Wars*, 277-278; *Examiner*, March 5, 1904; *Miners' Magazine*, March 10, 1904.
16. *Examiner*, March 5, 1904.
17. Andrew Floaten in *Miners' Magazine*, March 10, 1904.
18. *State of Idaho v. William D. Haywood*, Henry Maki testimony, trial transcripts, Vol. 6, pp. 3288-3298 (Jencks, CU Archives).
19. Ibid.
20. Ibid. Witness accounts also from *Miners' Magazine*, March 10, 1904.
21. *State of Idaho v. William D. Haywood*, Maki testimony (Jencks, CU Archives); Miners' Magazine, March 10, 1904; Examiner, March 5, 1904.
22. *Miners' Magazine*, March 10, 1904.
23. *Examiner*, March 5, 1904.
24. *Denver Post*, October 22, 1901.
25. *Examiner*, March 5, 1904; *Miners' Magazine*, March 10, 1904.
26. *Miners' Magazine*, March 10, 1904.
27. *Miners' Magazine*, March 10, 1904, *Examiner*, March 5 and 12, 1904; *Journal*, March 7, 1904; Miller, "The Telluride Strike," in Langdon, *The Cripple Creek Strike, Industrial Wars*, 277-278.
28. *Miners' Magazine*, March 10, 1904.

29. *Miners' Magazine*, March 10, 1904, *Examiner*, March 5 and 12, 1904; *Journal*, March 7, 1904; Miller, "The Telluride Strike" in Langdon, *The Cripple Creek Strike, Industrial Wars*, 277-278.
30. *Examiner*, March 12, 1904.
31. *Journal*, March 7, 1904.
32. Every detail of the assault was recorded in the *Journal*, March 7, 1904 and *Examiner*, March 12, 1904, as well as *Miners' Magazine*, March 10, 1904.
33. *Journal*, March 7, 1904.
34. *Journal*, March 7, 1904; *Examiner*, March 12, 1904; *Miners' Magazine*, March 10, 1904.
35. *Examiner*, March 12, 1904.
36. Ibid.
37. Ibid.
38. Ibid.
39. *Examiner*, March 19, 1904.
40. Kinikin, "Early Days in Telluride."
41. The details of the assault on Andrew Floaten are from his own account, which appeared in several newspapers in the following days, and was included verbatim in Miller, "The Telluride Strike," as published in Langdon, *The Cripple Creek Strike, Industrial Wars*, 280-282.
42. *Examiner*, March 19, 1904.
43. Ibid.
44. *Labor Disturbances*, 201; *Examiner*, March 19, 1904; *Rocky Mountain News*, March 17, 1904; *Miners' Magazine*, March 24 and 31, 1904; Miller, "The Telluride Strike," as published in Langdon, *The Cripple Creek Strike, Industrial Wars*; *Journal*, March 15, 1904.
45. *Rocky Mountain News*, March 17, 1904.
46. Ibid.
47. Murphy to Peabody, March 18, 1904, printed verbatim in Langdon, *The Cripple Creek Strike, Industrial Wars*, 282.
48. *Rocky Mountain News*, March 19, 1904.
49. *Miners' Magazine*, March 31, 1904.
50. *Rocky Mountain News*, March 23, 1904.
51. *Labor Disturbances*, 201.
52. The finest example of these editorials was *Rocky Mountain News*, March 20, 1904.
53. *Journal*, March 15, 16, 17, 19, and 24, 1904.
54. *Journal*, March 19, 1904.
55. Ibid.
56. Edwin Thomas's testimony, *People of Colorado v. Vincent St. John*, preliminary hearing, March 21, 1906; *Journal*, March 22, 1906.
57. *Examiner*, March 19, 1904.
58. *Journal*, March 31, 1904.
59. Stewart Forbes, "Notes From Telluride," *Miners' Magazine*, March 10, 1904.
60. *Rocky Mountain News*, March 17, 1904; *Miners' Magazine*, March 31, 1904; Langdon, *The Cripple Creek Strike, Industrial Wars*, 282-287.
61. *Journal*, March 24, 1904.
62. *Miners' Magazine*, March 31, 1904.
63. Petition printed verbatim, *Labor Disturbances*, 202.
64. *Labor Disturbances*, 203.
65. *Rocky Mountain News*, March 24, 1904.
66. *Labor Disturbances*, 202-203.
67. Bell to Wells, military order, March 23, 1904 (CNG Papers, CSA).
68. *Journal*, March 24, 1904.
69. Langdon, The Cripple Creek Strike, Industrial Wars, 284; *Miners' Magazine*, March 31, 1904.
70. *Rocky Mountain News*, March 17, 1904.
71. *Rocky Mountain News*, March 24, 1904.
72. *Miners' Magazine*, March 31, 1904.
73. Peabody to Bell, telegram, March 25, 1904 (Peabody Papers, CSA).

CHAPTER 21: **THE GOOSE IS COOKED** (Notes for pages 250-266)

1. *Labor Disturbances*, 204; CNG Initial Report, dated March 26, 1904 (CNG Papers, CSA); *Biennial Report*, "Report of Telluride Expedition No. 2," 202; The initial report dated this incident on March 26, although the *Biennial Report* puts it on the 25th, which is incorrect, as Gen. Bell was in the actual process of arriving on the 25th at 6:40, his staff, troops, horses, etc. being unloaded at the depot most of that evening.
2. *Biennial Report*, "Report of Telluride Expedition No. 2," 202.
3. *Miners' Magazine*, March 31, 1904.

4. *Denver Post*, December 6, 1902
5. Court Martial records (CNG Papers, CSA).
6. *Journal*, March 28, 1904.
7. *Labor Disturbances*, 204; *Journal*, March 31, 1904; *Miners' Magazine*, March 31, 1904.
8. *Labor Disturbances*, 204-205; *Journal*, March 28 and 31, 1904.
9. *Rocky Mountain News*, March 27, 1904; *Labor Disturbances*, 204.
10. *Labor Disturbances*, 204.
11. *Rocky Mountain News*, March 28, 1904; *Examiner*, April 4, 1904; *Journal*, March 31, 1904; *Labor Disturbances*, 229-230.
12. *Miners' Magazine*, March 31, 1904.
13. *Biennial Report*, "Report of Telluride Expedition No. 2," 204-205.
14. This and the preceding quotation from *Labor Disturbances*, 231.
15. *Journal*, April 7, 1904.
16. *Examiner*, April 2, 1904; *Journal*, April 2, 1904.
17. Wenger, "My Recollections of Robert Livermore and events in his Journal," Robert Livermore Papers (AHC-UW).
18. *Journal*, April 7, 1904; *Examiner* April 9, 1904; *People of Colorado v. Robert D. Meldrum*, Case #618, District Court Record, Book 19, pp. 274, 277-278 (SMCC).
19. *Examiner*, April 9, 1904.
20. *Labor Disturbances*, 202.
21. *Journal*, July 7, 1904.
22. Wenger, "My Recollections of Robert Livermore and events in his Journal," Robert Livermore Papers (AHC-UW). Wenger was one of those high school cadets and he could not "understand just why we were called into action especially with unloaded guns."
23. *Journal*, April 9, 1904; *Examiner*, April 9, 1904.
24. "My Recollections of Robert Livermore and events in his Journal," Robert Livermore Papers (AHC-UW). Wenger was an eyewitness to this.
25. *Journal*, April 9, 1904.
26. "My Recollections of Robert Livermore and events in his Journal," Robert Livermore Papers (AHC-UW).
27. *Journal*, April 9, 1904.
28. *Labor Disturbances*, 205; Police Docket book (SMCC).
29. For the lengthy Moyer habeas corpus case, see *Labor Disturbances*, 229-246; *Anarchy in Colorado*, 78-79 and 92-99.
30. *Labor Disturbances*, 243-245; *Anarchy in Colorado*, 98-99.
31. *Journal*, April 28, 1904; *Examiner*, April 30, 1904.
32. *Journal*, May 5, 1904.
33. *Journal*, April 28, 1904; *Examiner*, April 30, 1904.
34. *Examiner*, April 16, 1904.
35. Charles Lewis, May 1978; Langdon, *The Cripple Creek Strike, Industrial Wars*, 287; *Labor Disturbances*, 206, 287-288; *Examiner*, June 18, 1904.
36. *Examiner*, April 30, 1904.
37. *Journal*, May 5 and 19, 1904.
38. *Journal*, May 9, 1904.
39. *Examiner*, May 14, 1904.
40. Judge Stevens's ruling, District Court Record, May 10, 1904, Book 19 (SMCC); *Labor Disturbances*, 206.
41. *Examiner*, May 14, 1904.
42. Ibid.
43. *Journal*, May 12, 1904.
44. *Examiner*, May 14, 1904.
45. *WFM Proceedings*, Twelfth Annual Convention, 1904; 198-205.
46. *Labor Disturbances*, 206.
47. From Harry Floaten's own account as published in Langdon, *The Cripple Creek Strike, Industrial Wars*, 289.
48. Hogg to Peabody, June 6, 1904 (Peabody Papers, CSA).
49. *Labor Disturbances*; Langdon, *The Cripple Creek Strike, Industrial Wars*; *Journal*, June 9, 1904; *Examiner*, June 11, 1904.
50. *Examiner*, June 18, 1904.
51. *Examiner*, June 25, 1904.
52. *Examiner*, July 2, 1904.
53. Floaten, in Langdon, *The Cripple Creek Strike, Industrial Wars*, 289.
54. *Industrial Union Bulletin*, April 20, 1907.

55. *Examiner*, June 18, July 9 and 16, 1904; *Labor Disturbances*, 288; charges brought against Floaten, November 23, 1907, were later dropped.
56. All cases stemming from the 1902 grand jury indictments and illegally refiled by Howe were closed with this statement, District Court Record, November 18, 1904, Book 19 (SMCC).
57. Wells's own statement published in *Examiner*, July 2, 1904, indicated this low-grade property was not putting out the dividends to satisfy the stockholders, thus he had to blame someone for the decision to shut down.
58. *Examiner*, July 16 and 23, 1904.
59. *Examiner*, May 21, 1904; Benson wrote under the penname "Ben Ballou."
60. *Examiner*, July 23, 1904; Miller, "The Telluride Strike" in Langdon, *The Cripple Creek Strike, Industrial Wars*, 290.
61. Miller, "The Telluride Strike" in Langdon, *The Cripple Creek Strike, Industrial Wars*, 290-291.
62. *Labor Disturbances*, 289-290.
63. *Labor Disturbances*, 290.
64. *Labor Disturbances*, 292; *Examiner*, November 19, 1904.
65. Miller, "The Telluride Strike," in Langdon, *The Cripple Creek Strike, Industrial Wars*, 292-293.
66. Miller, "The Telluride Strike," in Langdon, *The Cripple Creek Strike, Industrial Wars*, 293; *Examiner*, December 3, 1904.
67. *Examiner*, December 3, 1904.
68. For details of how the card system worked throughout the MOA-run mining companies, as blacklisting of WFM members, see *Labor Disturbances*, 224-228.
69. *Denver Times*, November 26, 1904.

CHAPTER 22: **THE FIRE-EATER** (Notes for pages 267-279)

1. The state auditor's report of November 1904 from *Labor Disturbances*, 360.
2. "Gov. Peabody to the Voters, The Colorado Situation Discussed and Misstatements Refuted," pamphlet, 1904 (from Doris Gregory).
3. *Examiner*, August 6, 1904.
4. Jason Brockman, "Biography of Alva Adams," printed sheet (CSA).
5. Siringo, *Two Evil Isms*.
6. Langdon, *The Cripple Creek Strike, Industrial Wars*, 409-423; Suggs, *Colorado's War on Militant Unionism*, 186-187.
7. Alva Adams's statement of March 17, 1905, printed verbatim in Langdon, *The Cripple Creek Strike, Industrial Wars*, 423-427.
8. Red Book (*Criminal Record of the Western Federation of Miners*), 14-15.
9. Ballard's account, *Durango Democrat*, July 5, 1901; *People of Colorado v. Vincent St. John*, Case #686, District Court Record, preliminary hearing, March 21, 1906 (SMCC); *Journal*, March 22, 1906.
10. Red Book, 15.
11. *People of Colorado v. Vincent St. John*, Case #686, District Court Record, preliminary hearing, March 21, 1906 (SMCC); *Journal*, March 22, 1906.
12. Red Book, 16.
13. McParland to Gooding, April 6, 1906 (PNDAC, CHS).
14. Red Book, 16.
15. *Samuel Servis v. Elizabeth H. Servis*, Case # 987, August 6, 1901, County Court Record, Book 7, p. 501 (SMCC); Case #987, September 28, 1901, p. 517; San Miguel County Fee Book #2, September 1901, Samuel Servis paid docket fee, revenue, and "misc. for service, etc."
16. *People of Colorado v. Harry L. Servis*, Cases #591, November 11, 1903, District Court Record, Book 19, p. 225, (SMCC).
17. Red Book, 17.
18. Ibid.
19. *Daily Sentinel*, July 13, 1908, from testimony in the Steve Adams case.
20. *Mary A. Mahoney v. John Mahoney*, Case #1088, June 8, 1903, County Court Record, Book 8, pp. 243-244 (SMCC).
21. Mary Mahoney became McParland's enemy when she plotted with Adams's wife during the time Adams recanted the confession McParland had fashioned for him. See *Big Trouble*. McParland's disgust with the woman was well documented in his letters to Gooding and others in 1906-07 (PNDAC, CHS).
22. The booklet was published by the *Miners' Magazine*.
23. *Labor Disturbances*, 7-34.
24. *Journal*, May 18, 1905.
25. *Journal*, October 20, 1905; *Examiner*, October 21, 1905; Frederick N. MacLean, unpublished notes, (Frederick N. MacLean file, AHC-UW).
26. *Journal*, May 18, 1905.

27. Adams, "My Association With A Glamorous Man, Bulkeley Wells."
28. Contract between E. L. Oakes and Bulkeley Wells, March 27, 1905 (Idarado Papers). This contract was for two years, Oakes having to abandon his bid for county attorney as part of the agreement.
29. Wells to Bell, March 17, 1905 (CNG Papers, CSA).
30. As Adjutant General, Wells was a main player in the plot to kidnap the WFM leaders from Denver in order to avoid extradition, taking them to Idaho to face trial on charges of conspiracy to murder Steunenberg. See *Big Trouble*.
31. Polly Pry in the *Denver Post*, reprinted in the *Journal*, May 18, 1905.
32. Ibid.
33. Numerous brief accounts of Wells's hunting trips peppered both the *Journal* and the *Examiner* from 1903 on.
34. *Big Trouble* thoroughly analyzes and details the case of Harry Orchard, thus only those parts of Orchard's confession and information that have bearing on the Telluride-related cases will be mentioned.
35. McParland's reports to Gov. Gooding and his statements to the press consistently illustrated he believed "Divine Providence" was working through him.
36. McParland to Gooding, February 17 and 18, 1906 (PNDAC, CHS).
37. *Miners' Magazine*, September 27, 1906; St. John's presentation in Cincinnati from the *Cincinnati Post*, July 30, 1907; Langdon, *Labor's Greatest Conflicts*; Flynn, notes (IWW 130, Wayne State).
38. In McParland's reports to Gooding and his statements to the press, he presented St. John as vermin that deserved to hang. See reports, February-October, 1906. In an effort to keep the WFM cases secret, McParland developed a "Cipher Code," assigning what he considered derogatory code names to all the WFM members. Haywood was "Viper," Moyer was "Copperhead," Pettibone was "Rattler," and St. John was "Coyote." Other WFM members were "Jackal," "Buzzard," "Skunk," "Weasel," etc. McParland gave himself the code name, "Owl," the wise one; McParland to Hawley, February 8, 1906 (PNDAC, CHS).
39. *Industrial Union Bulletin*, April 20, 1907.
40. *Journal*, February 19, 1906.
41. From Steve Adams's confession as read before the court in *State of Idaho v. Stephen W. Adams*, District Court, Kootenai County, Idaho, 1907, p. 876 (CU Archives).
42. McParland to Gooding, March 8, 1906 (PNDAC, CHS).
43. McParland to Gooding, February 5, 1906 (PNDAC, CHS).
44. McParland to Gooding, February 18, 1906 (PNDAC, CHS), spells out the plot to get St. John and Carpenter on the Barney and Collins charges, before Adams ever "confessed."
45. McParland to Gooding, June 4, 1906 (PNDAC, CHS).
46. Steve Adams's confession as summarized by Idaho prosecutor James H. Hawley in 1906, reprinted in *Idaho Yesterdays*, Winter Issue 1963-64, "Steve Adams' Confession and the State's Case Against Bill Haywood."
47. *People of Colorado v. Vincent St. John*, preliminary hearing, March 21, 1906; *Hilton, Motion to Quash* (Jencks, CU Archives).
48. *Rocky Mountain News*, February 28, 1906.
49. Mrs. Mary C. St. John to Vice President Williams, March 8, 1906, as recorded in the *WFM Proceedings*, Fourteenth Annual Convention, Denver, May 28-June 13, 1906; Acting President Williams's report, 26.
50. *WFM Proceedings*, Thirteenth Annual Convention, Salt Lake City, May 22-June 9, 1905; Flynn, notes (IWW 130, Wayne State).
51. Mrs. St. John to Williams, March 8, 1906 *WFM Proceedings*, Fourteenth Annual Convention, Denver, May 28-June 13, 1906; Williams's report, 26.
52. Ibid. Williams to Mrs. St. John.
53. McParland to Gooding, March 12, 1906 (PNDAC, CHS); *Journal*, March 19, 1906.
54. All quotes in this and the preceding paragraph from *Daily Sentinel*, March 17, 1906.
55. Stone, *History of Colorado*, vol. 2, p. 287.
56. *People of Colorado v. Vincent St. John, Hilton, Motion to Quash* (Jencks, CU Archives).
57. In the short time between a 1904 photo of Rutan and a 1906 photo, he appeared in the latter in extreme poor health, his face bloated, his eyelids drooping, his eyes oddly glassy and puffy. He was known to have rheumatism and to be a heavy drinker. At the end of 1903 and thereafter, many of the deported men filed complaints of false imprisonment against him. Some sought damages of $10,000 because of business losses resulting from the arrests. Rutan was unable to shoulder the stress. He died in delirium in a Denver hospital, September 25, 1907.
58. *Durango Democrat*, May 2, 1906.
59. *Miners' Magazine*, June 14, 1906.
60. *Journal*, May 10, 1906.

61. All quotes in this and the four preceding paragraphs are from *People of Colorado v. Vincent St. John, Hilton, Motion to Quash* (Jencks, CU Archives).
62. *Daily Sentinel*, September 11, 1906.
63. *Rocky Mountain News*, February 28, 1906.

Chapter 23: GENERAL WELLS WANTS A HANGING (Notes for pages 280-292)

1. For a comprehensive history of the IWW, see Renshaw, *The Wobblies*; Dubofsky, *We Shall Be All*; and *Heritage of Conflict*.
2. The Socialist Labor Party was a different political entity than the Socialist Party of America.
3. *Heritage of Conflict*, 160-196; Renshaw, *The Wobblies*, 50-73.
4. St. John to Editor John M. O'Neill of the *Miners' Magazine*, October 30, 1906, published in *Miners' Magazine*, November 8, 1906.
5. Ibid.
6. For a detailed study of the WFM's break from the IWW, and Moyer's and Haywood's falling out, see *Heritage of Conflict, The Wobblies,* and *Big Trouble*. This was also played out in the issues of *Miners' Magazine*.
7. *Industrial Union Bulletin*, March 16, 1907, from St. John's speech at San Francisco, March 3, 1907, to a full-capacity crowd at the Walton Pavilion.
8. Vincent St. John, *The I.W.W., Its History, Structure, and Methods* (IWW 130, Wayne State).
9. *Industrial Union Bulletin*, May 4-July 20, 1907; Goldfield News, April 27, November 9 and 16, 1907; Pinkerton Operative reports, April 29, 1907 (PNDAC, CHS).
10. *Goldfield News*, November 9, 1907.
11. *Industrial Union Bulletin*, August 8 and 10, 1907; *Cincinnati Post*, July 30, 1907; Renshaw, *The Wobblies*.
12. *Industrial Union Bulletin*, August 8 and 10, 1907; *Cincinnati Post*, July 30, 1907.
13. Document as reproduced in *Miners' Magazine*, September 13, 1906.
14. *Industrial Union Bulletin*, June 22, 1907.
15. A detailed history of Warren Lillard was reported in the *Daily Sentinel*, June 29, 1908. In *Big Trouble*, Lukas identified Lillard as Annie's uncle, but he was clearly Steve's uncle.
16. Steve Adams's statement was reproduced in full in *Miners' Magazine*, September 13, 1906.
17. *State of Idaho v. Stephen W. Adams*, transcripts of trial, pp. 1038-1040; (CU Archives).
18. *Industrial Union Bulletin*, May 18, 1907.
19. From Idaho state records of the expenses of the various WFM trials, as reproduced in *Miners' Magazine*, September and October 1907.
20. Wells to F. W. Bradley, May 23, 1907 (PNDAC, CHS).
21. McParland to Gooding, September 12, 1906 (PNDAC, CHS).
22. The "McPartland" spelling was neither from Adams nor the court record, but from the newspaper in which this was reported, *Industrial Union Bulletin*, March 9, 1907.
23. *Industrial Union Bulletin*, March 9, 1907.
24. *State of Idaho v. Stephen W. Adams*, District Court, Kootenai County, Idaho, 1907, p. 809 (CU Archives); *Industrial Union Bulletin*, March 23, 1907; *Miners' Magazine*, September 13, 1907.
25. *Journal*, May 16, 1908, reprinted in the *Daily Sentinel*, May 18, 1908.
26. San Miguel County Sheriff's Office, Jail record.
27. *Rocky Mountain News*, September 24 and 25, 1907; medical record courtesy C. H. McVay.
28. *State of Idaho v. William D. Haywood*, 1907, transcripts, testimony of Cal Rutan, Vol. 13, pp. 4561-4581 (Jencks, CU Archives).
29. *Journal*, March 28, 1908.
30. This is Livermore's recollection from *Bostonians and Bullion*, 149.
31. *Journal*, March 28, 1908.
32. Ibid. The report of the bomb was in the *Journal*, March 28, 1908; follow-up reports and photos appeared March 30-April 16, 1908.
33. Explosives experts James Crippin, Director, Western Forensic Law Enforcement Training Center, Pueblo, Colorado; John R. Trujillo, Consulting Mining Geologist and mine safety consultant; P. Dan Martin, former Army explosives engineer; said the damage in the photos was too minimal for even one stick of dynamite and appeared to be staged. None of these experts believed anyone was in the bed when the blast occurred. They were given detailed descriptions of the damage as well as photos taken at the scene in 1908. Each came to the same conclusion independent of the others. Crippin stated: "I think the manager staged the incident for some reason. Even a single stick of dynamite should have completely destroyed that room and severely injured or killed whoever was in it at the time. The damage is wrong. Explosions take the path of least resistance. If all the force was directed downward into the floor, it should have completely destroyed the room downstairs. Why was he blown just to the ceiling? There should have been marks on the ceiling where he impacted. The gun

is broken, and he isn't, so to speak. Unlikely." (September 8, 2003) Documentary notes, photos, and correspondence in possession of the author.
34. *Journal*, March 28, 1908.
35. Wenger, "My Recollections of Robert Livermore and events in his Journal," Robert Livermore Papers (AHC-UW).
36. *Journal*, April 2, 1908.
37. Ibid.
38. *Daily Sentinel*, June 22, 1908.
39. McParland to Gooding, March 8, 1907 (PNDAC, CHS). McParland hated Mrs. Mahoney because the feisty woman had plotted against him, bolstered both Annie and Steve Adams, and stuck with them as a fierce ally that got in McParland's hair so badly that he started looking for the means to have her arrested for something-anything.
40. Certificate of Marriage between Stephen W. Adams and Mrs. Anna G. Dickson, San Miguel County Marriage Records, November 15, 1903, (SMCC). Witnesses were Mrs. Mary A. Mahoney and Miners' Union member A. J. Tyner.
41. *Daily Sentinel*, July 6, 1908.
42. Ibid.
43. *Daily Sentinel*, June 20, 1908.
44. *State of Idaho v. Stephen W. Adams*, District Court, Kootenai County, Idaho, 1907, p. 876 (CU Archives).
45. *People of Colorado v. Steve Adams*, Case #1875, District Court record, Mesa County; trial testimony from the *Daily Sentinel*, Telluride *Daily Journal, Denver Post, Denver Republican, Rocky Mountain News*, June 22-July 18, 1908.
46. *Daily Sentinel*, July 8, 1908.
47. McParland to Hawley, May 19, 1906 (PNDAC, CHS).
48. Adams's confession as summarized by Hawley in April, 13, 1906, reprinted in *Idaho Yesterdays*, "Steve Adams' Confession and the State's Case Against Bill Haywood."
49. *Daily Sentinel*, July 10, 1908.

CHAPTER 24: **IF BONES COULD SPEAK** (Notes for pages 293-305)
1. *Daily Sentinel*, July 9, 1908. Court testimony from *People of Colorado v. Steve Adams*, Case #1875; *Daily Sentinel, Journal, Denver Post, Denver Republican, Rocky Mountain News*, June 22-July 18, 1908.
2. *Daily Sentinel*, July 9, 1908.
3. *Daily Sentinel*, June 29, 1908; *State of Idaho v. Stephen W. Adams*, District Court, Kootenai County, Idaho, 1907 (CU Archives).
4. *Biennial Report* "Report of Telluride Expedition No. 2", 139 and 175.
5. BLM Homestead patent, Stephen W. Adams, Serial No. ORLGAA 097484. Steve assigned this to his brother Joseph, January 11, 1909.
6. *State of Idaho v. Stephen W. Adams*, testimony of Annie Adams, p. 1991 (CU Archives).
7. *Daily Sentinel*, July 10, 1908.
8. McParland to Gooding, February 5, 1906; McParland to Gooding, March 8, 1907 (PNDAC, CHS); *The Pinkerton Labor Spy*, 52-64. These are only a few examples of McParland's scripting exactly what should be said to certain persons by others for particular results. He even scripted a conversation for a sheriff to use on Moyer to get him to turn against Haywood: the sheriff was supposed to tell Moyer of a plot where Haywood hired Adams to bump off Moyer (entirely falsified information); McParland to Gooding, December 15, 1906 (PNDAC, CHS).
9. Adams's confession as summarized by Hawley in April, 13, 1906, reprinted in *Idaho Yesterdays*, "Steve Adams' Confession and the State's Case Against Bill Haywood."
10. *Daily Sentinel*, July 10, 1908.
11. *Denver Republican*, November 20, 1902.
12. *Daily Sentinel*, July 10, 1908.
13. This information also comes from the original newspaper reports of the shooting, November 1902.
14. *Journal*, July 11, 1908.
15. *Daily Sentinel*, July 9, 1908; *Journal*, July 11, 1908.
16. *Daily Sentinel*, July 10, 1908.
17. Ibid.
18. *Examiner*, April 2, 9, and 23, 1904; Kinikin, "Early Days in Telluride."
19. *Journal*, March 31, 1908.
20. *Daily Sentinel*, July 10-11, 1908.
21. *Journal*, July 11, 1908.
22. Ibid.
23. *Journal*, July 14, 1908.
24. *Daily Sentinel*, July 13, 1908.

25. All quotes in this and the preceding five paragraphs from *Daily Sentinel*, July 13, 1908.
26. *Daily Sentinel*, July 14, 1908.
27. Ibid.
28. Grand jury record, December 15, 1902, District Court Record, Book 19, (SMCC).
29. *Daily Sentinel*, July 14, 1908.
30. Ibid.
31. Ibid.
32. *Daily Sentinel*, July 15, 1908.
33. Ibid.
34. Ibid.
35. Ibid.
36. Experiments with both a Winchester Model 1897 pump-action shotgun with a sawed-off barrel (20 inches) and no choke, and Winchester with a full-length barrel and full-choke, proved the shot pattern in the Collins window had to be made at twenty feet from the window with the sawed-off shotgun. Shotgun cartridges used were 12-gauge, 12-pellet loads. The full-length shotgun made a foot-wide pattern at 70 feet, which was too wide, and the wad didn't hit the glass to create the thirteenth hole. The sawed-off shotgun at 70 feet made a pattern 18 inches or more. The sawed-off shotgun at 20 feet matched the Collins details, while the other shotgun at 20 feet made a pattern so small the pellet holes could not be counted in the glass. Experiments were conducted by the author, Peter P. Martin, Carl V. Button, Jr., and in consultation with Paul Miller of the San Juan Rifle Range and Gary Koverman, retired Colorado Bureau of Investigation agent. The experiments replicated the Collins murder scene, with the shooter firing through windows, oak chairs, and the target in the chair; from positions directly in line with the window and at an angle up a hill.
37. *Daily Sentinel*, July 15, 1908.

CHAPTER 25: **SUNDOWN** (Notes for pages 306-325)

1. *Daily Sentinel*, July 15, 1908; *Denver Post*, July 15, 1908.
2. *Denver Republican*, July 17, 1908.
3. *Daily Sentinel*, July 15, 1908.
4. Suggs, *Colorado's War on Militant Unionism*, 189.
5. *Daily Sentinel*, July 15, 1908.
6. Hilton to Elizabeth Gurley Flynn, September 16, 1918 (IWW 130, Wayne State).
7. Boulder County probate records, Estate of Sarah Rannells, May 15, 1911, Book 337, pp. 326-327.
8. Ralph B. Barney, unpublished manuscript in possession of Ralph C. Barney.
9. Photograph in possession of C. H. Lamb in 1991. Lamb's family had been the closest friends of Ella Barney Johnson's family.
10. BLM Homestead patent, William Barney, Serial No. MTLTN 0005380.
11. *Miners' Magazine*, September 13, 1906.
12. Adams said this in court during his habeas corpus hearing. Friedman, *The Pinkerton Labor Spy*, 214.
13. McParland to Gooding, June 4, 1906 (PNDAC, CHS).
14. In the rubble of boarding house dumps, the author has found coats, stockings, belts, shoes, and tobacco pipes that had not rotted in 100 years at the same altitude.
15. Thomas M. Canfield, MD, PC, Forensic Pathologist and Chief Medical Examiner, Office of Medical Investigation Southwestern Region, Montrose, Colorado; interviews, August and November 2002, regarding the detailed description of the body supposed to be William J. Barney (descriptions from James McParland's reports, Stephen Adams's confession, court trial transcripts, and newspaper accounts).
16. McParland to Gooding, April 6, 1906 (PNDAC, CHS).
17. *Journal*, July 2, 1907. During Haywood trial testimony of Max Malich, Max said Gratias, whose spy status was unknown to him at the time, had introduced Harry Orchard to Max as "Tom Hogan" and Tom proceeded to try to get Max to blow up a hotel full of strikebreakers. Max denounced him, so Harry said he was just joking.
18. *State of Idaho v. William D. Haywood*, District Court, Ada County, Idaho, June-July 1907, testimony of Harry Orchard (CU Archives).
19. McParland disliked Orchard from the start, saying he had "the most cold, cruel eyes I remember having seen" (January 26, 1906 report to Gooding) and calling Orchard "one of the greatest criminals on earth" (August 14, 1907 letter to Borah). He found him egotistical and disgusting, and only pretended to be Orchard's friend in order "to use him." McParland's reports from January 1906 to August 1907 were peppered with such remarks (PNDAC, CHS).
20. *Journal*, November 26, 1903. Detectives in Cripple Creek traced the revolver used to trigger the Vindicator bomb to the San Juan Hardware store in Telluride. The gun had been bought a few weeks earlier. Later Orchard confessed to placing the bomb in the mine with a gun he had bought himself.
21. Grimmett, *Cabal of Death*, 268.

22. Friedman, *The Pinkerton Labor Spy*, 23.
23. Siringo, *Two Evil Isms*, "Introduction" by Charles D. Peavy, pp. iii-xi.
24. *Denver Post*, May 19, 1919.
25. C. R. Waters, report to WFM headquarters, October 14, 1909; Waters to Charles Moyer, October 11, 1909; San Juan District Union meeting, November 20, 1907; Transfer of deed to Moyer, December 1909 (Jencks, CU Archives).
26. *Goldfield News*, October 25, November 2 and 9, 1907; *Goldfield Daily Tribune*, November 6, 1907; *Industrial Union Bulletin*, November 2, 1907.
27. *Goldfield Daily Tribune*, November 7, 1907.
28. *Goldfield News*, November 9, 1907.
29. *Nevada Workman*, November 9, 1907. This periodical was the official organ of Goldfield WFM Local 220.
30. Resolutions of the McCabe Miners' Union, December 7, 1907, against the "unwarranted, unprovoked and murderous attack on Vincent St. John."
31. The details of the Mullaney assault on St. John are from the *Goldfield News*, November 9, 1907; *Goldfield Daily Tribune*, November 6-7, 1907; *Industrial Union Bulletin*, November 16-23, 1907; St. John's own statement in the *Nevada Workman*, November 9, 1907; St. John's report at the Fourth Annual Convention of the IWW, November 1908.
32. The MOA *Goldfield Gossip*, as reprinted verbatim in the *Industrial Union Bulletin*, December 14, 1907.
33. Trautmann, autobiography (typed MS), "Fifty Years War. Book #2 1905 to 1920 The Rise and Fall of the Industrial Workers of the World," William Trautmann Collection I.D. #1660, Box 1 (hereafter cited as Trautmann 1660) Wayne State.
34. St. John's report to the Fourth Annual Convention of the IWW, November 1908.
35. *Journal*, July 17, 1907.
36. Rockwell, "Gentleman of Fortune," *Denver Westerners' Monthly Roundup*, May 1966, 8.
37. Adams, "My Association With a Glamorous Man, Bulkeley Wells."
38. Ibid.
39. Ibid.
40. *Bostonians and Bullion*, 187.
41. *San Francisco Chronicle*, January 21, 1923; State of Utah Marriage Application and License No. 41419, for Wells and Schmidt; *San Francisco Examiner*, May 27, 1931; *Bostonians and Bullion*; Adams, "My Association With a Glamorous Man, Bulkeley Wells."
42. This is the actual text of the note taken from a photo of it in the *San Francisco Chronicle*, May 27, 1931. The news article itself printed the note slightly differently.
43. *San Francisco Chronicle*, May 27, 1931; *San Francisco Examiner*, May 27, 1931.
44. Renshaw, *The Wobblies*; Dubovsky, *We Shall Be All*.
45. Lens, *The Labor Wars*.
46. Foster, *Bryan to Stalin*, as quoted in the *Daily Worker*, July 4, 1938 (IWW 130, Wayne State).
47. Lens, *The Labor Wars*, 183.
48. McGuckin, *Memoirs of a Wobbly*, 6.
49. Trautmann 1660 (Wayne State).
50. Ibid. Trautmann wrote his autobiography in his late 60s or early 70s, thus his references to Hitler and Mussolini as having "a peer in Vincent St. John" were made long after St. John was dead. Trautmann was the first general secretary of the IWW and, during his tenure, he was accused of fraud and mis-handling of funds, but St. John defended him. In 1908, the delegates voted St. John to replace him, rankling Trautmann. By the time Trautmann wrote his autobiography, he reviled St. John as a "cracked-brain" maniac bent on revenge and engaged in every crime from forgery to murder, who sent his thugs after Trautmann "with intent to kill." Trautmann claimed the U.S. Department of Justice asked him to testify against St. John so they could send him to prison. No evidence remotely backs these claims. Trautmann clearly had an axe to grind.
51. Cannon, *The I.W.W.* (31-page booklet).
52. Ibid.
53. Ibid.
54. Renshaw, *The Wobblies*, 138.
55. Statement of Vincent St. John, 1918 (IWW 130, Wayne State).
56. Renshaw, *The Wobblies*, 169-194.
57. Statement of Vincent St. John (IWW 130, Wayne State).
58. George, *The I.W.W. Trial*, 179-180.
59. *The New York Times*, May 27, 1919.
60. Federal Bureau of Prisons, USP Leavenworth record of V. St. John, BOP Archives.
61. All quotes in this paragraph are from Vincent St. John, "Experience of a First-timer," unpublished MS (IWW 130, Wayne State).

62. Lanier to President Wilson, April 19, 1919 (IWW 130, Wayne State).
63. Hilton to Flynn, September 16, 1918 (IWW 130, Wayne State).
64. Statement of Vincent St. John (IWW 130, Wayne State).
65. "Caroline" to Flynn, August 31, 1921 (IWW 130, Wayne State).
66. St. John to Weinberger, October 25, 1921 (IWW 130, Wayne State).
67. Flynn's appeal circular sent out to unions and individuals, dated April 29, 1921 (IWW 130, Wayne State).
68. Official federal warrant releasing St. John from prison, signed by Warren G. Harding, June 23, 1922 (Department of Justice).
69. St. John to Flynn, June 6, 1929 (IWW 130, Wayne State).
70. Letters in Flynn's papers show St. John had support from across the nation (IWW 130, Wayne State).
71. Witthoft to Flynn, June 21, 1929 (IWW 130, Wayne State).
72. Ibid.
73. J. Underwood Hall, M.D. letter, June 26, 1929, report on his care during St. John's illness, June 17-21, 1929 (IWW 130, Wayne State).
74. Witthoft to Flynn, June 26, 1929 (IWW 130, Wayne State).
75. Ibid.
76. *Industrial Worker*, June 29, 1929.
77. Quotes in this paragraph and the one preceding from undated letters and news clippings in Flynn's file on St. John (IWW 130, Wayne State).
78. Flynn, June 21, 1954, a handwritten note she titled, "To Saint" (IWW 130, Wayne State).

BIBLIOGRAPHY

GOVERNMENT PUBLICATIONS:

UNITED STATES:

"Capital and Labor Employed in the Mining Industry." *Report of the Industrial Commission, 1901, Vol. XII.* Washington, D.C.: GPO, 1901.

Cross, W., and C. W. Purington. *Description of the Telluride quadrangle, Colorado.* U.S. Geological Survey Geologic Folio 57, 1899.

Henderson, C. W. *Mining in Colorado.* U.S. Geological Survey Professional Paper 138, 1926.

Lord, Eliot. *Comstock Mining and Miners.* Monographs of the U.S. Geological Survey, Vol. 4, Department of Interior. Washington,

D.C.: Government Printing Office (GPO), 1883.

Papers Relative to Labor Troubles at Goldfield, Nevada. U.S. Congress. House, 60th Congress, 1st Session, Document 607. Washington, D.C.: GPO, 1908.

Purington, Chester W. *Preliminary Report on the Mining Industries of the Telluride Quadrangle.* Eighteenth Annual Report of the U.S. Geological Survey, 1898.

"Report of the Federal Mediation Commission to the President of the United States, January 9, 1918." *United States Official Bulletin,* Vol. 11, Feb. 11, 1918.

U.S. Commission on Industrial Relations. *Final Report and Testimony on Industrial Relations, Vol. IV.* Senate, 64th Congress, 1st Session, Document 22. Washington, D.C.: GPO.

U.S. Congress. *Coeur d'Alene Labor Troubles.* Senate, 56th Congress, 1st Session, Document 142. Washington, D.C.: GPO.

U.S. Congress. *Review of the Labor Troubles in the Metalliferous Mines of the Rocky Mountain Region.* Senate, 58th Congress, 2nd Session, Document 86. Washington, D.C.: GPO, 1904.

U.S. Congress. *Statement of the Secretary of the Mine Owners and Operators' Association.* Senate, 58th Congress, 2nd Session, Document 86. Washington, D.C.: GPO, 1904.

U.S. Congress. *Statement of the Western Federation of Miners. Presented by Senator Thomas M. Patterson.* Senate, 58th Congress, 2nd Session, Document 163. Washington, D.C.: GPO, 1904.

U.S. Dept. of Commerce and Labor, Bureau of the Census, Special Reports, Religious Bodies: 1906, Washington, D.C.: GPO, 1910.

U.S. Government Census records 1870-1920, National Archives.

U.S. Government Homestead records, Bureau of Land Management.

U.S. Government Mine Patent records, Bureau of Land Management.

Wright, Carroll D., Commissioner of Labor. *A Report on Labor Disturbances in the State of Colorado, From 1880 to 1904, Inclusive, with Correspondence Relating Thereto.* U.S. Congress. Senate, 58th Congress, 3rd Session, Document 122. Washington, D.C.: GPO, 1905.

COLORADO:

Attorney General Nathan C. Miller. *Biennial Report of the Attorney General of Colorado, 1903-1904.* Denver: Smith Brooks Printing Company, 1904. Colorado State Archives (CSA).

Biennial Report of the Adjutant General of Colorado, 1903-1904. Denver: Smith-Brooks Printing Co., 1904. CSA.

Biennial Report of the Adjutant General of Colorado, 1905-1907, Denver: Smith-Brooks Printing Co., 1907. CSA.

Colorado General Assembly *Session Laws* (Denver, 1889-1905). CSA.

Colorado House and Senate *Journals* (Denver, 1903-1904). CSA.

Colorado National Guard Papers: Military Affairs, Office of the Adjutant General, Records of General Sherman Bell, Box 10644, and Telluride Campaign, Box 10643. CSA.

Colorado State Auditor, *Biennial Report, 1903-1904*, Denver: Smith-Brooks Printing Co., 1904.

Colorado State Bureau of Labor Statistics *Biennial Report[s]* Denver: Smith-Brooks Printing Co., 1894-1908.

Colorado State Penitentiary Records, CSA.

Colorado Census Records 1885.

Colorado Marriage Records.

Military records, Civil War Book O; Transcript of Records Third Regiment Cavalry, CSA.

Republican State Central Committee of Colorado. *The Facts as to the Moyer and Other Decisions of the Supreme Court of Colorado*. Denver, Oct. 8, 1906. CSA.

SPECIAL COLLECTIONS:

Bishop Nicholas C. Matz Papers, Records of the Chancery Office, Archdiocese of Denver, Roman Catholic Church.

George Collins Collection. Includes ledgers and minutes of the Colorado Mine Operators Association; clippings, telegrams on the death of Arthur L. Collins; family history. Denver Public Library, Western History and Genealogy (DPL/WH).

Elizabeth Gurley Flynn Collection, I.D. Number 1742, Archives of Labor and Urban Affairs, Wayne State University (Wayne State).

Governor Alva Adams Papers, 1887-1889, 1897-1899, 1905 (CSA).

Governor Jesse F. McDonald Papers, 1905-1907 (CSA).

Governor James H. Peabody Papers, 1903-1905 (CSA).

James H. Hawley Papers, Idaho State Historical Society Library.

Idarado Papers, Idarado Mining Company records include original Smuggler-Union Mining Company documents and contracts with Telluride Miners' Union, inquest records, some employee records. Idarado Mining Company, Ouray, Colorado.

Industrial Workers of the World Collection, I.D. Number 130 (Wayne State).

Clinton Jencks Collection, Records of the Western Federation of Miners, including Records of Telluride Local #63 and San Juan District Union; University of Colorado at Boulder, Archives (CU Archives).

Robert Livermore Papers, American Heritage Center, University of Wyoming (AHC-UW).

Frederick N. MacLean file (AHC-UW).

Kenneth A. McLean file (AHC-UW).

Pinkerton's National Detective Agency Collection MSS #1037, Stephen H. Hart Library, Colorado Historical Society.

C. W. "Doc" Shores Collection, typed MSS, memoirs, clippings (DPL/WH).

Smuggler-Union Mining Company file (AHC-UW).

William Trautmann Collection, I.D. Number 1660 (Wayne State).

Bulkeley Wells file (AHC-UW).

COURT DOCUMENTS:

In Re Eight-Hour Bill, 21 Colorado 29 (1895), Colorado State Archives.

W. J. Barney v. Sarah E. Barney, Case #975, April 4, 1902, County Court Record, Book 8, San Miguel County Courthouse (SMCC).

James J. Blazer v. 16 to 1 Miners Union No. 63, Case #569, County Court Record, Book 8; District Court Record, Book 18 (SMCC).

Grand Jury Report December 15, 1902, District Court Record, Book 19 (SMCC).

Tony Langeri v. J. C. Rutan, Sheriff, Case #1147, County Court Record, Book 8 (SMCC).

Tony Langeri v. J. C. Rutan, Sheriff, Case #657, December 19, 1903, District Court Record, Book 19 (SMCC).

Louis Macari v. J. C. Rutan, Sheriff, Case #658, December 19, 1903, District Court Record, Book 19 (SMCC).

Mary A. Mahoney v. John Mahoney, Case #1088, June 8, 1903, County Court Record (SMCC).

Martin Marchiano v. J. C. Rutan, Sheriff, Case #656, December 19, 1903, District Court Record, Book 19 (SMCC).

Guy E. Miller v. J. C. Rutan, Sheriff, Case #655, December 19, 1903, District Court Record, Book 19 (SMCC).

Moyer v. Peabody, et al. (148 Fed. Rep. 870) and *Moyer v. Peabody* (212 U.S. 78).

In Re Moyer: Brief and Argument for Respondents, Denver: Smith-Brooks Printing Co. 1904.

In Re Moyer: Brief and Oral Argument of Attorney General and the Opinion of the Court, Denver: Smith-Brooks Printing Co., 1904.

People of Colorado v. Steve Adams, Case #714, November 1907-June 1908, District Court Record, Book 20 (SMCC).

People of Colorado v. Steve Adams, Case #1875, June 10, 1908, District Court Record, Mesa County Courthouse.

People of Colorado v. William Jordan, Case #384, May 23-June 6, 1901, District Court Record, Book 18 (SMCC).

People of Colorado v. Walt Kenley, Case #516, December 4, 1902, District Court Record, Book 19 (SMCC).

People of Colorado v. Robert D. Meldrum, Case #618, San Miguel County, District Court Record, Book 19 (SMCC).

People of Colorado v. Joe Mengoni, Case #395, District Court Record (SMCC).

People of Colorado v. John Miller, June 5, 1902, District Court Record, Book 18 (SMCC).

People of Colorado v. James Roner, May 16, 1903, District Court Record (SMCC).

People of Colorado v. Vincent St. John, Case #586, 587, 603, May 18, 1903-July 1903; District Court Record, Book 19 (SMCC).

People of Colorado v. Vincent St. John, Case #5686, preliminary hearing, March 21, 1906, San Miguel County Court record. Argument of O. N. Hilton, "Motion to quash." Mesa County District Court, 1906.

People of Colorado v. Harry L. Servis, Cases #396, #591, November 1901-November 1903, District Court Record, Books 18 and 19 (SMCC).

People v. Kenneth McLean, Case #951, San Miguel County Court Record, Book 8 (SMCC).

People v. F. Pheasey, Case #954, San Miguel County Court Record, Book 8 (SMCC).

People v. Fred Pheasey and K. A. McLean, Case #429, May 1902, San Miguel County Court Record, Book 8 (SMCC).

People v. Charles A. Smith, Case #1017, San Miguel County Court Record, Book 8 (SMCC).

Sarah Rannells estate (died March 17, 1896), settled in 1911; *Decree of Final Settlement*, Decrees, Book 337, pp. 326-327, Boulder County Clerk and Recorder.

Samuel Servis v. Elizabeth H. Servis, Case #987, August 6, 1901, San Miguel County Court Record, Book 7 (SMCC).

State of Idaho v. Stephen W. Adams, District Court, Kootenai County, Idaho, 1907.

State of Idaho v. William D. Haywood, District Court, Ada County, Idaho, June-July 1907.

Naomi Sumner v. Charles G. Sumner, Case #1077, February 5, 1903, San Miguel County Court Record, San Miguel County Courthouse.

Norman Young v. J. C. Rutan, Sheriff, Case #1146, San Miguel County Court Record, Book 8 (SMCC).

BOOKS AND ARTICLES:

Adamic, Louis. *Dynamite: The Story of Class Violence in America*. New York: The Viking press, 1931.

Adams, E. B. *Gio Oberto of Telluride, Colorado*. Grand Junction: Colorado Printing Co., (no date).

—-. *My Association With a Glamorous Man, Bulkeley Wells.* Booklet, privately published, 1961. (American Heritage Center, University of Wyoming)

Adams, Graham, Jr. *Age of Industrial Violence, 1910-1915: The Activities and Findings of the United States Commission on Industrial Relations.* New York: Columbia University Press, 1966.

"An Address Setting Forth the Objectives of the Citizens' Alliance of Helena." (Pamphlet) Helena, Mont., 1903.

Adler, Stephen J. *The Jury: Trial and Error in the American Courtroom.* New York: Times Books, 1994.

Aiken, Katherine G. "It May Be Too Soon to Crow: Bunker Hill and Sullivan Company Efforts to Defeat the Miners' Union, 1890-1900" *Western Historical Quarterly* XX1, No.3, August 1993.

Atkinson, Rev. H. A. *The Church and Industrial Warfare. A Report on the Labor Troubles in Colorado and Michigan.* Prepared by Federal Council of the Churches of Christ in America, 1914.

Aurand, Harold W. *From the Molly Maguires to the United Mine Workers, 1869-1897.* Philadelphia: Temple University Press, 1971.

Backus, Harriet Fish. *Tomboy Bride.* Boulder: Pruett Publishing Co., 1977.

Barger, H., and S. H. Schurr. *The Mining Industries, 1899-1939.* New York: National Bureau of Economic Research, 1944.

Bartholomew, H. E. *Anarchy in Colorado, Who is to Blame?* Colorado: Bartholomew Publishing Co., 1905.

Belsey, George. *When to Telluride, to Helluride.* Unpublished manuscript, 1962.

Bonnett, Clarence E. *Employers' Associations in the United States: A Study of Typical Associations.* New York: Macmillan Company, 1922.

Brissenden, Paul F. *Launching of the Industrial Workers of the World.* Berkeley: University of California Press, 1913.

—-. *The I.W.W., A Study of American Syndicalism.* New York: Columbia University Press, 1919.

Broehl, Wayne G. *The Molly Maguires.* Cambridge: Harvard University Press, 1964.

Brooks, J. G. *American Syndicalism, The I.W.W.* New York: Macmillan Co., 1913.

Bruns, Roger A. *The Bandit Kings.* New York: Crown Publishers, 1995.

Buchanan, J. R. *The Story of a Labor Agitator.* New York: Outlook Co., 1903.

Cannon, James P. *America's Road to Socialism.* New York: Pioneer, 1953.

—-. *The I.W.W.* New York: Merit Publishers, 1967.

Clark, M. R. *Labor Movement in America.* New York: W. W. Norton and Co., 1938.

Colorado Families: A Territorial Heritage. Denver: Colorado Genealogical Society, Inc., 1981.

Colorado Wage Slave. *Behind the Scenes.* New York: Labor News Press, 1904.

Constantine, Robert, J., editor. *Letters of Eugene V. Debs, 1874-1912,* 3 vols. Urbana and Chicago: University of Illinois Press, 1990.

Cornelius, Olive Frazier. *Pioneer History and Reminiscences of the San Juan Basin.* Unpublished, 1933. Colorado State Historical Society.

Creelman, James. *On The Great Highway, The Wanderings and Adventures of a Special Correspondent.* Boston: Lothrup Publishing Company, 1901.

Criminal Record of the Western Federation of Miners - Coeur d'Alene to Cripple Creek-1894-1904. (Red Book) Colorado Springs: Colorado Mine Operators Association, 1904.

Crown, H. T., and Mark Major. *A Guide to the Mollie Maguires.* Pennsylvania: Kelly Press, 1995.

Crown, H. T. *A Molly Maguire On Trial: The Thomas Munley Story.* Frackville, PA: Broad Mountain Publishing Co., 2002.

Crum, Josie Moore. *The Rio Grande Southern Railroad.* Durango: San Juan History, Inc., 1961.

Cummins, James. *Jim Cummins' Book, Written by Himself,* Denver: Reed Publishing Company, 1903.

Darrow, Clarence. *The Story of My Life.* New York: Charles Scribner's Sons, 1934.

Davie, John. *CWA Interviews,* c.1934 Colorado State Historical Society.

Dawson, Thomas J. *Scrapbooks.* Colorado State Historical Society.

Debs, Eugene V. *Writings and Speeches of Eugene V. Debs.* New York: Hermitage Press, 1948.

Debs: His Life, Writings and Speeches, Girard, Kansas: The Appeal to Reason, 1908.

Derickson, Alan. *Workers' Health, Workers' Democracy: The Western Miners' Struggle, 1891-1925.* Ithaca: Cornell University Press, 1988.

Dos Passos, John. *Nineteen Nineteen (U.S.A.I. The 42nd Parallel. II. Nineteen Nineteen. III. The Big Money).* New York: Modern Library, 1939.

Dubofsky, Melvyn. *We Shall Be All.* Urbana and Chicago: University of Illinois Press, 2000.

Elliott, W. Y., Elizabeth S. May, J. W. F. Rowe, Alex Skelton, and Donald H. Wallace. *International Control in the Non-Ferrous Metals.* New York: Macmillan Co., 1937.

Ferril, William C. *Sketches of Colorado.* Denver: Western Press, 1911.

—-. Scrapbook (Denver Public Library, Western History/Genealogy DPL/WH)

Foner, Philip. *History of the Labor Movement in the United States.* New York: International Publishers, 1964.

Foster, William Z. *American Trade Unionism, Principles and Organization, Strategy and Tactics.* New York: International Publishers, 1947.

Friedman, Morris. *The Pinkerton Labor Spy.* New York: Wilshire Book Co., 1907.

Gabbert, Kingly Prerogative. Denver: Smith-Brooks Press, 1906.

George, Harrison. *The I.W.W. Trial, Story of the Greatest Trial in Labor's History by one of the Defendants.* Chicago: Industrial Workers of the World, 1918 (?).

Gibbons, Rev. James J. *In the San Juan Colorado, Sketches,* 1898. Reprinted by St. Patrick's Parish, Telluride, 1972.

Glazer, Joe. *The Labor Movement in Folklore.* (Sound recording) DeLand, Florida: Everett & Edwards, 1979.

Gompers, Samuel. *American Labor and the War.* New York: G. H. Doran, c.1919.

—-. *Seventy Years of Life and Labor.* New York: E. P. Dutton and Co., 1925.

Gressley, Gene M., editor. *Bostonians and Bullion, The Journal of Robert Livermore 1892-1915.* Lincoln: University of Nebraska Press, 1968.

Grimmett, Robert G. *Cabal of Death.* Moscow: University Press of Idaho, 1977.

Griswold, Don. *Colorado's Century of Cities.* Denver: Smith Brooks Printing Co., 1958.

Hafen, LeRoy R. *Colorado, The Story of a Western Commonwealth.* Denver: Pierlin Publishing Co., 1933.

—-. *Colorado and Its People.* New York: Lewis Historical Publications, 1948.

Hall, Frank. *History of the State of Colorado.* Chicago: Blakely Printing Co., 1889-1895.

Harriman, J. *The Class Struggle in Idaho.* New York: Labor Publishing Assoc., 1904.

Harris, Herbert. *American Labor.* New Haven: Yale University Press, 1931.

Hawley, James H. "Steve Adams' Confession and the State's Case Against Bill Haywood." *Idaho Yesterdays,* Winter Issue 1963-1964.

Haywood, William D. *Bill Haywood's Book: The Autobiography of William D. Haywood.* New York: International Publishers, 1929. (This book was not written by Haywood in its entirety).

—-. *The General Strike.* Chicago: IWW Publishing Co., not dated.

Hill, Alice Polk. *Tales of the Colorado Pioneers.* Denver: Pierson & Gardner, 1884.

Hillebrand, J. R. *The Idarado Mine.* In Guidebook 8: Southwestern San Juan Mountains. New Mexico Geological Society, 1957.

Hillquit, Morris. *History of Socialism in the United States.* New York: Funk and Wagnalls, 1910.

Historical Encyclopedia of Colorado. Denver: Colorado Historical Association, 1968.

Holbrook, Stewart H. *The Rocky Mountain Revolution.* New York: Henry Holt & Co., 1956.

Hurt, W. *The Scarlet Shadow.* Girard, KS: Appeal to Reason, 1907.

Hutton, May Arkwright. *The Coeur d'Alenes-Or a Tale of the Modern Inquisition in Idaho.* Wallace, Idaho, 1900.

Industrial Workers of the World. *Twenty-five Years of Industrial Unionism.* Chicago: IWW Publishing Co., not dated.

Ingersoll, Ernest. "Silver San Juan," *Harper's New Monthly Magazine,* April 1882.

—-. *Crest of the Continent.* Chicago: R. R. Donnelley & Sons, 1885.

Jensen, Vernon H. *Heritage of Conflict, Labor Relations in the Nonferrous Metals Industry up to 1930.* Ithaca: Cornell University Press, 1950.

Jocknick, Sidney. *Early Days on the Western Slope of Colorado.* Glorieta, NM: Rio Grande Press, 1913.

Jones, Charles A. *The Early History of Rico.* Dolores, Colorado: The Dolores Star Press, 1964.

Jones, Rev. William H. *The History of Catholic Education in the State of Colorado.* Washington: Catholic University of America Press, 1955.

Josephson, Matthew. *The Robber Barons, The Great American Capitalists, 1861-1901.* New York: Harcourt, Brace and Co., 1934.

Kemp, D. C. *Colorado's Little Kingdom.* Golden: Sage Books, 1949.

Kenny, Kevin. *Making Sense of the Molly Maguires.* New York: Oxford University Press, 1998.

Kinikin, Luther C. "Early Days in Telluride," *Colorado Magazine,* June 1949.

Kirk, W. *National Labor Federations in the United States.* Baltimore: Johns Hopkins Press, 1906.

Koelsch, C. F. *The Haywood Case.* Boise: Idaho Mining Assoc., 1946.

Langdon, Emma F. *Labor's Greatest Conflicts.* Denver: Langdon, 1908.

—-. *The Cripple Creek Strike 1903-1904,* Victor, CO: Victor Daily Record press, 1904.

—-. *The Cripple Creek Strike: A History of Industrial Wars in Colorado, 1903-4-5.* Denver: Western Publishing Co., 1905.

Lavender, D. C. *A Rocky Mountain Fantasy: Telluride, Colorado.* Telluride: San Miguel County Historical Society, 1964.

Lawson, T. W. *Frenzied Finance.* New York: Ridgway Thayer, 1905.

Livermore, Robert. "Prolonging the Life of the Smuggler Union Enterprise." *Engineering and Mining Journal,* March 31, 1928.

Lens, Sidney. *The Labor Wars, From the Molly Maguires to the Sitdowns.* New York: Doubleday, 1973.

LeVeque, E. M. *The Romance of Cripple Creek.* Denver: Gleason and Co., 1928.

Lorwin, L. L. *The American Federation of Labor.* Washington: Brookings Institution, 1933.

Lukas, J. Anthony. *Big Trouble.* New York: Touchstone, 1997.

Lyman, G. D. *The Saga of the Comstock Lode.* New York: Charles Scribner's Sons, 1946.

Marshall, John. *Mining the Hard Rock in the Silverton San Juans.* Silverton: Simpler Way Book Co., 1996.

McGrath, Maria Davis. *The Real Pioneers of Colorado.* Denver: The Denver Museum, 1934.

McGuckin, Henry E. *Memoirs of a Wobbly.* Chicago: Charles H. Kerr Publishing Company, 1987.

McLean, Evalyn Walsh. *Father Struck It Rich.* 1907. (Reprint) Ouray, CO: Bear Creek Publishing Co., 1981.

Miller, Guy E. "The Telluride Strike," January 1904, later published in Emma Langdon's book, *The Cripple Creek Strike 1903-1904;* and an updated version in Langdon, *The Cripple Creek Strike: A History of Industrial Wars in Colorado, 1903-4-5.*

Mills, J. Warner, editor. *Mills' Annotated Statutes of the State of Colorado.* Denver: Mills' Publishing Co., 1891-1897, 3 vols.

—-. *Annotated Codes of the State of Colorado.* Denver: The Mills Publishing Company, 1896.

—-. *Mills' Annotated Statutes of the State of Colorado.* Denver: The Mills Publishing Company, 1891-1903.

Mines and Mining Men of Colorado. Denver: John G. Canfield, 1893.

"The Mining Industry, History, Production and Future Prospects of the Rich Mines of Rico." *Rico News,* special edition, 1892.

Misleaders of Labor. Trade Union Educational League, 1927.

Monroe, Arthur W. *San Juan Silver.* Montrose, CO: Arthur W. Monroe, 1940.

Morgan, H. Wayne. *William McKinley and His America.* Syracuse, NY: Syracuse University Press, 1963.

Myers, G. *History of Great American Fortunes.* Chicago: Charles H. Kerr and Co., 1910.

Nankivell, Major J. H. *History of the Military Organizations of the State of Colorado 1860-1935.* Denver: Kistler, 1935.

Noel, Thomas J. "William D. Haywood." *Colorado Heritage*, Issue 2, Colorado Historical Society, 1984.

—-. *Colorado Catholicism and the Archdiocese of Denver, 1857-1989.* Denver: University Press of Colorado, 1989.

Nossaman, Allen. *Many More Mountains, Volume I: Silverton's Roots.* Denver: Sundance Publications, Ltd. 1989.

—-. *Many More Mountains, Volume II: Ruts into Silverton.* Denver: Sundance Publications, Ltd. 1993.

—-. *Many More Mountains, Volume III: Rails into Silverton.* Denver: Sundance Publications, Ltd. 1998.

Olin, Karl G. (translated by S. Forsman). "Vörå Man Martyred at Telluride." Jakobstads Tidning, Jakobstad, Finland, Nov. 6, 1993.

Orchard, Harry. *The Confessions and Autobiography of Harry Orchard.* New York: McClure Co., 1907.

—-. (In collaboration with LeRoy Edwin Froom). *The Man God Made Again.* Nashville: Southern Pub. Assoc., 1952.

Parker, C. H. *The Casual Laborer and Other Essays.* New York: Harcourt, Brace and Howe, 1920.

Parkhill, Forbes. *The Wildest of the West.* Denver: Sage Books, 1957.

Peabody, J. H. *Governor Peabody to the Voters — The Colorado Situation Discussed and Misstatements Refuted.* Denver, Republican Party pamphlet, 1904.

Penman, Selig, and Philip Taft. *History of Labor in the United States, 1896- 1932.* New York: Macmillan Co., 1935.

Pinkerton's National Detective Agency and Its Connection With the Labor Troubles at Homestead, Penn., July 6th, 1892. New York: 1892.

Pioneers of the San Juan Country. Daughters of the American Revolution, Sarah Platt Decker Chapter. Durango: Big Mountain Printing, 1961.

Porter, J. A. *Eighth Annual Report of the Smuggler-Union Mining Company, January 1, 1899.* Denver: C. J. Kelley, 1899.

Portrait and Biographical Record of the State of Colorado. Chicago: A. W. Brown & Co., 1899.

Preston, William. *Aliens and Dissenters: Federal Suppression of Radicals, 1903-1933.* Urbana and Chicago: University of Illinois Press, 1994.

Progressive Men of Western Colorado. Chicago: A. W. Bowen & Co., 1905.

Ramsey, Howard Verne. *My Early Recollections of Telluride, Colorado.* Typewritten manuscript, 1988.

Rastall, Benjamin M. *The Labor History of the Cripple Creek District: A Study in Industrial Evolution.* Madison: University of Wisconsin Press, 1908.

Rathmell, Ruth. *Of Record and Reminiscence: Ouray and Silverton.* Ouray: Ruth Rathmell, 1976.

"The Reign of Lawlessness, Anarchy and Despotism in Colorado," *McClure's Magazine*, May 1904.

Renshaw, Patrick. *The Wobblies.* Chicago: Ivan R. Dee, 1999.

Reynolds, L. G., and C. C. Killingsworth. *Trade Union Publications.* Baltimore: Johns Hopkins Press, 1944-1945.

Rickard, T. A. *Across the San Juan Mountains.* New York: The Engineering & Mining Journal, 1903.

Rockwell, Wilson. "Gentleman of Fortune." *Denver Westerners' Monthly Roundup*, May 1966.

Rosen, George. *The History of Miners' Diseases, a Medical and Social Interpretation.* New York: Schuman's, 1943.

Rowan, Richard Wilmer. *The Pinkertons, a Detective Dynasty.* Boston: Little, Brown and Co., 1931.

San Juan! A History. Unknown publisher, 1880s or 1890s.

Saposs, D. *Left-wing Unionism.* New York: International Publishers, 1926.

Shinn, C. H. *Mining Camps: A Study in American Frontier Government.* New York: Charles Scribner's Sons, 1885.

Siringo, Charles A. *A Cowboy Detective, A True Story of Twenty-two Years with a World-Famous Detective Agency.* Chicago: W. B. Conkey Co., 1912.

—-. *Riata and Spurs.* Boston: Houghton Mifflin Co., 1927.

—-. *Two Evil Isms, Pinkertonism and Anarchism.* Austin: Steck-Vaughn Company, 1967 (Facsimile reprint of the 1915 original).

Sloan, Robert, and Carl Skowronski. *The Rainbow Route.* Denver: Sundance Publications, Ltd., 1984.

Smiley, Jerome C. *Semi-Centennial History of the State of Colorado.* Chicago: Lewis Publishing Co., 1913.

Smith, Duane A. *Rocky Mountain Boom Town: A History of Durango.* Albuquerque: University of New Mexico Press, 1980.

—-. *Silverton.* Ouray: Western Reflections, 1997.

—-. *Song of the Hammer and Drill.* Golden: Colorado School of Mines Press, 1982.

Smith, G. H. *History of Comstock Lode, 1850-1920.* University of Nevada Bulletin, Vol. XXXVII, No. 3. Reno: State Printing Office, 1943.

St. John, Vincent. *The I.W.W., Its History, Structure, and Methods.* Chicago: IWW Publishing Bureau, 1917 (Revised from 1912).

Stone, I. *Clarence Darrow-For the Defense.* New York: Garden City Publishing Co., 1941.

Stone, Wilbur Fisk. *History of Colorado.* Chicago: S. J. Clarke Publishing Co., 1919.

Suggs, George G., Jr. *Colorado's War on Militant Unionism.* Norman and London: University of Oklahoma Press, 1990.

Telluride Board of Trade. *Resources and Mineral Wealth of San Miguel County, Colorado, Past, Present and Future.* Denver: Publishers' Press Room Co., 1894.

Thayer, William Makepeace. *Marvels of the New West.* Norwich, CT: Henry Bill Publishing Co., 1887.

The Smuggler-Union Mines and Mills. Telluride: Telluride Journal special edition, 1899.

The Story of the World's Greatest Gold Camp-The Labor Troubles of 1903-04. Cripple Creek: Cripple Creek Times, 1904.

Van Wagenen, Theodore Francis. *Notes on Colorado,* personal journal. (DPL/WH)

Walling, William English. *Socialism as It Is: A Survey of the Worldwide Revolutionary Movement.* New York: Macmillan Co., 1912.

Warner, Matt, (as told to Murray E. King). *The Last of the Bandit Riders.* Caldwell: Caxton Printers, Ltd., 1940.

Warshaw, Herman T. *Representative Industries in the United States.* New York: Henry Holt and Co., 1928.

Waters, Frank. *Midas of the Rockies.* Denver: Sage Books, 1949.

Wellman, Walter. *Indictment of Moyer, Haywood and the Western Federation of Miners.* Published by Walter Wellman, c. 1905.

Wenger, Martin G. "My Recollections of Robert Livermore and events in his Journal," handwritten MS, December 1968, Robert Livermore Papers (AHC-UW). Later privately printed as *Recollections of Telluride 1895-1920.*

Western Federation of Miners. *Category of Crime of the Mine Operators' Association, A Partial List Showing 851 Men Murdered in Less Than Four Years.* Denver: Miners Magazine, 1904.

Western Federation of Miners. *Official Proceedings of Annual Conventions, 1900-1908.* (CU Archives).

Wilder, Walter L. Robert *Wilbur Steele: Defender of Liberty.* Denver: Carson-Harper Co., 1913.

Wood, Frances. *I Hauled These Mountains in Here.* Caldwell, Idaho: Caxton Printers, Ltd., 1977.

Workers of the World-Manifesto. Chicago: Conference of Industrial Unionists, January, 1905.

Wright, M. *Takes More than Guns.* Denver: IUMMSW, 1944.

Wyman, Louis W. *Snowflakes and Quartz.* Silverton: San Juan County Book Co., 1977.

Wyman, Mark. *Hard Rock Epic: Western Miners and the Industrial Revolution, 1860-1910.* Berkeley: University of California Press, 1989.

Yellen, Samuel. *American Labor Struggles.* New York: Harcourt, Brace and Co., 1936.

NEWSPAPERS & PERIODICALS:

Alamosa Journal
American Federationist
American Labor Union Journal (Butte)
Appeal to Reason (Kansas)
Aspen Daily Times
Boise Capitol News
Boise Idaho Statesman
Boston Transcript
Brooklyn Eagle (New York)
Carbonate Chronicle (Leadville)
Catholic World
Central City Register-Call
Chaffee County Record (Salida)
Chicago Daily Tribune
Chicago Mining Review
Cincinnati Post
Colorado Magazine (Denver)
Colorado Springs Gazette
Colorado Tribune (Denver)
Colorado Weekly Chieftain
Creede Candle
Cripple Creek Times
Daily Democrat (Denver)
Daily Herald (Silverton)
Daily Journal (Telluride)
Daily Sentinel (Grand Junction)
Daily Southwest (Durango)
Delta Chief and Independent
Denver Catholic Register
Denver Daily News
Denver Post
Denver Republican
Denver Times
Denver Tribune
Denver Westerners Monthly Roundup
Dolores News (Rico)
Durango Democrat
Durango Herald
Durango Record
Durango Telegraph
Engineering and Mining Journal (New York)
Evening Chronicle (Leadville)
Galveston News
Glenwood Post
Globe-Democrat (St. Louis)
Goldfield Daily Tribune (Nevada)
Goldfield News (Nevada)
Grand Junction News
Gunnison Daily News-Democrat
Gunnison Democrat
Gunnison News
Gunnison News-Champion
Gunnison Republican
Gunnison Review Press
Harper's Weekly (New York)
Hinsdale Phonograph (Lake City)
Idaho Yesterdays
Industrial Solidarity (Chicago)
Industrial Union Bulletin (Chicago)
Industrial Worker (Seattle)
Industrial Worker (Spokane)
International Socialist Review (Chicago)
Journal (Hannibal, MO)
Kansas City Times
Labor World (Butte)
La Plata Miner (Silverton)
Lake City Mining Register
Lake City Silver World
Lake City Times
Lee's Summit Journal (Missouri)
London Illustrated News
Magazine of Western History
McClure's Magazine (New York)
Miners' Magazine (Denver)
Mines and Minerals
Mining and Scientific Press (New York)
Mining Exchange Journal (Denver)
Mining News (Telluride)
Montezuma Journal (Cortez)
Montrose Daily Press
Montrose Enterprise
Mountain Mail (Salida)
Nevada Workman
New York Evening Post

The New York Times
Ouray Herald
Ouray Times
Plaindealer (Ouray)
Portland Oregonian
Pueblo Chieftain
Pueblo Courier
Railroad Red Book
Rico Democrat
Rico News
Rico News-Sun
Rico Record
Rocky Mountain Herald
Rocky Mountain News
Salida Herald
Salida Record
San Francisco Chronicle
San Francisco Examiner
San Juan Democrat (Silverton)
San Juan Herald (Silverton)
San Miguel Examiner (Telluride)
St. Louis Star (Missouri)
Seattle Post
Shenandoah Herald (Pennsylvania)
Silverite Plaindealer (Ouray)
Silverton Democrat
Silverton Miner
Silverton Standard
Solidarity (Chicago)
Solid Muldoon (Ouray)
Telluride Journal (weekly)
Telluride Miner
Telluride Republican
Victor Daily Record
White Pine News (Ely, Nevada)

ACKNOWLEDGEMENTS

So many people have helped to make this book possible that if I listed all their names here, I might cause the demise of another forest. Although I would like to personally thank all of them, space forbids it. Do know that everyone who contributed to this book, even in a small way, has my deepest thanks, for I could not have done this without you. Some, whose generous assistance went into the realm of soul, must be named here, for they deserve songs and parades in their honor. My gratitude is as big as the mountains and as eternal.

Many thanks to the most wonderful archivists a researcher can take advantage of, men and women who sought the last scrap of information for me without complaint or hesitation: William W. LeFevre (who should be canonized some day!), Archivist, Archives of Labor and Urban Affairs, Walter P. Reuther Library, Wayne State University; David M. Hays (faithful guardian at the gate), Archivist, Norlin Library, University of Colorado at Boulder; Leslie C. Shores (eternally kind), Archivist, American Heritage Center, University of Wyoming; Anne Diestel, Archivist, Federal Bureau of Prisons; Bruce Hanson, Denver Public Library, Western History/Genealogy; Tamara Vidos, University of Oregon; Linda Morton-Keithley, Idaho State Historical Society; Rose Mary Lankford and Donald R. Hale, William C. Quantrill Society; and Roger C. Adams, U.S. Department of Justice.

Many thanks to my forensics team, a great collection of criminally minded brains: Thomas M. Canfield, MD, PC, Forensic Pathologist and Chief Medical Examiner, Office of Medical Investigation Southwestern Region, Montrose, Colorado, (is it socially acceptable to have a favorite pathologist?); James Crippin, Director, Western Forensic Law Enforcement Training Center, Pueblo, Colorado; John R. Trujillo, Consulting Mining Geologist and mine safety consultant; P. Dan Martin (who likes to blow things up); San Miguel County Sheriff Bill Masters; Gary Koverman, retired Colorado Bureau of Investigation agent; Peter P. Martin (who loves to blow things up and can hit his target with any sort of gun, with his eyes closed); Carl V. Button, Jr. (cowboy shotgun enthusiast); and Paul Miller, of the San Juan Rifle Range.

Many thanks to my most dedicated, incredibly encouraging researcher in Ely, Nevada, Donna Frederick; and Deanna L. LaBonge, Nevada State Library Archives.

Enormous thanks to my legal team: Rene D. Tegtmeyer, attorney, who carefully reviewed my court cases and translated legalese to English (is that feasible?) and made years of research possible; Jane Thompson, Assistant Director for Faculty Services, University of Colorado Law Library; Judge P. David Smith; and Lena Martinez and the staff of the San Miguel County Combined Courts, for giving me free reign for the last thousand years to worm through the dusty court records.

A very special thanks to Alta Cassietto, long-time Postmaster of Telluride, whose love for her hometown spans nine decades. Her every microscopic detail remains so sharp that I could find hidden and demolished places, houses, and streets in the old town because of her.

Many thanks to my mining team: Richard River, Vice President and General Manager and later Director of Reclamation for the Idarado Mining Company (last owners of the Smuggler-Union), Ouray, Colorado; William "Senior" Mahoney, long time miner of Telluride, who shared his memories and historical materials as if I were an old friend; Ernie Kuhlman, and the delightful lads of the Old Hundred Mine at Silverton.

Many thanks to my wonderful (and sometimes wicked!) Pinkerton Detective Agency/James McParland expert team: Howard C. Crown, Patrick Campbell, and Wayne Kirchgasser.

A huge hug and deep thanks goes to Dorothy Patton, unflagging Interlibrary Loan Specialist of the Montrose Library District, for miracles performed.

Deepest thanks to Ralph C. Barney, C. H. Lamb, Charles Lewis, Marianne Wells Glaser, E. N. Wright, Albert Fuentes, Ann Hoffman, Bob Curvey, Charles Hosner, Davine Pera, Dawna & Jerry O'Rourke, Doris H. Gregory, Dr. Jim Parker, Frank Massard, George & Gay Cappis, Ginni Bosse, Harry B. Wright, Joel Swank, John Motter, John Walter "Shorty" Larson, Lillie Padilla, Oran David "Shorty" Campbell, Peggy Jacobson, Perk Vickers, Rosie Hall, Verena Jacobson, Stan Zuege, Elvira F. Wunderlich, Irene R. Visitin, M. L. Shaffer, and Linda Paulsen.

Eternal gratitude goes to the wonderful people of the Center of Southwest Studies, Ft. Lewis College; Chicago Historical Society; Colorado State Archives; Colorado State Historical Society; Denver Public Library, Western History Dept.; Illinois State Historical Library; Library of Congress; Pinkerton Western Office, San Diego, Calif.; Idaho State Historical Society; Nevada State Library and Archives; New York Historical Society; National Archives and Records Administration, Great Lakes Region; Ouray County Historical Society; Pennsylvania State Historical Society; San Juan County Historical Society; San Miguel County Historical Society; University of Colorado at Boulder Archives; Boulder Historical Society; Indiana State University; Diocese of Pueblo; Archdiocese of Denver; Public Library of Cincinnati; LDS Family History Library; Western State College, Leslie Savage Library, Gunnison, Colo.; and the Wilkinson Library, Telluride.

My fondest thanks to my dedicated proofreader, Joyce; and to my guardian angel, Rebecca, for keeping my computer in working order; and to my publisher, whose enthusiasm for this project has kept me pinned to the keyboard for twelve-and fourteen-hour days for more than a year, and for his *wonderful* patience.

To all of you, named and unnamed: my love, my gratitude, and my joy.

INDEX